Lecture Notes in Computer Science

Lecture Notes in Bioinformatics   **14881**

The series Lecture Notes in Bioinformatics (LNBI) was established in 2003 as a topical subseries of LNCS devoted to bioinformatics and computational biology.

The series publishes state-of-the-art research results at a high level. As with the LNCS mother series, the mission of the series is to serve the international R & D community by providing an invaluable service, mainly focused on the publication of conference and workshop proceedings and postproceedings.

De-Shuang Huang · Qinhu Zhang · Jiayang Guo
Editors

Advanced Intelligent Computing in Bioinformatics

20th International Conference, ICIC 2024
Tianjin, China, August 5–8, 2024
Proceedings, Part I

 Springer

Editors
De-Shuang Huang
Eastern Institute of Technology
Ningbo, China

Qinhu Zhang
Eastern Institute of Technology
Ningbo, China

Jiayang Guo
Xiamen University
Xiamen, China

ISSN 0302-9743   ISSN 1611-3349 (electronic)
Lecture Notes in Bioinformatics
ISBN 978-981-97-5688-9   ISBN 978-981-97-5689-6 (eBook)
https://doi.org/10.1007/978-981-97-5689-6

LNCS Sublibrary: SL8 – Bioinformatics

This Springer imprint is published by the registered company Springer Nature Singapore Pte Ltd.
The registered company address is: 152 Beach Road, #21-01/04 Gateway East, Singapore 189721, Singapore

If disposing of this product, please recycle the paper.

Preface

The International Conference on Intelligent Computing (ICIC) was started to provide an annual forum dedicated to emerging and challenging topics in artificial intelligence, machine learning, pattern recognition, bioinformatics, and computational biology. It aims to bring together researchers and practitioners from both academia and industry to share ideas, problems, and solutions related to the multifaceted aspects of intelligent computing.

ICIC 2024, held in Tianjin, China, August 5–8, 2024, constituted the 20th International Conference on Intelligent Computing. It built upon the success of ICIC 2023 (Zhengzhou, China), ICIC 2022 (Xi'an, China), ICIC 2021 (Shenzhen, China), ICIC 2020 (Bari, Italy), ICIC 2019 (Nanchang, China), ICIC 2018 (Wuhan, China), ICIC 2017 (Liverpool, UK), ICIC 2016 (Lanzhou, China), ICIC 2015 (Fuzhou, China), ICIC 2014 (Taiyuan, China), ICIC 2013 (Nanning, China), ICIC 2012 (Huangshan, China), ICIC 2011 (Zhengzhou, China), ICIC 2010 (Changsha, China), ICIC 2009 (Ulsan, South Korea), ICIC 2008 (Shanghai, China), ICIC 2007 (Qingdao, China), ICIC 2006 (Kunming, China), and ICIC 2005 (Hefei, China).

This year, the conference concentrated mainly on the theories and methodologies as well as the emerging applications of intelligent computing. Its aim was to unify the picture of contemporary intelligent computing techniques as an integral concept that highlights the trends in advanced computational intelligence and bridges theoretical research with applications. Therefore, the theme for this conference was "Advanced Intelligent Computing Technology and Applications". Papers that focused on this theme were solicited, addressing theories, methodologies, and applications in science and technology.

ICIC 2024 received 2189 submissions from 15 countries and regions. All papers went through a rigorous single-blind peer-review procedure and each paper received at least three review reports. Based on the review reports, the Program Committee finally selected 863 high-quality papers for presentation at ICIC 2024, included in twenty-one volumes of proceedings published by Springer: thirteen volumes of Lecture Notes in Computer Science (LNCS), six volumes of Lecture Notes in Artificial Intelligence (LNAI), and two volumes of Lecture Notes in Bioinformatics (LNBI).

In addition, this year we selected 134 Poster papers from the remaining papers, which will be made accessible on the open access website http://poster-openaccess.com/.

This volume of LNBI_14881 includes 39 papers.

The organizers of ICIC 2024, including Eastern Institute of Technology, Ningbo, China; Tianjin University of Science and Technology, China; China University of Mining & Technology (Beijing), China; China University of Mining and Technology (Xuzhou), China; and North China University of Science and Technology, China, made an enormous effort to ensure the success of the conference. We hereby would like to thank the members of the Program Committee and the referees for their collective effort in reviewing and soliciting the papers. In particular, we would like to thank all the authors for contributing their papers. Without the high-quality submissions from the authors, the success of the conference would not have been possible. Finally, we are especially grateful to the International Neural Network Society and the National Science Foundation of China for their sponsorship.

De-Shuang Huang
Fuping Lu

Organization

General Co-chairs

De-Shuang Huang   Eastern Institute of Technology, China
Fuping Lu   Tianjin University of Science and Technology, China

Program Committee Co-chairs

Prashan Premaratne   University of Wollongong, Australia
Xiankun Zhang   Tianjin University of Science and Technology, China
Chuanlei Zhang   Tianjin University of Science and Technology, China
Wei Chen   China University of Mining and Technology, China
Jair Cervantes Canales   Autonomous University of Mexico State, Mexico
Yijie Pan   Eastern Institute of Technology, China
Qinhu Zhang   Eastern Institute of Technology, China
Jiayang Guo   Xiamen University, China

Organizing Committee Co-chairs

Zhanjun Si   Tianjin University of Science and Technology, China
Xiaoyue Liu   North China University of Science and Technology, China
Fan Zhang   China University of Mining and Technology (Beijing), China

Organizing Committee Members

Yarui Chen   Tianjin University of Science and Technology, China
Jing Su   Tianjin University of Science and Technology, China

Shuo Yang	Tianjin University of Science and Technology, China
Jing Han	Tianjin University of Science and Technology, China
Yiying Zhang	Tianjin University of Science and Technology, China
Jucheng Yang	Tianjin University of Science and Technology, China
Qian Long	Tianjin University of Science and Technology, China
Yongjun Ma	Tianjin University of Science and Technology, China
Lin Sun	Tianjin University of Science and Technology, China
Guoliang Gong	Tianjin University of Science and Technology, China

Award Committee Chair

Kang-Hyun Jo	University of Ulsan, South Korea

Tutorial Co-chairs

Abir Hussain	Liverpool John Moores University, UK
Michal Choras	Bydgoszcz University of Science and Technology, Poland

Publication Co-chairs

Jair Cervantes Canales	Autonomous University of Mexico State, Mexico
Chenxi Huang	Xiamen University, China

Special Session Co-chairs

Valeriya Gribova	Far Eastern Branch of Russian Academy of Sciences, Russia
M. Michael Gromiha	Indian Institute of Technology Madras, India

Special Issue Co-chairs

Yu-Dong Zhang   University of Leicester, UK
Yoshinori Kuno   Saitama University, Japan
Phalguni Gupta   Indian Institute of Technology Kanpur, India

International Liaison Chair

Prashan Premaratne   University of Wollongong, Australia

Workshop Co-chairs

Kyungsook Han   Inha University, South Korea
Laurent Heutte   Université de Rouen Normandie, France

Publicity Co-chairs

Chun-Hou Zheng   Anhui University, China
Dhiya Al-Jumeily   Liverpool John Moores University, UK
Han Huang   Nanjing University of Information Science and
                         Technology, China

Program Committee Members

Antonio Brunetti   Polytechnic University of Bari, Italy
Bin Liu   Beijing Institute of Technology, China
Bin Qian   Kunming University of Science and Technology,
                         China
Bin Yang   Zaozhuang University, China
Bing Wang   Anhui University of Technology, China
Bingqiang Liu   Shandong University, China
Binhua Tang   Hohai University, China
Bo Li   Wuhan University of Science and Technology,
                         China
Caihong Mu   Xidian University, China
Changqing Shen   Soochow University, China
Chao Song   University of South China, China
Cheng Tang   Kyushu University, Japan

Chin-Chih Chang	Chung Hua University, Taiwan, RoC
Chuanlei Zhang	Tianjin University of Science and Technology, China
Chunhou Zheng	Anhui University, China
Chunmei Liu	Howard University, USA
Chunquan Li	University of South China, China
Cong Shen	Tianjin University of Technology, China
Daowen Qiu	Sun Yat-sen University, China
Delong Yang	First People's Hospital of Foshan, China
Dian Ding	Shanghai Jiao Tong University, China
Dong Wang	University of Jinan, China
Duo Chen	Nanjing University of Chinese Medicine, China
Eros Gian Alessandro Pasero	Politecnico di Torino, Italy
Fa Zhang	Beijing Institute of Technology, China
Fei Guo	Central South University, China
Fei Luo	Wuhan University, China
Fei Shen	Nanjing University of Science and Technology, China
Feng Liu	East China Normal University, China
Feng Zou	Huaibei Normal University, China
Fengfeng Zhou	Jilin University, China
Fudong Nian	Hefei University, China
Fuxue Li	Yingkou Institute of Technology, China
Gang Li	Qilu University of Technology, China
Gaoxiang Ouyang	Beijing Normal University, China
Guanghui Gong	Eastern Institute of Technology, Ningbo, China
Guohui Ding	Shenyang Aerospace University, China
Guoliang Li	Huazhong Agricultural University, China
Han Zhang	Nankai University, China
Hao Huang	Hubei University, China
Hao Lin	University of Electronic Science and Technology of China, China
Haodi Feng	Shandong University, China
Haodong Zhu	Zhengzhou University of Light Industry, China
Heng Li	Southern University of Science and Technology, China
Hoang-Anh Ngo	The University of Waikato, New Zealand
Hongjie Wu	Suzhou University of Science and Technology, China
Hongmin Cai	South China University of Technology, China
Hulin Kuang	Central South University, China
Jiahui Pan	South China Normal University, China

Jian Huang	University of Electronic Science and Technology of China, China
Jian Shen	Beijing Institute of Technology, China
Jiang Xie	Shanghai University, China
Jianrong Li	Tianjin University of Science and Technology, China
Jiawei Luo	Hunan University, China
Jiayang Guo	Xiamen University, China
Jing Chen	Suzhou University of Science and Technology, China
Jing Hu	Wuhan University of Science and Technology, China
Jintian Lu	Jishou University, China
Jin-Xing Liu	University of Health and Rehabilitation Sciences, China
Jipeng Wu	First People's Hospital of Foshan, China
Joaquin Torres-Sospedra	Universidade do Minho, Portugal
Juan Liu	Wuhan University, China
Junfeng Xia	Anhui University, China
Jungang Lou	Huzhou University, China
Junqing Li	Yunnan Normal University, China
Junyi Li	Harbin Institute of Technology (Shenzhen), China
Ka-Chun Wong	City University of Hong Kong, China
Kangning Zhang	Academy of Mathematics and Systems Science, CAS, China
Ke Niu	Beijing Information Science and Technology University, China
Laurent Heutte	Université de Rouen Normandie, France
Le Zhang	Sichuan University, China
Lei Wang	Guangxi Academy of Sciences, China
Lejun Gong	Nanjing University of Posts and Telecommunications, China
Liang Gao	Huazhong University of Science & Technology, China
Lida Zhu	Huazhong Agriculture University, China
Lin Wang	University of Jinan, China
Lin Yuan	Qilu University of Technology, China
Liqiang Liu	Xi'an Technological University, China
Li-Wei Ko	National Yang Ming Chiao Tung University, Taiwan, RoC
Long Shao	Beijing Institute of Technology, China
Long Xu	Ningbo University, China
Meiyan Xu	Minnan Normal University, China

Meng Liu — National University of Defense Technology, China

Michael Gromiha — Indian Institute of Technology Madras, India

Michal Choras — Bydgoszcz University of Science and Technology, Poland

Mingyong Li — Chongqing Normal University, China

Mohd Helmy Abd Wahab — Universiti Tun Hussein Onn Malaysia, Malaysia

Nicola Altini — Polytechnic University of Bari, Italy

Nier Wu — Inner Mongolia University of Technology, China

Peipei Gu — Zhengzhou University of Light Industry, China

Peng Chen — Anhui University, China

Pengjiang Qian — Jiangnan University, China

Pengwei Hu — Xinjiang Technical Institute of Physics and Chemistry, CAS, China

Prashan Premaratne — University of Wollongong, Australia

Pu-Feng Du — Tianjin University, China

Qi Sun — Hangzhou Nuowei Information Technology Co., Ltd., China

Qi Zhao — University of Science and Technology Liaoning, China

Qifang Luo — Guangxi University for Nationalities, China

Qinhu Zhang — Eastern Institute of Technology, Ningbo, China

Qiuzhen Lin — Shenzhen University, China

Quan Zou — University of Electronic Science and Technology of China, China

Rong Wang — Sichuan Normal University, China

Rong-Qiang Zeng — Chengdu University of Information Technology, China

Rui Wang — National University of Defense Technology, China

Saiful Islam — Aligarh Muslim University, India

Shanfeng Zhu — Fudan University, China

Shitong Wang — Jiangnan University, China

Shixiong Zhang — Xidian University, China

Sungshin Kim — Pusan National University, South Korea

Taisong Jin — Xiamen University, China

Tian Wu — Nanchang University, China

Tieshan Li — University of Electronic Science and Technology of China, China

Valeria Gribova — Far Eastern Branch of Russian Academy of Sciences, Russia

Wangren Qiu — Jingdezhen Ceramic University, China

Waqas Haider Bangyal — Kohsar University Murree, Pakistan

Wei Chen	China University of Mining and Technology (Xuzhou), China
Wei Chen	Chengdu University of Traditional Chinese Medicine, China
Wei Jiang	Fujian Medical University, China
Wei Wang	Henan Normal University, China
Wei Xu	East China Normal University, China
Weichao Wu	Beijing Institute of Technology, China
Weiwei Kong	Xi'an University of Posts and Telecommunications, China
Weixiang Liu	Shenzhen University, China
Wen Jiang	Ctrip Computer Technology (Shanghai) Co., Ltd., China
Wen-Sheng Chen	Shenzhen University, China
Wenzheng Bao	Xuzhou University of Technology, China
Xiangtao Li	Jilin University, China
Xiaodi Li	Shandong Normal University, China
Xiaofeng Wang	Hefei University, China
Xiaoke Ma	Xidian University, China
Xiaolei Zhu	Anhui Agricultural University, China
Xiaoli Lin	Wuhan University of Science and Technology, China
Xiaoqing Li	Capital University of Economics and Business, China
Xin Zhang	Jiangnan University, China
Xingjian Xu	Inner Mongolia Normal University, China
Xingquan Cai	North China University of Technology, China
Xingtao Wang	Harbin Institute of Technology, China
Xinguo Lu	Hunan University, China
Xingyu Feng	City University of Hong Kong, China
Xinlu Li	Hefei University, China
Xinzheng Xu	China University of Mining and Technology (Xuzhou), China
Xiufen Zou	Wuhan University, China
Xiujuan Lei	Shaanxi Normal University, China
Xiwei Liu	Tongji University, China
Xiyuan Chen	Southeast University, China
Xizhao Luo	Soochow University, China
Xulong Zhang	Ping An Technology (Shenzhen) Co., Ltd., China
Yang Yang	Hubei University, China
Yansen Su	Anhui University, China
Yijie Pan	Eastern Institute of Technology, Ningbo, China
Yiming Tang	Hefei University of Technology, China

Yizhang Jiang	Jiangnan University, China
Yong Wang	Academy of Mathematics and Systems Science, CAS, China
Yong Wu	Anhui Normal University, China
Yonggang Lu	Lanzhou University, China
Yu Lu	Shenzhen Technology University, China
Yu Xue	Huazhong University of Science and Technology, China
Yunxia Liu	Zhengzhou Normal University, China
Yupei Zhang	Northwestern Polytechnical University, China
Yushan Qiu	Shenzhen University, China
Yuyan Zheng	Shandong Normal University, China
Zhan-Li Sun	Anhui University, China
Zhen Shen	Nanyang Institute of Technology, China
Zhendong Liu	Shandong Jianzhu University, China
Zhenran Jiang	East China Normal University, China
Zhenyi Shen	Zhejiang University, China
Zhi-Hong Guan	Huazhong University of Science and Technology, China
Zhi-Ping Liu	Shandong University, China
Zhong-Qiu Zhao	Heifei Institute of Technology, China
Zhuangzhuang Chen	Hong Kong University of Science and Technology, China
Zhuo Lei	City Cloud Technology China Co., Ltd., China
Zixiao Kong	University of International Relations, China

Contents – Part I

Biomedical Informatics Theory and Methods

Pattern Recognition

Contents – Part II

**Medical Image Intelligent Diagnosis and Precision Analysis for
Human Health**

Biomedical Data Modeling and Mining

Alzheimer's Disease Diagnosis via Specific-Shared Representation Learning in Multimodal Neuroimaging

Mingxia Wang[1], Yun Yang[1], Jun Qi[2], Fengtao Nan[3], Shunbao Li[4], and Po Yang[4(✉)]

[1] National Pilot School of Software, Yunnan University, Kunming 650091, Yunnan, China
[2] Department of Computing, Xi'an Jiaotong-Liverpool University, Suzhou 215123, Jiangsu, China
[3] School of Information Science and Engineering, Northwest A&F University, Xianyang 712100, Shanxi, China
[4] Department of Computer Science, University of Sheffield, Sheffield S10 2TN, UK
poyangcn@gmail.com

Abstract. Mild cognitive impairment (MCI) is a potential prodromal state of Alzheimer's disease (AD), an irreversible progressive neurodegenerative disease. Early diagnosis and intervention of AD are crucial. Recent studies have shown that deep learning methods exhibit excellent performance in predicting AD from multimodal neuroimaging. However, previous studies tend to focus on one aspect of the analysis of intra-modal information or inter-modal shared information, while ignoring the integration of the two. To solve this problem, this paper proposes a fusion network that combines modality-specific representations with shared representations between modalities to assist AD diagnosis. The network first extracts modality-specific representations from each modality data, then introduces a modality shared representation extraction network to supplement modality shared information, and finally performs modality representation fusion and performs AD prediction. In addition, in order to reduce the interference of non-disease-related regions between modalities on shared information, the network combines the shallow representation extracted by the single-modal network to guide the extraction process of shared representation. Combining these representations can provide a comprehensive view of multimodal data, thereby more accurately assisting in AD prediction tasks. The experimental results of this article on the Alzheimer's Disease Neuroimaging Initiative (ADNI) database prove the effectiveness and excellent performance of this model in assisting in the diagnosis of AD.

Keywords: Multimodal Neuroimaging · Alzheimer's Disease Diagnosis · Multimodal Representation Learning

D.-S. Huang et al. (Eds.): ICIC 2024, LNBI 14881, pp. 3–14, 2024.
https://doi.org/10.1007/978-981-97-5689-6_1

4 M. Wang et al.

1 Introduction

Alzheimer's disease (AD) is an irreversible neurodegenerative disease that has a significant negative impact on the physical and mental health of the elderly [1, 2]. The medical field uses various imaging technologies to assist diagnosis, aiming to improve the understanding of the pathogenesis and progression of these diseases [3, 4], and then conduct early screening, intervention and treatment of AD. With the continuous advancement of medical neuroimaging technology, deep learning methods based on neuroimaging analysis have gradually improved, showing excellent performance in disease diagnosis tasks [5, 6].

In recent years, the introduction of multimodal learning has brought significant changes to deep learning-based medical imaging diagnosis methods. These methods no longer rely solely on a single imaging technology, but instead integrate multi-source neuroimaging data to achieve a more comprehensive diagnostic perspective. Recent research results [7–10] further confirm that deep learning methods using multi-source medical neuroimaging (such as considering MRI and PET neuroimaging simultaneously) are more effective in disease diagnosis than considering only a single source. This advantage mainly arises from the complementarity of functional and structural information provided by different modalities [8]. Effective integration of information across different modalities has become an important focus of current research.

Some disease prediction networks based on deep learning introduce decision-level fusion methods in the prediction stage [11]. They perform weighted fusion based on different single-modal disease prediction results to form the final prediction result, which fully considers the specific information of different modalities for disease diagnosis, but ignores the complementary information between modalities. Some methods use simple representation fusion methods [7, 12, 13], such as concatenation, addition, multiplication, etc. in the multi-modal information fusion stage of AD diagnosis. This strategy considers information sharing between different modalities, but weakens the consideration of specific information within a modality. [14, 15] introduced a novel representation fusion method to generate inter-modal information while retaining intra-modal information to achieve better disease diagnosis. However, this representation fusion method causes the multi-modal representation dimension to expand exponentially, leading to a sharp increase in model parameters and introducing more noise, thus affecting the extraction of effective representations. [16] alleviated the parameter redundancy problem through low-rank decomposition, but the information loss caused by hyperparameter selection and the noise brought by this method are still problems.

Based on the above analysis of the advantages and disadvantages of other methods, in this paper, we propose a multimodal deep learning framework for intra-modal specific representation and inter-modal shared representation learning and disease diagnosis (Fig. 1). Our framework is divided into three stages: the first stage, neuroimaging shallow feature learning and single-modality specific representation learning, the second stage, multi-modality shared representation learning, and the third stage, modality representation fusion and disease prediction. Experimental results on two ADNI public datasets show that our method achieves better results than other methods in AD recognition and MCI conversion prediction, and also proves that our method effectively learns intra-modal information and inter-modal information.

The main contributions of our proposed approach can be summarized as follows:

(1) We consider the important role of intra-modal specific information and inter-modal shared information in disease diagnosis. Based on the disease diagnosis task, we extract representations of these two types of information based on the diagnosis task to ensure the relevance and effectiveness of the extracted representations to the task.
(2) Considering the interference of modality-disease diagnosis irrelevant areas on the learning of shared information between modalities, the prior knowledge learned by the single-modal network is introduced to assist the learning of the shared representation extraction module between modalities.
(3) We propose a multimodal specific-shared representation learning framework (MSSRL) aimed at assisting the diagnosis of AD. This approach considers the integration of intra-modality specific information and cross-modality shared information, providing a more comprehensive and accurate perspective for disease prediction.

2 Method

2.1 Overview

We denote D as the dataset that comprises N subjects. Each subject in this dataset has complete MRI (represented by M) and PET (represented by P) medical imaging scans. The dataset D can be expressed as $D = \{M_i, P_i, y_i\}_N^i$, where M_i and P_i correspond to the MRI and PET scans of the ith subject, respectively. Furthermore, y_i is a binary disease label. $y_i = 0$ means that the subject's diagnosis result is CN. Depending on the task, $y_i = 1$ means that the subject's diagnosis result is AD or MCI. Our proposed framework aims to predict disease labels $\hat{y}_i$ by considering intra-modal specific representation as well as inter-modal shared representation. To achieve this goal, the framework consists of two shallow feature extraction $\{SFE_m, SFE_p\}$, two single-modality specific representation learning networks $\{E_m, E_p\}$ and a multi-modal shared representation learning network $\{E_{sh}\}$. The modality-specific representation and the inter-modality shared representation are respectively obtained by the corresponding representation extractors, represented as f_m, f_p and f_{sh}. Subsequently, f_1 and f_2 are obtained through the representation fusion module $\{Fu_1, Fu_2\}$. Finally, f_1 and f_2 as the input of the classifier L to obtain the prediction result of AD.

In the following, we will delve into the specific details of each network.

2.2 Shallow Feature Learning

Relevant research shows that only certain areas in the entire medical image are relevant to disease diagnosis [17–19]. Single-modality diagnostic networks can focus on disease-related regions in images from different modalities. In order to reduce the interference of irrelevant areas of the image on modality sharing representation learning, we embed the knowledge learned by the single-modality network to assist in guiding the training of the modality sharing module.

Modality Shallow Feature. We consider the joint training of the shallow feature extraction module and the single-modal representation extraction module, and guide the

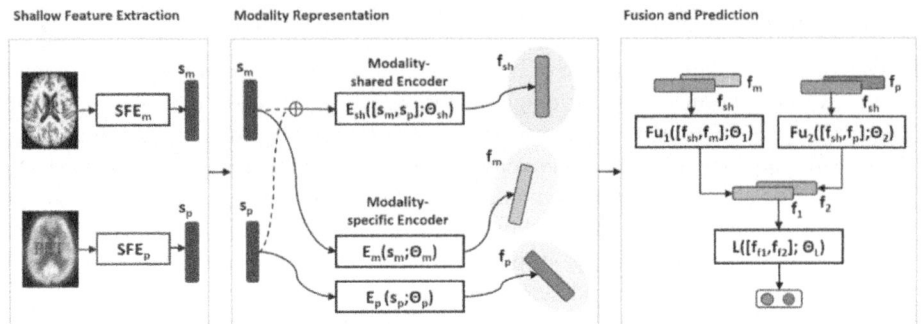

Fig. 1. The figure above shows the framework diagram of our network

training of the modal shared representation extraction module in the multi-modal shared feature learning stage, but do not participate in the gradient optimization of the model. We will mention the specific details below. Shallow features can be expressed as

$$s_{mi} = SFE_m(M_i; \theta_{sm}), \quad s_{pi} = SFE_p(M_i; \theta_{sp}), \tag{1}$$

where θ_{sm} and θ_{sp} are parameters that can be learned by the corresponding shallow feature extractor.

Network Architecture. Shallow convolutional layers SFE_m and SFE_p have the same structure, with only one convolution block, which contains the $Conv3d - Instancenorm3d - ReLU - MaxPooling$ component. The number of output channels of the convolutional layer is 16, the kernel size is set to $3 \times 3 \times 3$, stride and padding are set to 1, while the kernel size of the pooling layer is set to 2.

2.3   Modality Specific Representation Learning

Specific brain diseases are closely related to changes in certain brain regions, which has been confirmed by numerous studies [17–19]. Considering that MRI and PET scans use different imaging methods, the disease-related regions are not exactly the same [20, 21]. We introduce two independent single-modal representation extractors $\{E_m, E_p\}$ to learn specific information related to disease diagnosis in neuroimaging (see Fig. 1). These representation extractors can directly extract disease-relevant information from whole-brain images. It is worth noting that the single-modal feature extractor extracts task-related representations based on tasks, ensuring that the extracted representations are highly relevant to the disease prediction task, and the module can implicitly identify disease-related regions.

Modality Specific Representation. Given a pair of subject images (M_i, P_i) with category labels y_i, we adopt a joint training method and apply the two-part combined network to MRI and PET respectively: $\{SFE_m, E_m\}$ and $\{SFE_p, E_m\}$. These networks aim to focus on high-dimensional raw image data for disease diagnosis-relevant regions and convert them into low-dimensional visual representations rich in semantic information to perform the classification task more efficiently, its visual representation is expressed as follows:

$$f_{mi} = E_m(SFE_m(M_i, \theta_{sm}), \theta_{em}), f_{pi} = E_p(SFE_p(P_i, \theta_{sp}), \theta_{ep}), \tag{2}$$

where $\theta_{em}, \theta_{ep}, \theta_{sm}, \theta_{sp}$ are the corresponding parameters that can be learned.

Then, we apply linear classification layers L_m and L_p corresponding to MRI and PET to the extracted low-dimensional semantic representations to classify the disease accurately. The network focuses on extracting important information from these low-dimensional representations to achieve accurate assessment of the subject's state. At the same time, the representation extraction layer and classification network are jointly optimized to ensure the effectiveness of the overall network. The overall process can be expressed as

$$\hat{y}^{M_i} = L_m(f_{mi}; \theta_{lm}), \hat{y}^{P_i} = L_p(f_{pi}; \theta_{lp}), \tag{3}$$

where θ_{lm} and θ_{lp} are parameters that can be learned by the linear classification layers.

Network Architecture. The modality-specific representation extractor E_m and E_p contains four convolutional blocks. Among them, the first convolution block processes the input, while the subsequent three blocks process the representation maps output by the previous block. The components of the first three blocks are $Conv3d - InstanceNorm - ReLU - Maxpooling$ and the component of the last block is $Conv3d - InstanceNorm - ReLU - Avgpooling$. The kernel size of all convolutional layers is $3 \times 3 \times 3$, the stride is 1, and the padding is 1. The pooling layer has parameter settings of kernel size 2 and stride 2. Finally, the linear classification module performs category prediction on the extracted final representation map through a fully connected layer.

2.4 Modality Shared Representation Learning

Not all areas in MRI/PET scans are associated with specific brain diseases [23, 24]. We consider combinations of $\{SFE_m, E_m, L_m\}$ and $\{SFE_p, E_p, L_p\}$ as experts in diagnosing these diseases, and they are able to effectively identify key disease-related regions in MRI/PET. Therefore, we can use the knowledge of these experts to guide the learning process of shared information between modalities, thereby reducing the interference of disease-irrelevant regions on model learning, reducing the learning difficulty of the model, and indirectly improving the learning performance of the model.

Modality Shared Representation. Considering that the shallow convolution layer in the model mainly focuses on the position and detailed information of the image, its semantics is low, while the deep convolution layer pays more attention to the overall information of the image and has higher semantics, but its ability to perceive details is weak. In order to retain more spatial and detailed information during inter-modal fusion to achieve more refined information sharing, we choose $\{SFE_m\}$ and $\{SFE_p\}$ as our final guidance models to reduce the difficulty of shared information learning.

We concatenate the representation maps $[s_{mi}, s_{pi}]$ output from the shallow convolutional layer and use $[s_{mi}, s_{pi}]$ as the input of the shared representation extraction module E_{sh}, the shared representation f_{sh} can be expressed as

$$f_{sh_i} = E_{sh}([s_{mi}, s_{pi}], \theta_{esh}), \tag{4}$$

where θ_{esh} is the parameter to be learned, θ_{sm} and θ_{sp} do not appear in the formula, which means that they do not participate in the optimization process of the model.

Then, f_{sh} is used as the input of linear classifier L_{sh} to achieve the disease prediction task

$$\hat{y}_i^{sh} = L_{sh}(f_{shi}, \theta_{lsh}), \tag{5}$$

where θ_{lsh} is the parameter to be learned.

Network Architecture. The modality-shared representation extractor contains four convolutional blocks. Among them, the first convolution block processes the input, while the subsequent three blocks process the representation maps output by the previous block. The components of the first three blocks are $Conv3d - InstanceNorm - ReLU - Maxpooling$ and the component of the last block is $Conv3d - InstanceNorm - ReLU - Avgpooling$. The output channels of the convolutional layer are all 64, the kernel size of the convolutional layer is $3 \times 3 \times 3$, stride is 1, and padding is 1. The parameters of the pooling layer are set to kernel size 2 and stride 2. Finally, the linear classification module performs category prediction on the extracted final representation map through a fully connected layer.

2.5 Modality Specific-Shared Representation Learning

Considering only inter-modal information, we may lose intra-modal specific information. Similarly, if we only consider intra-modal information, we will lose the complementary relationship between different functional and structural information between modalities [8]. In this regard, we consider both parts of information comprehensively to obtain more comprehensive and accurate prediction results.

Modality Specific-Shared Representation. The structure of the MSSRL model is shown in the Fig. 1. The model has two input parts, which are MRI and PET images. Subsequently, the corresponding features $\{s_{mi}, s_{pi}\}$ are obtained through a shallow feature extractor $\{SFE_m, SFE_p\}$. $\{s_{mi}, s_{pi}\}$ are used as input to $\{E_m, E_p, E_{sh}\}$ and feature $\{f_{mi}, f_{pi}, f_{shi}\}$ is extracted. These modules are trained based on disease diagnosis tasks, ensuring the effectiveness of the extracted representation $\{f_m, f_p, f_{sh}\}$ for disease diagnosis. Then, we further fuse the modality-specific representation and the modality-shared representation to obtain MRI-shared representation f_{1i} and PET-shared representation f_{2i}. These representations contain richer and more comprehensive semantic information, they can be expressed as

$$f_{1i} = Fu_1([f_{mi}, f_{shi}], \theta_1), f_{2i} = Fu_2([f_{pi}, f_{shi}], \theta_2), \tag{6}$$

where θ_1 and θ_2 is a learnable parameter.

Then f_1 and f_2 is used as the input of the linear classifier L to perform the classification task and realize the diagnosis of the disease

$$\hat{y}_i = L([f_{1i}, f_{2i}], \theta_l), \tag{7}$$

where θ_l is learnable parameter.

Table 1. Simplified Demographic and Clinical Information from Two Datasets

Dataset	Cat	M/F	Age	Edu	MMSE
ADNI-1	CN	55/38	76 ± 5	16 ± 3	29 ± 1
	AD	48/26	76 ± 7	14 ± 3	23 ± 2
	MCI	121/62	76 ± 7	16 ± 3	28 ± 2
ADNI-2	CN	114/131	73 ± 6	17 ± 3	29 ± 1
	AD	78/58	74 ± 8	16 ± 3	23 ± 2
	MCI	182/146	72 ± 7	16 ± 3	28 ± 2

Network Architecture. The representation extraction module of MSSRL consists of $\{SFE_m, SFE_p\}$ and $\{E_m, E_p, E_{sh}\}$. The structure of these modules has been explained previously. The representation fusion module $\{Fu_1, Fu_2\}$ and classification module L contain a linear fusion layer and a linear classification layer respectively.

3 Experiments

3.1 Materials and Image Pre-processing

We selected two sub-datasets of ADNI for experiments, including ADNI-1 and ADNI-2. Subjects in these two datasets were divided into four categories: (1) AD (2) CN (3) sMCI (4) pMCI. sMCI and pMCI can be combined into the MCI category [13]. According to whether the modality is missing, subjects in ADNI-1 and ADNI2 can be divided into three parts (1) subjects with only MRI images (2) subjects with only PET images (3) subjects with MRI and PET images of subjects. In the experiment, we only selected subjects who had both MRI and PET images (see Table 1).

In the data preprocessing stage, we first perform skull removal processing on MRI images using FreeSurfer. Subsequently, each PET scan was linearly calibrated to its matching MRI image. Each MRI image was then affine transformed according to the standard MNI template using SPM, and the PET image matching the MRI was also adjusted using the same transformation parameters. This approach ensures that MRI and PET images of the same participant are spatially consistent. After completing pre-processing, all MRI and PET images were resized to be consistent with the MNI template ($182 \times 218 \times 182$), with a voxel size of $1 \times 1 \times 1$ mm. During the training phase of the model, we again adjusted the image size to $112 \times 128 \times 112$, and the corresponding voxel size was $1.625 \times 1.704 \times 1.625$ mm to adapt to the network structure and reduce model parameters.

We performed two groups of experiments on the ADNI-1 and ADNI-2 datasets: (1) Normal people and AD patients (CN versus AD classification) (2) Normal people and cognitive impairment patients (CN versus MCI classification).

Table 2. Performance comparison of various methods for AD vs. CN and CN vs. MCI

Method	CN vs. AD					CN vs. MCI				
	ACC	AUC	SPE	SEN	F1S	ACC	AUC	SPE	SEN	F1S
CNN-M	87.57	93.16	90.32	84.52	86.59	61.96	66.07	64.52	60.66	67.89
CNN-P	88.14	95.94	86.02	90.48	87.86	69.93	72.98	58.06	**75.96**	77.01
FF	92.09	97.02	**93.55**	90.48	91.57	67.39	71.85	62.37	69.96	73.99
DF	91.35	96.90	**93.55**	89.29	90.91	69.57	73.22	62.37	73.22	76.14
TFN	91.67	95.37	93.24	89.84	90.91	64.13	70.76	64.52	63.93	70.27
LMF	92.09	97.07	92.27	91.67	91.67	63.41	64.02	66.67	61.75	69.11
Ours	**93.79**	**97.12**	**93.55**	**94.05**	**93.49**	**73.55**	**76.34**	**70.97**	74.86	**78.96**

3.2 Comparison Methods

We compare the following methods on different binary classification tasks to evaluate the effectiveness and robustness of our method.

CNN-M. MRI images are input into a 3D convolutional network to predict diagnostic results.

CNN-P. PET images are input into a 3D convolutional network to predict diagnostic results.

Feature-Level Fusion [13]**.** FF considers the information interaction between modalities at the representation level, and further obtains prediction results through the information interaction between modalities.

Decision-Level Fusion [13]**.** DF comprehensively considers the prediction results of different modalities, and the final prediction result is obtained by weighting the prediction results of different models.

Tensor Fusion Network [15]**.** TFN obtains intra-modal specificity information and inter-modal interaction information through vector outer product operations.

Low-rank Multi-modal Fusion [16]**.** LMF utilizes the low-rank structure of the high-dimensional information matrix to obtain intra-modal specificity information and inter-modal interaction information.

3.3 Experimental Setup

We calculated five statistical metrics for model evaluation, including accuracy (ACC), area under the receiver operating characteristic curve (AUC), sensitivity (SEN), specificity (SPE) and F1-score (F1S). In the experiment, ADNI-2 was selected as the training set and ADNI-1 as the testing set.

Table 3. Performance comparison of diagnosis with and without prior knowledge.

Method	CN vs. AD			CN vs. MCI		
	ACC	AUC	F1S	ACC	AUC	F1S
Ours(without)	93.22	97.06	92.86	67.03	70.85	73.62
Ours(with)	**93.79**	**97.12**	**93.49**	**73.55**	**76.34**	**78.96**

Table 4. Performance comparison of different representation combinations for diagnosis

Method	CN vs. AD			CN vs. MCI		
	ACC	AUC	F1S	ACC	AUC	F1S
$[f_m]$	87.57	93.16	86.96	61.96	66.07	67.89
$[f_p]$	88.14	95.94	87.86	69.93	71.98	77.01
$[f_m, f_{sh}]$	92.66	96.88	92.40	68.12	72.94	74.12
$[f_p, f_{sh}]$	90.40	96.54	89.94	72.46	75.84	78.77
$[f_m, f_p, f_{sh}]$	**93.79**	**97.12**	**93.49**	**75.55**	**76.34**	**78.96**

3.4 Evaluation of Automated Diseases Diagnosis

As shown in Table 2, the ADNI-1 test results show that in two binary classification diagnosis tasks, our proposed method leads other fusion methods on most evaluation indicators. In the first classification task CN vs AD, the diagnostic performance of the multi-modal fusion method is better than that of the single-modal diagnostic method in all aspects, indicating that the multi-modal fusion diagnostic method has advantages in certain tasks. Experimental results show that the performance of TFN and LMF methods for disease diagnosis is lower than that of FF and DF methods. Although TFN and LMF have intra-modal and inter-modal representations, their performance is not superior enough due to the acquisition of more information but also the acquisition of noise or the loss of disease-related information due to the setting of hyperparameter.

On the second classification task, except for our proposed method, other multi-modal fusion methods are better than the single-modal MRI network, but not as good as the single-modal PET network. Experimental results show that the second task is more difficult than the first task, which indirectly shows that it is difficult for a single-modal network to extract a pure representation with high discriminative ability to distinguish CN and MCI. The advantage of multimodal fusion networks is that they directly or indirectly consider information from different modalities. The insufficient performance of the single-modal network shows that the single-modal representation extracted by the single-modal extraction module contains a large amount of useless information, which seriously affects the effectiveness of the fusion of multi-modal information, thereby reducing the accuracy of disease diagnosis. Contrary to other methods, the basic network performance of our method on CN vs MCI tasks far exceeds that of single-modality networks. The difference from TFN and LMF is that we train the multi-modal shared representation extraction network based on a specific diagnostic task. This approach not only ensures the effectiveness of the extracted inter-modal information, but also reduces the generation of noise. The experimental results fully verify the effectiveness and generalization of our method.

3.5 Ablation Study

Introducing the Knowledge Gained from Single-Modal Learning Is of Great Significance for Guiding Multi-modal Learning. In MRI/PET scans, considering that not all areas are related to the disease, the knowledge learned by the single-modal network can be used to guide the representation learning between modalities in the multi-modal feature extraction stage, effectively reducing the pairs of disease-irrelevant information. Interference in model learning. As shown in Table 3, ablation experiments conducted in two classification tasks compared the introduction of single-modal learning knowledge with and without the introduction. Experimental results show that on these two tasks, the performance of the network incorporating single-modal learning knowledge surpasses that of the unincorporated network, verifying the effectiveness of this method, especially on more challenging tasks. This performance improvement is closely related to the efficiency of single-modal networks in filtering disease-irrelevant region information to enhance the feature learning efficiency of multi-modal networks.

The Consideration of Specificity and Sharing of Multimodal Features is Crucial. In the process of interaction and sharing of different modal information, there is a possibility that the multi-modal network may ignore single-modal specific information that is closely related to the disease, but this information is important and complementary. To test this hypothesis, ablation experiments were performed. As shown in Table 4, on two classification tasks, different information combinations are used to predict the two-classification task. The experimental results reveal that combining different information (see Fig. 1) can achieve better classification performance than a single modality. Especially when single-modality specific information and multi-modality shared information are fully utilized, the optimal disease classification effect can be achieved. These results not only confirm our hypothesis but also demonstrate its validity.

4 Conclusion

This paper delves into the significance of intra-modal and inter-modal information in disease diagnosis through the lens of multi-modal neuroimaging and proposes a Multi-Modal Specific Shared Representation Learning framework (MSSRL) aimed at assisting the diagnosis of Alzheimer's disease. The framework extracts intra- and inter-modal representations in a task-oriented manner, ensuring their relevance and effectiveness for disease prediction. To mitigate potential interference from modality-irrelevant regions in disease diagnosis, this study integrates prior knowledge from single-modality network learning during the multi-modal network learning phase to guide the extraction of shared representations between modalities. Experimental results on the ADNI public dataset demonstrate the superiority of our method over others, effectively enhancing the diagnostic performance for AD and confirming the importance of combining intra-modal specific information with inter-modal shared information to improve the accuracy of disease diagnosis.

Acknowledgments. The authors acknowledge the support from the National Natural Science Foundation of China (62301452), the National Natural Science Foundation of China (No. 62061050), and supported by the Yunnan University Professional Degree Graduate Practice Innovation Fund Project (ZC-2323516).

References

1. Petersen, R.C., Stevens, J.C., Ganguli, M., Tangalos, E.G., Cummings, J.L., DeKosky, S.T.: Practice parameter: early detection of dementia: mild cognitive impairment (an evidence-based review). Report of the quality standards subcommittee of the American academy of neurology. Neurology **56**, 1133–1142 (2001)
2. Baumgart, M., Snyder, H.M., Carrillo, M.C., Fazio, S., Kim, H., Johns, H.: Summary of the evidence on modifiable risk factors for cognitive decline and dementia: a population-based perspective. Alzheimers Dement. **11**, 718–726 (2015)
3. Kapoor, V., McCook, B.M., Torok, F.S.: An introduction to PET-CT imaging. Radiographics **24**, 523–543 (2004)
4. Weishaupt, D., Kochli, V.D., Marincek, B., Kim, E.E.: How does MRI work? An introduction to the physics and function of magnetic resonance imaging. J. Nucl. Med. **48**, 1910 (2007)
5. Shen, D., Wu, G., Suk, H.I.: Deep learning in medical image analysis. Annu. Rev. Biomed. Eng. **19**, 221–248 (2017)
6. Litjens, G., et al.: A survey on deep learning in medical image analysis. Med. Image Anal. **42**, 60–88 (2017)
7. Pan, Y., Liu, M., Xia, Y., Shen, D.: Disease-image-specific learning for diagnosis-oriented neuroimage synthesis with incomplete multi-modality data. IEEE Trans. Pattern Anal. Mach. Intell. **44**, 6839–6853 (2022)
8. Xiang, S., Yuan, L., Fan, W., Wang, Y., Thompson, P.M., Ye, J.: Bi-level multi-source learning for heterogeneous block-wise missing data. Neuroimage **102**, 192–206 (2014)
9. Pan, Y., Liu, M., Lian, C., Zhou, T., Xia, Y., Shen, D.: Synthesizing missing PET from MRI with cycle-consistent generative adversarial networks for Alzheimer's disease diagnosis. Med. Image Comput. Comput. Assist. Interv. **11072**, 455–463 (2018)
10. Zhou, J., Yuan, L., Liu, J., Ye, J.: A multi-task learning formulation for predicting disease progression. ACM (2011)
11. Sleeman, W.C., Kapoor, R., Ghosh, P.: Multimodal classification: current landscape, taxonomy and future directions. ACM Comput. Surv. **55**, 1–31 (2023)
12. Liu, Y., Yue, L., Xiao, S., Yang, W., Shen, D., Liu, M.: Assessing clinical progression from subjective cognitive decline to mild cognitive impairment with incomplete multi-modal neuroimages. Med. Image Anal. **75**, 102266 (2022)
13. Guo, Z., Li, X., Huang, H., Guo, N., Li, Q.: Medical image segmentation based on multi-modal convolutional neural network: study on image fusion schemes. IEEE (2018)
14. Zadeh, A., Chen, M., Poria, S., Cambria, E., Morency, L.-P.: Tensor fusion network for multimodal sentiment analysis. In: Association for Computational Linguistics (2017)
15. Chen, R.J., et al.: Pathomic fusion: an integrated framework for fusing histopathology and genomic features for cancer diagnosis and prognosis. IEEE Trans. Med. Imaging **41**, 757–770 (2022)

16. Liu, Z., Shen, Y., Lakshminarasimhan, V.B., Liang, P.P., Bagher Zadeh, A., Morency, L.-P.: Efficient low-rank multimodal fusion with modality-specific factors. In: Association for Computational Linguistics (2018)
17. Lian, C., Liu, M., Zhang, J., Shen, D.: Hierarchical fully convolutional network for joint atrophy localization and Alzheimer's disease diagnosis using structural MRI. IEEE Trans. Pattern Anal. Mach. Intell. **42**, 880–893 (2020)
18. Wachinger, C., Salat, D.H., Weiner, M., Reuter, M.: Whole-brain analysis reveals increased neuroanatomical asymmetries in dementia for hippocampus and amygdala. Brain **139**, 3253–3266 (2016)
19. Liu, M., Zhang, J., Adeli, E., Shen, D.: Landmark-based deep multi-instance learning for brain disease diagnosis. Med. Image Anal. **43**, 157–168 (2018)
20. Zhang, D., Wang, Y., Zhou, L., Yuan, H., Shen, D.: Alzheimer's disease neuroimaging, I.: multimodal classification of Alzheimer's disease and mild cognitive impairment. Neuroimage **55**, 856–867 (2011)
21. Zhang, D., Shen, D., Alzheimer's disease neuroimaging, I.: multi-modal multi-task learning for joint prediction of multiple regression and classification variables in Alzheimer's disease. Neuroimage **59**, 895–907 (2012)
22. Baron, J.C., et al.: In Vivo mapping of gray matter loss with voxel-based morphometry in mild Alzheimer's disease. Neuroimage **14**, 298–309 (2001)
23. Cui, R., Liu, M.: Hippocampus analysis by combination of 3-D DenseNet and shapes for Alzheimer's disease diagnosis. IEEE J. Biomed. Health Inform. **23**, 2099–2107 (2019)
24. Durazzo, T.C., Mattsson, N., Weiner, M.W., Alzheimer's disease neuroimaging, I.: smoking and increased Alzheimer's disease risk: a review of potential mechanisms. Alzheimers Dement **10**, S122–145 (2014)

An Activity Graph-Based Deep Convolutional Neural Network Framework in Symptom Severity Diagnosis Towards Parkinson's Disease Using Inertial Sensors

Mingchang Xu[1], Xiyang Peng[2], Po Yang[2(✉)], Jun Qi[3], and Yun Yang[1]

[1] National Pilot School of Software, Yunnan University, Kunming 650500, China
[2] Department of Computer Science, University of Sheffield, Sheffield S10 2TN, UK
po.yang@sheffield.ac.uk
[3] Department of Computing, Xi'an Jiaotong-Liverpool University, Suzhou 215123, China

Abstract. Parkinson's disease (PD) is a neurodegenerative disorder diagnosed and assessed primarily through the subjective Hoehn and Yahr (H-Y) staging system, which can be limited by doctor's subjectivity, particularly in classifying subtle motor symptoms, leading to potential misclassification. Previous research has predominantly relied on machine learning algorithms that incorporated hand-crafted feature extraction techniques. However, these approaches are constrained by domain-specific knowledge, which restricts the complexity of feature extraction, subsequently impacting algorithmic performance. To address these challenges, we propose a novel approach: a PD diagnosis assistance framework based on convolutional neural networks (CNNs) for automatic feature extraction and PD severity classification. In this paper, we collaborated with the First People's Hospital of Yunnan Province to collect motor data from 70 PD patients using wearable sensors equipped with an accelerometer and gyroscope. Neurologists assessed the PD severity on the Unified Parkinson's Disease Rating Scale (UPDRS) from simultaneously recorded video footages. The measured time data were transformed into activity graphs using recurrence transform, and two-dimensional images were constructed for training the network. The CNN model was trained by convolving images representing H-Y staging with kernels. The proposed symptom severity diagnosis of PD framework based on CNN was compared to previously studied machine learning algorithms and found to outperform them (accuracy = 84.52, recall = 80.18, f1-score = 80.69).

Keywords: Parkinson's Disease · Convolutional neural network · activity graphs · Inertial sensor

1 Introduction

Parkinson's Disease (PD) is a degenerative disorder of the central nervous system that affects around 1% of people over 60 in industrialised countries. People affected by PD show a variety of motor features that gain in severity with the progression of the disease,

D.-S. Huang et al. (Eds.): ICIC 2024, LNBI 14881, pp. 15–26, 2024.
https://doi.org/10.1007/978-981-97-5689-6_2

which include rigidity, slowness of motion, shaking and problems with gait. The severity and nature of these motor features vary over the course of the day, which has a significant impact on the quality of life of people with PD [1]. Objective, automated means to assess PD in people's daily lives are therefore much desired. Accurate classification of disease staging can help clinicians and patients make the best decisions [2]. To achieve this goal, various prediction methods based on machine learning algorithms are being developed, such as regression algorithms, kernel methods, and deep learning.

The Hoehn and Yahr Staging Scale [4] of Parkinson's disease is a widely used system to classify the severity of Parkinson's disease based on clinical observations of motor symptoms. Traditionally, H-Y Staging assessment relies on patient self-reporting or clinician examination. However, these methods are limited by their reliance on human input, making continuous and frequent monitoring outside clinical settings challenging. Additionally, the accuracy of measurements can vary significantly between different raters. To address these limitations, various alternative methods utilizing IoT-based sensor technologies, such as electromagnetic motion trackers, electromyography, touch sensors, and inertial sensors, have been explored. Among these, accelerometers and gyroscopes are particularly favored due to their compact size and affordability [3].

Previous research has explored machine learning algorithms for diagnosing Parkinson's disease, achieving high-level classification beyond linear predictive functions [5]. G. Ferrari et al. [6] proposed motion features in time and frequency domains for automatic UPDRS score evaluation during sit-to-stand tasks, employing a support vector machine model for classification. Parisi F et al. [7] suggested kinematic features for three tasks (Leg agility, Sit-to-Stand, and Gait) and designed an automatic scoring system using K-nearest neighbor and support vector machine classifiers. Machine learning classifiers outperform linear regression in rating scale estimation, yet their efficacy may be limited by manual feature extraction. Researchers often operate within domain-specific knowledge constraints when conducting handcrafted feature extraction, restricting feature complexity and impacting algorithm performance.

Grover et al. [8] proposed a deep learning framework to extract features from voice signals of Parkinson's disease patients and predict disease severity. In contrast, our study focuses on analyzing motion data. Kim HB et al. [9] introduced a method using a convolutional neural network to estimate tremor severity in PD patients. Tuan D. Pham et al. [10] developed a method for feature dimension creation and utilized deep learning models for classifying computer-key hold time series. However, their study only examines typing activity and overlooks other relevant motor signs for PD diagnosis and monitoring.

At present, the majority of Parkinson's movement datasets focus on specific actions such as finger tapping, walking, and drawing spirals. In contrast, our dataset selects 14 movements from the Unified Parkinson's Disease Rating Scale (UPDRS) [11], aiming to capture a more comprehensive range of motor symptoms in Parkinson's patients. This addresses the limitations of current datasets, providing us with a more comprehensive set of movement data for better support in comprehensive analysis and assessment.

In this paper, we present a novel activity graph-based deep convolutional neural network framework for symptom severity diagnosis in Parkinson's Disease. To the best of our knowledge, this represents the first application of activity graphs and CNN in the

Table 1. Patient characteristics.

Characteristics	Observations
Number of patients	**70**
Sex (male/female)	**34/36**
Age (y)	**66.5 ± 9.6**
Height (cm)	**160.3 ± 7.3**
Weight (kg)	**59.1 ± 9.1**
Hoehn and Yahr (H&Y) stage	**2.4 ± 0.9**

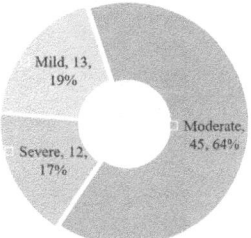

Fig. 1. Distribution of Parkinson's Disease Patients' Symptom Severity

analysis of PD symptom severity. Our key contributions include the introduction of a specialized CNN architecture, incorporating activity graph-based features, and composed of convolutional layers that autonomously extract symptom severity features from raw data. Additionally, we collaborated with the First People's Hospital of Yunnan Province to collect motor data from 70 PD patients using wearable motion sensors equipped with an accelerometer and gyroscope. The proposed framework demonstrated superior performance compared to established machine learning and deep learning algorithms.

2   Subjects and Data Collection

2.1   Participants

Participants were recruited from individuals exhibiting symptoms of Parkinson's disease who visited the First People's Hospital of Yunnan Province between August 2020 and July 2022. A total of 70 patients voluntarily participated in the experiments, providing written informed consent prior to their inclusion. The final analysis included only those patients diagnosed with PD. Demographic and clinical data for the 70 PD patients are summarized in Table 1, while Fig. 1 illustrates the distribution of symptom severity among Parkinson's disease patients.

2.2   Data Collection

Inertial Measurement Unit (IMU) data were acquired through the use of accelerometers and gyroscopes integrated into the Shimmer3 platform. All study participants were equipped with Shimmer sensors for the purpose of collecting exercise-related data. Each participant wore a total of two sensors positioned in different locations, namely the right hand and left hand as illustrated in Fig. 2. Subsequently, the sensors were connected to a laptop via Bluetooth connectivity. Following this, the Shimmer sensors were configured using ConsensysPRO software, utilizing a data collection and storage frequency of 201.3Hz for 3D accelerometer, 3D gyroscope, and 3D magnetometer data. The collected data from multiple sensors was displayed and recorded in real-time through video

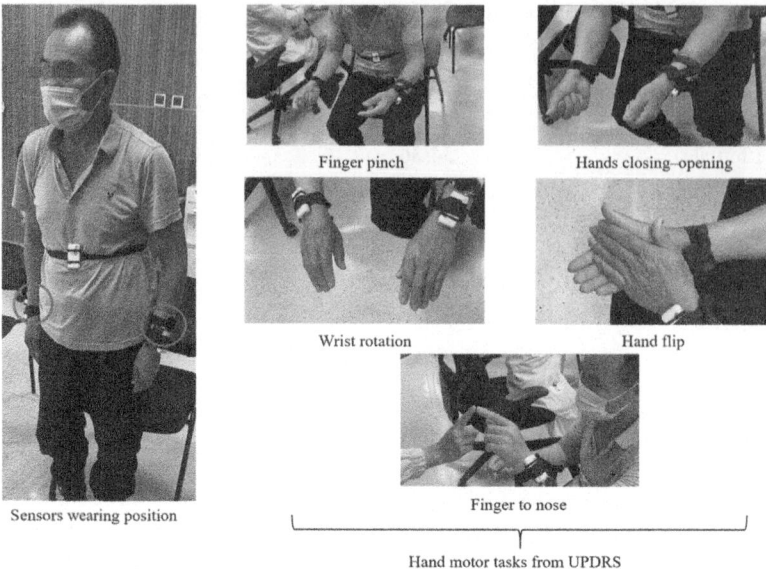

Sensors wearing position

Finger pinch

Hands closing–opening

Wrist rotation

Hand flip

Finger to nose

Hand motor tasks from UPDRS

Fig. 2. Sensors wearing position and examples of hand motor tasks

recording. Skilled medical professionals then assessed the patient's H-Y staging grade based on the recorded video footage.

The exercise data of both healthy individuals and PD patients were gathered following identical standards. This study encompassed a total of basic activity types (Finger Tapping, Clench Fist and Open, Hands Rotation, Hand Alternating-right/left, Finger to nose-right/left, Standing with arms held) and daily life (Walk Back and Forth, Arising from a Chair, Drinking Water, Pick Things, Sitting, Standing), which included tasks from the third segment of UPDRS. Participants were briefed on the execution of each activity, with each task lasting between 20 to 50 s.

3   Methodology

This paper aims to improve the accuracy achieved in earlier works in distinguishing the presence of Parkinson's Disease. The main innovation of this approach lies in utilizing a convolutional neural network to classify activity graphs using the recurrence transform of motion time series signals from Parkinson's patients. Another point of enhancement is a data augmentation technique developed specifically for activity graphs using recurrence transform, considering the nature of the image and the structures forming the textures. An overview of the steps adopted in this approach is shown in Fig. 3.

First, activity graphs using recurrence transform were generated from the wearable motion signals. Then, a data augmentation technique was applied to increase the number of samples. The sub-images were then used as independent inputs to the CNN. The network output was three classes of PD H-Y Staging: Mild, Moderate, and Severe.

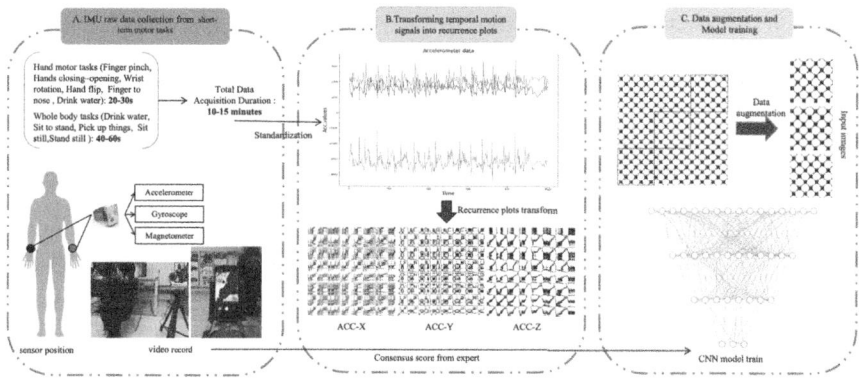

Fig. 3. Overall experimental flow in this study. Sub-Fig.A shows the data acquisition forms. Sub-Fig.B shows how to transform temporal motion signals into an activity graph using the recurrence transform. Sub-Fig.C shows the data augmentation applied to the characteristics of recurrence plots and the training of the convolutional neural network model.

3.1 Activity Graph Generation

Due to the nature of our collected dataset, which involves Parkinson's patients performing the same repetitive actions within a specified time, the inherent repetitiveness serves as an indicator of bradykinesia symptoms in these patients. The activity graphs using recurrence transform illustrate the repetitive or recursive relationships between data points in the time series [13]. This transformation is valuable for capturing patterns, periodicity, and repetitiveness in the movement time series of Parkinson's patients, facilitating a more in-depth analysis of symptom severity. Therefore, we utilize recurrence transformation to generate activity graphs that represent the motion patterns of Parkinson's patients.

The core concept in applying nonlinear dynamics methods to analyze the trajectory of a dynamic system is that the fundamental structure, containing all information about the system, can be reconstructed using only one component of the system. The commonly employed Packard-Takens procedure facilitates the reconstruction of the phase trajectory of a dynamic system from a single realization.

$$F(t) = [x(t), x(t + \tau), \dots, x(t + m\tau)] \tag{1}$$

where $F(t)$ is m-dimensional pseudo-phase space, $x(t)$ is time realization of the system, τ is delay period.

Recurrence is a fundamental property of dynamical systems, which can be exploited to characterise the system's behaviour in phase space [12]. Recurrence plots are two-dimensional matrices of order N that allow the visualization of the recurring behavior of dynamic systems, regardless of their dimensionality. In this matrices, black or white points are marked. A black point indicates the recurrence of a specific state at time i at another time j [13].

According to Marwan [14], a recurrence plot can be defined mathematically by Eq. 2:

$$R_{i,j}^{m,\epsilon} = \theta(\varepsilon - \|x_i - x_j\|), x_i, x_j \in R_m, i, j = 1 \dots N \tag{2}$$

where N is the number of states x_j formed by the system, obtained through the method of time delays [15]. And ε is neighborhood size of a point x_i, $\|x_i - x_j\|$ is distance between points, θ is Heaviside function, denoted as $H(t)$, is defined as

$$H(t) = \begin{cases} 0 \; if \; t < 0, \\ 1 \; if \; t \geq 0. \end{cases} \tag{3}$$

Numerical analysis of recurrence plots makes it possible to calculate the complexity measures of recurrence structures [14]. Consider the main quantitative characteristics of recurrence plots.

Recurrence rate (RR)

$$RR = \frac{1}{N^2} \sum\nolimits_{i,j=1}^{N} RP_{i,j}^{\varepsilon} \tag{4}$$

RR shows the density of recurrent points, simply by counting them. This measure shows the probability of location a point in the recurrence plot (probability of state repetition).

The probability, P_t, that the system returns to the ε- neighborhood of a previous point x_i in the trajectory after τ time steps can be directly estimated from the recurrence plot.

$$P_\tau = \frac{1}{N - \tau} \sum\nolimits_{i,j=1}^{N-\tau} \theta(\varepsilon - \|x_i - x_{i+\tau}\|) \tag{5}$$

The next measure considers diagonal lines. Let the $P(l) = \{l_i; i = 1 \ldots N_l\}$ is frequency distribution of the lengths of the diagonal lines in RP, where l_i is length of i - th diagonal line, N_l is number of diagonal lines. Stochastic time series may generate very short diagonals or none at all, whereas deterministic processes produce long diagonals and a small number of isolated recurrence points.

Recurrence points relation

$$DET = \frac{\sum_{l=l_{min}}^{N} lP^{\varepsilon}(l)}{\sum_{i,j}^{N} R_{i,j}^{\varepsilon}(l)} \tag{6}$$

is referred to as a measure of determinism (DET) or system predictability. It is important to note that this measure does not reflect the actual determinism of the process.

We convert the motion time-series data from the three-axis accelerometer and three-axis gyroscope worn by Parkinson's disease patients into a recursive graph, as shown in Fig. 4.

3.2 Data Augmentation

Due to the difficulty for mild patients to perceive Parkinson's disease and seek medical attention, as well as the challenge for severe patients to visit hospitals, our dataset lacks data from mild and severe patients, leading to a certain imbalance in classes, as illustrated in Fig. 1. To address this issue, we employ a data augmentation technique.

Convolutional Neural Network (CNN) algorithms typically demand a substantial quantity of images to achieve generalization capabilities that comprehend the potential

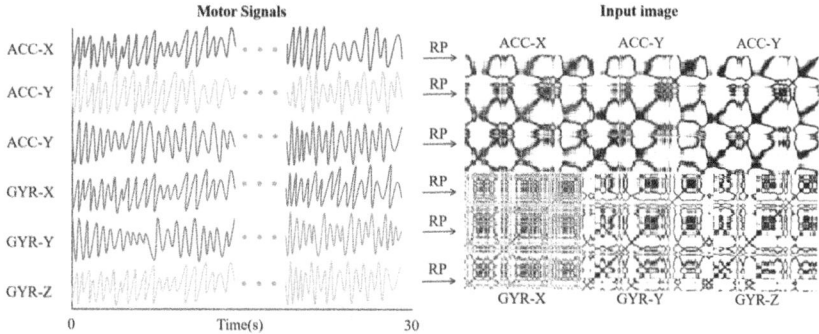

Fig. 4. Input image for training and testing of CNN. All signals from the accelerometer and gyroscope have been converted into the recurrence plots.

variations within the dataset. However, in the case of activity graphs using recurrence transform derived from wearable motion signals, despite a total of 5880 samples, their recurrent nature enables them to be treated as texture images. This characteristic allows for an increase in the effective sample size by subdividing the original images.

In this research, a data augmentation technique was devised for activity graphs using recurrence transform by employing a strategy of splitting the original dataset's images. To preserve the texton concept, the orientation of pattern structures along the main diagonal and a minimum scale were taken into account. The original activity graphs using recurrence transform have dimensions of 500×500 pixels. To obtain sub-images, a mask measuring 167×167 pixels was utilized, as illustrated in sub-figure C of Fig. 3. By generating three non-overlapping sub-images, independent inputs were created for training the CNN model.

A significant benefit of the data augmentation technique developed in this study is that it preserves the original features of the images without modification. The proposed method ensures that the symmetry of activity graphs using recurrence transform about the main diagonal is maintained. This is crucial because certain operations could introduce new features into the database, potentially leading to erroneous attribution of non-existent anomalies.

3.3   Convolutional Neural Network

1) Neural Network Architecture: Convolutional Neural Networks (CNNs) are a type of Artificial Neural Network (ANN) specifically designed for data with a grid topology, such as images and videos. Convolutions play a crucial role in these networks and are considered the most important operation. In this study, a CNN model was developed to classify activity graphs using recurrence transform of wearable motion signals into three categories based on PD H-Y Staging: Mild, Moderate, and Severe.

The CNN model's feature extraction component is crucial for capturing meaningful information from input images. It starts with a 3-channel input, followed by convolutional operations using 16 filters sized 3×3. ReLU activation is then applied to introduce non-linearity, enhancing the network's ability to detect complex patterns. Subsequently, max pooling with a kernel size of 2 × 2 and stride of 2 is employed to reduce the spatial dimensions of the feature maps. The process continues with a second convolutional layer comprising 32 filters, each 3 × 3 in size. ReLU activation is again utilized, followed by max pooling operations. This sequence of convolution and pooling operations facilitates hierarchical feature extraction. The third and final convolutional layer includes 64 filters of size 3 × 3. ReLU activation is applied similarly to the previous layers, and max pooling is employed to downsample the feature maps. This layer further enhances the extracted features, capturing more intricate patterns.

Following the convolutional layers, the feature maps are flattened into a one-dimensional vector using the flatten layer. This transformation allows for the transition from spatial information to a linear representation. The classifier component of the CNN model is responsible for the final classification. It starts with a dropout layer, where 50% of the input units are randomly set to zero during training, reducing overfitting risks. This layer is followed by a fully connected layer with 256 neurons, employing ReLU activation. Finally, another linear layer maps the extracted features to the desired number of classes (in this case, three classes corresponding to PD patients' severity).

2) Training: Because of the small number of Severe PD samples, an imbalance in the data. Thus, 13 samples of mild PD, 45 of moderate PD and 12 of severe PD were randomly selected. Consequently, the final set used for testing and training the model was made 70 Parkinson's patient samples. The training group was split according to the technique of data augmentation exposed in this study. The images were then used as independent inputs to train the model. A back-propagation algorithm was used to estimate the network parameters, which is a method based on the gradient descent optimization. More specifically, the Adam optimizer algorithm was used, with a learning rate $\eta = 0.0001$, $\beta_1 = 0.9$, $\beta_2 = 0.999$ and $\varepsilon = 10^{-7}$. During training, the stop conditions were if the number of epochs reached 100, or, to avoid overfitting, if the validation loss increased five consecutive times. To ensure robust generalization of the network across diverse datasets, we adopted a 70:30 split between the training and test sets.

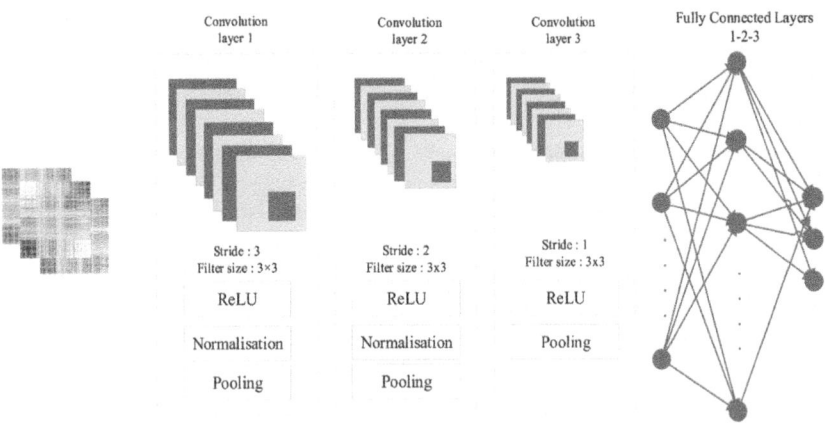

Fig. 5. CNN architecture.

4 Results

We transformed the motion time-series data of Parkinson's patients into activity graphs using recurrence transform and observed distinct features corresponding to different symptomatic stages, as illustrated in Fig. 6. It can be observed that there are significant differences in the activity graphs between different disease severity levels, while activity graphs between similar disease severity levels exhibit similar structural features.

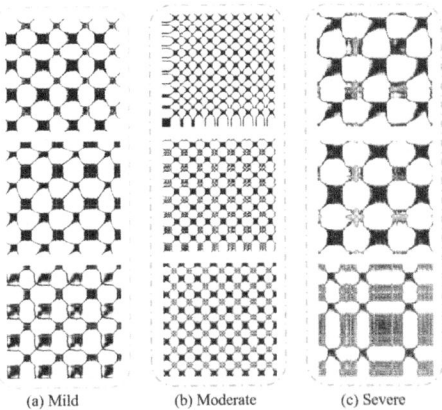

(a) Mild   (b) Moderate   (c) Severe

Fig. 6. Activity graphs using recurrence transform for patients with Parkinson's disease in mild, moderate, and severe stages

Table 2 summarizes the predictions generated by proposed framework and other classifiers through the analysis of accelerometer and gyroscope data on the wrist. Our proposed framework demonstrates significant improvement compared to traditional machine learning algorithms, achieving an accuracy of 84.52%, recall of 80.18%, precision of 82.51% and F1-score of 80.69%. The accuracy is 7.57% higher than the best-performing machine learning algorithm with the decision tree and 4.36% higher than the best-performing deep learning model with AlexNet. This demonstrates superior performance in comparison.

Table 3 summarizes a more detailed comparison of our proposed framework's predictive results in patients with mild, moderate, and severe stages of Parkinson's disease, as compared to traditional machine learning methods. Our approach achieves the best performance in predicting patients with moderate symptoms, with a precision of 86.57%. However, its recall is lower than some traditional machine learning algorithms such as KNN and SVM, which is attributed to the imbalance in data caused by the largest number of patients with moderate symptoms. Particularly, there is a significant improvement in recall and F1-score for other severity levels, indicating the robustness and generalization capability of our model.

Table 2. Prediction results of the proposed framework and conventional machine learning classifiers.

Method	Accuracy	Precision	Recall	F1-Score
Proposed Framework	**84.52**	**82.51**	**80.18**	**80.69**
AlexNet	80.16	76.50	74.59	74.91
ResNet	67.41	65.11	49.07	52.15
Support Vector Machine	70.86	67.37	49.58	52.64
Random Forest	72.39	65.72	61.82	63.30
Decision Tree	76.95	70.36	70.16	70.25
Logistic Regression	72.43	65.39	58.69	61.00
k-Nearest Neighbors	74.52	74.01	55.97	60.41
Multilayer Perceptron	74.73	75.66	56.99	59.62

Table 3. The predicted results of the proposed framework and traditional machine learning methods in patients with mild, moderate, and severe stages of Parkinson's disease.

Method	Mild			Moderate			Severe		
	Prec	Recall	F1	Prec	Recall	F1	Prec	Recall	F1
Proposed Framework	**80.72**	**70.53**	**75.28**	**86.57**	91.19	**88.82**	80.23	75.82	**77.97**
AlexNet	61.11	64.76	67.25	77.78	86.94	87.36	78.60	62.07	70.13
ResNet	39.08	26.36	31.48	70.53	89.28	78.81	75.71	31.58	46.15
Support Vector Machine	58.99	20.45	30.37	71.61	94.17	81.36	71.50	34.13	46.20
Random Forest	47.92	48.88	48.40	78.95	84.02	81.41	70.29	52.51	60.11
Decision Tree	60.92	62.59	61.75	84.14	84.41	84.28	66.00	63.48	64.72
Logistic Regression	50.59	31.29	39.14	76.19	87.36	81.39	63.39	56.80	62.47
k-Nearest Neighbors	67.86	33.17	44.56	74.65	**94.89**	79.52	83.56	39.86	53.10
Multilayer Perceptron	79.80	19.70	31.60	74.90	93.98	72.29	83.36	57.28	63.91

Finally, we trained the proposed network with input images from different sensor signals and observed the quantification performance that shown in Table 4. It can be observed that the model, which incorporates both accelerometer and gyroscope data into 2D input images, achieves the best performance, outperforming the individual use of the accelerometer and gyroscope.

Table 4. Classification results using different 2D input images.

Method	Input	Accuracy	Recall	F1-score
Our Framework	Acc. And Gyro. (6 axes)	**84.52**	**80.18**	**80.69**
	Accelerometer (3 axes)	74.21	63.51	64.88
	Gyroscope (3 axes)	75.59	65.09	67.62

5 Discussion and Conclusion

In this study, this paper introduced a novel activity graph-based deep convolutional neural network (CNN) framework to accurately assess H-Y staging severity by analyzing raw data directly. Unlike traditional methods, our framework autonomously extracts relevant patterns from minimally processed raw data, eliminating the need for manual feature extraction. To achieve comprehensive diagnosis akin to a doctor, the approach integrates global information for a holistic assessment, extending beyond analyzing local features.

For the first time, this paper explored using an activity graph-based CNN framework for clinical score estimation. We evaluated its effectiveness against prior methods documented in the literature for extracting distinguishing features from both time and frequency domains.

Dealing with larger and more diverse datasets presents a formidable challenge in discerning effective features for symptom severity classification. The observed enhancements across all performance metrics signify that automatic feature extraction within our proposed framework surpasses traditional manual feature engineering approaches. Our proposed framework architecture outperformed most conventional machine learning methods in precision, recall, and F1-score, as shown in Table 3.

Unlike conventional approaches relying on handcrafted features, our framework was trained using activity graphs generated from minimally processed raw signals. Results showed noticeable enhancement in estimation accuracy through network design optimization. We believe this methodology can be adapted for precise and continuous monitoring of H-Y staging in everyday activities.

Acknowledgments. This research was supported by the National Natural Science Foundation of China (No. 62061050), the Young Scientists Fund of the National Natural Science Foundation of China (No.62301452) and the Scientific Research Fund of the Yunnan Provincial Department of Education (No. 2024Y034).

References

1. Twelves, D., Perkins, K.S., Counsell, C.: Systematic review of incidence studies of Parkinson's disease. Mov. Disord. **18**, 19–31 (2003)

2. Kubler, D., Schroll, H., Buchert, R., Kuhn, A.A.: Cognitive performance correlates with the degree of dopaminergic degeneration in the associative part of the striatum in non-demented Parkinson's patients. J. Neural Transm. (Vienna) **124**, 1073–1081 (2017)

3. Qi, J., Yang, P., Hanneghan, M., Tang, S., Zhou, B.: A hybrid hierarchical framework for gym physical activity recognition and measurement using wearable sensors. IEEE Internet Things J. **6**, 1384–1393 (2019)

4. Hoehn, M.M., Yahr, M.D.: Parkinsonism: onset, progression, and mortality. 1967. Neurology **57**, S11–26 (2001)

5. Mei, J., Desrosiers, C., Frasnelli, J.: Machine learning for the diagnosis of Parkinson's disease: a review of literature. Front Aging Neurosci. **13**, 633752 (2021)

6. Giuberti, M., et al.: Automatic UPDRS evaluation in the sit-to-stand task of parkinsonians: kinematic analysis and comparative outlook on the leg agility task. IEEE J. Biomed. Health Inform. **19**, 803–814 (2015)

7. Parisi, F., et al.: Body-sensor-network-based kinematic characterization and comparative outlook of UPDRS scoring in leg agility, sit-to-stand, and gait tasks in Parkinson's disease. IEEE J. Biomed. Health Inform. **19**, 1777–1793 (2015)

8. Grover, S., Bhartia, S., Akshama, Yadav, A., Seeja, K.R.: Predicting severity of Parkinson's disease using deep learning. Procedia Comput. Sci. **132**, 1788–1794 (2018)

9. Kim, H.B., et al.: Wrist sensor-based tremor severity quantification in Parkinson's disease using convolutional neural network. Comput. Biol. Med. **95**, 140–146 (2018)

10. Pham, T.D., Wårdell, K., Eklund, A., Salerud, G.R.: Classification of short time series in early Parkinson s disease with deep learning of fuzzy recurrence plots. IEEE/CAA J. Automatica Sinica **6**, 1306–1317 (2019)

11. Martinez-Martin, P., et al.: Parkinson's disease severity levels and MDS-Unified Parkinson's Disease Rating Scale. Parkinsonism Relat. Disord. **21**, 50–54 (2015)

12. Marwan, N., Carmenromano, M., Thiel, M., Kurths, J.: Recurrence plots for the analysis of complex systems. Phys. Rep. **438**, 237–329 (2007)

13. Eckmann, J.P., Kamphorst, S.O., Ruelle, D.: Recurrence plots of dynamical systems. Europhys. Lett. (EPL) **4**, 973–977 (1987)

14. Marwan, N.: Encounters with neighbours: current developments of concepts based on recurrence plots and their applications. Norbert Marwan (2003)

15. Takens, F.: Detecting strange attractors in turbulence. pp. 366–381. Springer, Heidelberg (1981)

An Optimization Method for Drug Design Based on Molecular Features

Xuan Liu[1,2], Xiaoli Lin[1,2(✉)], and Fengli Zhou[3]

[1] School of Computer Science and Technology, Wuhan University of Science and Technology, Wuhan 430081, China
linxiaoli@wust.edu.cn
[2] Hubei Province Key Laboratory of Intelligent Information Processing and Real-Time Industrial System, Wuhan 430081, China
[3] Department of Information Engineering, Wuhan City College, Wuhan 430081, China

Abstract. Computer-aided drug design can reduce the cost and improve the efficiency of drug development. Due to the fact that the features of a molecule are closely related to its structure, the commonly used SMILES representation can only characterize the relative arrangement order of atoms to atoms in the horizontal direction, missing the relative positional relation between atoms in space. Therefore, this paper proposes an optimization generation model based on AIS encoding and drug-target interaction constraints. The model employs AIS coding for feature representation of drug molecules. The local structural information is further enhanced on the basis of obtaining global information by aggregating the information of neighboring atoms or fragments of each atom. To further validate the effect of the targeted drug generation model, the main protease 3CLpro was selected as the targeted protein. Experiments show that the proposed optimization method can generate the molecules with high affinity.

Keywords: Drug design · Drug generation · Optimization · Molecular features

1 Introduction

With the development of bioinformatics field and the deepening understanding of disease mechanism, the drug design has become one of the research topics in the field of medicine [1]. Drugs are chemical substances or biological products that are used to prevent, diagnose, treat diseases or adjust physiological functions. They can be naturally occurring or synthetic molecules with the ability to produce specific biological effects on the organism through various mechanisms [2]. Although existing drug libraries already contain a wide range of drugs for the treatment of different diseases, there is a relative lack of targeted drug therapies at this stage [3]. The development of novel drugs is urgently needed for chronic diseases such as cardiovascular diseases, oncology and diabetes. Therefore, designing drugs for diseases that are currently difficult to overcome [4] and enhancing existing drug treatment strategies based on the designed drugs [5, 6] are of significant research value for clinical treatment practice. This not only has the potential to provide patients with more effective treatment options, but also may further promote the implementation of individualized treatment plans [7, 8].

© The Author(s), under exclusive license to Springer Nature Singapore Pte Ltd. 2024
D.-S. Huang et al. (Eds.): ICIC 2024, LNBI 14881, pp. 27–36, 2024.
https://doi.org/10.1007/978-981-97-5689-6_3

Advances in computational methods and technologies are driving a fundamental shift in the drug discovery and design paradigm [9, 10]. In traditional drug discovery [11], high-throughput screening techniques [12] are commonly used to identify candidate compounds with potential medicinal value. Then, these compounds are prepared by means of chemical synthesis, and their pharmacological effects are evaluated by a series of biological activity tests. Conventional drug discovery is often time-consuming and the experimental process is fraught with uncertainty. Despite the historical success of traditional method, its limitations in terms of efficiency and cost remain key issues that need to be addressed.

The design of drug molecules that bind to receptor proteins is an important challenge in the current field of drug design for the design of rationally targeted drugs [13]. Masuda et al. [14] proposed to convert complexes of receptor proteins and ligand molecules into three-dimensional images, and then, to convert the drug design task into generating three-dimensional images. However, the use of 3D convolutional neural networks on atomic density networks is spatially incapable of achieving degeneracies such as translation and rotation. Luo et al. [15] proposed a new autoregressive sampling method to generate molecules in the spatial extent of the binding pockets of receptor proteins, which combines a message-passing neural network as well as a multilayered perceptron classifier using a masking pattern training method. The method is only capable of generating atoms in space, and lacks the ability to generate chemical bonds corresponding to the connections, such that some of the generated molecules, which are heavily distorted in space, lacking guidance to the structure of the drug itself.

With the increasing availability and variety of biological data in relevant databases such as the ChEMBL database [16], the DrugBanK database [17], and the Uniprot database [18], the drug design has made a remarkable progress. In addition, the protein interaction prediction helps to identify key proteins associated with diseases [19, 20], which often become potential targets for drug design and provide a clear direction for drug discovery. Moreover, the emergence of deep learning techniques has opened up new opportunities for drug design [21]. Currently, the task of designing drug molecules capable of targeting specific protein binding pockets in conjunction with structural information about the receptor remains challenging. Currently, there are still major challenges in the task of designing drug molecules for specific targeted protein binding pockets in combining the structural information of the receptor.

Based on our previous study [22, 23], this study proposes an optimization method for drug design based on molecular features. The remaining of this paper is organized as follows. Section 2 introduces the proposed method. Experiments are presented and discussed in Sect. 3. Finally, Sect. 4 gives conclusion and perspectives for the future.

2 Methods

2.1 Pocket of Targeted Protein

In this paper, the Main Protease [24] (3C-like protease, 3CLpro) of a Corona Virus Disease was used as the targeted protein for the study. Multiple active pockets may exist in a protein. It is general most convenient to select by default the active pocket in the targeted protein where the ligand is carried without any processing and screening. However, for the same protein structure, due to different methods of resolving the structure, different protein structures carrying ligands are obtained, and the positions of the active pockets where the ligands are located are different.

There are several active pockets in a resolved protein structure. It cannot be ensured that the ligand carried by a particular resolved protein structure is located in the best active pocket of that protein. To obtain the best active pocket, the ligand carried by the selected protein is stripped from its resolved file. Then, the ligand is blindly docked to that protein. Finally, the best active pocket of that protein is determined by combining the position of the pocket where the ligand is located, as shown in Fig. 1.

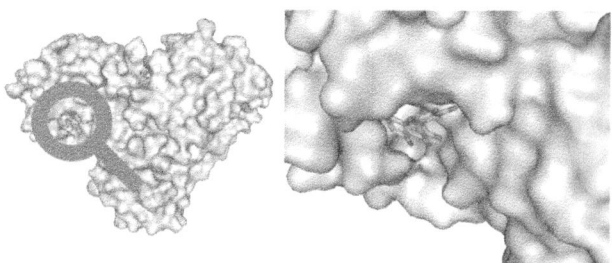

Fig. 1. Targeted protein with Ligand

2.2 Feature Extraction of Targeted Protein

The aim of this paper is to generate drugs with specific features and the generated drug has the potential to treat a specific disease. Drugs and the targeted protein produce biological effects through intermolecular interactions [25]. In this process, the drug binds to the targeted protein and forms a stable complex in the targeted protein pocket, thus exerting the therapeutic effect of the drug. The specific region of the targeted protein that can interact with the drug and perform a biological function is known as the active site of the targeted protein. This region usually includes a specific set of atoms or residues. Therefore, it is of great interest to rationally characterize the active site of the targeted protein. In addition, the biological function of proteins usually involves interactions based on amino acid residues, so it is more appropriate to study the active sites of a targeted protein at the level of amino acid residues to meet the needs of drug design.

The process of obtaining the active site in the targeted protein in this paper is shown in Fig. 2. The process of obtaining the active site of the targeted protein is divided into three steps:

- Construct an atomic dictionary of the entire targeted protein.
- The distance between the atoms of the targeted protein and the atoms of the drug is used as a measure to determine the atoms of the protein corresponding to the active site of the targeted protein.
- From the complete dictionary of atoms of the targeted protein, the amino acid residues where the atoms of the identified proteins are located are selected.

The structure of constructing the atomic dictionary corresponding to the targeted protein is shown in Fig. 3. Each atom in the targeted protein is constructed according to the format of the atom dictionary, which makes it easy to find the information about the corresponding amino acid residue after searching for the information of the atom in the targeted protein.

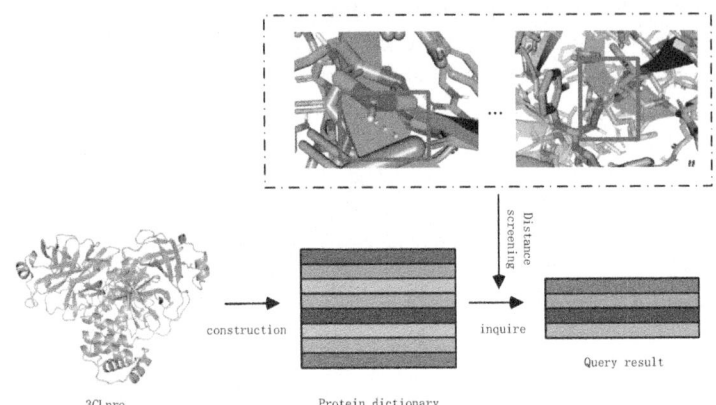

Fig. 2. Process of obtaining the active site of a targeted protein

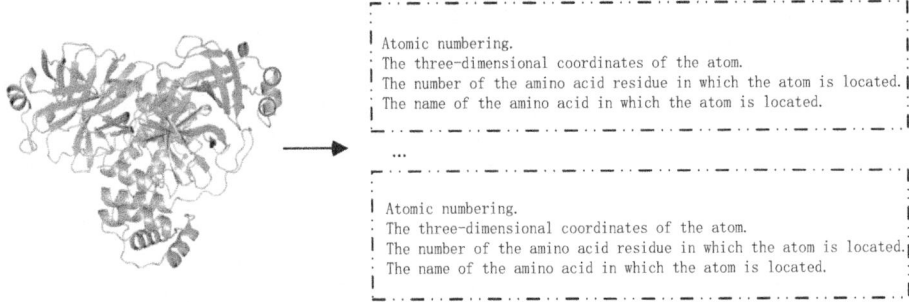

Fig. 3. Structure of the atomic dictionary of the targeted protein

2.3 Feature Representation of Drug Molecule

Compounds can usually be represented using a one-dimensional linear sequence. Among the common representation methods are Simplified Molecular Input Line Entry System (SMILES) representation, International Chemical Identifier (InChI), DeepSMILES representation, and so on.

The SMILES representation is a simple and compact description. It is very easy to understand and is also the most widely used drug representation. The InChI identifier is a standardized representation of chemical substances introduced by the International Union of Pure and Applied Chemistry. It represents each compound uniquely, precisely and without ambiguity. This is due to the fact that it is designed with multiple layers of information and sublayer compositions in its syntax, but this complexity makes it very difficult for models to learn and generate drugs efficiently. The DeepSMILES representation simplifies the syntactic representation of the rings and branches in the SMILES representation by making syntactic modifications. However, the conversion between DeepSMILES sequences and molecular structures is not necessarily fully reversible. Therefore, there may be some pitfalls in learning and generating drugs based on DeepSMILES representations.

To better help the model learn the structure of the molecule, the AIS (atom-in-SMILES) algorithm [26] is used, which builds on the SMILES representation by aggregating information about the neighboring atoms around it for each atom, enriching the local structural information contained in that atom. As shown in Table 1, the SMILES representation, the InChI identifier, the DeepSMILES representation, and the AIS encoded representation are performed for the same chemical substance.

Table 1. Examples of different drug representations

Methods	Representations
SMILES	O(C)c1ccc(C(=O)O)c(O)c1OC
InChI	InChI = 1S/C9H10O5/c1–13-6–4-3–5(9(11)12)7(10)8(6)14–2/h3–4,10H,1-2H3,(H,11,12)
DeepSMILES	OC)ccccC = O)O))cO)c6OC
AIS	[CH3;!R;O] [O;!R;CC] [c;R;CCO] 1 [cH;R;CC] [cH;R;CC] [c;R;CCC] ([C;!R;COO] (= [O;!R;C]) [OH;!R;C]) [c;R;CCO] ([OH;!R;C]) [c;R;CCO] 1 [O;!R;CC] [CH3;!R;O]

2.4 Model

To better help the model learn the drug features and its relation with the target, the drug-target interaction features are used as the generative constraints. The AIS coding algorithm is used to characterize the feature vectors of the drug. A targeted drug generation model is built by combining the Linear layer, the Batch Normalization layer and the Gated Recurrent Unit (GRU). It can generate small molecules that are more suitable for the specific targeted protein by learning the basic laws of drug-target interactions.

The GRU layer controls the flow of information by employing a gating mechanism, and this flexible flow of information enables the GRU layer to improve the model's ability to capture long term dependencies, which further enhances the model's ability to learn from sequential data. The gating mechanism in the GRU layer consists of a reset gate r_t and an update gate z_t. The reset gate is responsible for the extent to which past information is forgotten and retained. The update gate is responsible for selectively merging current information with past information.

First, the hidden state of the previous time step is set to h_{t-1} and the input of the current time step is set to x_t. The information of reset gate r_t and update gate z_t is obtained through h_{t-1} and x_t. The reset gate r_t is defined as

$$r_t = sigmoid\left(W_r[h_{t-1}, x_t] + b_r\right) \tag{1}$$

where W_r is the weight matrix of the reset gate and b_r is the bias term of the reset gate. $sigmoid(\cdot)$ Denotes the sigmoid activation function. The update gate z_t is defined as

$$z_t = sigmoid\left(W_z[h_{t-1}, x_t] + b_z\right) \tag{2}$$

where W_z is the weight matrix of the update gate and b_z is the bias term of the update gate. Next, the hidden state of the previous time step is reset by the reset gate, and the candidate hidden state $\tilde{h}_t$ is obtained by combining the current information. $\tilde{h}_t$ is defined as

$$\tilde{h}_t = tanh\left(W[r_t \times h_{t-1}, x_t] + b\right) \tag{3}$$

where W is the weight matrix of the candidate hidden state, and b is the bias term. $tanh(\cdot)$ Denotes the tanh activation function. Finally, the final hidden state h_t, of the current time step is obtained by combining the candidate hidden states, the information obtained from updating the gate and the hidden state of the previous time step. h_t is defined as

$$h_t = (1 - z_t) \times h_{t-1} + z_t \times \tilde{h}_t \tag{4}$$

where $(1 - z_t)$ and z_t determine how much the hiding state and candidate hiding state of the previous time step affect the final hiding state of the current time step, respectively.

The model uses multivariate cross-entropy loss as a loss function during training to achieve minimizing the difference between the predicted and actual distributions. In addition, the Adam optimizer is used to adaptively and dynamically adjust the learning rate. The multivariate cross-entropy loss is defined as

$$L(y, f(x)) = -\frac{1}{N}\sum\nolimits_{i=1}^{N}\sum\nolimits_{c=1}^{C} y_{i\in c}\log(p_{\in c}(f(x_i))) \tag{5}$$

where $f(x_i)$ denotes the model prediction result for the i th sample x_i. $p_{\in c}(\cdot)$ denotes the probability that the value belongs to category c. $y_{i\in c}$ denotes whether or not the true observation corresponding to x_i in the dataset belongs to category c, and if it does, the value is 1, and otherwise the value is 0. N denotes the total number of samples, and C denotes the total number of categories.

3 Experimental Results

3.1 Datasets

The dataset consists of two parts: the training set and the active set. The training set contains the drug set and the drug-target interaction set. The active set contains the specific active drug and its drug-target interaction with the master protease. The specific active drug is derived from a collection of drugs currently known to have a therapeutic effect on the master protease. The ChEMBLdb1 dataset was derived from the ChEMBL database, which consists of 917,821 drug molecules. The CMPair dataset is a drug-target interaction dataset, in which drug molecules were derived from the ChEMBLdb1 dataset.

Experimental parameter setting is shown in Table 2. Due to the specificity of the task of generating molecules, it is difficult to judge the best generating model only by the loss function value of the model. Therefore, in this paper, during the training process, the validity of the molecules obtained from the sampling of the training model is combined as a judgement condition, which in turn determines the best training model to be saved.

Table 2. Experimental parameter setting

Name	Parameter value
Total number of rounds	10
Initial learning rate	0.001
Batch size	256
Activation function	ReLu
Optimiser	Adam
Number of GRU layers	3

3.2 Comparison of Experiments

In this paper, the ability of the model to generate molecules is evaluated based on four metrics (validity, uniqueness, novelty and average affinity). To demonstrate the superiority of the proposed model, it was compared with other models on the ChEMBLdb1 dataset and CMPair dataset. Table 3 lists the results of ablation experiments. It can be seen that the Interaction Constraint Module (ICM) helps to improve the validity and affinity, while uniqueness and novelty decline slightly. The Affinity of AIS coding module was slightly better than that of the baseline model. Uniqueness of AIS, although decreasing relative to the baseline model, is higher than that of the ICM. When the ICM and the AIS coding module are combined, the special qualities of both modules are integrated. Compared to the baseline model, there is a greater increase in affinity.

Table 4 shows the comparative results of validity, uniqueness and novelty of the molecules generated by three models based on the ChEMBLdb1 dataset. It can be seen that our model achieves the best result in the validity metric at 0.8836. Uniqueness and novelty exceed 0.99, and although it is slightly worse relative to the 3CLpro2mol model. Table 5 demonstrates the results of the comparison of the average affinity and the percentage of high-quality affinity of the molecules generated by each model based on the CMPair dataset, and the proposed model achieved the best results in both metrics, with an average affinity of -8.144 and a percentage of high-quality affinity of 0.2933. In addition, the interaction information for the four best affinities is shown in Table 6.

Table 3. Results of ablation experiments

Model	Validity (↑)	Uniqueness (↑)	Novelty (↑)	Average affinity (↓)
Baseline	0.8810	**1.0**	0.9959	-7.780
ICM	**0.9116**	0.9978	0.9912	-8.0391
AIS	0.8642	0.9993	0.9958	-7.9513
ICM+AIS	0.8836	0.9995	**0.9972**	**-8.144**

Table 4. Comparative results of different methods based on ChEMBLdb1 dataset

Model	Validity (↑)	Uniqueness (↑)	Novelty (↑)
DeepTarget [27]	0.8083	0.997	0.9989
3CLpro2mol [22]	0.8623	**1.0000**	**0.9998**
Our Model	**0.8836**	0.9995	0.9972

Table 5. Comparative results of different methods based on CMPair dataset

Model	Average affinity (↓)	Percentage of high-quality affinity (↑)
CVAE [28]	−6.144	0.238
AR [15]	−6.215	0.2670
Our Model	**−8.144**	**0.2933**

Table 6. The interaction information for the four best affinities

Number of Drug	DTI	Residue	Distance (Å)
D_b01	Hydrogen bonding	Chain A-LYS-5	2.7
D_b01	Hydrogen bonding	Chain B -LYS-5	2.1
D_b02	Hydrogen bonding	Chain A -LYS-5	2.2
D_b03	Hydrogen bonding	Chain A-LEU-282	2.1
D_b03	Hydrogen bonding	Chain B-SER-284	2.3
D_b04	Hydrogen bonding	Chain A-ALA-285	2.6
D_b04	Hydrogen bonding	Chain B-LYS-5	2.1
D_b04	Hydrogen bonding	Chain B-ARG-4	2.7
D_b04	Hydrogen bonding	Chain A- LYS-5	2.8

4 Conclusion

In this paper, we propose a targeted drug generation model based on the AIS encoding and drug-target interaction. To obtain the structural information inside the drug molecule, the AIS encoding is used to aggregate the information of neighboring atoms or fragments of neighboring atoms of an atom in the drug molecule. It can enrich the relative positional relation between atoms in the drug molecule in space. Thereby, the representation of local features of the drug molecule is further enhanced on the basis of the overall representation of the drug molecule. In addition, to effectively correlate the relation between drug features and targeted protein features, this paper uses a drug-target interaction constrained generation strategy, which precisely guides the generation process of drug molecules at the level of atomic-amino acid residue with five types of drug-target interactions as the constraints, so that the generated drugs can have stronger interactions with the targeted protein, thus enhancing the quality of the generated molecules. Comparing with other models, it can be demonstrated that the proposed model has the ability to generate high-quality targeted drugs efficiently.

Acknowledgement. The authors thank the members of Machine Learning and Artificial Intelligence Laboratory, School of Computer Science and Technology, Wuhan University of Science and Technology, for their helpful discussion within seminars. This work was supported by National Natural Science Foundation of China (No. 61972299).

References

1. Blanco-Gonzalez, A., Cabezon, A., Seco-Gonzalez, A., et al.: The role of AI in drug discovery: challenges, opportunities, and strategies. Pharmaceuticals **16**(6), 891 (2023)
2. Moreira-Filho, J.T., Silva, A.C., Dantas, R.F., et al.: Schistosomiasis drug discovery in the era of automation and artificial intelligence. Front. Immunol. **12**, 642383 (2021)
3. Sun, D., Gao, W., Hu, H., et al.: Why 90% of clinical drug development fails and how to improve it? Acta Pharmaceutica Sinica B **12**(7), 3049–3062 (2022)
4. Desai, S.R., Baldwin, H., Del Rosso, J.Q., et al.: Microencapsulated Benzoyl Peroxide for Rosacea in Context: A Review of The Current Treatment Landscape. Drugs, 1–10 (2024)
5. Heo, Y.A.: Apadamtase Alfa: First Approval. Drugs 1–6 (2024)
6. Bharatam, P.V.: Drug discovery and development: from targets and molecules to medicines, 137–210 (2021)
7. Asif, F., Zaman, S.U., Arnab, M.K.H., et al.: Antimicrobial peptides as therapeutics: confronting delivery challenges to optimize efficacy. Microbe 100051 (2024)
8. Rohall, S.L., Auch, L., Gable, J., et al.: An artificial intelligence approach to proactively inspire drug discovery with recommendations. J. Med. Chem. **63**(16), 8824–8834 (2020)
9. Abdul Raheem, A.K., Dhannoon, B.N.: Automating drug discovery using machine learning. Curr. Drug Discov. Technol. **20**(6), 79–86 (2023)
10. Chen, L., Fan, Z., Chang, J., et al.: Sequence-based drug design as a concept in computational drug design. Nat. Commun. **14**(1), 4217 (2023)
11. He, X., You, C., Jiang, H., et al.: AlphaFold2 versus experimental structures: evaluation on G protein-coupled receptors. Acta Pharmacol. Sin. **44**(1), 1–7 (2023)
12. Berdigaliyev, N., Aljofan, M.: An overview of drug discovery and development. Future Med. Chem. **12**(10), 939–947 (2020)

13. Biala, G., Kedzierska, E., Kruk-Slomka, M., et al.: Research in the field of drug design and development. Pharmaceuticals 16(9), 1283 (2023)
14. Masuda, T., Ragoza, M., Koes, D.R.: Generating 3D molecular structures conditional on a receptor binding site with deep generative models. arXiv preprint arXiv:2010.14442 (2020)
15. Luo, S., Guan, J., Ma, J., et al.: A 3D generative model for structure-based drug design. Adv. Neural. Inf. Process. Syst. 34, 6229–6239 (2021)
16. Mendez, D., Gaulton, A., Bento, A.P., et al.: ChEMBL: towards direct deposition of bioassay data. Nucleic Acids Res. 47(D1), D930–D940 (2019)
17. Wishart, D.S., Feunang, Y.D., Guo, A.C., et al.: DrugBank 5.0: a major update to the drugbank database for 2018. Nucleic Acids Res. 46(D1), D1074–D1082 (2018)
18. UniProt Consortium. UniProt.: A Worldwide Hub of Protein Knowledge. Nucleic Acids Res. 47(D1), D506–D515 (2019)
19. Lin, X.L., Zhang, X.L., Xu, X.: Efficient classification of hot spots and hub protein interfaces by recursive feature elimination and gradient boosting. IEEE/ACM Trans. Comput. Biol. Bioinf. 17(5), 1525–1534 (2020)
20. Lin, X.L., Zhang, X.L.: Prediction of hot regions in PPIS based on improved local community structure detecting. IEEE/ACM Trans. Comput. Biol. Bioinf. 15(5), 1470–1479 (2018)
21. Bai, Q., Liu, S., Tian, Y., et al.: Application advances of deep learning methods for de novo drug design and molecular dynamics simulation. Wiley Interdiscipl. Rev. Comput. Mol. Sci. 12(3), e1581 (2022)
22. Lin, X.L., Liu, X., Zhang, X.L.: Anti-3CLpro molecular design based on the model constrained by specific DTIs. In: 2023 IEEE International Conference on Bioinformatics and Biomedicine (BIBM), pp. 3796–3803 (2023)
23. Lin, X.L., Zhu, Q.L., Zhang, X.L.: Generating molecules conditional 3D protein pockets with HGAF. In: 2023 IEEE International Conference on Bioinformatics and Biomedicine (BIBM), pp.446–449 (2023)
24. Jin, Z., Du, X., Xu, Y., et al.: Structure of Mpro from SARS-CoV-2 and discovery of its inhibitors. Nature 582(7811), 289–293 (2020)
25. Lin, X.L., Xu, S., Liu, X., Zhang, X.L., Hu, J.: Detecting drug-target interactions with feature similarity fusion and molecular graphs. Biology 11(7), 967 (2022)
26. Ucak, U.V., Ashyrmamatov, I., Lee, J.: Improving the quality of chemical language model outcomes with atom-in-SMILES tokenization. J. Cheminformatics 15(1), 55 (2023)
27. Chen, Y., Wang, Z., Wang, L., et al.: Deep generative model for drug design from protein target sequence. J. Cheminform. 15(1), 38 (2023)
28. Masuda, T., Ragoza, M., Koes, D.R.: Generating 3D molecular structures conditional on a receptor binding site with deep generative models. arXiv preprint arXiv:2010.14442 (2020)

Application of Machine Learning and Large Language Model Module for Analyzing Gut Microbiota Data

Jianhua Cao$^{(\boxtimes)}$, Xiaoyu Xu, and Xiaomeng Zhang

College of Artificial Intelligence, Tianjin University of Science and Technology, Tianjin 300457, China
caojh@tust.edu.cn

Abstract. In this paper, we aim to explore the applications of machine learning and large language model for analyzing gut microbiota data, particularly attempting to investigate large language model module to intelligently conduct data analysis just through prompts. The data of gut microbiota is from 16S rRNA sequencing result of obese mouse for obesity experimental research, and our primary task focus on identifying differentially expressed genes and uncovering biomarker microbiota associated with obesity. Statistical methods, including diversity analysis and principal component analysis, have been firstly conducted, revealing significant differences between experimental and control groups. Then different machine learning algorithms including random forest, SVM-RFE, Lasso, and XGBoost have been selected to discover distinguished genes through feature importance ranking. Several types of microbial genes have been identified by both of the machine learning methods, and these biomarkers exhibit significant abundance changes between groups. Some of them have been confirmed with related research literature, and the remains are worthy of attention for more investigation. Furthermore, we have developed a large language model module named Chat2GM, which is based on the LangChain framework, python Flask, and API of OpenAI's GPT model, particularly for this data analysis task. Upon uploading the original microbiota data, data exploration and analysis can be conducted merely by using prompts. It has been demonstrated to achieve satisfactory performance in completing the aforementioned analysis tasks with precise natural language instructions, indicating that applications powered by large language models with more expertise knowledge have great potential in gut microbiota research field.

Keywords: Gut microbiota · Machine learning · Large language models · LangChain framework

1 Introduction

Gut microbiota play crucial roles in maintaining the functionality of the human digestive system, regulating the immune system, and controlling metabolism, thus exerting significant effects on human health [1]. But the data has typically features of large scale and high dimensionality, which pose challenges for effectively analyzing such data using

D.-S. Huang et al. (Eds.): ICIC 2024, LNBI 14881, pp. 37–48, 2024.
https://doi.org/10.1007/978-981-97-5689-6_4

traditional statistical methods. Machine learning has shown great advantages at processing datasets of varying scales and discovering underlying patterns and regularities, and identifying microbial features associated with specific health conditions, providing support for personalized medicine and offering new perspectives and methods for the prevention, diagnosis, and treatment of relevant diseases [2]. However, the application of machine learning may be difficult for researchers in bioengineering. It not only requires understanding the algorithms themselves but may also entail proficiency in programming development.

Large language models (LLMs), owing to their rich knowledge and strong reasoning abilities, have demonstrated the powerful abilities in executing a wide range of tasks such as translation, creative writing, data analysis, software development, etc. [3]. And applications are solely through prompts rather than programming. ChatGPT is the most famous large language model application, and its interactive working way is just chatting. It has sparked a global wave of LLM applications and simultaneously brought about new paradigms to research in various industries [4]. Currently, there are few large language model applications or publications in the specific field of gut microbiota data analysis.

In this paper we have chosen gut microbiota data output from 16S rRNA gene sequencing of target OB mice and aimed to investigate obesity-related microbial markers using machine learning. Additionally, we have developed Chat2GM, a large language model module based on python Flask and LangChain framework to explore its capability of analyzing gut microbiota data solely by using instructions.

2   Methodology

2.1   Overview

Using the abundance data of gut microbiota at the genus level obtained from 16S rRNA sequencing of OB mice as the raw data, we conduct data modeling and analysis using statistical methods and machine learning algorithms. Additionally, we have developed a large language model application based on the LangChain framework by invoking the API of OpenAI large model and creating agents, enabling completion of data analysis tasks through natural language instructions.The architecture overview of this study is shown as Fig. 1.

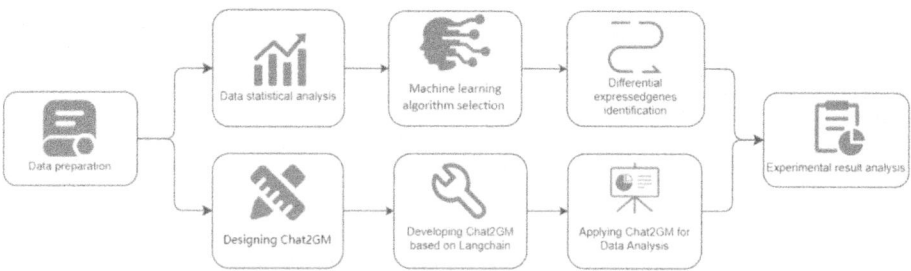

Fig. 1. Gut microbiota data analysis overview using machine learning and LLM module

2.2   Machine Learning Algorithms

The main goal is to identify differentially expressed genes between experimental group and control group, and uncovering biomarkers associated with obesity.Considering the large-scale and high-dimension features of the raw data, several machine learning algorithms, including Random Forest, SVM-RFE, and Lasso regression and XGBoost have been selected in this study.

Random Forest. Random Forest is an Ensemble Learning model composed of multiple Decision Trees, with each tree constructed from randomly selected features and samples [5]. During the training process, multiple samples are randomly drawn from the original dataset through bootstrapping to train each decision tree. Random Forest exhibits strong predictive performance and robustness, making it effective in handling high-dimensional and complex gut microbiota data.

SVM-RFE. The SVM-RFE algorithm is a feature selection method based on Support Vector Machine (SVM), which stands for Support Vector Machine - Recursive Feature Elimination (SVM-RFE). This algorithm combines the principles of SVM and Recursive Feature Elimination (RFE), iterating through training SVM models repeatedly and selecting features based on their importance [6]. SVM-RFE recursively trains SVM models and ranks and filters features based on their weights.

LASSO. LASSO (Least Absolute Shrinkage and Selection Operator) regression is a linear regression method used for feature selection and regularization [7]. It identifies representative features and shrinks the coefficients of features that have little or no impact on the target prediction to zero. This achieves the effect of feature selection, reducing the complexity of the model and mitigating the influence of collinearity. Consequently, it enhances the interpretability and generalization ability of the model.

XGBoost. XGBoost is an integrated learning algorithm that uses gradient boosting technology to train multiple decision tree models to form a powerful integrated model. During the training process, multiple decision trees are iteratively trained and the data is weighted based on the prediction results of the previous round of models to gradually improve model performance [8]. It flexibly discovers non-linear relationships in data and reveals the importance between features.

2.3   Chat2GM - a LLM Module Based on Langchain Framework

In this study, Chat2GM - a LLM module has been developed particularly for gut microbiota data analysis using the LangChain framework and API of the OpenAI GPT model. The module is composed of webpage templates and back-end python service using Python Flask framework. Users can upload data and input prompts, and it has an aesthetic interface design and clear layouts(shown as Fig. 2). Left part is for starting a new chat and listing the chat history, and Question-Answer pairs are shown in the middle zone. An input box is set for giving instructions or prompts on the bottom of the middle part. On the right side, the file-upload button is for uploading data with Excel format, and some simple charts could be shown for the analysis. The source code is available at https://github.com/zxm-a11y/Chat2GM.

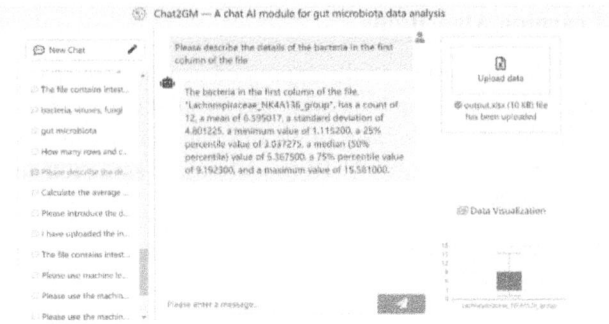

Fig. 2. User interface of Chat2GM module

The overall design architecture of Chat2GM service is based on a combination of a large language model, Prompt templates, Agent proxies, and LLM Chain processing tools (shown as Fig. 3). This architecture enables functions such as handling user input questions, keyword searching, data analysis, and displaying results. And Google service is also embedded as a tool in the chains. When data with Excel format is ready, it can be uploaded through a file-upload button on the webpage templates. And then users can start to input some instructions or request prompts through a message input box. A create_csv_agent and Google service agent have been set on the back-end server side. When receiving the instructions, the server directly request the large language model through using API to start data analysis based on the text instructions. The task will be completed through agent chains, sequentially from thought, action to observation and final answer. Users can overview the details of agent execution on the server terminal (shown as Fig. 4). Ultimately, the final answer of the large language model's data analysis can be promptly sent back to the front-end page via the asynchronous AJAX request, where they can be reviewed by the user.

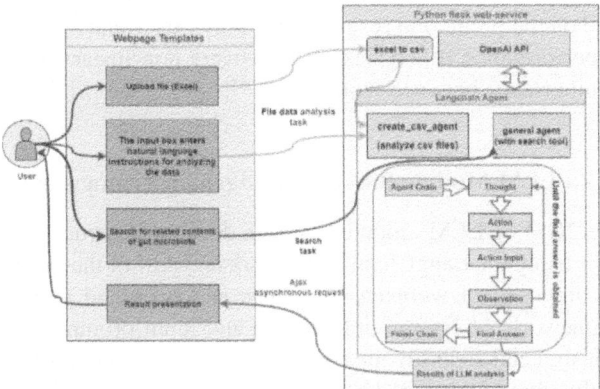

Fig. 3. Chat2GM module schema based on python Flask and LangChain framework

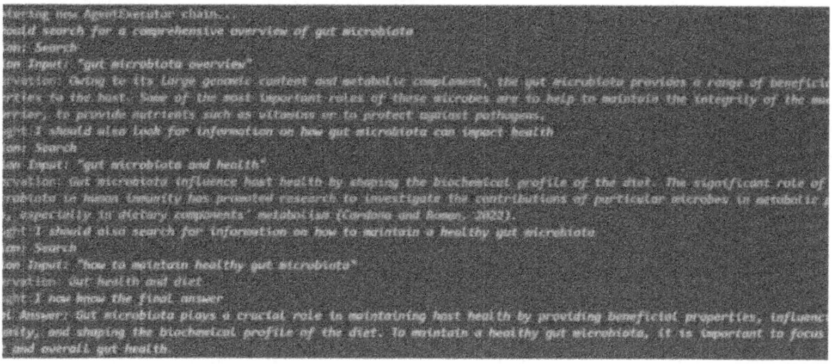

Fig. 4. Logs of LangChain agent chains execution on the server side

3   Applications and Analysis

3.1   Data

The data is collected from the obesity intervention experiment of obese mouse. The obese mouse, also named as OB mouse, is a mutant mouse that eats excessively due to mutations in the gene responsible for the production of leptin and becomes profoundly obese. It is an animal model of type II diabetes. Identification of the gene mutated in OB led to the discovery of the hormone leptin, which is important in the control of appetite. During the experiments, the OB mice are classified as control group and experimental group. Usually, multiple sets of comparative experiments are designed to observe the changes in relevant microbiota obtained from 16S rRNA genetic sequencing to discover the biomarkers that affecting obesity.

Samples of microbial abundance data at the genus level is shown as Table 1. The first column "Genus" represents the names of microbial genera at this level, such as Alloprevotella, Bifidobacterium, etc. Actually in this genus level, there has more than 150 types of genes. The first row header includes "NC" for the healthy control group and "MC" for the experimental group, followed by sample numbers. For instance, "NC1" refers to the first sample in the control group, and "MC1" refers to the first sample in the experimental group.There are totally 12 samples, with 6 for each group. The cell values represent the abundance levels of microbial genera in a specific sample.

The abundance matrix data mentioned above needs to be transposed for further analysis, with the columns representing microbiota species and the rows representing samples. If each microbial is considered as a feature, the transposed matrix data has more than 150 kinds of features, and many of the cells have an abundance value of 0. This is a typical high-dimensional, sparsely distributed matrix data, and we need to evaluate these complex features and uncover the meaning of their abundance changes, which is obvious challenging. The raw data could also be available at https://github.com/zxm-a11y/Chat2GM.

3.2   Species Diversity Analysis with Statistical Methods

α-**diversity analysis.** α-diversity analysis is used to assess the species diversity and richness between groups. In this study, three indices, Observed-OTUs, Shannon, and

Table 1. samples of gut microbiota data at the genus level

Genus	NC1	NC2	NC3	NC4	NC5	NC6	MC1	MC2	MC3	MC4	MC5	MC6
Alloprevotella	0.401	0.268	1.791	0.4	0.773	0.335	9.502	12.668	10.219	9.1005	8.6337	5.7857
Bifidobacterium	0.866	0.612	1.013	4.429	0.398	0.643	3.907	6.595	3.466	1.2781	0.5483	0.3842
Lactobacillus	1.311	2.001	1.689	7.224	0.405	11.88	2.904	2.991	3.511	2.291	1.6222	1.3972
Bacteroides	1.077	0.703	0.818	2.415	1.104	4.864	1.759	0.541	0.903	0.876	1.8921	1.0902
Helicobacter	2.389	0.552	3.518	1.705	3.861	0.678	2.537	0.738	0.734	2.92	3.1228	6.9807
Alistipes	0.688	0.512	0.636	1.277	0.604	0.936	2.195	2.033	1.507	1.5735	1.6151	2.2503

PD-whole-tree, have been calculated, and the comparative results are shown in Fig. 5. There are significant differences between NC group and MC group. Among them, the MC group samples exhibit fewer intestinal microorganisms compared to the NC group according to the Observed-OTUs index, while they show significantly higher diversity in species according to the Shannon index, indicating richer species diversity. It is evident that the range of PD-whole-tree values in the MC group is larger than that in the NC group, indicating significant differences between the two groups of samples.

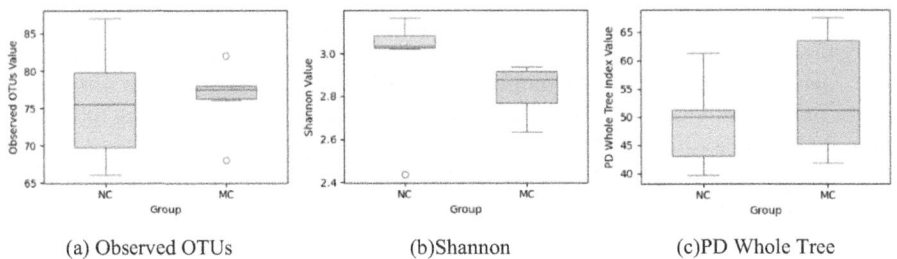

(a) Observed OTUs   (b)Shannon   (c)PD Whole Tree

Fig. 5. α-diversity analysis diagrams

β-diversity Analysis. β-diversity analysis is used to compare the differences in microbiota composition between different samples. In this study, Unweighted UniFrac distance, Weighted UniFrac distance, and Bray-Curtis distance have been calculated, as shown in Fig. 6. In the figure, darker colors indicate smaller distance values, indicating greater similarity in microbiota structure between two samples, while lighter colors indicate greater differences in microbiota structure between them. The results of the calculations for all three indices show differences in microbiota composition between samples, with some differences being quite significant.

PCA Analysis. PCA analysis is used to reduce data dimensionality and highlight the differences. By mapping high-dimensional microbiota data into a lower-dimensional principal component space, PCA helps us discover the main directions of variation and patterns in the data, thus providing a more intuitive understanding of the structure and characteristics. The PCA analysis results for this data are shown in Fig. 7, where the

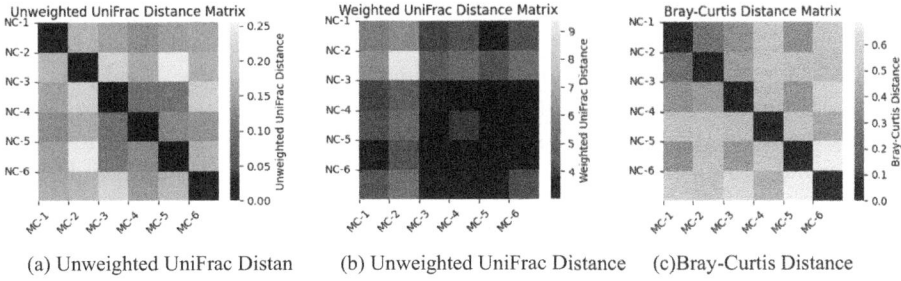

(a) Unweighted UniFrac Distan   (b) Unweighted UniFrac Distance   (c)Bray-Curtis Distance

Fig. 6. β-diversity analysis diagrams

positions of the dots reflect the projection of each sample along these two principal component directions. The results show significant differences between MC and NC group data along the direction of Principal Component 1 (PC1).

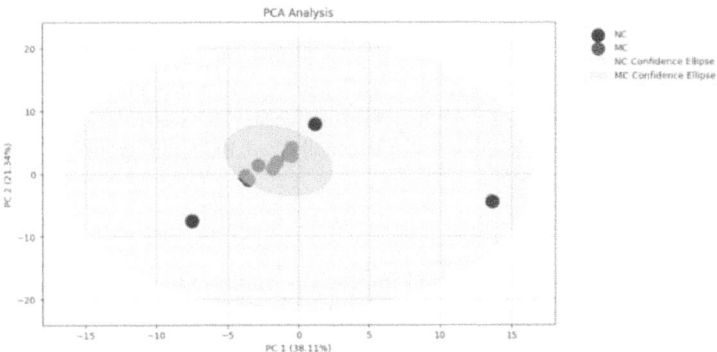

Fig. 7. PCA analysis diagrams

3.3 Identification of Obesity-Related Biomarkers via Machine Learning

Machine learning has been used to find meaningful biomarkers that distinguished the experimental group from the control group using the abundance matrix of the gut microbiota at the selected genus level. Random Forest can directly provide importance ranking for each feature, helping to identify colonies that have a high impact on the target variable. Lasso regression, which automatically selects the most important features and sets the coefficients of other irrelevant features to 0, can help to exclude colonies that have a low impact on the target variable. SVM-RFE combines the ideas of the Support Vector Machine and Recursive feature elimination, and can automatically select the most important features through an iterative process. The XGBoost algorithm provides attributes for directly obtaining feature importance rankings, and supports visualization tools to help identify features that have a greater impact on the target variables.Since the raw data has more than 150 kinds of genes at the genus level, the target is to find out differentially expressed genes between the samples, which are just the obesity-related biomarkers. Each microbial is considered as one feature for both NC and MC group. And the microbial assessment becomes an effort to rank the importance of features. Therefore, four algorithms including Random Forest, SVM-RFE, Lasso and XGBoost have been chosen for the study. We use the machine learning package provided by the Scikit-Learn library for program development, and their hyperparameters and some of genes identified are listed in Table 2.

Table 2. Model hyperparameters and identified genes using machine learning

Ml algorithms	Model hyperparameters	Differentially expressed genes identified
Random Forest	n_estimators:50	coprostanoligenes_group,Alloprevotella, Family_XIII_AD3011_group,Mucispirillum…
SVM-RFE	C:10	Alloprevotella, Bifidobacterium, Lactobacillus, Bacteroides, Alistipes, Blautia…
Lasso	alpha:0.1;max_iter:100	Alloprevotella, Lachnospiraceae_NK4A136_group, Alistipes, Phaseolus_acutifolius, Anaerotruncus…
XGBoost	L1: alpha-0.1;ratio:0.6 L2: lambda:0.1, max_depth:3, n_estimators:50	Turicibacter, Alloprevotella, Alistipes, Prevotellaceae_UCG-001, Rikenellaceae_RC9_gut_group, Akkermansia…

Figure 8 demonstrates the microbial importance ranking bars using Random Forest and XGBoost algorithms respectively. In the bar chart, the height of each bar represents the difference in abundance values of the corresponding microbiota. The taller the bar, the greater the difference in abundance between the two groups of multiple samples. Fig. 9a is the Venn diagram used to visualize the intersection and differences of microbial selection results between SVM-RFE and Lasso Regression methods. 14 types of microbial are selected by both methods. Therefore at the genus level, 5 types of genes with the largest abundance changes have been identified by all the four methods, including Alloprevotella, Alistipes, Mucispirillum, Blautia, and Prevotellaceae_UCG-001.Their abundance values show very distinct variation between NC and MC group (Shown as Fig. 9b). For instance, Alloprevotella abundance in MC group is evidently increasing comparing with that in NC group, with ratio more than 800%. While Blautia abundance has the opposite trend. It has been noted in research publications that the changes of Alloprevotella in intestinal microbial community do affect the obesity and metabolic problems of mice [9]; Mucispirillum prevents colitis by interfering with the expression of pathogen invasion mechanism [10]; Probiotic therapy can partially reduce obesity caused by diet by regulating the gut microbiota such as Anaerotruncus [11];The Family_XIII_AD3011_group in the gut microbiota of obese children is different from that of normal children, and the relative abundance of the Family_XIII_AD3011_group in obese children is reduced [12]. These scientific publications confirm of the reliability of the data analysis. Genes of Blautia and Prevotellaceae_UCG-001 are worthy of attention, with abundance downregulated, which indicates the potential of these microbiota to inhibit obesity level growth, and can provide some basis for subsequent regulation and treatment for obesity. Referring to the research procedures for biologists, more animal experiments are to be designed to validate these genes.

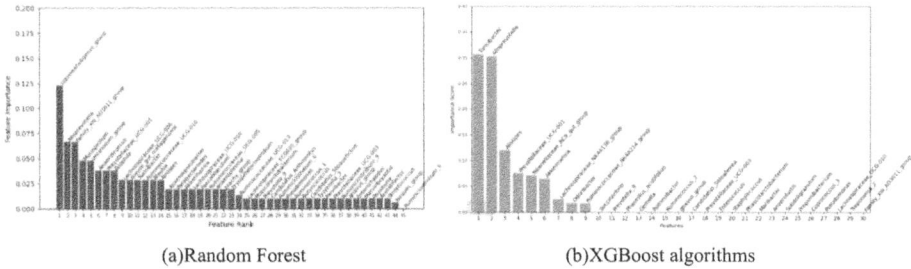

(a)Random Forest   (b)XGBoost algorithms

Fig. 8. Microbial importance ranking bars

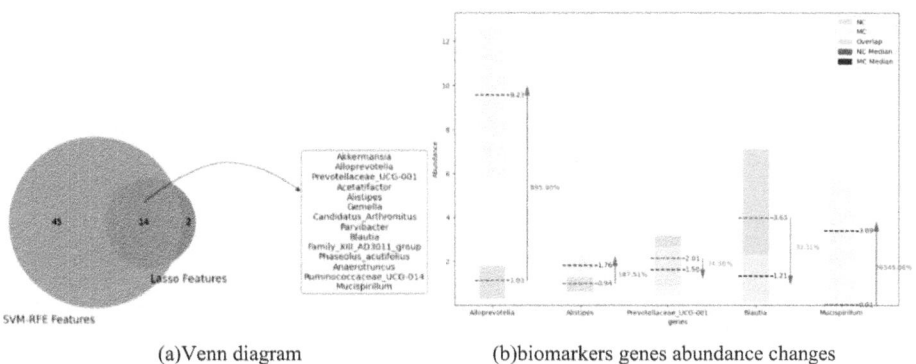

(a)Venn diagram   (b)biomarkers genes abundance changes

Fig. 9. Microbial identification diagrams

3.4   Gut Microbiota Data Analysis with Chat2GM Module

Chat2GM module has been evaluated and applied in analyzing the same gut microbiota data mentioned above. The working pattern for using Chat2GM is shown as Fig. 10. When the data is uploaded into the module, users can then prepare the natural language instructions for data inspection and analysis. As shown in Fig. 10, the request is "Please help me analyze the α diversity of gut microbiota data, calculate the Observed OTUs index and output the result". Chat2GM will then try to understand the requests and call the relevant functions of large language models. The action chain is Thought-Action-Action Input-Final Answer, with details shown in the right part of Fig. 10. The user finally get the response from Chat2GM, which is the Final Answer generated from chains.

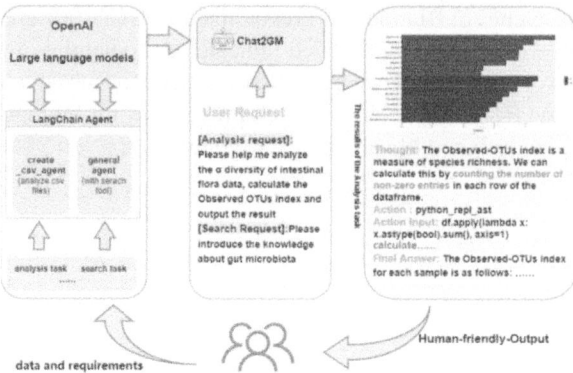

Fig. 10. working mode for using Chat2GM

Data Inspection Evaluation. The prompt is "please give the basic information of the data". More prompts, like "list the first five columns of the file" can be entered to further explore the data. The interactions are shown as Fig. 11a. Diversity analysis has also been tested using Chat2GM as shown as Fig. 11b.

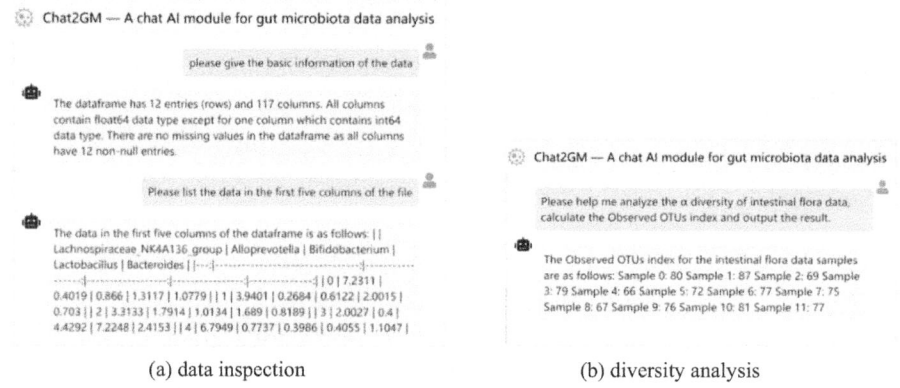

(a) data inspection   (b) diversity analysis

Fig. 11. Data analysis request-answers using Chat2GM

Experiment of Differentially Expressed Genes Identification. Two types of identification strategies using prompts have been applied in the experiments.The first one is to give the requests of analysis without specified method (Shown as Fig. 12). Responses show that it performs the analysis task by using the Random Forest algorithm, with 20 types of differentially expressed genes in the answers. The second one is to specify the machine learning methods and request of recognizing top 10 targets (shown as Fig. 13). The outputs are almost identical When comparing using specified Random Forest and none-specified schemes. And some of the biomarkers also exist in the outcomes of using programs shown in Fig. 8. Therefore, once the algorithm is specified, the module also give feedback that make sense, which can be verified by comparison with the research result using machine learning in Chapter 4.3.

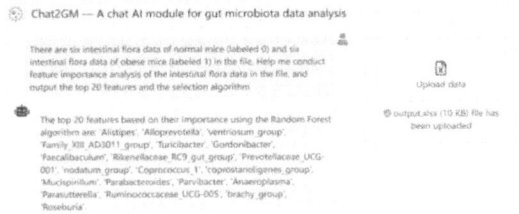

Fig. 12. Chat2GM for genes identification using prompts with no specified algorithms (a) request in the same prompts (b) request respectively with prompts

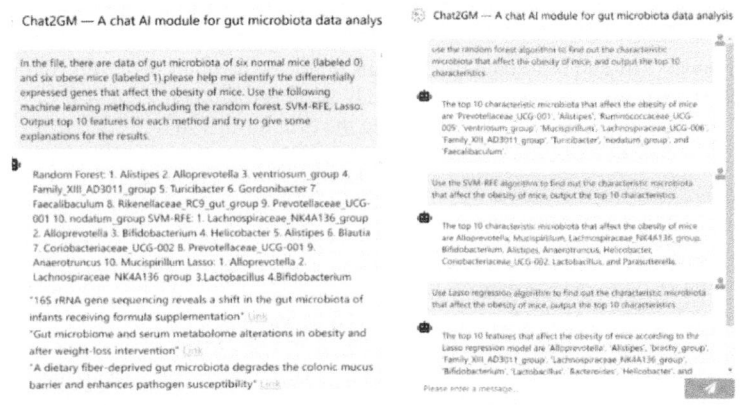

(a)request in the same prompts   (b)request respectively with prompts

Fig. 13. Chat2GM for genes identification using prompts with specified algorithms

Discussions. The experiments show the possibilities of data analysis with Chat2GM using instructions rather than programming, and also the capabilities of generating reliable analysis results providing precise instructions combined with expertise knowledge. It could be an useful tool for those researchers who are not proficient programmers or having few knowledge about algorithms. It is also an meaningful attempt for LLMs application in the gut microbiota research domain. But due to the inherently generative nature of the LLMs, the same request may yield non-deterministic results with minor difference. Improvements are indeed urgent for the practical research. One effective strategy is to incorporate more expertise literature and then fine-tune the model based on RAG strategy. Publications such as articles, reports and patents on microbiota analysis need to be integrated as external reference data, so that a professional gut microbiota knowledge base could be set up. In addition to textual generations, incorporating data visualization such as charts would enhance the analysis performance.

4 Conclusions

In order to identify differentially expressed genes from 16S rRNA genetic sequencing microbiota data of OB mice, machine learning algorithms have been applied in this work. Several types of microbial biomarkers have been successfully identified using the Random Forest, Lasso Regression, SVM-REF and XGBoost machine learning algorithms. These microbiota markers do influence obesity and metabolic issues referring to public research publications. And Genes of Blautia and Prevotellaceae_UCG-001 are worthy of attention for further research. Additionally, a module named Chat2GM based on the LangChain framework has been developed to enable data analysis just by giving prompts other than programming. Experiments demonstrate satisfactory performance could be achieved giving expertise instructions, and several kinds of differentially expressed genes related to the obesity have been discovered from the same gut microbiota data, which is almost identical to the machine learning programming outputs. Improvements have been suggested for further research. Complete RAG strategy based LangChain framework will be put forward for expertise gut microbiota data analysis in future study.

Acknowledgments. The authors would like to thank all the anonymous reviewers for their valuable comments to improve this paper. And sincerely thank Prof. Zhou Zhongkai from Tianjin University of Science and Technology for providing the data for this study.

References

1. Zhang, P., Yang, C., Ji, S., et al.: Gut microbiota and associated diseases. World Chin. J. Digestol. **24**(15), 2355–2360 (2016)
2. Cao, H., Zhu, J., Ma, Y., et al.: Application of machine learning in predicting host phenotypes based on gut microbiota. Biotechnol. Adv. **13**(05), 671–680 (2023)
3. Alberts, I.L., Mercolli, L., Pyka, T., et al.: Large language models (LLM) and ChatGPT: what will the impact on nuclear medicine be? Eur. J. Nucl. Med. Mol. Imaging **50**(6), 1549–1552 (2023). https://doi.org/10.1007/s00259-023-06172-w
4. Yin, H., Gu, Z., Wang, F., et al.: An evaluation of large language models in bioinformatics research. arXiv preprint arXiv:2402.13714 (2024)
5. Breiman, L.: Random forests. Mach. Learn. **45**, 5–32 (2001)
6. Sanz, H., Valim, C., Vegas, E., et al.: SVM-RFE: selection and visualization of the most relevant features through non-linear kernels. BMC Bioinformatics **19**, 1–18 (2018)
7. Hans, C.: Bayesian lasso regression. Biometrika **96**(4), 835–845 (2009)
8. Chen, T., Guestrin, C. Xgboost: a scalable tree boosting system. In: Proceedings of the 22nd ACM SIGKDD International Conference on Knowledge Discovery and Data Mining, pp. 785–794 (2016)
9. Li, S., You, J., Wang, Z., et al.: Curcumin alleviates high-fat diet-induced hepatic steatosis and obesity in association with modulation of gut microbiota in mice. Food Res. Int. **143**, 110270 (2021)
10. Herp, S., Durai Raj, A.C., et al.: The human symbiont Mucispirillum Schaedleri: causality in health and disease. Med. Micro Biol. Immunol. **210**(4), 173–179 (2021)
11. Kong, C., Gao, R., Yan, X., et al.: Probiotics improve gut microbiota Dysbiosis in obese mice fed a high-fat or high-sucrose diet. Nutrition **60**, 175–184 (2019)
12. Sun, J., Jin, X., Yang, L., et al.: Gut microbiota associated with obesity and high carotid intima-media thickness among Chinese children. Res. Square (2022). https://doi.org/10.1007/s00259-023-06172-w

CVAE-Based Hybrid Sampling Data Augmentation Method and Interpretation for Imbalanced Classification of Gout Disease

Xiaonan Si[1,2], Yifan Fu[3], Xinran Liu[4], Rulin Wang[5], Wenchang Xu[2], and Lei Wang[2(✉)]

[1] Institute of Software, Chinese Academy of Sciences, Beijing 100190, China
sxnyyds@mail.ustc.edu.cn
[2] CAS Key Lab of Bio-Medical Diagnostics, Suzhou Institute of Biomedical Engineering and Technology, Chinese Academy of Sciences, Suzhou 215163, China
wanglei@sibet.ac.cn
[3] State Key Laboratory for Strength and Vibration of Mechanical Structures, Xi'an Jiaotong University, Xi'an 710049, China
[4] School of Internet, Anhui University, Hefei 230039, China
[5] School of Forestry, Northeast Forestry University, Harbin 150040, China

Abstract. Gout disease is a highly painful condition worldwide. The use of clinical data for staging gout is crucial in aiding experts with rapid diagnosis and treatment. However, current research has not given sufficient consideration to the accurate staging of gout. On the other hand, imbalanced data presents a challenge in disease diagnosis, which hinders accurate predictive data analysis in real-world medical application. Hence, this study proposes a new predictor model with interpretation for the stages of gout disease (termed PIGD). To address the issue of imbalance, a new hybrid sampling method CBHS that utilises the generative network is proposed. Improved TabNet was used as the foundation for gout detection. The PIGD used the SHAP to propose an interpretable model for the detection of gout, which reveals causal relationships between features to experts and overcomes the limitations of 'black-box'. The results indicate that proposed PIGD performed the best in the clinical dataset for gout, which get a precision of 86.65%, an AUC of 82.96%, an accuracy of 80%, and a G-means of 77.51%.

Keywords: Gout · Hybrid sampling · Imbalanced data · CVAE · SHAP · TabNet

1 Introduction

Gout is a chronic metabolic disorder that causes joint inflammation. It has become increasingly prevalent in recent decades [1, 2]. In addition to joint damage, gout is associated with a number of other conditions such as hypertension, diabetes, kidney disease and dyslipidaemia [3]. Timely diagnosis of gout is crucial in preventing rapid deterioration of the condition, which can be fatal. Machine learning is increasingly being used to simplify disease diagnosis, saving significant time and improving the validity of diagnoses. For example, machine learning has been utilized to diagnose heart disease, detect stress levels, and diagnose diabetes [4–6].

© The Author(s), under exclusive license to Springer Nature Singapore Pte Ltd. 2024
D.-S. Huang et al. (Eds.): ICIC 2024, LNBI 14881, pp. 49–60, 2024.
https://doi.org/10.1007/978-981-97-5689-6_5

However, few studies have been dedicated to the use of machine learning for accurately diagnosing the stages of gout patients. In addition, medical datasets are frequently imbalanced, leading to models that do not treat all samples equally. Previous. studies have used the resampling method to balance data distribution like SMOTE and CVAE [7]. Although these oversampling algorithms have been widely used in many real-world applications, they are still somewhat ill-conceived for rebalanced datasets [8, 9]. First, SMOTE oversamples less informative minority samples [10]. Figure 1 (d), (e) and (g) show the samples generated by SMOTE and variants, focuses on less informative samples that have limited contributions to the classifier. Second, oversampling algorithms often generate noisy samples, particularly due to outlier samples. Last, oversampling algorithms fit the global data distribution well, but ignores class overlap near the boundary and will result in performance deterioration in the presence of class overlap [11]. As shown in Fig. 1(b), (f) and (h), this result suggests that synthetic data is generated in the overlapping region. More importantly, machine learning models for disease diagnosis are often considered 'black-box' as they do not provide an interpretation to aid the diagnostic process. This can lead doctors to question the diagnostic models.

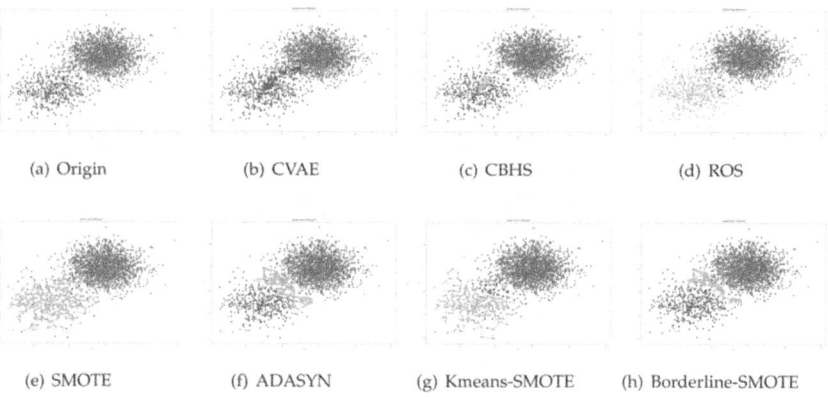

Fig. 1. (a) An imbalanced dataset where green represents the majority class and red represents the minority class, generated the minority points (orange). (b) A balanced dataset achieved by CVAE with an equal number of samples. (c) Using CBHS, (d) using ROS. (e) using SMOTE. (f) using ADASYN. (g) using Kmeans SMOTE. (h) using Borderline SMOTE.

To address the limitations of previous work, this study proposes a stable and interpretable predictor of gout staging (PIGD). PIGD employs a CVAE-based hybrid sampling (CHBS) for data balancing and improved TabNet for prediction. The performance of PIGD was evaluated on a clinical dataset, and a 5-fold cross-test was used to validate the performance difference between the proposed model (PIGD) and other models. Additionally, the Shapley Additive Explanation algorithm (SHAP) framework and the model's outputs were utilized to enhance model interpretation. The experimental results demonstrate that the method proposed in this paper outperforms other balancing algorithms and alleviates the imbalance problem, as shown in Fig. 1(c). The classification accuracy achieves the best classification results, improves the interpretability of the model, and outperforms existing methods. As far as we know, this is the first attempt at staging gout. A summary of the paper's contributions is as follows:

1. A new hybrid sampling method for data imbalance, CHBS, based on conditional variational autocoder and adaptive undersampling is proposed and compared with traditional methods. The results show that the proposed CHBS method outperforms the traditional method and gets higher evaluation of classification.
2. A prediction model for gout (PIGD) has been proposed, which combines CHBS and improved TabNet to achieve accurate predictions of gout. Compared to previous methods, PIGD achieved higher accuracy on clinical gout datasets by improving feature extraction and addressing the issue of imbalanced data.
3. An interpretable model for gout was proposed by combining SHAP and TabNet feature importances. This model demonstrates the most important attributes in the decision-making process, enabling PIGD to produce interpretability.

2   Materials and Methods

2.1   CVAE-Based Hybrid Sampling

To address both class imbalance and class overlap, a hybrid CVAE-based sampling technique called CBHS was proposed. As explained in the related work, generative model-based sampling can effectively solve the class imbalance problem. However, the data generated by CVAE exacerbates the degree of class overlap. To deal with this problem, the CBHS is processed in three stages as shown in Fig. 2. In the first phase, outlier samples from all classes are cleaned by evaluating the neighbors of the majority and minority class samples. In the second stage, the CVAE technique is used to generate data to balance the different classes based on the global distribution. In the final stage, a neighbour-based adaptive weighted undersampling method was proposed to reduce overlap between the generated data and the original majority data. Figure 1(a) and (c) shows how the three stages can effectively prevent noisy samples, oversampling of less-informative minority samples, and class overlap resulting from global sampling. The steps of the method are presented in Algorithm 1.

In the initial phase, the KNN algorithm is used to identify the K nearest neighbours of each instance from majority class, and the number of minority class samples (m) within these neighbours is determined. If $Wmid \leq m \leq K$, then the instance is considered an outlier sample. Using this instance to synthesise a new instance may result in noise and boundary blurring. Therefore, this instance should be classified as a noise sample and removed. Similarly, the same operation is performed for instances in the minority class, and the number of samples in the majority class within the neighbourhood is calculated and compared. Lines 4 to 17 of Algorithm 1 demonstrate the specific operations mentioned above. Prior to implementing the clean algorithm, the appropriate value of K is determined by consulting relevant literature, with a value of 5 being selected. Additionally, $Wmid$ is set to 4 [12].

Algorithm 1 CVAE-based hybrid sampling (CBHS)

1: **Input:** Initial imbalanced dataset $T_{Origin_{im}}$ parameters k and W_{mid}
2: **Output:** Balanced data T_{ba}.
3: $T_{Origin_{maj}}/T_{Origin_{min}}$ majority/minority instances in $T_{Origin_{im}}$
4: **Start of clean noise instance:** Identification and clean of noise instance
5: **for** each $x \in T_{Origin_{maj}}$ **do**
6:　　Find the k nearest neighbors of x from $T_{Origin_{maj}}$
7:　　**if** Neighbors of minority $< W_{mid}$ **then**
8:　　　Put into T_{maj}
9:　　**end if**
10: **end for**
11: **for** each $x \in T_{Origin_{min}}$ **do**
12:　　Find the k nearest neighbors of x from $T_{Origin_{min}}$
13:　　**if** Neighbors of majority $< W_{mid}$ **then**
14:　　　Put into T_{min}
15:　　**end if**
16: **end for**
17: **End of clean**
18: **Start of CVAE-based hybrid sampling**
19: Generated instances T_{gen} by CVAE with T_{min}
20: **for** each $x \in T_{gen}$ **do**
21:　　Find the k nearest neighbors of x from T_{gen}
22:　　**if** Majority neighbors $N_{maj} \leq [k/2]$ **then**
23:　　　Put into T_{ba}
24:　　**end if**
25: **end for**
26: Clean $= len(T_{gen}) - len(T_{ba})$
27: **for** each $x \in T_{maj}$ **do**
28:　　Find the k nearest neighbors of x from T_{maj}
29:　　$N_{min} \leftarrow$ numbers of minority
30:　　$N_{gen} \leftarrow$ numbers of generated
31:　　$Weight_x = N_{min} * \frac{N_{min}}{N_{maj}} + N_{gen}$
32: **end for**
33: $T_{maj} \leftarrow T_{maj} -$ Clean instances with largest $Weight_x$ in T_{maj}
34: Put T_{min} into T_{ba}
35: Put T_{maj} into T_{ba}
36: **End of CVAE-based hybrid sampling**

During the second stage, the CBHS utilized the CVAE model to produce synthetic minority samples. It involves two main processes, encoding and decoding [13]. Encoding: Original data and class label information are input to encoder to obtain hidden variable variance probability distribution. Decoding: The decoder uses the variational probability distribution of the hidden variable to generate data that approximates the original data. The loss function used in CVAE is presented in Eq. 1. The first part of the equation serves as a reconstruction term, which guarantees the accuracy of the reconstruction. The second part of the equation is a constraint that improves the multivariate nature of the hidden variable. To enhance the variety and representativeness of the generated samples, the coefficient of KL loss is defined as IR. The IR is the ratio of the total number of samples in the majority class to the total number of samples in the minority class, as shown in Eq. 2. Figure 3 illustrates the structure and information flow of the CVAE. The encoder network takes inputs X and labels y, which generate the intermediate latent space Z through the mean (μ) and standard deviation (σ) vectors. The decoder uses both the latent variable Z and the labels y to generate new data X'.

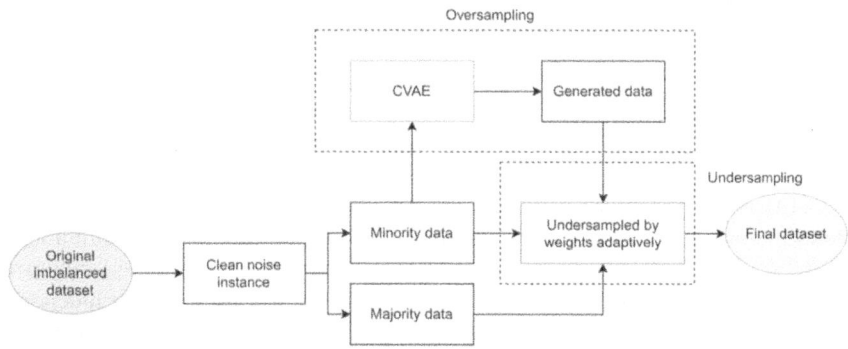

Fig. 2. General overview of CVAE-based Hybrid Sampling (CBHS)

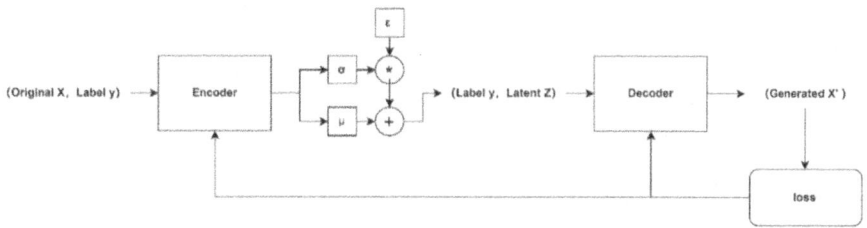

Fig. 3. Diagram of the proposed CVAE model.

Adaptive weighted undersampling was proposed in the final stage to process the oversampling based on CVAE. As demonstrated in Fig. 1(a) and (b), the CVAE effectively generates data that fits the global data distribution. However, it fails to address the issue of class overlap and may even worsen it. As shown in Fig. 1(b), this result suggests that some synthetic data were generated in the overlap region. Thus, the proposed method addresses class overlap by undersampling the generated and majority class samples. The input consists of three parts: the dataset for the minority class (*Tmin*), the dataset for the majority class (*Tmaj*), and the dataset generated for the mi-nority class by CVAE (*Tgen*). The first thing to consider is the undersampling of the generated data, as you can see lines 20 to 25 in Algorithm 1. For each instance in Tgen, decide whether to delete if more than half its k nearest neighbours is *Tmaj*. The next step is the undersampling of the original majority samples *Tmaj*. Line 31 of Algorithm 1 shows how the weight of each instance from the majority class is calculated based on its nearest neighbours. The weight takes a weighted sum of the minority and generated samples because the minority is more important than the majority. According to line 33 of Algorithm 1, instances with higher weights in *Tmaj* are deleted afterwards. The number of majority instances deleted is equal to the number of generated data deleted to ensure that the final dataset is balanced. As demonstrated in Fig. 1(b) and (c), the method effectively removes the data in the class overlap region, reducing the extent of class overlap while ensuring class balance.

$$minE_{q(z|x,y)}[logp(x|z, y)] - IR * KL(q(z|x, y)||p(z|y)) \tag{1}$$

$$IR = \frac{N_{maj}}{N_{min}} \tag{2}$$

2.2 Detection Model

TabNet [14] is a deep neural network specifically designed for tabular data, Fig. 4 illustrates the overall structure of the classifier. The architecture of it is continuous and multistep, at each step, a D-dimensional feature vector is processed and output to a Feature Transformer block. The Feature Transformer is composed of four independent blocks that may be shared across decision steps or unique to a decision step. Each independent block comprises three layers: fully connected layers, a batch normalization layer, and a Gated Linear Unit (GLU) activation. Additionally, the GLU is connected to a normalization residual connection, which helps to stabilize the variance throughout the network. Based on this, fully connected layers that enhance the input of the model has been added to the Feature Transformer in this study. Moreover, a normalization residual connection is also added, which can reduce the loss of information during transmission and protect the integrity of information. In this way, Feature Transformer assists in the selection of features and improves the parameter efficiency of the network, as shown in Fig. 5. In addition, the activation function has been changed from the Relu function to the Softmax function to better suit multiclassification tasks.

The Feature Transformer is connected to the Attentive Transformer and Mask. The Attentive Transformer is a multi-layered block that includes fully connected batch normalization layers, Prior Scales, and Sparsemax. The formulation of the Attentive Transformer and masking procedure is Eq. 3

$$a[i - 1] : M[i] = sparsemax(P[i - 1].h_i[a - 1]) \tag{3}$$

h_i refers to the FC and BN trainable functions, P $[i - 1]$ is the a priori scale and $a[i - 1]$ represents the previous step. Sparsemax reduces dimensionality by introducing sparsity into feature vectors and then projecting these features onto a probability map in Euclidean space. Each projected feature vector has an associated probability, making the model easier to interpret. The prior scale term, $P[i]$, is a measure of the saliency of a feature in the previous steps and is defined as Eq. 4

$$P[i] = \prod_{j=1}^{i} \{\gamma-\}M[j])\tag{4}$$

$M[j]$ is the trainable mask chosen by the Attentive Transformer. where γ defines the relationship between the enforcement of a feature in one decision step or in several steps.

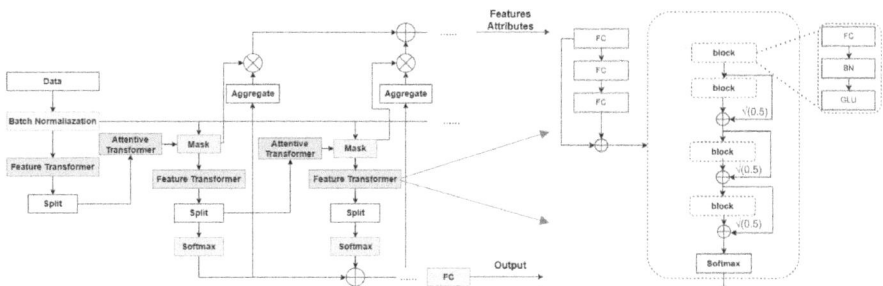

Fig. 4. Architecture of TabNet, each step contains an attentive transformer, mask, feature transformer, split node and SoftMax activation, right is improved feature transformer

2.3 Interpretation

To demonstrate the interpretability of the algorithm, this paper combines SHAP [15] (Shapley Additive Explanation) and TabNet. When using TabNet for prediction, a co-efficient is generated to determine the importance of features in the final decision, achieved by combining masks at each step. In addition, SHAP uses the principles of cooperative game theory to assign importance scores to each input feature for a given prediction. Each feature in the dataset can be seen as a player, and training a model on that dataset and obtaining a prediction can be seen as the payoff of many players cooperating on a project. And, the contribution of each player to the game is determined by calculating the SHAP value, which helps to identify the impact of a particular feature on the overall prediction. The SHAP is defined bas Eq. 5

$$\varphi_i(f) = \sum_{S \in Fi} \frac{|S|!}{|F|!} x(|S| - |F| - 1)! \times \left[f(X_{S \cup i}) - f(X_s) \right]\tag{5}$$

$\varphi_i(f)$ represents the weighted average of SHAP values for a given feature when all other features are excluded. F represents the total number of features, S represents a subset of feature F, $f(X_{S \cup i})$ represents the model prediction when feature i is included, and $f(X_s)$ represents the model prediction without feature i.

3 Experiment and Result

3.1 Datasets

The experimental dataset, which includes 39 physical examination characteristics and clinical laboratory indicators of gout patients, was provided by the hospital in Shandong. The patients were classified into four categories: hyperuricaemia acute gout stage, intermittent period, and chronic gouty arthritis [16]. The dataset comprised 4,362 cases, including 133 cases of hyperuricemia, 281 cases of attacks of acute gout, 1993 cases of intermittent period, and 1,955 cases of chronic gouty arthritis. To enhance data quality, it is recommended to use data preprocessing techniques, especially to handle missing values, prior to making predictions. The experiment randomly selects 20% of the initial training data as the test set and uses the remaining 80% as the training set. For the sake of ensuring experiment accuracy, the final classification results were averaged using a 5-fold cross-validation.

3.2 Classification Results

Figure 5 presents the visualization results for Accuracy, Precision, F1 Score, AUC, Recall, and G-mean for each model. The results suggest that the existing methods may be effective, but this is due to indiscriminately grouping the minority class with the majority class. In all metrics, the method presented in this paper outperforms the suboptimal algorithm. It is worth noting that the AUC, recall, and G-mean all improve by 5% compared to the suboptimal model. The precision is also higher, reaching 86.58%. The F1 score also improves by 5%, reaching 75.2%. Furthermore, the accuracy improves by 2.5%, reaching 80%. The results indicate that the CBHS method handles imbalanced data to reduce classification difficulty, and the improved feature transformer can more efficiently capture multilevel high-dimensional features, leading to improved model performance.

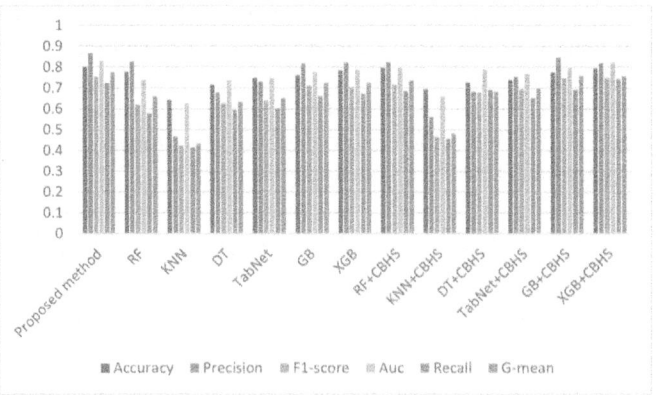

Fig. 5. Visualization results for of each model in the gout dataset.

3.3 Comparison of Balancing Strategies

This section discusses the effectiveness of the CHBS in addressing data imbalances.

Balanced datasets are generated using other techniques such as RandomOver, SMOTE, Borderline SMOTE, ADASYN and Kmeans SMOTE, and will then be classified using the proposed model. The main measures considered were recall, specificity, G-mean, and AUC. The experimental results are presented in Fig. 6. Compared to the original data, the CHBS proposed in this paper improves the recognition rate of minority classes from 32.3% to 55.44%, the recall rate from 65.58% to 72.27%, and the AUC value from 76.6% to 82.96%, while maintaining specificity. The reason for the improved performance compared to the oversampling methods RandomOver, SMOTE, Borderline SMOTE, ADASYN and Kmeans-SMOTE is that these methods increase the number of minority class samples through simple replication or linear interpolation. This introduces redundant information that increases noise, ultimately reducing overall performance and weakening generalization ability in certain cases. In contrast, this paper proposes the CHBS, which can reduce the impact of dataset imbalance and maximally reflect the characteristics of medical data. Therefore, the method proposed in this paper effectively addresses the problem of data imbalance, as all four indicators show improved performance

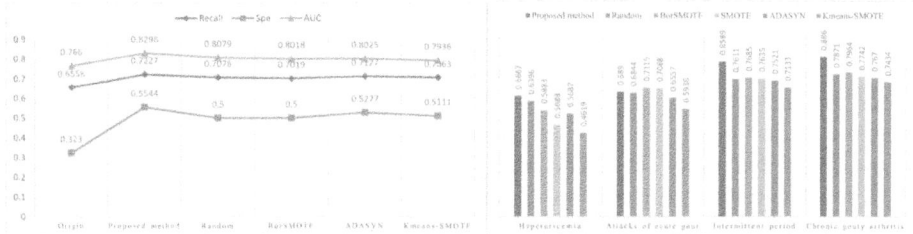

Fig. 6. Comparison of different balanced strategies

3.4 Model Interpretation

Explainable learning has become a popular research topic in recent years, particularly in medical applications. PIGD will be able to explain how the decision was made in each case. Figure 7 shows that PIGD offers instance feature selection with feature weights for each step. The features selected for each step are indicated by each Masks. The figure displays 50 samples vertically and 39 feature entries horizontally. Brighter areas indicate that the feature carries significant weight in the process and has higher impact on the final result. The masks that are used for each step are referred to in this paper as Mask0, Mask1 and Mask2. The proposed PIGD focuses on enhancing the performance of gout by addressing these features. After consulting with gout specialists, PIGD was developed to align with clinical findings, demonstrating the algorithm's credibility.

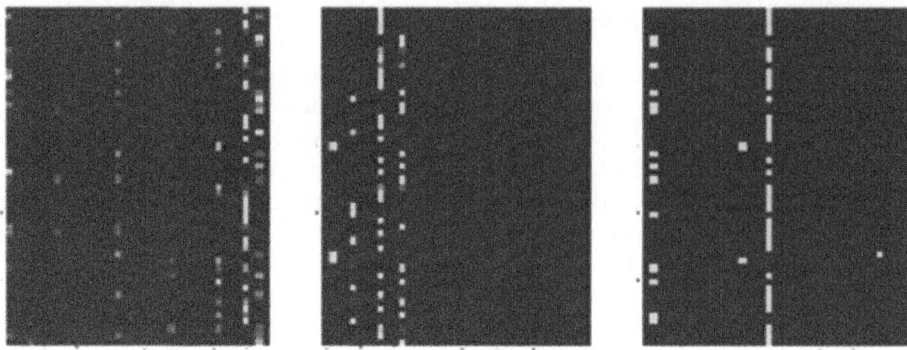

Fig. 7. Masks from PIGD.

In addition, the SHAP technique is useful for generating interpretable predictions and assigning a relevance score to each feature in complex models which was used to determine the essential PIGD attributes. Summary plot provides an overview of the overall and local feature importance of PIGD. The SHAP analysis presents a ranking of feature importance for different features on the gout dataset. The impact of each feature on the staging of gout is visible in Fig. 8(c). For instance, gout stone age, and CREA (creatinine clearance) have a greater impact on the model output, while UA (Uric acid) and others contribute less. The feature importance for different classes in the gout dataset is shown in the Fig. 8(a). It is evident that each feature has a varying impact on different classes. For instance, Ccr and Joint swelling assessment have a greater im-pact on the diagnosis of the minority class, while gout stone and vas score contribute more to the diagnosis of the majority class. The bee swarm plot displays the distribution of feature values and their relationship with the target variable. Each point represents a sample. The colour of the point indicates the value of the feature, with red indicating larger values and blue indicating smaller values. The horizontal axis represents the feature value, and the vertical axis represents the target variable. The Fig. 8(b) illustrates that the distribution of gout stone, age, and CREA are more dispersed, indicating a larger influence, while the distribution of other features is more compact and the influence is smaller. Waterfall plots are utilized to display the effect of features on the prediction results for a single sample. Each horizontal bar represents a feature, and its length indicates the contribution of feature to the prediction. Starting from zero, the values are added (in red) or subtracted (in blue) from the bottom of the waterfall plot to obtain the final prediction. The Fig. 8(d) below illustrates the prediction for the first observation in the gout dataset. The waterfall plot provides a sequential explanation of how the model adjusts its predictions based on each feature value, aiding in the comprehension of the significance of each feature for a single sample

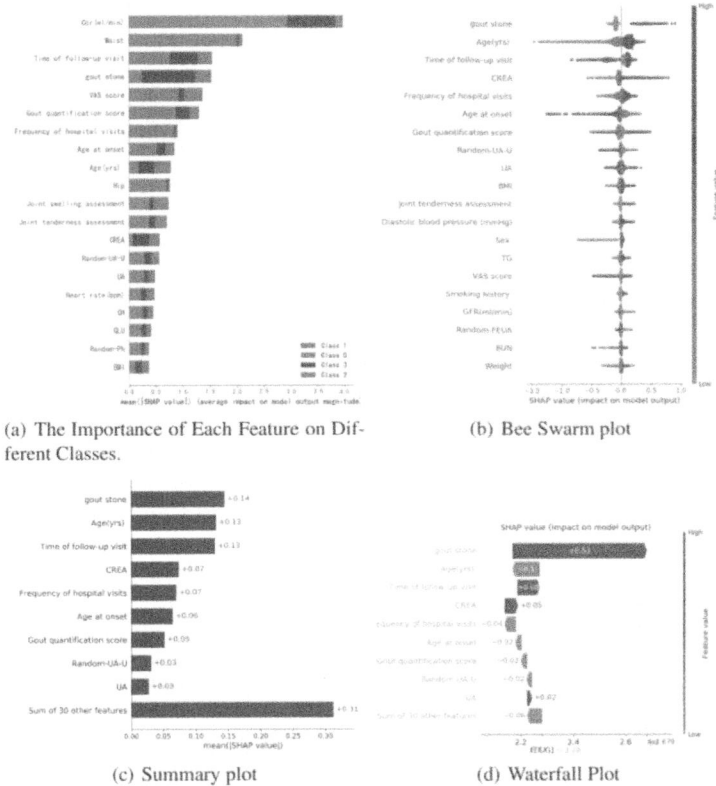

(a) The Importance of Each Feature on Different Classes.

(b) Bee Swarm plot

(c) Summary plot

(d) Waterfall Plot

Fig. 8. Plot of Shap

4 Conclusion

This study presents an interpretable, ML-based approach to accurately predict gout progression. This study proposes a hybrid sampling method based on CVAE to address the issue of dataset imbalance. It outperforms other sampling methods by fully reflecting the characteristics of the minority class. Furthermore, the enhanced TabNet model was utilized to classify gout, resulting in improved feature extraction and enhanced recognition accuracy for the minority class. This method has the best overall effect compared to the existing model, with better generalization performance and superior results in gout progression classification. Additionally, the interpretation of SHAP and TabNet are utilized to identify the significant features for predicting the progression of gout. Improvement in the interpretability of the model can be a means of reducing the potential risks inherent in its practical application. The study results indicate that the proposed model outperformed existing methods on gout dataset, achieving a precision of 86.65%, an AUC of 82.96%, an accuracy of 80%, and a G-means of 77.51%.

Even though the performance of the proposed method in predicting gout is better. But, the training dataset for gout is limited, the number of patients and features included in the study is relatively limited. As a result, the predictive performance for patient status is not entirely satisfactory. Therefore, the focus of future work will be on collecting additional data, including more features, from a larger patient population. It could also be easily applied and extended to other chronic tasks, such as diabetes.

Acknowledgments. This work was supported by the National Key Research and Development Plan Project of China 2022YFC2503300 and the Key Research and Development Plan of Shandong Province (2021CXGC011103, 2021SFGC0104).

References

1. Wortmann, R.L.: Gout and hyperuricemia. Current opinion in rheumatol-Ogy **14**(3), 281–286 (2002)
2. Punzi, L., et al.: One year in review 2020: Gout. Clin. Exp. Rheumatol. **38**(5), 807–821 (2020)
3. Beunza, J.J., Puertas, E., et al.: Comparison of machine learning algorithms for clinical event prediction (risk of coronary heart disease). J. Biomed. Inform. **97**, 103257 (2019
4. Asif, S., et al.: Improving the accuracy of diagnosing and predicting coronary heart disease using ensemble method and feature selection techniques. Cluster Comput. **27**(2), 1927–1946 (2023). https://doi.org/10.1007/s10586-023-04062-2
5. Xue, X., et al.: Effect of clinical typing on serum urate targets of benzbromarone in Chinese gout patients: a prospective cohort study. Front. Med. **8**, 806710 (2022)
6. Wang, C., et al.: Profiling of serum oxylipins identifies distinct spectrums and potential biomarkers in young people with very early onset gout. Rheumatology **62**(5), 1972–1979 (2023)
7. Abdellatif, A., et al.: Computational detection and interpretation of heart disease based on conditional variational auto-encoder and stacked ensemble-learning framework. Biomed. Signal Process. Control **88**, 105644 (2024)
8. Elreedy, D., Atiya, A.F.: A comprehensive analysis of synthetic minority oversampling technique (SMOTE) for handling class imbalance. Inf. Sci. **505**, 32–64 (2019)
9. Soltanzadeh, P., Hashemzadeh, M.: RCSMOTE: range-Controlled synthetic minority oversampling technique for handling the class imbalance problem. Inf. Sci. **542**, 92–111 (2021)
10. Yuan, X., Chen, S., Zhou, H., Sun, C., Yuwen, L.: CHSMOTE: convex hull-based synthetic minority oversampling technique for alleviating the class imbalance problem. Inf. Sci. **623**, 324–341 (2023)
11. Zhu, B., Pan, X., vanden Broucke, S., Xiao, J.: A GAN-based hybrid sampling method for imbalanced customer classification. Inform. Sci.**609**, 1397–1411 (2022)
12. Zhou, H., Wu, Z., Xu, N., Xiao, H.: PDR-SMOTE: an imbalanced data processing method based on data region partition and K nearest neighbors. Int. J. Mach. Learn. Cybern. **14**(12), 4135–4150 (2023)
13. Wang, Y.R., Sun, G.D., Jin, Q.: Imbalanced sample fault diagnosis of rotating machinery using conditional variational auto-encoder generative adversarial network. Appl. Soft Comput. **92**, 106333 (2020)
14. Arik, S.Ö., Pfister, T.: TabNet: attentive interpretable tabular learning. In: Proceedings of the AAAI Conference on Artificial Intelligence, vol. 35, no.8, pp. 6679–6687 (2021)
15. Lundberg, S.M., Lee, S.I.: A unified approach to interpreting model predictions. In: Advances in Neural Information Processing Systems, vol. 30 (2017)
16. Ragab, G., Elshahaly, M., Bardin, T.: Gout: An old disease in new perspective–a review. J. Adv. Res. **8**(5), 495–511 (2017)

DepthParkNet: A 3D Convolutional Neural Network with Depth-Aware Coordinate Attention for PET-Based Parkinson's Disease Diagnosis

Maoyuan Li[1,2], Ling Chen[3], Jianmin Chu[3], Xinchong Shi[4], Xiangsong Zhang[4], Gansen Zhao[1,2(✉)] ⓘ, and Hua Tang[1,2(✉)] ⓘ

[1] School of Computer Science, South China Normal University, Guangzhou, China
{gzhao,tanghua}@m.scnu.edu.cn
[2] Key Lab on Cloud Security and Assessment Technology of Guangzhou, Guangzhou, China
[3] Department of Neurology, First Affiliated Hospital, Sun Yat-Sen University, Guangzhou, China
[4] Department of Nuclear Medicine, First Affiliated Hospital, Sun Yat-Sen University, Guangzhou, China

Abstract. Positron Emission Tomography (PET) imaging is essential for accurately diagnosing Parkinson's Disease (PD). With the advancement of deep convolutional networks, 3D convolutional neural networks (CNNs) have been extensively utilized to analyze PET images. However, the limited availability of PET imaging data represents a significant bottleneck in this field. Training on small datasets could restrict the performance of 3D CNNs. Moreover, models trained on small datasets tend to exhibit weak generalizability, making it challenging to handle variations in PET images caused by clinical factors. This paper proposes depth-aware coordinate attention that integrates clinical prior knowledge. It's composed of two modules: the coordinate attention module and the depth-aware attention module. The coordinate attention module could preserve positional information while capturing long-range dependencies. Meanwhile, the depth-aware attention module aims to reveal the relationships among various depths within a PET image. These innovations enable our DepthParkNet to extract critical features from limited PET imaging data, thus enhancing its performance. Additionally, this paper proposes an augmentation pipeline named PDaug, which includes three transformations aimed at improving the generalizability of the model. The proposed model is validated on two datasets, demonstrating outstanding performance with balanced accuracy of 99.17% and 98.13%.

Keywords: Parkinson's disease · Deep learning · 3D convolutional neural network · Attention mechanism · Class imbalance

1 Introduction

Parkinson's Disease (PD) is a neurodegenerative disorder. Although there is currently no cure for PD, early detection and appropriate treatment could alleviate symptoms and improve the quality of life. Positron Emission Tomography (PET) is currently considered a crucial modality for assessing dopaminergic degenerative changes in subjects and plays a vital role in the early diagnosis of PD [1].

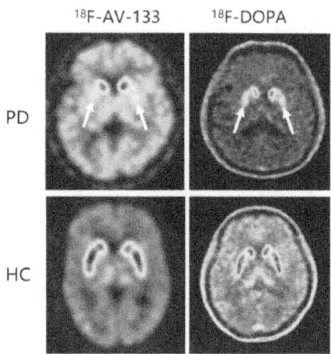

Fig. 1. Comparative view of PET images from PD patients and healthy controls (HC) using ^{18}F-DOPA and ^{18}F-AV-133 radiotracers.

Convolutional Neural Networks (CNNs) have been extensively employed in medical imaging tasks, including image classification [2], and segmentation [3], achieving promising results. Utilizing CNNs for the automated, early diagnosis of PD has been reported in [4–12]. However, the development of these models has not fully considered the distinctive features of PET images in the diagnosis of PD. Instead, they rely on generic methodologies for network development, which may not be entirely suitable for specific nuances of PET images. The limited availability of PET imaging data further complicates the development of a model with sufficient generalizability for clinical application. In clinical settings, PET images can exhibit a range of characteristics influenced by various clinical factors. These include the type of scanners and radiotracers employed, differences in individual patients' uptake of radiotracers, as well as variations in imaging angles and positions. A comparative view of PET images in PD and healthy control (HC) using ^{18}F-DOPA and ^{18}F-AV-133 radiotracers is presented in Fig. 1. It is observable that PET images with different radiotracers exhibit certain differences, yet both are capable of reflecting the distinct dopaminergic changes characteristic of the disorder. Therefore, it is essential for models to have strong generalizability to effectively deal with the image variations resulting from clinical factors. Furthermore, existing studies commonly neglect the issue of class imbalance [13]. The majority of studies that utilize deep learning for diagnosing PD come across a significant imbalance in the number of samples between healthy individuals and patients.

To tackle the aforementioned challenges, this work proposes DepthParkNet, a 3D CNN for PD diagnosis. DepthParkNet embeds a carefully crafted attention module, named depth-aware coordinate attention, to extract critical features from limited PET images. Simultaneously, this work proposes an augmentation pipeline, called PDaug, specifically designed for PET images used in PD diagnosis, to enhance the model generalizability. Additionally, this work introduces the class-balanced loss [14] to alleviate issues related to class imbalance.

The main contributions of this work are as follows:

- This work proposes DepthParkNet, a 3D CNN with an attention module, called depth-aware coordinate attention. This design integrates prior knowledge tailored to PD diagnosis, efficiently preserving positional details and long-range dependencies, merging multi-dimensional information, and minimizing redundancies in the depth dimension. It allows the proposed network to focus more precisely on lesion areas, thereby facilitating the extraction of critical features from PET images.
- This work proposes the PDaug, a specialized augmentation pipeline designed for PET image analysis. This method significantly improves the generalizability of the proposed model, allowing the model to effectively analyze PET images across diverse imaging conditions.
- The proposed DepthParkNet achieves balanced accuracy of 99.17% on the DOPA dataset and 98.13% on the AV133 dataset. These results substantiate the efficacy of the proposed model in the accurate diagnosis of PD, making it potentially a reliable tool in clinical settings.

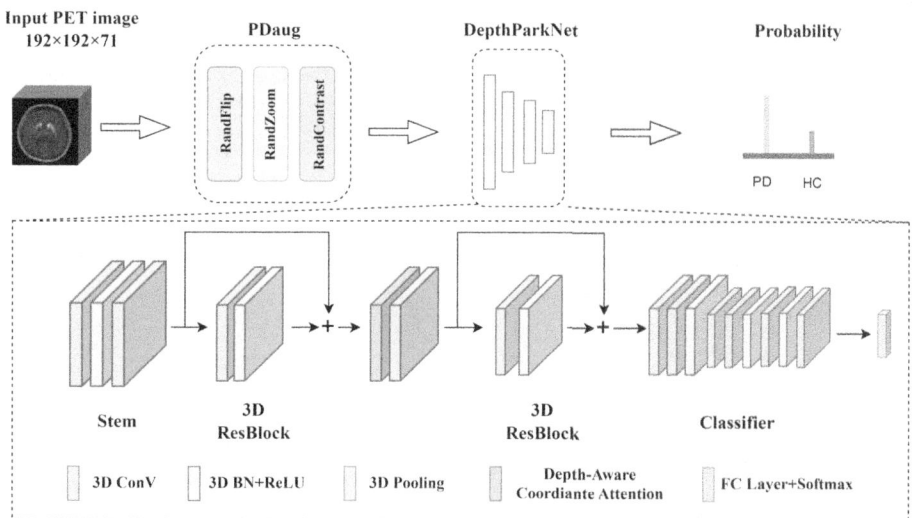

Fig. 2. The entire workflow of the proposed method.

2   Method

The complete workflow of the proposed method is illustrated in Fig. 2. PET images initially undergo enhancement through PDaug, followed by feature extraction and classification via our proposed 3D CNN, DepthParkNet. DepthParkNet, adapted from DeCoV-Net [15] which was initially designed for detecting COVID-19 in CT scans, incorporates depth-aware coordinate attention right after the first ResBlock. This design is driven by the intent to amplify features before excessive pooling or convolution operations, which tend to reduce resolution and depth, leading to a loss of information.

2.1   Depth-Aware Coordinate Attention

This work proposes the depth-aware coordinate attention that enhances the ability of model to capture neurodegenerative signatures associated with PD in PET images. As depicted in the Fig. 3, the architecture of our attention module comprises two modules: the coordinate attention module and the depth-aware attention module.

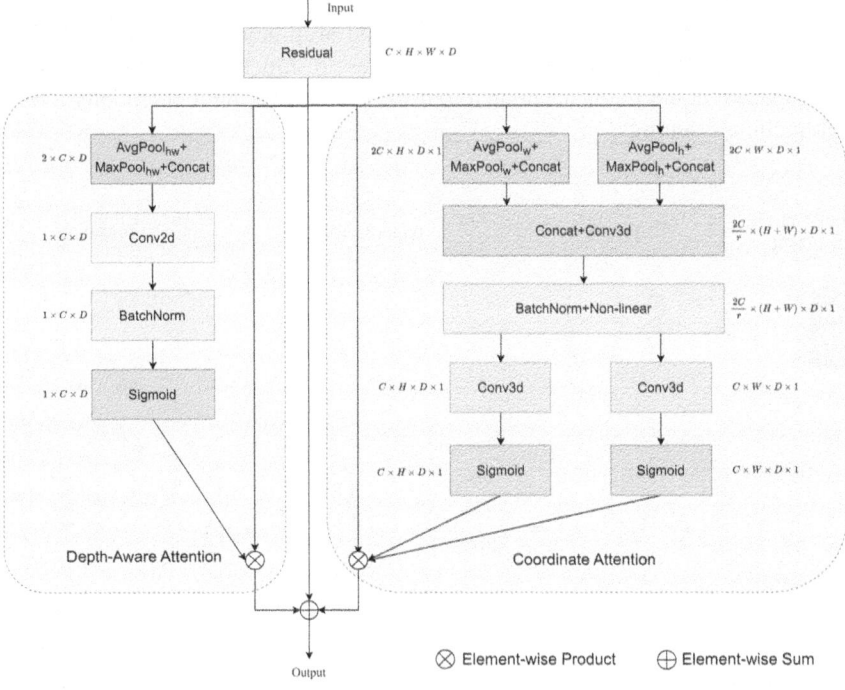

Fig. 3. Structure of Depth-Aware Coordinate Attention, where $AvgPool_w$ and $AvgPool_h$ denote average pooling using pooling kernels of size $(1, W, 1)$ and $(H, 1, 1)$, respectively. $AvgPool_{hw}$ represents average pooling using a pooling kernel of size $(H, W, 1)$. MaxPool operates in a similar manner but utilizes max pooling.

Coordinate Attention. In PET images, the regions relevant to PD diagnosis are often of limited extent. It is therefore important to develop a network that can better preserve positional information and enable full interaction with channel information. Coordinate attention [16] achieves this by first aggregating the input features along the vertical and horizontal directions using two 1D global pooling operations. This creates two separate direction-aware feature maps, each of which captures long-range dependencies along one spatial direction. The positional information is thus preserved in the generated attention maps.

This work adapts this concept from 2D CNNs to 3D CNNs, with a key distinction in the pooling process to retain depth dimension information. Specifically, for feature maps $X \in \mathbb{R}^{H \times W \times D \times C}$, pooling kernels with dimensions $(1, W, 1)$ and $(H, 1, 1)$ are utilized to encode each channel along the horizontal and vertical coordinates, respectively. This allows the decomposition of the feature maps into horizontal and vertical maps. To retain more information, both max pooling and average pooling are employed in the process. The feature maps generated from each pooling method are then concatenated along the channel dimension. These can be formulated as:

$$X_h = \left[\text{MaxPool}_w(X), \text{AvgPool}_w(X)\right], \tag{1}$$

$$X_v = \left[\text{MaxPool}_h(X), \text{AvgPool}_h(X)\right]. \tag{2}$$

Here, X_h and X_v denote the horizontal and vertical feature maps. $\text{MaxPool}_w(\cdot)$ and $\text{AvgPool}_w(\cdot)$ denote the application of max pooling and average pooling, respectively, each using pooling kernels of size $(1, W, 1)$. Likewise, $\text{MaxPool}_h(\cdot)$ and $\text{AvgPool}_h(\cdot)$ represent the execution of max pooling and average pooling, each employing a pooling kernel with the dimensions of $(H, 1, 1)$. The notation $[\cdot, \cdot]$ indicates concatenation along the channel dimension.

The generated horizontal and vertical feature maps are concatenated and then processed through a shared $1 \times 1 \times 1$ convolutional transformation function F_1, which integrates channel and spatial information, yielding:

$$f = \delta(F_1([X_h, X_v])). \tag{3}$$

In this formula, $[\cdot, \cdot]$ indicates concatenation along the spatial dimension, while δ refers to the Hardswish activation function, and $f \in \mathbb{R}^{\frac{2C}{r} \times (H+W) \times D \times 1}$ represents the intermediate feature map that captures spatial information in both the horizontal and vertical directions. Here, r is the reduction ratio similar to the SE block [17]. The tensor f is then divided into two separate tensors along the spatial dimension: $f^h \in \mathbb{R}^{\frac{2C}{r} \times H \times D \times 1}$ and $f^v \in \mathbb{R}^{\frac{2C}{r} \times W \times D \times 1}$. Subsequently, two $1 \times 1 \times 1$ convolutional transformations, F_h and F_v are applied to f^h and f^v, aligning their channel numbers with the input X. The outputs are subsequently multiplied with X to yield the result:

$$X_{ca} = X \times \sigma\left(F_h\left(f^h\right)\right) \times \sigma\left(F_v\left(f^v\right)\right), \tag{4}$$

where σ denotes the sigmoid function, and X_{ca} denotes the output of coordinate attention module.

Depth-Aware Attention. Clinically, particularly in the diagnosis of PD, not all brain imaging data are equally informative. Clinicians primarily focus on a few key slices in a PET image that contain the striatum to assess the condition. To maximize the extraction of critical information and minimize redundancy, a depth-aware attention module is proposed to capture relationships between different depths within a PET image.

For feature maps X, two pooling kernels, each with dimensions of $(H, W, 1)$, are utilized to perform max pooling and average pooling, respectively. The resulting feature maps are then squeezed and concatenated to form a feature map $X_d \in \mathbb{R}^{2 \times C \times D}$. This process can be written as:

$$X_d = \left[\text{MaxPool}_{\text{hw}}(X), \text{AvgPool}_{\text{hw}}(X) \right], \tag{5}$$

where $\text{MaxPool}_{\text{hw}}(\cdot)$ and $\text{AvgPool}_{\text{hw}}(\cdot)$ denote the application of max pooling and average pooling, respectively, each using pooling kernels of size $(H, W, 1)$. Then a 3×3 convolutional transformation F_2 is utilized to extract the mutual information between channels and depth dimensions, while also capturing the relationships between different depths. This is followed by a Batch Normalization (BN) layer and a sigmoid activation function δ to derive the depth-aware attention. The resultant attention weights are then multiplied with X to obtain the final output X_{da}. These can be formulated as:

$$X_{da} = X \times \delta(\text{BN}(F_2(X_d))). \tag{6}$$

Finally, residual connections are employed to integrate the outputs of both the coordinate attention module X_{ca} and depth-aware attention module X_{da} with the original feature maps X. This integration results are the final output of the depth-aware coordinate attention module X_{out}. The process can be mathematically expressed as follows:

$$X_{out} = X_{ca} + X_{da} + X. \tag{7}$$

2.2 PDaug

Various clinical factors can affect PET images characteristics including scanner types, radiotracers, individual patient uptake, etc. Therefore, models need to have a certain degree of generalizability to handle these variations. This work utilizes data augmentation to improve model generalizability. The proposed augmentation pipeline named *PDaug* including three transformations: *RandFlip*, *RandZoom*, and *RandContrast*.

RandFlip. To encourage the model to focus on striatal asymmetries, essential in diagnosing PD, we apply random flips (horizontal and vertical) to each image with a 50% probability, thereby enhancing the adaptability of model to structural variances.

RandZoom. Considering the importance of preserving complete striatal information, we don't use random cropping in our augmentation strategy. Instead, we employ a zooming technique with a probability of 0.5 and a zoom range of $[0.9, 2.0]$. This approach allows the model to focus on varying scales of striatal details without risking the loss of crucial information.

RandContrast. The contrast and intensity in PET images can vary significantly depending on the radiotracer used. To ensure that the model remains focused on key features, we incorporate a random adjustment of image contrast using the formula:

$$x = \left(\frac{x - x_m}{r}\right)^{\gamma} \times r + x_m, \tag{8}$$

where x_m is the minimum intensity in the image, r denotes the intensity range, and γ is a hyperparameter, which we randomly sampled from a uniform distribution in the range of [0.5,4.5].

2.3 Class-Balanced Loss

To mitigate class imbalance in the dataset, we utilize the class-balanced loss [14], which assigns weights inversely proportional to class frequencies, thereby emphasizing minority classes. Let the output logits of the model be denoted as $z = [z_1, z_2, \ldots, z_C]$, where C is the total number of classes. Considering a specific class y with y_n samples, the class-balanced loss introduces a weight factor $\frac{1-\beta}{1-\beta^{n_y}}$ to the standard cross-entropy loss. This factor, incorporating a hyperparameter $\beta \in [0,1)$, is calibrated to balance each class's contribution to the loss function. The class-balanced (CB) loss function is formulated as:

$$CB(z, y) = -\frac{1 - \beta}{1 - \beta^{n_y}} \log\left(\frac{\exp(z_y)}{\sum_{j=1}^{C} \exp(z_j)}\right). \tag{9}$$

3 Experiments

3.1 Datasets and Preprocessing

The experimental data for this study consists of PET images using ^{18}F-DOPA and ^{18}F-AV-133 radiotracers. Based on the type of radiotracer used, the data are divided into two groups: the DOPA dataset and the AV133 dataset, with their statistical properties detailed in Table 1.

Table 1. The statistical number of subjects on DOPA, AV133 datasets

Dataset	Diagnosis	No. Cases
DOPA	PD	613
	HC	84
AV133	PD	73
	HC	39

In the DOPA dataset, 20% of subjects from both the Parkinson's Disease (PD) and Healthy Control (HC) groups are allocated to the test set using a stratified sampling approach, ensuring balanced representation. The remaining 80% form the training set. To validate the robustness of model performance, the partitioning and training processes are repeated five times, with each iteration involving retraining and re-optimizing all parameters from scratch. The final results are determined by calculating the average values obtained from these five iterations. The AV133 dataset is reserved for testing to assess model generalizability. All PET images are resized to $192 \times 192 \times 71$ and normalized using the z-score method to achieve a standard intensity range for consistency.

3.2 Implementation Details

All models are developed using the Pytorch framework [18], while all augmentations are achieved using the MONAI framework [19]. Four metrics are employed to evaluate the performance of the models: accuracy (ACC), sensitivity (SEN), specificity (SPE), and balanced accuracy (BAC). Due to the class imbalance present within our dataset, balanced accuracy is particularly adopted to provide a more nuanced reflection of the performance of the model. The training process extends for 100 epochs with a batch size of 32, and all experiments are conducted using a Tesla A800 GPU.

3.3 Comparison

We conduct a comparative analysis of the proposed DepthParkNet against other models [20–27] on the DOPA dataset, as delineated in Table 2. DepthParkNet achieves outstanding performance across various metrics, surpassing the baseline DeCoVNet by 1.4% in balanced accuracy. This outcome underscores the efficacy of our model in capturing critical features. It is observed that direct transformation of 2D CNN architectures into 3D variants often results in suboptimal performance. Similarly, applying models originally designed for video recognition to Parkinson's diagnosis exhibits inherent limitations. Furthermore, the UniFormer [26] and Swin Transformer [27] models, based on the transformer architecture, exhibit poor performance. This could be attributed to the inherent requirement of transformer architectures for large datasets to achieve optimal training.

In assessing the generalizability of the proposed model, we conduct a direct test on the AV133 dataset using models previously trained on the DOPA dataset. As demonstrated in Table 2, DepthParkNet exhibits excellent performance, particularly notable in its balanced accuracy, which reaches 98.13%. In contrast, some models exhibit a notable decrease in balanced accuracy. This suggests that the proposed model is better equipped to capture crucial information in PET images of PD patients, thereby demonstrating enhanced generalizability. While models such as MobileNetV2 achieve 100% sensitivity, its relatively lower balanced accuracy of 70.51% indicates a deficiency in discerning critical features between PD and HC subjects, hinting at a potential bias.

We employ LayerCAM [28] for visualizations on the AV133 dataset. Figure 4 shows that DepthParkNet consistently concentrates on both sides of the caudate nucleus, a critical area associated with dopaminergic dysfunction in early-stage PD. In contrast, DeCoVNet primarily focuses on just one side of the caudate nucleus. This distinction

Table 2. Comparison of proposed model with other models

Method	Dopa				AV133			
	ACC	SEN	SPE	BAC	ACC	SEN	SPE	BAC
ResNet-50 [20]	99.28%	99.51%	97.65%	98.58%	90.18%	99.45%	72.82%	86.14%
EfficientNet [21]	92.25%	94.62%	74.85%	84.73%	86.25%	98.90%	62.57%	80.73%
MobileNetV2 [22]	94.24%	94.26%	94.12%	94.19%	79.46%	**100%**	41.03%	70.51%
I3D [23]	99.28%	99.51%	97.65%	98.58%	85.36%	99.73%	58.46%	79.09%
TSM [24]	99.14%	99.19%	98.82%	99.00%	94.29%	98.36%	86.67%	92.47%
TANet [25]	97.13%	97.23%	96.32%	96.78%	78.22%	88.22%	59.49%	73.85%
UniFormer [26]	82.36%	85.31%	79.41%	82.36%	65.18%	65.48%	64.61%	65.05%
Swin-T [27]	95.00%	98.91%	66.67%	82.79%	89.58%	98.17%	73.50%	85.84%
DeCoVNet [15]	98.71%	99.02%	96.47%	97.75%	96.78%	96.99%	96.41%	96.70%
Ours	**99.43%**	**99.51%**	**98.82%**	**99.17%**	**98.03%**	97.80%	**98.46%**	**98.13%**

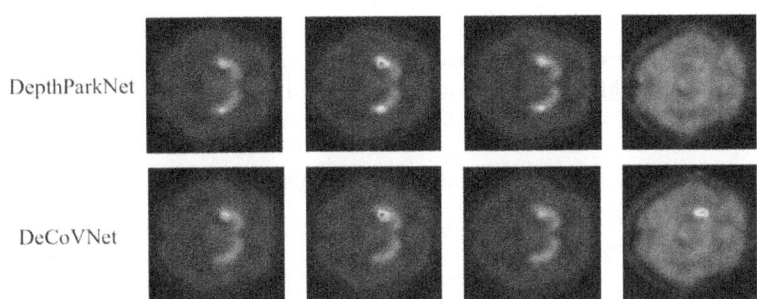

Fig. 4. Visualization using LayerCAM of the outputs from the second ResBlock in both DepthParkNet and DeCoVNet.

highlights the ability of DepthParkNet to more accurately extract and learn pathologically significant features from PD PET. Furthermore, as illustrated in the fourth column of Fig. 4, in comparison to DeCoVNet, the class activation maps of our proposed network mainly highlight regions in images containing the striatum, while images lacking this structure show minimal highlighted areas. This finding suggests the capability of our depth-aware attention module to effectively suppress redundant information.

3.4 Ablation Study

Effectiveness of Depth-Aware Coordinate Attention. Ablation experiments on the DOPA dataset validate the effectiveness of our proposed module. We remove each component of depth-aware coordinate attention—coordinate attention and depth-aware attention—individually, with results shown in Table 3. The findings reveal that each module individually contributes to an improvement over the DeCoVNet baseline, but optimal results are achieved only when both modules are incorporated. These experimental results demonstrate that our depth-aware coordinate attention module enhances the classification of PET images more effectively.

Table 3. Ablation study on attention modules

Method	ACC	SEN	SPE	BAC
Baseline	98.71%	99.02%	96.47%	97.75%
+ Coordinate Attention	99.14%	99.19%	98.82%	99.01%
+ Depth-Aware Attention	99.00%	99.02%	98.82%	98.92%
+ Depth-Aware Coordinate Attention	**99.43%**	**99.51%**	**98.82%**	**99.17%**

Effectiveness of PDaug. We perform ablation experiments on the AV133 dataset to evaluate the efficacy of PDaug. We systematically investigate the impact of each augmentation independently, with the results detailed in Table 4. Although each augmentation individually contributes to performance enhancement, the most significant improvement is observed when all three augmentations are applied in conjunction.

Table 4. Ablation study on augmentations

Method	ACC	SEN	SPE	BAC
No augmentation	75.00%	100%	28.00%	64.10%
+RandContrast	84.82%	98.63%	58.97%	78.80%
+RandFlip	77.68%	**100%**	35.90%	67.95%
+RandZoom	92.86%	90.41%	97.44%	93.92%
+RandZoom+RandContrast	80.36%	73.97%	92.31%	83.14%
+PDaug	**98.21%**	97.26%	**100%**	**98.63%**

4 Conclusion

This work proposes an innovative method for the automated diagnosis of PD using PET images. The developed model, DepthParkNet, embeds a depth-aware coordinate attention tailored for PD diagnosis. This attention design, incorporating clinical insights, efficiently captures positional information and long-range dependencies. It also integrates information across various dimensions and reduces redundancy within the depth dimension. Consequently, it enables the model to identify key features for PD diagnosis from PET images, thus enhancing the performance of the model. Besides, this work proposes PDaug, a specialized augmentation pipeline for PET images, which significantly improves the generalizability of the model across varied datasets. Furthermore, this work tackles the frequently underestimated issue of class imbalance by utilizing class-balanced loss. Future work will focus on developing interpretability techniques to provide insights into the decision-making processes within the model, thereby increasing the validity of the model.

Acknowledgments. This work was financially supported by the National Natural Science Foundation of China (No. 82271267), the National Social Science Fund of China (No. 19ZDA041), Guangzhou Science and Technology Plan Program (No. 2012224–12).

References

1. Kreisl, W.C., Kim, M.J., Coughlin, J.M., Henter, I.D., Owen, D.R., Innis, R.B.: PET imaging of neuroinflammation in neurological disorders. Lancet Neurol. **19**(11), 940–950 (2020)
2. Yadav, S.S., Jadhav, S.M.: Deep convolutional neural network based medical image classification for disease diagnosis. J. Big Data **6**(1), 1–18 (2019)
3. Tajbakhsh, N., Jeyaseelan, L., Li, Q., Chiang, J.N., Wu, Z., Ding, X.: Embracing imperfect datasets: a review of deep learning solutions for medical image segmentation. Med. Image Anal. **63**, 101693 (2020)
4. Wenzel, M., et al.: Automatic classification of dopamine transporter SPECT: deep convolutional neural networks can be trained to be robust with respect to variable image characteristics. Eur. J. Nucl. Med. Mol. Imaging **46**(13), 2800–2811 (2019)
5. Xiao, B., et al.: Quantitative susceptibility mapping based hybrid feature extraction for diagnosis of Parkinson's disease. NeuroImage: Clinical **24**, 102070 (2019)
6. Kiryu, S., et al.: Deep learning to differentiate parkinsonian disorders separately using single midsagittal MR imaging: a proof of concept study. Eur. Radiol. **29**(12), 6891–6899 (2019)
7. Magesh, P.R., Myloth, R.D., Tom, R.J.: An explainable machine learning model for early detection of Parkinson's disease using LIME on DaTSCAN imagery. Comput. Biol. Med. **126**, 104041 (2020)
8. Nazari, M., et al.: Explainable AI to improve acceptance of convolutional neural networks for automatic classification of dopamine transporter SPECT in the diagnosis of clinically uncertain parkinsonian syndromes. Eur. J. Nucl. Med. Mol. Imaging **49**(4), 1176–1186 (2022)
9. Zhao, Y., et al.: Decoding the dopamine transporter imaging for the differential diagnosis of parkinsonism using deep learning. Eur. J. Nucl. Med. Mol. Imaging **49**(8), 2798–2811 (2022)
10. Sun, X., et al.: Use of deep learning-based radiomics to differentiate Parkinson's disease patients from normal controls: a study based on [18F]FDG PET imaging. Eur. Radiol. **32**(11), 8008–8018 (2022)
11. Wang, Y., et al.: An automatic interpretable deep learning pipeline for accurate Parkinson's disease diagnosis using quantitative susceptibility mapping and T1-weighted images. Hum. Brain Mapp. **44**(12), 4426–4438 (2023)
12. Zhao, L., et al.: MetaViT: metabolism-aware vision transformer for differential diagnosis of Parkinsonism with 18F-FDG PET. In: Frangi, A., de Bruijne, M., Wassermann, D., Navab, N. (eds.) IPMI 2023. LNCS, vol. 13939, pp. 132–144. Springer, Cham (2023). https://doi.org/10.1007/978-3-031-34048-2_11

13. Johnson, J.M., Khoshgoftaar, T.M.: Survey on deep learning with class imbalance. J. Big Data **6**(1), 1–54 (2019)
14. Cui, Y., Jia, M., Lin, T.Y., Song, Y., Belongie, S.: Class-balanced loss based on effective number of samples. In: 2019 IEEE/CVF Conference on Computer Vision and Pattern Recognition (CVPR), pp. 9260–9269 (2019)
15. Wang, X., et al.: A weakly-supervised framework for COVID-19 classification and lesion localization from chest CT. IEEE Trans. Med. Imaging **39**(8), 2615–2625 (2020)
16. Hou, Q., Zhou, D., Feng, J.: Coordinate attention for efficient mobile network design. In: 2021 IEEE/CVF Conference on Computer Vision and Pattern Recognition (CVPR), pp. 13708–13717 (2021)
17. Hu, J., Shen, L., Sun, G.: Squeeze-and-excitation networks. In: 2018 IEEE/CVF Conference on Computer Vision and Pattern Recognition, pp. 7132–7141 (2018)
18. Paszke, A., et al.: Pytorch: an imperative style, high-performance deep learning library. In: Advances in Neural Information Processing Systems, vol. 32 (2019)
19. Cardoso, M.J., et al.: Monai: an open-source framework for deep learning in healthcare. arXiv preprint arXiv:2211.02701 (2022)
20. He, K., Zhang, X., Ren, S., Sun, J.: Deep residual learning for image recognition. In: 2016 IEEE Conference on Computer Vision and Pattern Recognition (CVPR), pp. 770–778 (2016)
21. Tan, M., Le, Q.: EfficientNet: rethinking model scaling for convolutional neural networks. In: International Conference on Machine Learning, pp. 6105–6114. PMLR (2019)
22. Sandler, M., Howard, A., Zhu, M., Zhmoginov, A., Chen, L.-C.: MobileNetV2: inverted residuals and linear bottlenecks. In: 2018 IEEE/CVF Conference on Computer Vision and Pattern Recognition, pp. 4510–4520 (2018)
23. Carreira, J., Zisserman, A.: Quo Vadis, action recognition? A new model and the kinetics dataset. In: 2017 IEEE Conference on Computer Vision and Pattern Recognition (CVPR), pp. 4724–4733 (2017)
24. Lin, J., Gan, C., Han, S.: TSM: temporal shift module for efficient video understanding. In: Proceedings of the IEEE/CVF International Conference on Computer Vision, pp. 7083–7093 (2019)
25. Liu, Z., Wang, L., Wu, W., Qian, C., Lu, T.: TAM: temporal adaptive module for video recognition. In: 2021 IEEE/CVF International Conference on Computer Vision (ICCV), pp. 13688–13698 (2021)
26. Li, K., et al.: UniFormer: unifying convolution and self-attention for visual recognition. IEEE Trans. Pattern Anal. Mach. Intell. **45**(10), 12581–12600 (2023)
27. Liu, Z., et al.: SWIN transformer: hierarchical vision transformer using shifted windows. In: Proceedings of the IEEE/CVF International Conference on Computer Vision (ICCV), pp. 10012–10022 (2021)
28. Jiang, P.T., Zhang, C.B., Hou, Q., Cheng, M.M., Wei, Y.: LayerCAM: exploring hierarchical class activation maps for localization. IEEE Trans. Image Process. **30**, 5875–5888 (2021)

Gene Selection and Classification Method Based on SNR and Multi-loops BPSO

Baofang Chang[1,2,3], Fuxiang Sun[1,2,3(✉)], Maodong Li[1,2,3], Peiyan Yuan[1,2,3], Hecang Zang[4], and Hu Jin[5]

[1] College of Computer and Information Engineering, Henan Normal University, Xinxiang 453007, China
sun307157171@outlook.com
[2] Big Data Engineering Lab of Teaching Resources and Assessment of Education Quality, Xinxiang 453007, China
[3] Key Laboratory of Artificial Intelligence and Personalized Learning in Education of Henan Province, Xinxiang 453007, China
[4] Institute of Agricultural Information Technology, Henan Academy of Agricultural Sciences, Zhengzhou 450002, China
[5] Department of Electronics and Communication Engineering, Hanyang University, Ansan 15588, South Korea

Abstract. For gene expression data with a massive amount of redundant data and noise, gene selection methods based on binary particle swarm optimization algorithm (BPSO) is an important method to improve classification performance. However, most BPSO-based methods can only handle finite sets, resulting in more selected genes and lower classification accuracy. This paper proposes a hybrid method SNR-MBPSO that utilizes signal-to-noise ratio (SNR) combined with multi-loops BPSO and various classifiers. It is important to point out that the multi-loops BPSO structure in this paper is first proposed and used to solve gene selection problems. And the multi-loops BPSO shows a remarkable improvement in the selection when it is combined with SNR. What is more, the traditional BPSO is modified by using adaptive weights and an improved bit-value changing strategy. To verify the performance of the proposed method, the SNR-MBPSO is compared with the other seven recently published algorithms in the literature. Experimental results based on nine publicly available gene expression datasets have shown that the proposed method significantly outperforms the state-of-the-art methods in terms of classification accuracy and the number of key genes.

Keywords: binary particle swarm optimization algorithm · signal-to-noise ratio · support vector machine

1 Introduction

Advances in DNA microarray technology and the need of analyzing gene expression have stimulated a shining road of research in biology and medical fields [1]. DNA microarray studies are used to discover which specific genes are important for the development of

D.-S. Huang et al. (Eds.): ICIC 2024, LNBI 14881, pp. 73–84, 2024.
https://doi.org/10.1007/978-981-97-5689-6_7

a disease, and they are used to analyze gene expression data associated with a specific diagnosis. However, gene expression data are high-dimensional data produced by DNA microarray experiments. The challenge of high-dimensional data analysis is to select information about the disease from a huge amount of redundant data and noise. As such, the selection and classification of gene data is a great hotspot in the medical and computer science sector [2]. Many solutions have been proposed for the above challenges and can be arranged into three general categories: filter, wrapper, and embedded [3].

In recent years, most advanced methods have adopted a hybrid strategy to improve and enhance performance [4]. In general, most methods use the filter method for data screening in the first stage. It means that the filter methods used in the first stage need to be able to ensure that key genes are not discarded. There is an additional requirement that the method in the first stage can reduce the time complexity. We are inspired to construct a hybrid feature selection method that includes preprocessing, selection, and classification [5]. In the first stage, namely the pre-processing stage, non-parameter filters such as Pearson Correlation coefficient (PCC) and signal-to-noise ratio (SNR) are relatively widely used due to their better performance. Pearson Correlation Coefficient (PCC) is used to determine the correlation between the features and gene classification, but it cannot guarantee that key genes will not be removed, especially since it probably eliminates key genes when combined with Wrapper methods [6]. In contrast, SNR can achieve ideal performance when it is combined to form a hybrid method. Golub et al. [1] use SNR for selection and prove that SNR is an efficient algorithm that ensures key features are not lost. It is worth noting that the proposed method also uses SNR in the pre-processing phase. After the first stage is completed, the processed results generally show a steep drop in the number of gene scales compared to the original data. However, the processed results still contain many redundant genes and difficult-to-process genes. Therefore, it is necessary to continuously optimize in subsequent steps to achieve satisfactory results. In other words, it is desirable to select a smaller subset of gene data for higher accurate classification [7].

Due to multiple constraints such as highly correlated genes, better classification performance, and a number of genes, researchers have proposed many bio-inspired optimization methods in recent years [8]. Particle swarm optimization algorithm (PSO) is one of the bio-inspired methods that quickly became well-known for its superior performance in solving many optimization problems [9]. Kennedy et al. [10] propose a particle swarm optimization algorithm not only solves continuous space problems but also solves discrete space problems. Although the BPSO is used for feature selection and results consistent with the concept that BPSO is beneficial to get a higher classification accuracy, the selected subsets are still having redundant genes that hinder the further improvement of classification accuracy. Collectively, BPSO is probably to be trapped in local optimal as it is generally accepted that caused by inertia weight and bit value change strategy [11]. Shi Y et al. [11] utilized the adaptive inertia weight which achieved more superiority in global searching ability and accelerated particle convergence. Jian-hua Liu et al. [12] provided a new bit value change strategy that ensured global optimal convergence and strengthened local searching performances. However, it ignored the influences of inertia weights. Taken together, the reason for the unsatisfactory performance may be that it falls into the local area too quickly in the early search process, or the search results may become unstable due to the lack of constraints on the global search. In general, the solution of gene selection optimization problems is a complex process that requires sophisticated and fine-grained control. Meanwhile, we found that both inertia weights and bit value change have important effects on the search process. Hence, the inertia weights and the bit value change are used alternately in the search process to control the optimization performance. Therefore, by modifying the inertia weight and position value change strategy of the traditional BPSO algorithm, the global and local optimal problems can be solved as best as possible.

In addition, some advanced algorithms in the field of gene selection in recent years have used heuristic-based strategies for optimization. Salem H et al. [13] exported IG-SGA which combined both Information Gain (IG) and Standard Genetic Algorithm (SGA) for selection. Dabba A et al. [14] proposed MIM-MFA using Mutual Information Maximization (MIM) and modified Moth Flame Algorithm (MFA). Due to the characteristics of gene datasets, few scholars use deep learning methods for gene selection [15]. And deep learning methods do not show great superiority in gene selection [16]. But a recent method based on deep learning has attracted widespread attention. That is Shah S H et al. [17] presented LS-CNN based on Laplacian Score (LS) and Convolutional Neural Network (CNN) for the classification. Algamal Z Y et al. [18] exported TSLR, including two-stage sparse logistic regression, which combined the screening approach as a filter and adaptive lasso with a new weight as an embedded method. Hameed et al. [19] developed the combination of PCC and BPSO (PCC-BPSO) for selection in high-dimensional data. And it is confirmed by the results that BPSO-based methods can achieve relatively high classification accuracy. In these hybrid methods, there is a common problem that the selected genes still remain a large number of redundant genes. And they face a critical challenge in practical applications when there are high correlations among genes. To address this problem, we first proposed the multi-loops structure and used it in the selection stage to solve the problem that redundant genes are difficult to remove. In this paper, a three-phase hybrid method is proposed, to obtain an efficient subset of genes with high classification capabilities by combining the SNR as a pre-processing approach and multi-loops BPSO. In the pre-processing phase, SNR is employed to eliminate redundant genes for high-dimensional gene datasets. To our knowledge, this is the first attempt at applying the multi-loops BPSO as a gene selection to guarantee smaller gene numbers and a high classification. Finally, three different classifiers are applied to evaluate the classification accuracy.

We briefly state the key contributions of the work presented in this paper below.

1. To address the problem that most of the BPSO-based feature selection methods failed to select a smaller subset of important genes efficiently, by combining multi-loops structure with BPSO, the multi-loops BPSO is proposed. The proposed method probably runs BPSO multiple times to obtain the smaller subset. When BPSO is terminated, the reserved genes are determined by the degree of SVM. Meanwhile, a new terminated condition is proposed to address the computational burden.

2. To improve particle search ability and accelerate particle convergence, the adaptive inertia weight and a new bit-value change strategy are instead. The adaptive inertia weight is updated by the current fitness and ensures to avoid trapping in local optimal.

Furthermore, the new bit-value change strategy with two parts is conducive to particle search and convergence.

2 Method

In this section, we present a new BPSO-based method, called SNR-MBPSO, which includes three phases as shown in Fig. 1, pre-processing, selection, and classification. In the pre-processing phase, the Min-Max method is utilized to normalize the gene data and guarantee a stable convergence of weights and biases [20]. Then, we use the Signal-to-noise ratio (SNR) to obtain pre-selected genes. During the selection phase, the proposed method probably runs modified BPSO several times to obtain the smaller subset. Instead of traditional BPSO, the modified BPSO raises adaptive inertia weight and bit-values change strategy to improve particle search and convergence. This positive impact is seen to be varied for different datasets based on the final applied classifier. In the last phase, three different classifiers are employed to classify the selected gene subsets. The results indicate that the accuracy rates are all in the acceptable rational regions. Especially, the proposed method with SVM classifier significantly outperforms the other state-of-the-art methods in terms of classification accuracy and the number of selected genes.

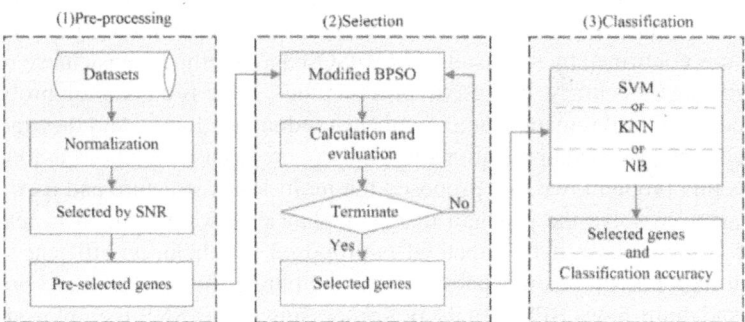

Fig. 1. The overall of proposed method

2.1 The Multi-loops BPSO

Though the BPSO methods are widely used in gene classification and have remarkable performance in feature selection, it is insufficient to deal with high correlation among the relevant genes. These lead us to find more options and better solutions with BPSO. Meanwhile, it is observed that there remain some irrelevant genes with BPSO running only once. If the BPSO runs again, the results prove that the selection is very effective in terms of classification accuracy. Thus, a new hybrid method with combines multi-loops structure based on BPSO is proposed for gene selection. This paper provides the terminated condition of BPSO and gives a complexity analysis in this section. Moreover, the traditional BPSO is modified by using the adaptive inertia weight and bit-value change strategy in this paper.

The Workflow of Multi-loops BPSO. After pre-processing, assume that we get a gene subset G_s^1. Thus, the G_s^1 will be evaluated by multi-loops BPSO in selection. After the first BPSO end, we can get a gene subset G_1. G_1 will be beneficial to classification and retained. The other genes, denoted as G_1', where $G_1 \bigcup G_1' = G_s^1$, which maybe has some relevant genes and need to be calculated. G_1' will be evaluated to pick up some relevant genes and generate a subset G_1''. The $G_1 \bigcup G_1''$ maybe remains irrelevant genes. Suppose that $G_s^2 = G_1 \bigcup G_1''$. The gene subset G_s^2 will be determined to whether run the loop or not by the terminate conditions. The loops will be executed repeatedly until there are no irrelevant or satisfy the terminate conditions.

According to the above assumptions, although the irreverent gens G_1' will be discarded, there are probably still some key genes which beneficial to classification in G_1'. The process of discarding some key genes usually occurs in the early loops [21]. To ensure the classification performance, we evaluated the genes G_1' further to avoid discarding several key genes.

If G_i' is the irreverent genes during i^{th} loop. $x(g_i) \subseteq G_i'$. Suppose that G_i' with m genes. Now, the individual g_i will be decided to whether remain or not by the following:

$$x(g_i) = \begin{cases} 1 & \omega_i^2 < \gamma \cdot \frac{1}{m}\sum_{j=1}^{m}\omega_j^2 \\ 0 & \omega_i^2 \geq \gamma \cdot \frac{1}{m}\sum_{j=1}^{m}\omega_j^2 \end{cases} \tag{1}$$

When $x(g_i) = 1$, the individual will remain. Otherwise, the individual will be discarded. Where γ within $(0,1]$.

By computation, a smaller subset G_i'' will be obtained. Then, the gene subset $G_i \bigcup G_i''$ will be evaluated to determine whether to restart a loop by the terminate condition. Because the loops depend on the terminate condition, the proper terminate condition contributes to decreasing the time complexity. In this study, the multi-loops are terminated when the selection results at the $D + 1^{th}$ and the Dth BPSO are similar. The formalization is given by the following:

$$Rj^{D+1} = Rj^D \tag{2}$$

where Rj^{D+1} is the number of eliminated genes at the $D + 1^{th}$ BPSO. Rj^D is the number of eliminated genes at the Dth BPSO.

The Modified BPSO. We have made deep improvements to traditional BPSO. And the main goal is to ensure classification performance and reduce computational complexity. In terms of ensuring classification performance, it is mainly accomplished by a combination of adaptive inertia weights and bit value change strategies. Since using the multi-loops method may increase the time overhead. We also use some tricks to reduce the time overhead and mainly achieve this by controlling the particle swarm size and the number of iterations.

The inertia weight affects the process of the particle swarm to find the optimal solution. The high weight is beneficial to the global search, but not conducive to the convergence of the particle swarm; On the contrary, it is beneficial to local search, but it is easy to fall into local optimum. In this paper, the adaptive inertia weight is used to update the weight at this time according to the current fitness of particles. This is beneficial for particles to avoid falling into local optimal and improve the search ability of particles. The bit value change strategy has effects on the search of particles. In the traditional strategy, although the particle has good global search ability, it is not conducive to particle convergence. Using the original strategy results in the search unstable and negative effect for selection. Based on this, we employ an improved bit value change strategy. This strategy speeds up particle convergence, to ensure the selection performance and reduce the time complexity.

3 Experiments and Results

3.1 Experiment Preparation

To demonstrate the superiority of the method proposed in this paper, the comprehensive results of the proposed method in nine gene datasets are performed with other comparative methods. The running environment of all the simulation experiments is provided by PYTHON 3.7.3, which is installed on a personal computer running Windows 10 on an Intel(R) i5-10210U at 2.11 GHz and 8.0 GB memory. During the experiments, three-quarters of the total sample is used for training and the rest is used for testing.

3.2 Experimental Design Principles

To fully demonstrate the advantages of this method, we employ three classifiers including SVM, KNN (K-Nearest Neighbor), and NB (Naive Bayes) and the results are mainly included in the following three parts:

First, to illustrate the screening effect of SNR, a comparable experiment is carried out between SNR and original data processing (ODP) on nine gene sets and three classifiers.

Second, we conducted experiments on BPSO under different loops. In the experiment, the performance is compared by setting one loop and multiple loops. According to the criteria of classification accuracy and the number of selected genes, the experiments clearly show the superior performance of multi-loops BPSO on all datasets.

Finally, to better verify the proposed method, the method is fully compared with the other seven similar methods in the above datasets. These methods include five state-of-the-art processing methods IS-SGA, MIM-mMFA, LS-CNN, TSLR (Two-stage Sparse Logistic Regression), and SIL (SNR ILasso) [23] in recent years, and two more classical processing methods PCC-BPSO and PCC-GA. The experiments show that the proposed method has better performance on all datasets.

3.3 Preprocessing by SNR

For the first experiment, we performed SNR preprocessing to address the excess of redundant genes. Meanwhile, the accuracy of the classifiers applied to the original datasets (ODP) is evaluated. Thus, we compare the performance of SNR and ODP on multiple classifiers. And the datasets generally be ranked and divided into five intervals with a step of 0.2. Each interval gene can be used as a result of gene screening. It is worth noting that the same pre-processing settings are used in the literature [22] in terms of screening. For most gene sets, key genes are concentrated in [0.8, 1], and the rest of the intervals are redundant genes. For the two gene sets of Colon and Prostate, the number of genes between [0.8, 1] is too small to be meaningful as key genes. Therefore, we selected genes between [0.6, 0.8] as key genes.

The information on screening genes in each gene set can be gained in Table 1 demonstrated that the number of genes in each gene set decreased significantly after SNR processing. Column 5 represents the proportion of selected genes in total genes. The proportion of selected genes is lower than 1% in most gene sets. It is observed that the SNR can remove relevant genes efficiently from Table 1. Specifically, the Colon shows the worst selection effect and the selected genes account for 2.9% of total genes and this is caused by a high correlation with redundant genes.

Table 1. Gene set information after SNR

Datasets	SNR interval	The size after SNR	Original size	Percentage
RAOA	(0.8, 1]	194	18433	0.0105
RAHC	(0.8, 1]	20	41057	0.0005
DLBCL	(0.8, 1]	28	5469	0.0051
SRBCT	(0.8, 1]	17	2308	0.0074
Leukemia	(0.8, 1]	46	7129	0.0065
Lung	(0.8, 1]	127	12533	0.0101
Ovarian	(0.8, 1]	84	15154	0.0055
Colon	(0.6, 0.8]	58	2000	0.029
Prostate	(0.6, 0.8]	84	10509	0.008

In a further experiment, three classifiers SVM, KNN, and NB are used to examine the impact of both ODP and SNR methods on classification performance as shown in Table 2. It can be seen that SNR has better classification accuracy on all classifiers. A lot of irrelevant and redundant genes are removed from the original genes by SNR, thus increasing the classification accuracy of most gene sets significantly. Hence, it is confirmed that SNR can ensure key genes while greatly reducing a lot of redundancy. Although results demonstrate the validity of the processing effect of SNR, the screened gene sets still include many redundant genes and need to be further processed.

3.4   The Comparison of One-Loop and Multi-loops on BPSO

We compared the results of one-loop and multiple-loops on BPSO in the terms of classification accuracy and the number of selected genes. For the convenience of description, the one-loop BPSO and the multi-loops BPSO are denoted as O-BPSO and M-BPSO respectively. Tables 3 and 4 display the comparative experiments for the number of classification accuracy and selection genes, respectively.

Table 2. Classification accuracy of ODP and SNR on three classifiers

Dataset	ODP				SNR			
	Gene	SVM	KNN	NB	Gene	SVM	KNN	NB
RAOA	18433	0.833	0.667	0.667	194	**1**	**0.833**	0.667
RAHC	41057	0.9	0.7	0.7	20	0.9	**0.8**	**0.8**
DLBCL	5469	0.938	0.812	0.688	28	**1**	**0.938**	**0.938**
SRBCT	2308	1	0.909	1	17	1	**1**	1
Leukemia	7129	1	0.933	**1**	46	1	**1**	0.933
Lung	12533	0.972	0.889	0.972	127	0.972	**0.972**	0.972
Ovarian	15154	1	0.941	0.941	84	1	**0.98**	**0.961**
Colon	2000	0.846	0.692	0.308	58	0.846	**0.846**	**0.615**
Prostate	10509	0.905	0.809	0.524	84	**0.952**	**0.952**	**0.857**

As can be seen from Table 3, the M-BPSO is more comprehensive and superior in classification performance generally. But there are individual cases where the accuracy of the M-BPSO is lower than that of O-BPSO. This may be caused by the M-BPSO losing some key genes during the execution. As far as these few genes are concerned, the M-BPSO can reach a high and acceptable level of classification accuracy with only a quarter of the key gene scale as shown in Table 4.

Table 3. The classification accuracy of O-BPSO and M-BPSO

Datasets	SVM(%)		KNN(%)		NB(%)	
	M-BPSO	O-BPSO	M-BPSO	O-BPSO	M-BPSO	O-BPSO
RAOA	1	1	1	0.833	1	0.817
RAHC	0.83	**0.9**	**0.87**	0.86	**0.9**	0.82
DLBCL	1	1	1	1	1	0.9
SRBCT	1	1	1	1	1	1
Colon	**0.962**	0.908	0.931	**0.946**	**0.9**	0.662
Prostate	**0.971**	0.967	0.967	**0.971**	**0.957**	0.952
Leukemia	1	0.973	0.987	1	**0.987**	0.973
Lung	**0.978**	0.972	**0.981**	0.972	0.989	**0.994**
Ovarian	1	1	1	0.984	1	0.982

Table 4 shows the number of selected key genes under different loop conditions. Where, the average number of genes selected by M-BPSO on the classifiers SVM, KNN, and NB is 5.5, 5.6, and 5.9. Corresponding to the results of M-BPSO, the results of O-BPSO are 36.8, 35.0, and 34.6 respectively. According to the results of the three classifiers SVM, KNN, and NB, the ratio of M-BPSO to O-BPSO is 0.2157, 0.2685, and 0.2414. In other words, according to the overall average percentage, the number of M-BPSO is about a quarter of O-BPSO or even less.

Generally, it is confirmed that the M-BPSO has a better performance in terms of both the classification accuracy and the number of selected genes than O-BPSO.

Table 4. The number of genes of O-BPSO and M-BPSO (Horizontal Layout)

Datasets	SVM			KNN			NB		
	M-BPSO	O-BPSO	Percentage	M-BPSO	O-BPSO	Percentage	M-BPSO	O-BPSO	Percentage
RAOA	**9.1**	98.4	0.0925	**8**	95.8	0.0835	**14.9**	97.1	0.1535
RAHC	**3.2**	10.2	0.3137	**8**	8.6	0.9302	**4.2**	8	0.525
DLBCL	**6.1**	21.5	0.2837	**6.6**	18.7	0.3529	**4.3**	20.1	0.2139
SRBCT	**2.8**	6.9	0.4058	**2.4**	8.4	0.2857	**3.1**	7.5	0.4133
Colon	**5.5**	27.7	0.1986	**5.9**	27.9	0.2115	**6.6**	25.6	0.2578
Prostate	**4.9**	43.9	0.1116	**4.8**	36.9	0.1301	**4.7**	34.6	0.1358
Leukemia	**7.2**	22.2	0.3243	**5.6**	24.5	0.2286	**5**	20.6	0.2427
Lung	**7.4**	61.7	0.1199	**4.9**	59.2	0.0823	**5.1**	59.6	0.0856
Ovarian	**3.5**	38.4	0.0911	**3.9**	34.8	0.1121	**5.5**	38	0.1447
Avg	5.5	36.8	0.2157	5.6	35	0.2685	5.9	34.6	0.2414

3.5   Comparative Experiment and Analysis

The experiments in this section demonstrate the classification performance of our SNR-MBPSO method and other advanced methods in terms of average selected genes and classification accuracy. Table 5 shows the comparison results of method SNR-MBPSO with other five state-of-the-art methods, these five state-of-the-art methods include IS-SGA, MIM-mMFA, LS-CNN, TSLR, and SIL. In the table, the symbol "-" denotes that no results are obtained using the corresponding methods.

Table 6 shows the comparison results of method SNR-MBPSO with two classic heuristics methods PCC-BPSO and PCC-GA in recent years. For a fairer experimental comparison, we use the same settings as in the literature [19] which employed the same classifier SVM, KNN, and NB with the same dataset Colon, Leukemia, Prostate, and Ovarian.

As can be seen from Table 5 and Table 6, for Classifier SVM, the SNR-MBPSO has the best accuracy performance on all datasets. For Classifier KNN and NB, the PCC-BPSO and PCC-GA achieve a slightly higher classification accuracy of Colon and Leukemia than the SNR-MBPSO, but the SNR-MBPSO is only about a quarter of both PCC-BPSO and PCC-GA or even less in term of the number of selected genes. In the rest cases, the SNR-MBPSO shows higher or equal classification accuracy than PCC-BPSO or PCC-GA. In all cases, the SNR-MBPSO can obtain fewer genes than the other two methods.

In summary, whether compared with the state-of-the-art five methods or compared with the traditional two methods, the SNR-MBPSO can be only about a quarter or even less in the number of the selected genes with better or equal classification accuracy.

Table 5. Comparison results of us with the other SOTA methods

Methods		DLBCL	Colon	Prostate	Leukemia	Lung	Avg.	Std.
SNR-MBPSO	Acc	1	0.967	0.971	1	0.978	0.983	0.01
	Genes	**6.1**	5.5	**4.9**	7.2	7.4	**6.2**	**0.96**
IS-SGA	Acc	0.948	0.855	1	0.971	1	0.955	0.05
	Genes	110	60	26	**3**	9	41.6	39.53
MIM-mMFA	Acc	1	**1**	1	1	–	1	**0**
	Genes	14.7	26.3	18.6	7.5	–	16.78	6.79
LS-CNN	Acc	0.98	**1**	1	0.99	–	0.993	0.01
	Genes	1500	**1000**	2500	1700	–	1675	540.25
TSLR	Acc	–	0.938	0.951	0.937	0.964	0.948	**0**
	Genes	–	5	10	7	9	7.8	1.92
SIL	Acc	–	0.903	0.961	0.986	1	0.963	0.04
	Genes	–	**4**	9	14	**7**	8.5	3.64

Table 6. Comparison results of ours' with two classical methods

	Method	Colon		Leukemia		Prostate		Ovarian	
		Acc.	Genes	Acc.	Genes	Acc.	Genes	Acc.	Genes
SVM	PCC-BPSO	0.919	25	1	18	**0.971**	33	1	17
	PCC-GA	0.919	29	1	35	0.961	26	1	22
	proposed	**0.962**	5.5	1	7.2	**0.971**	4.9	1	3.5
KNN	PCC-BPSO	**0.936**	36	1	17	0.961	15	1	15
	PCC-GA	**0.936**	47	1	25	0.951	29	0.996	11
	proposed	0.931	5.9	0.987	**5.6**	**0.967**	4.8	1	3.9
NB	PCC-BPSO	**0.919**	21	1	27	**0.961**	22	1	18
	PCC-GA	**0.919**	23	1	29	0.941	27	1	20
	proposed	0.9	**6.6**	0.987	5	0.957	**4.7**	1	**5.5**

4   Conclusion

Reducing the redundant or irrelevant genes of gene expression datasets can effectively decrease the cost of cancer classification. In this paper, a novel gene selection method based on SNR and multi-loops BPSO is proposed. The multi-loops BPSO shows an excellent performance in the selection when it is combined with SNR. What is more, the BPSO is modified by using adaptive weights and an improved bit-value changing strategy. In order to fully exhibit the good performance of the methods, nine high-dimensional data sets are used for experiments. Experimental results display that the proposed method is effective to select the most relevant genes with higher classification accuracy, as compared with representative algorithms. However, the proposed method cannot remove some redundant genes which are high correlation and has high time complexity. In further work, to more efficient discriminate against high correlation and reduce the time-consuming of our method, more efficient discriminate strategies and termination conditions should be explored to achieve a better classification performance.

Acknowledgments. This work is supported in part by the National Natural Science Foundation of China under Grant No.62072159 and No.61902112, the Postgraduate Education Reform and Quality Improvement Project of Henan Province No. YJS2024AL095.

References

1. Golub, T.R., Slonim, D.K., Tamayo, P., et al.: Molecular classification of cancer: class discovery and class prediction by gene expression monitoring. Science **286**(5439), 531–537 (1999)
2. Alizadeh, A.A., Eisen, M.B., Davis, R.E., et al.: Distinct types of diffuse large B-cell lymphoma identified by gene expression profiling. Nature **403**(6769), 503–511 (2000)
3. Sun, L., Wang, L., Xu, J., et al.: A neighborhood rough sets-based attribute reduction method using lebesgue and entropy measures. Entropy **21**(2) (2019)
4. Sun, L., Zhang, X., Qian, Y., et al.: Feature selection using neighborhood entropy-based uncertainty measures for gene expression data classification. Information Sciences **502**, 18–41 (2019)
5. Hambali, M., Oladele, T., Adewole, K.: Microarray Cancer Feature Selection: Review, Challenges and Research Directions. International Journal of Cognitive Computing in Engineering **1**(3), 78–97 (2020)
6. Hira, Z.M., Gillies, D.F.: A Review of feature selection and feature extraction methods applied on microarray data. Adv. Bioinform. (2015)
7. Nada, A., Alshamlan, H.: A survey on hybrid feature selection methods in microarray gene expression data for cancer classification. IEEE Access **7**, 78533–78548 (2019)
8. Dashtban, M., Balafar, M., Suravajhala, P.: Gene selection for tumor classification using a novel bio-inspired multi-objective approach. Genomics **110**, 10–17 (2018)
9. Wang, X., Yang, J., Teng, X., et al.: Feature selection based on rough sets and particle swarm optimization. Pattern recogn. lett. **28**(4), 459–471 (2007)
10. Kennedy, J., Eberhart, R.: Particle swarm optimization. In: Proceedings of ICNN 1995-International Conference on Neural Networks, vol.4, pp. 1942–1948. IEEE (1995)
11. Shi, Y., Eberhart, R.: A modified particle swarm optimizer. In: IEEE International Conference on Evolutionary Computation Proceedings. IEEE World Congress on Computational Intelligence (Cat. No. 98TH8360), pp.69–73. IEEE (1998)

12. Liu, J.H., Yang, R.H., Sun, S.H.: The analysis of binary particle swarm optimization. J. Nanjing Univ. Nat. Sci. Edition **47**(5), 11 (2011)
13. Salem, H., Attiya, G., El-Fishawy, N.: Classification of human cancer diseases by gene expression profiles. Applied Soft Computing **50**, 124–134 (2017)
14. Dabba, A., Tari, A., Meftali, S., et al.: Gene selection and classification of microarray data method based on mutual information and moth flame algorithm. Expert Systems with Applications **166** (2021)
15. Mori, Y., Yokota, H., Hoshino, I., et al.: Deep learning-based gene selection in comprehensive gene analysis in pancreatic cancer. Sci. Rep. **11**(1), 1–9 (2021)
16. Montesinos-López, O.A., Montesinos-López, A., Pérez-Rodríguez, P., et al.: A review of deep learning applications for genomic selection. BMC Genomics **22**(1), 1–23 (2021)
17. Shah, S.H., Iqbal, M.J., Ahmad, I., et al.: Optimized gene selection and classification of cancer from microarray gene expression data using deep learning. Neural Comput. Appli. **1–12** (2020)
18. Algamal, Z.Y., Lee, M.H.: A two-stage sparse logistic regression for optimal gene selection in high-dimensional microarray data classification. Adv. Data Anal. Classifi. **13**(3), 753–771 (2019)
19. Hameed, S.S., Muhammad, F.F., Hassan, R., et al.: Gene selection and classification in microarray datasets using a hybrid approach of PCC-BPSO/GA with Multi Classifiers **14**(6), 868–880 (2018)
20. Jain, A., Nandakumar, K., Ross, A.: Score normalization in multimodal biometric systems: Pattern Recogn. **38**, 2270–2285 (2005)
21. V. Vapnik.: The Nature of Statistical Learning Theory. Springer science & business media (1999)
22. Xu, J.C., Li, T., Sun, L., et al.: Feature gene selection based on SRN and neighborhood rough set. Data Collection Processing **30**(5), 973–981 (2015)
23. Gordon, G.J., Jensen, R.V., Hsiao, L.L., et al.: Translation of microarray data into clinically relevant cancer diagnostic tests using gene expression ratios in lung cancer and mesothelioma. Cancer research **62**(17), 493–4967 (2002)

Graph Convolutional Networks Based Multi-modal Data Integration for Breast Cancer Survival Prediction

Hongbin Hu[1], Wenbin Liang[2], Xitao Zou[3], and Xianchun Zou[1(✉)]

[1] College of Computer and Information Science, Southwest University,
Chongqing 400715, China
[2] Key Laboratory of Luminescence Analysis and Molecular Sensing, Ministry of Education,
College of Chemistry and Chemical Engineering, Southwest University,
Chongqing 400715, China
[3] School of Intelligent Technology and Engineering, Chongqing University of Science and
Technology, Chongqing 401331, China
zouxc@swu.edu.cn

Abstract. Recently, multi-modal breast cancer survival prediction (MBCSP) has been widely researched and made huge progress. However, most existing MBCSP methods usually overlook the structural information among patients. While certain studies may address structural information, they often ignore the abundant semantic information within multi-modal data, despite its significant impact on the efficacy of cancer survival prediction. Herein, we propose a novel method for breast cancer survival prediction, termed graph convolutional networks based multi-modal data integration for breast cancer survival prediction (GMBS). In essence, GMBS firstly defines a series multi-modal fusion module to integrate diverse patient data modalities, yielding robust initial embeddings. Subsequently, GMBS introduces a patient-patient graph construction module, aiming to delineate inter-patient relationships effectively. Lastly, GMBS incorporates a Graph Convolutional Network framework to harness the intricate structural information encoded within the constructed graph. Extensive experiments on two well-known MBCSP datasets demonstrate the superior performance of GMBS method compared to representative baseline methods.

Keywords: Breast Cancer · Survival Prediction · Graph Convolutional Networks

1 Introduction

Breast cancer, the second most prevalent cancer in women, recorded 357,200 new cases in China in 2022, posing a significant threat to women's health [1]. The heterogeneity within invasive breast cancer, encompassing various molecular subtypes and clinical presentations, poses challenges in treatment decisions and accurate prognostication [2]. Survival predictions are crucial for managing invasive breast cancer. They guide personalized treatment plans [3], optimize palliative or terminal care [4], streamline patient selection for clinical trials, and ultimately improve patient quality of life.

© The Author(s), under exclusive license to Springer Nature Singapore Pte Ltd. 2024
D.-S. Huang et al. (Eds.): ICIC 2024, LNBI 14881, pp. 85–98, 2024.
https://doi.org/10.1007/978-981-97-5689-6_8

Breast cancer invasion, a complex process, is closely linked to molecular mechanisms. While gene expression [5, 6], copy number variation [4, 7], and clinical data [4, 8–10] offer insights, each has limitations. Integrating multi-modal data comprehensively assesses these factors, enhancing survival prediction accuracy. Clinicians consider all relevant patient information, but traditional statistical methods may struggle with complex datasets, hindering nuanced relationship capture [9].

Deep learning methods are increasingly preferred over traditional statistical approaches for breast cancer survival prediction (BCSP) [3, 4, 11]. However, existing MBCSP methods barely consider the inclusion of structural information [3, 4, 11, 12], which refers to relationships that exist among patients. Integrating structural information not only captures intricate patient associations but also facilitates the extraction of fine-grained associations within multi-modal patient data, offering valuable support for prediction tasks. However, it remains challenging in incorporating structural information among patients for cancer survival prediction [13].

Graph neural networks (GNNs), inherits the characteristics of deep learning, excelling in processing multi-modal [7] and graph-structured [14, 15] data, offering a promising learning framework. Despite their wide application [16, 17], developing effective mapping models for structured data remains challenging. In BCSP, this involves addressing complexities in feature representation and graph construction.

In line with the above considerations, we propose a novel GMBS method to integrate multi-modal data and explore rich structural information among patients. Specifically, GMBS firstly defines a series multi-modal fusion module to integrate multi-modal data of patients to obtain robust initial embeddings. Subsequently, GMBS proposes a patient-patient graph construction module to construct a graph with the goal of bridging relations of patients. Then, a graph convolutional network [15] is introduced in GMBS to exploit the structural information. The final survival prediction is produced by integrating node representations from the network's last layer. The main contributions are outlined as follows.

1. We introduce an innovative GMBS framework tailored for survival prediction using gene expression, copy number alteration, and clinical data, where series multi-modal fusion module is built for multi-modal data to integrate data for each modality.
2. We build the sample feature matrix and embed it in the patient-patient graph, design the graph convolutional network module to explore the rich structural information among patients and obtain the final node representation to produce survival prediction.
3. We conduct extensive experiments on TCGA-BRCA dataset and validation experiments on METABRIC dataset to evaluate the effectiveness of our proposed model. Experimental results show that the proposed model achieves better performance than the existing methods.

2 Method

In this section, we outline the workflow of our GMBS framework designed for survival prediction, depicted in Fig. 1.

In typical scenarios, the life expectancy of cancer patients is classified into long-time (more than or equal to 5-year survival) and short-time (less than 5-year survival) survivors

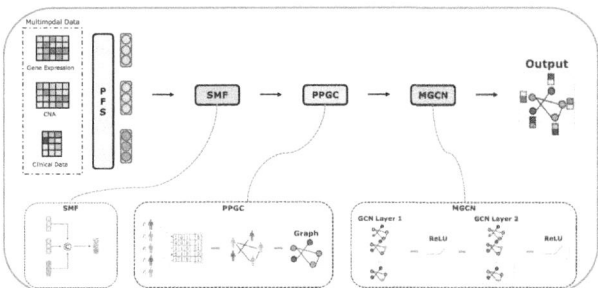

Fig. 1. A depiction of the GMBS method proposed herein. GMBS is composed of four modules: the preprocessing and feature selection module (PFS), the series multi-modal fusion module (SMF), the patient-patient graph construction module (PPGC) and the multi-modal graph convolutional networks module (MGCN). Concretely, the PFS is leveraged to obtain robust initial embeddings of patients, the SMF is utilized to integrate the data of each modality, the PPGC is employed to build efficient mapping models for highly structured multi-modal data, and the MGCN is employed to leverage the structural information among patients and obtain the node representation of the final layer for survival prediction.

[18]. The short-time survivors are labeled as 0, and the long-time survivors are labeled as 1, following the manipulation of previous research [4, 9]. We use notation $y_i \in \{0,1\}$ to represent whether a patient i is a short- or long-time survivor. We aim to develop a survival prediction model that takes gene expression G_j, copy number alteration N_j, and clinical data C_j of an unseen patient j (not included in the training set) as input to predict her or his survival status $\hat{y}$ as either short- or long-time survival.

2.1 Feature Selection and Fusion

The preprocessing strategies for the three modal data types of gene expression data, CNA data, and clinical data, are implemented as follows.

Gene Expression Data. In the process of obtaining various data of patients, mistakes may occur in the process of collection or collation, resulting in missing part of the obtained data. Following the previous works [3, 8], for missing data, we use K nearest neighbors (KNN) for interpolation. If more than 20% of a patient's data is missing, this patient is filtered out from the dataset. Likewise, if a feature of the single-modal data has more than 10% missing values among all patients, the feature is filtered out. Gene expression data were normalized and further classified into three categories based on two thresholds, which determine the levels of high expression and low expression respectively: 1 (high expression, the value is higher than or equal to high expression threshold), 0 (normal expression, between low expression threshold and high expression threshold), and -1 (low expression, less than or equal to the low expression threshold).

CNA Data. Consistent with gene expression data, we use the same ways to filter out both the patients and features. We directly exploited the raw data containing five discrete values: -2 (homozygous deletion), -1 (hemizygous deletion), 0 (no deletion), 1 (copy number gain), and 2 (high-level amplification).

Clinical Data. Consistent with gene expression data and CNA data, we use the same ways to filter out both the patients and features. As the clinical data have many non-numeric features, we use one-hot encoding to process them. In addition, we use the Z-score to normalize the single-modal data:

$$Z = \frac{x - \mu}{\sigma} \tag{1}$$

where x is the original value, μ is the average value, σ is the standard deviation, and Z represents the value after normalization.

Gene expression data is 20,530-dim, and copy number variation data is 24,776-dim. The high dimensionality can easily result in Curse of Dimensionality [19]. We apply the improved mRMR method [20] (Fast-mRMR) for dimensionality reduction.

The data of all patients after dimension reduction feature selection can be represented as:

$$X_{\text{selected}} = \mathcal{G}, \mathcal{N}, \mathcal{C} \in \mathbb{R}^{n \times d} \tag{2}$$

where $\mathcal{G}, \mathcal{N}, \mathcal{C}$ denotes the gene expression data, copy number alteration, and clinical data after preprocessing and feature selection for all patients, respectively. It's important to note that these variables now denote the data after the mentioned processing steps, not the original data. The dimension is defined as $d = m + n + c$, where m, n, c represents the dimension of the gene expression data, the CNA data, and the clinical data, respectively.

At present, most multi-modal data fusion methods use summation [21], multiplication [18], or concatenation [22]. In view of the different data dimensions of different modalities, the summation methods and the multiplication methods cannot be directly used for fusion. Besides, the concatenation method can take full advantage of multi-modal data. Therefore, we adopt the method of concatenation, using the SMF, all the modalities are connected serially. The next phase of input X_0 involves merging all multi-modal patient data through data processing and feature selection:

$$X_0 = \mathcal{G} \oplus \mathcal{N} \oplus \mathcal{C} \in \mathbb{R}^{n \times d} \tag{3}$$

where $\oplus$ represents concatenate. The feature vector X_0 can also be represented as:

$$X_0 = \{x_1, x_2, \ldots, x_n\} \tag{4}$$

where $x_1, x_2, \ldots x_n$ are feature vectors of each patient.

2.2 Patient-Patient Graph Construction

After we obtain the feature vector, we introduce how patient-patient graphs are constructed as the input to the networks. We construct the patient-patient graph as follows:

Each patient P_i is treated as a node, and its corresponding feature vector x_i (i-th row of X_0), is used as the node attribute. The nodes in the entire graph are categorized into two classes according to sample labels: long-term survival and short-term survival. We construct an undirected graph G, with edges defined as:

$$e_{ij} = \begin{cases} 0, & \text{if } y_i \neq y_j \\ 1, & \text{if } y_i = y_j \end{cases} \tag{5}$$

where e_{ij} represents whether an edge exists between node i and j (0: not exists; 1: exists) and y_i and y_j are the labels of these two nodes. This construction will establish edges between any two patients with short-term survival and any two patients with long-term survival.

This abstraction process incorporates patients, feature vectors, relationships, and labels to obtain a patient-patient graph (see Fig. 2).

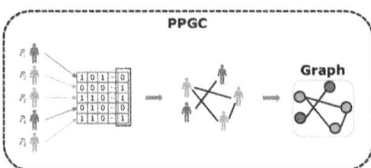

Fig. 2. An instance of PPGC, where 5 patients in a dataset are denoted as P_1, P_2, P_3, P_4, and P_5 with their feature vectors shown in the figure. The last column in Fig. 2 represents their labels. Patients P_1 and P_4 are short-term survivors, while patients P_2, P_3, and P_5 are long-term survivors.

2.3 Multi-modal Graph Convolutional Networks Module

We then introduce the graph neural network (GNN). The initial node representation is denoted as $H^0 = X_0$. The GNN module creates additional node representations $H^k \in \mathbb{R}^{n \times F}$ layer by layer where $k = 1, 2, \cdots, K$, $K = 2$ in our case, n denotes the total number of nodes and F is the dimension of node representation. Each layer involves two crucial functions: the AGGREGATE function and the COMBINE function.

The AGGREGATE function, represented by Eq. 6, plays a key role in summarizing information from the neighbors of each node. For instance, it gathers relevant data from neighboring nodes in the graph, providing a holistic view of the local information surrounding a particular node:

$$a_v^k = \text{AGGREGATE}^k \left(\left\{ H_u^{k-1} : u \in N(v) \right\} \right) \tag{6}$$

where $N(v)$ is the set of neighbors of the v-th node.

The COMBINE function, represented by Eq. 7, is responsible for updating the representation of a node. It combines the aggregated information obtained from the neighbors, denoted as a_v^k, with the current node representation H_v^{k-1}. This combination process refines the node representation in each layer, capturing the evolving relationships within the graph:

$$H_v^k = \text{COMBINE}^k \left(H_v^{k-1}, a_v^k \right) \tag{7}$$

The node representation of the last layer H^K can be regarded as the final node representation. We use H^K for downstream tasks as default since it is often more expressive and offers better performance.

As shown in Fig. 3, the proposed multi-modal GNN comprises two graph convolutional layers [15]. The node representations of each layer are updated according to the following propagation rules [23]:

$$H^k = \sigma\left(\widetilde{D}^{-\frac{1}{2}} \widetilde{A} \widetilde{D}^{-\frac{1}{2}} H^{k-1} W^{k-1}\right) \tag{8}$$

where $k = 1,2$, $\widetilde{A} = A + I$ is the self-connected adjacency matrix of the given undirected graph G, I is represented as the identity matrix, and $\widetilde{D}$ represents a diagonal matrix where each $\widetilde{D}_{ii} = \sum_j \widetilde{A}_{ij}$. The activation function $\sigma(\cdot)$ (ReLU in this paper) is applied according to ReLU$(x) = max(0, x)$. The matrix $W^{k-1} \in \mathbb{R}^{F \times F'}$ (F and F' are the dimensions of the node representation at layer $k - 1$ and layer k respectively) is a trainable transformation matrix, which is optimized during the training process.

We can further decompose Eq. 8 to understand the AGGREGATE and COMBINE functions defined in multi-modal graph convolutional networks. For a node i, how the node representation is updated can be reformulated as follows:

$$
\begin{aligned}
a_i^k &= \text{AGGREGATE}^k\left(\left\{H_j^{k-1} : j \in N(i)\right\}\right) \\
&= \sum_{j \in N(i)} \frac{A_{ij}}{\sqrt{\widetilde{D}_{ii}\widetilde{D}_{jj}}} H_j^{k-1} W^k
\end{aligned} \tag{9}
$$

$$
\begin{aligned}
H_i^k &= \text{COMBINE}^k\left(a_i^k, H_i^{k-1}\right) \\
&= \sigma\left(\sum_{j \in N(i) \cup i} \frac{\widetilde{A}_{ij}}{\sqrt{\widetilde{D}_{ii}\widetilde{D}_{jj}}} H_j^{k-1} W^k\right) \\
&= \sigma\left(\sum_{j \in N(i)} \frac{A_{ij}}{\sqrt{\widetilde{D}_{ii}\widetilde{D}_{jj}}} H_j^{k-1} W^k + \frac{1}{\widetilde{D}_i} H_i^{k-1} W^k\right)
\end{aligned} \tag{10}
$$

The AGGREGATE function is defined as a weighted average of neighbor node representations, where the weight of each neighbor node j is determined by the edge weight between node i and j (normalized by the degrees of the two nodes). The COMBINE function is then defined as the sum of the aggregated message and the node representation itself, with the node representation normalized by the node's own degree [23].

The predicted label of node i (represented by $\hat{y}_i$) is calculated using the Softmax function, defined as follows:

$$\hat{y}_i = \text{Softmax}\left(H_i^K W^{\mathsf{T}}\right) \tag{11}$$

where $W \in \mathbb{R}^{|\mathcal{L}| \times F}$, $|\mathcal{L}|$ is the number of survival labels in the output space (2 in our case), F is the dimension of node representation in the K-th layer. The parameter K represents the sequence number of the last layer of the graph convolutional network, and in this case, $K = 2$. T means transpose. The Softmax function is applied to produce a probability distribution over the possible labels, providing a normalized prediction for the survival status.

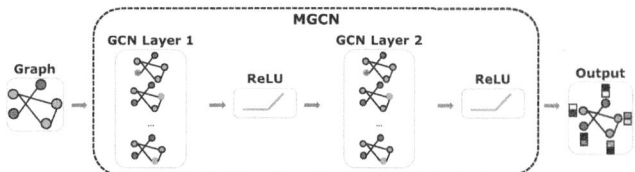

Fig. 3. Structure of MGCN. The whole MGCN has two graph convolutional layers, where each layer has two important functions: the AGGREGATE function and the COMBINE function. ReLU is used as the activation function in each layer. This module is designed to utilize the structural information among patients and acquire the node representation of the final layer for survival prediction.

2.4 Training Details

Loss Functions. The proposed model is trained in a supervised fashion. We employ the binary cross-entropy loss function for survival prediction classification. To mitigate the risk of overfitting, we incorporate L2 regularization into our loss function, a technique widely used in various deep learning studies [3, 4, 11, 24]. The loss function is defined as:

$$L(y_i, \hat{y}_i) = -\frac{1}{N} \sum_{i=0}^{N} [y_i \log(\hat{y}_i) + (1 - y_i)\log(1 - \hat{y}_i)] + \frac{\lambda}{2} \sum_{k=1}^{L} \sum_{j=1}^{n_k} \sum_{i=1}^{m_k} w_{ij}^{k\,2}$$

(12)

In this formulation, the first term represents the cross-entropy loss for binary classification, where y_i is the actual class label and $\hat{y}_i$ is the predicted probability. The second term introduces L2 regularization to prevent overfitting, with λ controlling the regularization strength, and w_{ij}^k is the element of W^k and W^k represents the k-th weight matrix in the model ($W^k = (w_{ij}^k)_{m_k \times n_k}$, L is the number of weight matrices).

Other Details. We implemented our model using Pytorch 1.9.0 and torch-geometric 2.1.0 on an NVIDIA RTX 2080Ti GPU server. The model was trained with Adam optimizer [25]. GMBS has a total of two graph convolutional layers with input feature dimension d and output feature dimension 16. The second convolutional layer has an input feature dimension of 16 and an output feature dimension of the number of classes (2 in our case). The hyperparameters in Sect. 2.1 were set as $m = 400, n = 200, c = 55, d = 655$, the learning rate was initialized as 0.01, and $\lambda = 5e - 6$ in Eq. 12.

3 Experiments

3.1 Datasets and Evaluation Metrics

Datasets. The TCGA (The Cancer Genome Atlas) database is a Cancer Gene mapping project jointly launched by the National Cancer Institute (NCI) and the National Human

Genome Research Institute (NHGRI) [26]. The TCGA database holds clinical and multi-omics data from over 11,000 cancer patients spanning 33 distinct tumor types [27].

We used the TCGA database as the source of experimental data to construct a dataset known as the TCGA-BRCA multi-modal dataset. Among them, the data types included in the dataset are gene expression, copy number variation, and clinical data. The TCGA database contains 1218 gene expression data, 1080 copy number variation data, and 1096 clinical data. We extracted 1068 valid breast cancer patient data from the TCGA database, denoted as the TCGA-BRCA multi-modal dataset, which included 823 short-term survivors and 245 long-term survivors. In the TCGA-BRCA dataset, the feature dimension of gene expression data is as high as 20530, the feature dimension of copy number variation data is as high as 24776, and the clinical data is 203. Building upon previous methodologies [18], we applied data preprocessing and feature selection techniques (detailed in Sec. 2.1) to curate a set of model input features. Specifically, we selected 400 genes from gene expression data, 200 genes from copy number variation data, and retained 55 clinical features from the clinical data. The evaluation of our model and other comparison methods is conducted using 10-fold cross-validation splits.

Evaluation Metrics. Following previous work [3, 8], Receiver Operating Characteristics curve (ROC) and Area Under the Curve (AUC) values are used as primary measures for hyper-parameter tuning and identifying the best model. The performance of our model is also evaluated in other performance indicators: Sn for sensitivity, Sp for specificity, Acc for accuracy, Pre for precision, and Mcc for Matthew's correlation coefficient. The detailed explanation of evaluation metrics can be found in [28].

3.2 Comparisons with State-of-The-Art

For a more comprehensive comparison, we use the same 10-fold cross-validation split to implement and evaluate some of the most recent survival prediction methods. As shown in Fig. 4, we compare the performance of the proposed model (GMBS) with other methods using the ROC curve as the primary index. Among them, the left figure represents the comparison between GMBS and traditional machine learning methods (LR [29], RF [30], and SVM [31]). GMBS demonstrates a significant advantage in the ROC curve, which can be explained by those traditional methods are not in place for multi-modal data mining. The right graph shows the comparison between GMBS and BCSP methods proposed in recent years (MDNNMD [4], Stacked RF [11], SiGaAtCNN [3], MAFN [8], Utility Kernel SVM [12], and ChoqFuzGCN [24]). GMBS holds a leading position in the ROC curve against recently proposed methods, e.g., it improves in AUC by 17.9% compared with MDNNMD [4]. The reason why GMBS is ahead of other state-of-the-art methods may be that graph neural networks can capture more fine-grained relationships between data. However, the listed recent models tend to ignore the use of structural information. Meanwhile, compared with ChoqFuzGCN [24], which also uses Graph Neural Network, our model also shows advantages. It might be because the edges in our graph are not constructed by similarity, while ChoqFuzGCN uses similarity to construct edges between patients. As mentioned earlier, patients with genetic abnormalities may still have a better prognosis. Therefore, when constructing edges by similarity, it may lead to wrong classification and ultimately affect the performance.

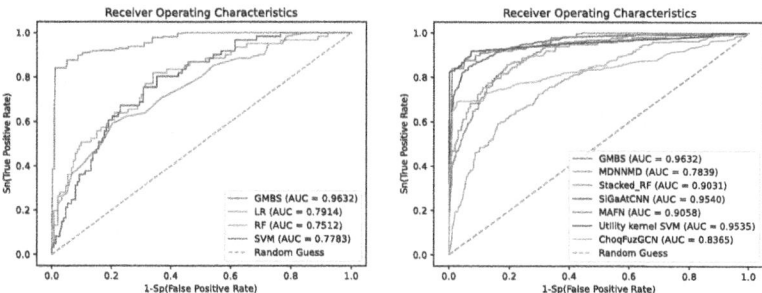

Fig. 4. The ROC curves of different approaches on TCGA-BRCA dataset. "GMBS" is the proposed approach. The left figure is the ROC curve comparison between our proposed GMBS and the traditional methods. Meanwhile, the right figure is a comparison between our proposed method and the methods proposed in recent years. Our model has obvious advantages.

Table 1. The secondary performance of different approaches. There are three traditional methods in the first part of the table (LR [29], RF [30], and SVM [31]). The second part consists of five deep learning survival prediction methods proposed recently (MDNNMD [4], Stacked RF [11], SiGaAtCNN [3], MAFN [8], and Utility Kernel SVM [12]). The third part of the table represents the survival prediction methods which use Graph Neural Network (ChoqFuzGCN [24], GMBS). The last method is the one proposed by us. Each method is evaluated under high specificity ($Sp = 99\%$) and medium specificity ($Sp = 95\%$). GMBS performs well in secondary performance.

Model	Acc	Pre	Sn	Mcc	Model	Acc	Pre	Sn	Mcc
($Sp = 99\%$) (Threshold $= 0.704$)					($Sp = 95\%$) (Threshold $= 0.446$)				
LR	0.790	0.674	0.508	0.451	LR	0.795	0.665	0.527	0.247
RF	0.720	0.931	0.143	0.109	RF	0.786	0.801	0.267	0.308
SVM	0.672	0.758	0.487	0.336	SVM	0.701	0.692	0.509	0.343
MDNNMD	0.799	0.862	0.156	0.327	MDNNMD	0.810	0.794	0.387	0.409
Stacked RF	0.909	0.857	0.682	0.735	Stacked RF	0.917	0.831	0.804	0.764
SiGaAtCNN	0.927	0.890	0.754	0.796	SiGaAtCNN	0.933	0.845	0.872	0.815
MAFN	0.850	0.872	0.839	0.572	MAFN	0.863	0.838	0.856	0.728
Utility SVM	0.932	0.901	0.830	0.837	Utility SVM	0.940	0.879	**0.877**	0.850
ChoqFuzGCN	0.715	0.396	0.440	0.230	ChoqFuzGCN	0.803	0.370	0.699	0.428
GMBS(Ours)	**0.939**	**0.946**	**0.840**	**0.859**	GMBS(Ours)	**0.943**	**0.898**	**0.877**	**0.862**

We also compared other metrics [28] when verifying our framework, as presented in Table 1. The left and right sections of the table compare the model's performance under high specificity ($Sp = 99\%$) and medium specificity ($Sp = 95\%$), respectively. Our model also achieves superior performance on these metrics. The general improvement demonstrates that our framework is effective in making reasonable predictions compared to both conventional methods and recently proposed state-of-the-art methods.

3.3 Ablation Studies

In this section, we conduct additional experiments to further explore the impact of various modalities and components of our models.

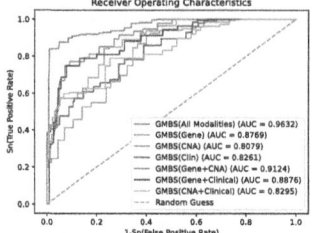 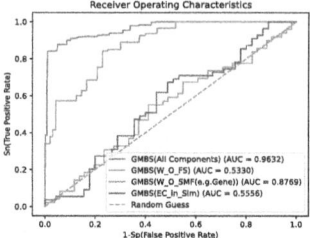

Fig. 5. The ROC curves of different modalities and components. The left part represents the ROC curves under different modalities, and the content of the brackets represents the modalities used by the model. The right part shows the ROC curves after removing or change one of three components, and the content of the brackets means that removing or changing a certain component. W_O represents without, FS refers to feature selection, EC means edge construction, and Sim is an abbreviation for Similarity. Both modalities and components are important factors affecting the performance of the model ROC curve.

Table 2. The secondary performance under different modalities. The content of the brackets represents the modalities used by the model.

Model	Acc	Pre	Sn	Mcc
GMBS (Gene)	0.824	0.806	0.873	0.665
GMBS (CNA)	0.657	0.743	0.498	0.510
GMBS (Clinical)	0.729	0.634	0.877	0.163
GMBS (Gene + CNA)	0.861	0.842	**0.883**	0.535
GMBS (Gene + Clinical)	0.833	0.830	0.881	0.701
GMBS (CNA + Clinical)	0.799	0.786	0.866	0.490
GMBS (All modalities)	**0.943**	**0.898**	0.878	**0.862**

Impacts of Modalities. The proposed GMBS model incorporates features from three modalities: Gene, copy number change (CNA), and Clinical data. In the left part of Fig. 5 and Table 2, we investigate how the model performance varies with different combinations of modalities. We employ the SMF method proposed in this paper for feature fusion when dealing with two or more modalities, and the results in Table 2 are obtained under moderate strictness ($Sp = 95\%$). Removing certain modalities leads to a severe decrease in model performance. For example, compared to GMBS with all modalities, the AUC of variant removing Gene modality, i.e., GMBS (CNA + Clinical), decreases by 13.4%, and Acc decreases by 14.4%. However, by removing the other modalities (GMBS (Gene + Clinical), GMBS (Gene + CNA)), we do not see a significant decrease in performance. This may be due to greater prognostic implications of gene-expression data. What we also need to see is that compared with the GMBS model using gene expression data alone, the performance of the model after adding CNA and Clinical modalities has been greatly improved. This further highlights the superiority of multi-modality over single-modality approaches in BCSP.

Table 3. The secondary performance after removing or changing one of three key components, and the content of the brackets means that removing or changing a certain component: 1) Removing SMF module; 2) Removing PFS module; 3) change edge construction methods in PPGC module. Removing any of the three components results in a significant decrease in secondary performance.

Model	Acc	Pre	Sn	Mcc
GMBS (W/O SMF (e.g. Gene))	0.824	0.806	0.873	0.665
GMBS (W/O Feature Selection)	0.648	0.578	0.069	0.126
GMBS (Edge Construction in Sim.)	0.710	0.558	0.379	0.227
GMBS (All Components)	**0.943**	**0.898**	**0.878**	**0.862**

Impacts of Components. The GMBS model comprises three critical components: the PFS module, SMF module, and PPGC module. The comparative analysis of three components is shown in the right section of Fig. 5 and Table 3. It is observed that the more modalities used, the greater the performance enhancement attributed to the SMF module. When no feature selection is applied, the model's AUC value decreases by 43.0% when 400, 200, and 55 dimensions are randomly selected from Gene, CNA, and Clinical data. This may be due to the random selection of features, which leads to the learning of a large amount of information that is not relevant to patient survival and thus severely affects the prediction performance. It also verifies the effectiveness of the Fast-mRMR algorithm [20] we used. Furthermore, constructing the edges of the patient-patient graph using cosine similarity results in a decrease in model performance. This suggests that computing cosine similarity between patient features may not yield optimal results. We hypothesize that the original representation of the patient needs to be further processed before computing cosine similarity. Otherwise, it might result in performance degradation. In summary, these findings serve to reinforce the efficacy of the proposed method.

3.4   Validation

To assess the generalization capability of our proposed model, we undertake a validation experiment utilizing the METABRIC dataset, employing the same 10-fold cross-validation split. Furthermore, we implement various state-of-the-art survival prediction methods and compare their performance with our proposed framework.

The METABRIC dataset comprises a total of 1980 patients, with 1489 classified as long-term survivors and 491 as short-term survivors. Notably, short-term patients are initially labeled as 1, while long-term patients are labeled as 0 in the METABRIC dataset.

To align with the conventions of the TCGA-BRCA dataset, we transform the labels accordingly: short-term patients are labeled as 0, and long-term patients are labeled as 1. Following what we did in Sec. 2.1, we preprocess the METABRIC dataset and extracts features. The resulting dimensions for the three modalities are 400, 200, and 25, denoted as $m = 400$, $n = 200$, and $c = 25$, respectively.

The performance comparison between the proposed model and other models is presented in Table 4, where the results are obtained under moderate strictness ($Sp = 95\%$). The comparison across models reveals significant improvements in various metrics besides Utility Kernel SVM [12] exhibiting a marginal lead of 0.9% in the Accuracy (Acc) value. We hypothesize that different models may favor different datasets. Beyond that, our GMBS outperforms other models. These results underscore the strong generalization ability of the proposed method and its robust performance on the METABRIC dataset.

Table 4. The experimental results on METABRIC dataset. The experiment was carried out under medium specificity (Sp = 95%). The same as before, the first three models in the table represent traditional models, the middle five models represent deep learning survival prediction models proposed in recent years, and the last two models use graph neural networks for survival prediction. Our proposed model is in the lead.

Model	AUC	Acc	Pre	Sn	Mcc
LR	0.864	0.830	0.880	0.855	0.493
RF	0.830	0.854	0.859	0.871	0.527
SVM	0.767	0.782	0.783	0.866	0.128
MDNNMD	0.845	0.827	0.760	0.466	0.492
Stacked RF	0.902	0.887	0.839	0.739	0.670
SiGaAtCNN	0.912	0.914	0.840	0.748	0.713
MAFN	0.917	0.879	0.901	0.793	0.571
Utility SVM	0.915	**0.925**	0.897	0.842	0.617
ChoqFuzGCN	0.830	0.820	0.630	0.666	0.528
GMBS(Ours)	**0.919**	0.916	**0.912**	**0.874**	**0.764**

4 Conclusion and Future Work

In this paper, we introduce GMBS, a novel framework for survival prediction, that integrates multi-modal data and harnesses rich structural information among patients through sample feature matrix and patient-patient graph construction. Our extensive experiments on the TCGA-BRCA dataset and the METABRIC dataset showcase that our proposed framework consistently outperforms state-of-the-art methods.

While our work has demonstrated excellent performance, it's important to note that the current public datasets, such as TCGA-BRCA with only 1068 samples, struggle to handle such small datasets using overly complex networks. We expect that future advancements will enable even greater improvements, particularly with access to larger datasets. Additionally, the development of cross-modal learning technology presents an exciting opportunity to enhance patient embedding learning and further improve survival prediction performance. Moreover, expanding the dataset to include additional modalities, such as gene methylation and miRNA, is also promising.

Acknowledgments. This study was funded by the National Natural Science Foundation (NNSF) of China (No. 22074122) and the Fundamental Research Funds for the Central Universities (No. SWU-KT22029), China.

References

1. Han, B., Zheng, R., Zeng, H., Wang, S., Sun, K., Chen, R., et al.: Cancer incidence and mortality in china, 2022. J. National Cancer Center **4**(1), 47–53 (2024)
2. Lukasiewicz, S., Czeczelewski, M., Forma, A., Baj, J., Sitarz, R., Stanis lawek, A.: Breast cancer—epidemiology, risk factors, classification, prognostic markers, and current treatment strategies—an updated review. Cancers **13**(17), 4287 (2021)
3. Arya, N., Saha, S.: Multi-modal advanced deep learning architectures for breast cancer survival prediction. Knowl.-Based Syst. **221**, 106965 (2021)
4. Sun, D., Wang, M., Li, A.: A multimodal deep neural network for human breast cancer prognosis prediction by integrating multi-dimensional data. IEEE/ACM Trans. Comput. Biol. Bioinf. **16**(3), 841–850 (2019)
5. Sun, Y., Goodison, S., Li, J., Liu, L., et al.: Improved breast cancer prognosis through the combination of clinical and genetic markers. Bioinformatics **23**(1), 30–37 (2007)
6. Van De Vijver, M.J., He, Y.D., et al.: A gene-expression signature as a predictor of survival in breast cancer. N. Engl. J. Med. **347**(25), 1999–2009 (2002)
7. Gao, J., Lyu, T., Xiong, F., Wang, J., Ke, W., Li, Z.: Predicting the survival of cancer patients with multimodal graph neural network. IEEE/ACM Trans. Comput. Biol. Bioinf. **19**(2), 699–709 (2021)
8. Guo, W., Liang, W., Deng, Q., Zou, X.: A multimodal affinity fusion network for predicting the survival of breast cancer patients. Front. Genet. **12** (2021)
9. Sun, D., Li, A., Tang, B., Wang, M.: Integrating genomic data and pathological images to effectively predict breast cancer clinical outcome. Comput. Methods Programs Biomed. **161**, 45–53 (2018)
10. Sun, W., Cai, Z., Li, Y., Liu, F., Fang, S., et al.: Data processing and text mining technologies on electronic medical records: a review. J. Healthcare Eng. **2018** (2018)
11. Arya, N., Saha, S.: Multi-modal classification for human breast cancer prognosis prediction: proposal of deep-learning based stacked ensemble model. IEEE/ACM Trans. Comput. Biol. Bioinf. **19**(2), 1032–1041 (2020)
12. Arya, N., Mathur, A., Saha, S.: Proposal of svm utility kernel for breast cancer survival estimation. IEEE/ACM Trans. Comput. Biol. Bioinf. **20**(2), 1372–1383 (2022)
13. Palmal, S., Arya, N., Saha, S., Tripathy, S.: A multi-modal graph convolutional network for predicting human breast cancer prognosis. In: International Conference on Neural Information Processing, pp. 187–198. Springer (2022)
14. Kipf, T.N., Welling, M.: Variational graph auto-encoders. CoRR abs/1611.07308 (2016)
15. Kipf, T.N., Welling, M.: Semi-supervised classification with graph convolutional networks. In: 5th International Conference on Learning Representations, ICLR 2017, Toulon, France, 24–26 April 2017, Conference Track Proceedings. OpenReview.net (2017)

16. Ma, T., Chen, J., et al.: Constrained generation of semantically valid graphs via regularizing variational autoencoders. Adv. Neural Inform. Process. Syst. **31** (2018)
17. Perozzi, B., Al-Rfou, R., Skiena, S.: Deepwalk: online learning of social representations. In: Proceedings of the 20th ACM SIGKDD International Conference on Knowledge Discovery and Data Mining, pp. 701–710 (2014)
18. Chen, R.J., Lu, M.Y., Wang, J., Williamson, D.F., Rodig, S.J., et al.: Pathomic fusion: an integrated framework for fusing histopathology and genomic features for cancer diagnosis and prognosis. IEEE Trans. Med. Imaging **41**(4), 757–770 (2020)
19. Subramanian, I., Verma, S., et al.: Multi-omics data integration, interpretation, and its application. Bioinform. Biol. Insights **14**, 1177932219899051 (2020)
20. Ramírez-Gallego, S., Lastra, I., Martínez-Rego, D., et al.: Fast-mrmr: Fast minimum redundancy maximum relevance algorithm for high-dimensional big data. Int. J. Intell. Syst. **32**(2), 134–152 (2017). https://doi.org/10.1002/int.21833
21. Liu, H., Kurc, T.: Deep learning for survival analysis in breast cancer with whole slide image data. Bioinformatics **38**(14), 3629–3637 (2022)
22. Li, R., Wu, X., Li, A., Wang, M.: HFBSurv: hierarchical multimodal fusion with factorized bilinear models for cancer survival prediction. Bioinformatics **38**(9), 2587–2594 (2022)
23. Wu, L., Cui, P., Pei, J., Zhao, L.: Graph Neural Networks: Foundations, Frontiers, and Applications. Springer Singapore (2022)
24. Palmal, S., Arya, N., Saha, S., Tripathy, S.: Breast cancer survival prognosis using the graph convolutional network with choquet fuzzy integral. Sci. Rep. **13**(1), 14757 (2023)
25. Kingma, D.P., Ba, J.: Adam: A method for stochastic optimization. In: 3rd International Conference on Learning Representations, ICLR 2015, San Diego, CA, USA, 7–9 May 2015, Conference Track Proceedings (2015), http://arxiv.org/abs/1412.6980
26. Wei, Y., Su, Y.: Using machine learning and rna to enhance the efficacy of anti-tumor immunotherapy. Evol. Intel. **16**(5), 1555–1563 (2023)
27. Howard, F.M., Dolezal, J., et al.: The impact of site-specific digital histology signatures on deep learning model accuracy and bias. Nat. Commun. **12**(1), 4423 (2021)
28. Tharwat, A.: Classification assessment methods. Applied computing and informatics **17**(1), 168–192 (2020)
29. Jefferson, M.F., Pendleton, N., Lucas, S.B., Horan, M.A.: Comparison of a genetic algorithm neural network with logistic regression for predicting outcome after surgery for patients with nonsmall cell lung carcinoma. Cancer: Interdisciplinary Inter. J. Am. Cancer Soc. **79**(7), 1338–1342 (1997)
30. Nguyen, C., Wang, Y., Nguyen, H.N.: Random forest classifier combined with feature selection for breast cancer diagnosis and prognostic. J. Biomed. Sci. Eng. **06**(05) (2013)
31. Xu, X., Zhang, Y., Zou, L., Wang, M., Li, A.: A gene signature for breast cancer prognosis using support vector machine. In: 2012 5th International Conference on Biomedical Engineering and Informatics, pp. 928–931. IEEE (2012)

IDHPre: Intradialytic Hypotension Prediction Model Based on Fully Observed Features

Yifan Yao[1,2], Zemin Kuang[3(✉)], Xiwen Yang[1], Baoquan Wang[1], Zhaomeng Niu[4], Jiaxin Yang[5], Lun Hu[1], Xi Zhou[1(✉)], and Pengwei Hu[1(✉)]

[1] Xinjiang Technical Institute of Physics and Chemistry, Chinese Academy of Sciences, Urumqi 830011, China
yaoyifan1@hotmail.com, {zhouxi,hpw}@ms.xjb.ac.cn
[2] The School of Software, Xinjiang University, Urumqi 830046, China
[3] Beijing Anzhen Hospital of Capital Medical University, Beijing 100029, China
kzmkk@foxmail.com
[4] Department of Health Informatics, Rutgers School of Health Professions, NJ 1709, USA
[5] School of Engineering and Applied Science, Columbia University, NY 10964, USA

Abstract. In the course of dialysis, hypotension is a common complication, known as Intradialytic Hypotension (IDH). To improve the predictive accuracy and practicality of IDH, we conducted research on machine learning techniques. To effectively impute missing values while maintaining the interpretability and confidence intervals of the data, we utilized the covariance structure of observed data to estimate missing values for imputation, preserving the correlation between data features and variables. We also introduced a feature dimensionality reduction module that supports nonlinear features, employing a tree-based method capable of capturing nonlinear relationships in the data. This method prioritizes basic functions with a small number of features to effectively describe the complex structure of the data. We used a weighted Lasso optimization criterion to select a sparse subset of features for the basic functions to ensure the retention of the most important features while reducing the complexity of the model. Experimental results demonstrate that our proposed model achieves high accuracy and stability in predicting IDH. Compared to traditional methods, it can more accurately identify patient risks, providing strong support for clinical decision-making. Therefore, our research provides a new approach and perspective for predicting and preventing IDH, with promising clinical application prospects.

Keywords: Intradialytic Hypotension · Machine Learning · Data Analysis

1 Introduction

Dialysis treatment is a therapeutic method used to replace kidney function, Blood pressure fluctuations between dialysis sessions affect approximately 50% of dialysis patients [1]. IDH happens often and is a serious problem. It occurs in about 30% of dialysis treatments, and even more in some kinds of patients [2]. IDH refers to a sudden drop in blood pressure to abnormally low levels during dialysis, leading to syncope and other severe consequences. Given cardiovascular disease mortality, especially sudden cardiac passing, is the primary reason of passing in dialysis patients, every effort ought to be made to avoid IDH [3]. Therefore, timely prediction and identification of risk factors for hypotension during dialysis are crucial for personalized treatment and intervention [4].

Most studies on IDH have adopted the definition from international guidelines [5]. Regular blood pressure measurements during dialysis, including pre-dialysis, intra-dialysis, and post-dialysis blood pressure, are essential for removing excess fluid from patients and maintaining fluid balance. Both too rapid and too slow ultrafiltration rates during dialysis can lead to hypotension [4–6, 8]. Therefore, doctors adjust the ultrafil-tration rate based on changes in the patient's weight and clinical symptoms to prevent hypotension. Although the instability of blood pressure during hemodialysis has long been recognized, we have limited knowledge about the factors that promote systemic IDH [7].

Medical work increasingly employs computer technology, including new record-keeping and data analysis tools, to study disease progression [9]. Digital healthcare has gradually shifted its focus from health management to treatment [10]. The combination of images and artificial intelligence in medical research has gained significant attention, such as Tong, Le's lightweight transformer for retinal vessel segmentation [11], which has clinical value in analyzing complex vascular conditions. Model interpretability poses a challenge, as demonstrated by Li et al. who utilized a model-free lesion generation and learning framework to investigate the interpretability of diabetic retinopathy detection [12]. Furthermore, the introduction of multimodality can enhance the performance of AI in medical applications [13]. Several IDH risk prediction models developed by Ma et al. have been systematically upgraded and updated to evaluate their performance in patient cohorts from new periods and environments [14]. Luis et al. associated data with sessions for multiple patients, where sessions comprised a set of analytical variables recorded close to the date of high-definition sessions and values of another set of clinical variables under the detection of "hour_0" and SBP throughout the entire dialysis session [15]. However, these methods still lack sufficient attention to specific event information in physiological data, which may affect predictive ability in terms of data dimensions.

Drawing inspiration from the above, we strengthened the data processing part of our model by enhancing imputation missing data through determining conditional distribu-tion of features based on completely observed features. This approach not only focuses on the feasibility of imputing missing data in real-world data but also partially addresses the interpretability and scalability issues of imputation methods, thereby improving the credibility of data analysis. Above all, the contributions of our paper are:

In order to enhance the explainability and extensibility of the whole prediction model, we optimize the whole process of IDH prediction and propose IDHPRE. We proposed a method by completely observed features to determine conditional distribution of features for imputing missing values, aiming to improve the interpretability and credibility of data analysis. By introducing a tree ensemble to separate important individual features and small feature subsets, and by solving a weighted Lasso optimization problem to select a subset of trees that collectively utilize a small number of features, we enhance the recognition of important feature combination.

2   Related Work

2.1   Imputation of Missing Values

The presence of missing values is a common and practical issue in machine learning [16]. Many learning algorithms don't work well for some types of information because they are made assuming that the data is complete and has no missing parts [17]. Traditional methods for imputing missing values include deletion, mean imputation, median imputation, Expert domain knowledge imputation and so on [18]. These method is simple but may result in the loss of valuable information, causing the data mining results to deviate from the actual data [19–21].

Abel et al. proposed a regression framework method, including TSR, KDR, KDR-PCR, and KDR-PLS [22]. These methods share a similar strategy when handles missing data. These methods judge whether the algorithm has converged based on specific tolerance values, facilitating iteration or stopping iteration during the missing value imputation process. Cahan et al. advanced a factor-based imputation mode [23]for handling missing values in high-dimensional panel data, where the covariance matrix plays a critical role in allocating weights to construct minimum variance data combinations.

2.2   Feature Selection

Although the number of features sometimes has a positive impact on machine learning models, in many cases, too many features can lead to confusion. In medical fields, there are often numerous features and variables involved, like gene expression and clinical indicators [24]. As a practical matter, extracting important gene tags from large volumes of redundant and noisy values facilitates accurate medical tumor/disease classification, survival prediction, and individual patient assessment [25–27].

Vinh et al. proposed an EMRMR function for feature selection via spectral relaxation and semidefinite programming, emphasizing proper handling of self-redundancy [28]. However, there is still controversy over how to address the issue of self-redundancy, which may result in a bias towards low-entropy features, thereby affecting the accuracy of feature selection. Wang et al. introduced feature equivalent partition, using MIGM to assess new feature discriminability [29]. But this method ignores inter-feature correlations, it requires exhaustive search when dealing with large-scale data features, leading to high computational complexity.

3   IDHPre

Figure 1 shows an overview of IDHPre. It has two main parts: data preprocessing and prediction. In the data preprocessing part, we focus on studying missing value imputation methods and feature selection methods to prepare structured data for the model adequately.

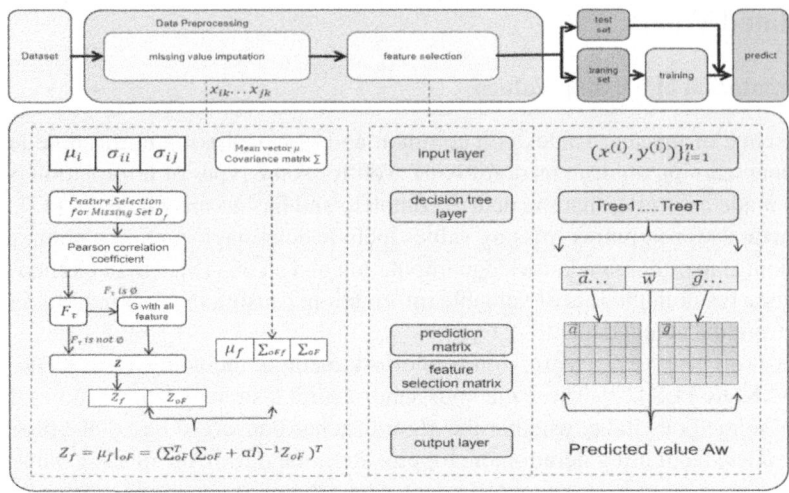

Fig. 1. The basic structure diagram of IDHPre

3.1 Imputation of Missing Values

In the imputation step, a shrinkage estimator is introduced in relation to Ridge regression [30] to reduce estimation variance and enhance the stability of the imputation results. Leveraging Gaussian conditional formulae, feature selection, and l2-norm regularization, it infers interpretability, scalability, and robustness to multicollinearity.

First, the algorithm utilizes parameterized expectation representation to estimate the covariance matrix, iteratively handling each feature with missing values in the test set for imputation. The imputation operation starts from an example s and searches for similar features with the same missing pattern for imputation. The imputation for each feature f is based on the following operations:

Compute the mean, sample variance, and covariance between each pair of features in the training data. Calculate the sample mean μ_i and variance σ_{ii} for each available feature , x_{ik} represents k-th sample value of i-th feature, n_i represents number of available entries for the i-th feature. Calculate the sample variance σ_{ii} for each available feature. Ultimately, mean vector μ and covariance matrix Σ, obtained in this way, constitute estimates for dataset.

$$\mu_i = \frac{1}{n_i}\sum\nolimits_{k=1}^{n_i} x_{ik} \tag{1}$$

$$\sigma_{ii} = \frac{1}{n_i}\sum\nolimits_{k=1}^{n_i} (x_{ik} - \mu_i)^2 \tag{2}$$

$$\sigma_{ij} = \frac{1}{n_i}\sum\nolimits_{k=1}^{n_i} (x_{ik} - \mu_i)(x_{jk} - \mu_j) \tag{3}$$

Create the initial matrix X_{input} by directly assigning the original test set to X_{input}. Look for samples with missing values in feature f, and assign the indices or actual data set containing these samples to D_f, so that D_f contains information about all missing values in feature f.

Handling the case where F_τ is empty, first identify the feature set G that is not in F_τ. If F_τ is an empty set, meaning no features highly correlated with the target feature f are found, then designate the features in the feature set q that are not in F_τ (including the target feature f) as set G. Next, set the feature subset F and add all features from set G to the feature subset F.

Processing missing values for samples with specific missing patterns, we construct a sample set Z with the same missing pattern as D_f (the sample set in feature f with missing values). Next, we extract submatrices Z_f and Z_{oF} from set Z, corresponding to the values of feature f and the selected feature subset F excluding f respectively. At the same time, we extract corresponding submatrices $\widehat{\Sigma}_{oFf}$ and $\widehat{\Sigma}_{oF}$ from the covariance matrix $\widehat{\Sigma}$. Using the following formula:

$$Z_f = \hat{\mu}_f|_{oF} = \left(\Sigma_{oF}^T(\Sigma_{oF} + \alpha I)^{-1} Z_{oF}\right)^T \tag{4}$$

Fill in the missing values in Z_f. Here, $\hat{\mu}_f|_{oF}$ represents the conditional mean of feature f given the values of other selected features F. α is a regularization parameter used to ensure that the denominator is not zero and may help improve numerical stability. Next, we update the set difference between D_f and Z, removing the samples for which missing values have already been handled in Z. Merge the Z with missing values processed back into the original test set X_{test} to obtain the final imputed test set.

The specific implementation principle for filling in missing values is as follows:

First, supplement the properties of the conditional distribution of multivariate normal distribution [31]: If a random vector y follows a multivariate normal distribution $N_p(\gamma, \Sigma)$, and is divided into two parts $\mathbf{y} = \left(\mathbf{y}_1^T, \mathbf{y}_2^T\right)$, then given $\mathbf{y}_2 = \mathbf{a}$, the contingent dispersal of $\mathbf{y}_1$ maintains its pattern of multivariate gaussianity.

We have the complete mean vector γ and covariance matrix Σ of the multivariate normal distribution, as well as the known part of the data, denoted as a or u_o. Next, the mean vector γ is divided into γ_1 and γ_2 (or γ_o and γ_m). The covariance matrix Σ is partitioned into corresponding sub-blocks, denoted as $\Sigma_{11}, \Sigma_{12}, \Sigma_{21}, \Sigma_{22}$ (or $\Sigma_o, \Sigma_{om}, \Sigma_{mo}, \Sigma_m$). For $\mathbf{y}_1$ given $\mathbf{y}_2 = \mathbf{a}$ conditionally, the conditional mean is $\overline{\gamma} = \gamma_1 + \Sigma_{12}\Sigma_{22}^{-1}(\mathbf{a} - \gamma_2)$. For $\mathcal{U}_m$ given $\mathcal{U}_o = u_o$ conditionally, the conditional mean is $\mu_{m|o} = \Sigma_{om}^T\Sigma_o^{-1}u_o$. For the conditional covariance matrix of $\mathbf{y}_1$ and $\mathcal{U}_m$, denoted as $\overline{\Sigma} = \Sigma_{11} - \Sigma_{12}\Sigma_{22}^{-1}\Sigma_{21}$ and $\Sigma_{m|o} = \Sigma_m - \Sigma_{om}^T\Sigma_o^{-1}\Sigma_{om}$ respectively, the resulting conditional mean is either $\overline{\gamma}$ or $\mu_{m|o}$, and the conditional covariance matrix is $\overline{\Sigma}$ either or $\Sigma_{m|o}$.

The above process allows us to estimate conditional distribution of missing part in multivariate based on the known partial data, by calculating the conditional mean and conditional covariance matrix, we can infer the expected value and variation of the missing data.

3.2 Feature Selection

We introduce a feature dimension reduction module that supports nonlinear features. The algorithm first constructs a large ensemble of trees, prioritizing basic functions with a small number of features. It then utilizes a weighted Lasso optimization criterion [32] to select a sparse subset of features from these basic functions. The resulting sub-forest is more interpretable than standard tree ensembles like random forests.

First, we are given n output responses $\{(x^{(i)}, y^{(i)})\}_{i=1}^{n}$, and $x^{(i)}$ is input feature and $y^{(i)}$ is the corresponding response (label). Then, we construct T trees, each of which maps the sample $x^{(i)}$ to a predicted value $a^{(i)}$. For each tree t, we define a prediction vector $a^{(t)} \in \mathbb{R}^{N}$ and a binary feature selection vector $g^{(t)} \in \{0,1\}^{P}$. Next, we construct the prediction matrix A and the feature selection matrix G, where each column of A is a prediction vector of a tree, and each column of G is a binary feature selection vector of a tree.

$$A = \left[a^{(1)}, \ldots, a^{(T)} \right] \in \mathbb{R}^{N \times T} \qquad (5)$$

$$G = \left[g^{(1)}, \ldots, g^{(T)} \right] \in \mathbb{R}^{N \times T} \qquad (6)$$

Then, we define a weight vector $w \in \mathbb{R}^{T}$, whose elements are non-negative, representing the contribution of each tree. On the optimization problem:

$$\begin{array}{l} minimize \ \frac{1}{N} \sum_{n=1}^{N} \ell(Aw_n, y_n) + \alpha u^T w \\ subject \ to \ w \geq 0 \end{array} \qquad (7)$$

where Aw_n is the dot product between the n-th row of matrix A and weight vector w, representing the prediction for the n-th sample. $\ell(Aw_n, y_n)$ is the loss between this prediction and the true response y_n. α represent regularization argument controlling the quantity of selected features. $u^T w$ is the weighted sum of feature selections, where u_t is the number of features used by tree t. For the case of squared loss:

$$\begin{array}{l} minimize \ \frac{1}{N} \|y - Aw\|_2^2 + \alpha u^T w \\ subject \ to \ w \geq 0 \end{array} \qquad (8)$$

where $\|y - Aw\|_2^2$ is the squared Euclidean distance between the predicted values Aw and the true response y.

Tree Construction: In random forests [33], if each tree uses almost all features for splitting, it becomes difficult for the lasso regularization method to find a sparse subset of features because all features are used frequently. However, if during ensemble construction, each tree is allowed to use only a subset of features, the lasso method can effectively select the subset of features that contribute the most to predictions.

we define a diagonal matrix D with elements $d_{tt} = u_t$, representing the number of features used by the t-th tree. Since $u_t > 0$, matrix D is positive definite, so it is invertible. The problem can be rewritten as:

$$\begin{array}{l} minimize \ \frac{1}{N} \|y - AD^{-1}Dw\|_2^2 + \alpha \|Dw\|_1 \\ subject \ to \ w \geq 0 \end{array} \qquad (9)$$

Here, $AD^{-1}Dw$ represents the model's predictions, $\|Dw\|_1$ is the L_1 norm of Dw, used for regularization to encourage sparsity. α is the regularization parameter, controlling the strength of regularization. By replacing variables with $x = Dw$, the problem can be further simplified to:

$$minimize \quad \frac{1}{N}\|y - AD^{-1}x\|_2^2 + \alpha\|x\|_1$$
$$subject\ to\ x \geq 0, D^{-1}x \geq 0 \tag{10}$$

After obtaining x through solving, we can retrieve the original weight vector $w = D^{-1}x$ by inverse solving. Since D is positive definite and invertible, this operation is feasible. The resulting w will be a sparse vector, with many elements possibly being zero, indicating that the corresponding tree's contribution in the final model is zero, thus achieving the goal of feature selection.

4 Experiment and Evaluation

In this section, we qualitatively and quantitatively compare our method with baseline methods. Additionally, we perform ablation study to demonstrate the availability of our approach.

4.1 Implementation Details

We implemented our method using PyTorch [34] and trained/tested our model on the dataset. The input size of our model is a matrix data of size 18309×95, split into training and testing sets with a ratio of 0.8 to 0.2, respectively. The information contained in this dataset is all clinical real data collected by medical staff for dialysis patients, and the patients have been anonymized, the information in the entire dataset is authentic and reliable.

4.2 Qualitative and Quantitative Comparison

We compared our method with SVC, Logistic Regression, Decision Tree, AdaBoost [35], LightGBM [36],CatBoost [37],HyperFast [38] and TabPFN [39] methods. Among them, AdaBoost combines multiple weak learners into a strong classifier by sequentially fitting a series of weak models to repeatedly modified versions of the information. LightGBM is an algorithm implemented through trees, which utilizes gradient-based one-side sampling to achieve the purpose of reducing memory usage and improving model speed.The HyperFast utilizes meta-trained hypernetworks to generate task-specific neural networks tailored for unseen datasets, thus eliminating the need for traditional model training. Throughout the process, it cleverly employs random features and Principal Component Analysis (PCA) to efficiently handle datasets with different dimensions. TabPFN as a Transformer model specifically designed for small-scale tabular classification tasks, cleverly integrating prior knowledge from SCM and BNN. During the inference stage, It demonstrates an approximation capability for the dataset's prior PPD, including a preference for concise and causal explanations of data, thereby ensuring the accuracy and reliability of prediction results.

For evaluation metrics, we chose to use accuracy, precision, recall, F1 score, ROC curve/AUC, PR curve, and AUPR, all of which are common metrics in classification tasks.

ROC Curve and AUC are important metrics for evaluating classifier performance, commonly used for handling imbalanced datasets at different thresholds.PR Curve demonstrates the relationship between precision and recall under different threshold values. AUPR is one of the important metrics for evaluating classification model performance, especially suitable for handling imbalanced datasets. Classification prediction results are shown in Table 1. Figure 2 shows the ROC curve and PR curve comparing the IDHPre model with the baselines.

Table 1. Comparing the quantitative results of the baseline and our method

Model	Acc	Precision	Recall	F1score	AUC	AUPR
SVC	0.816	0.847	0.828	0.838	0.895	0.922
Decision Tree	0.778	0.811	0.797	0.804	0.862	0.897
LR	0.812	0.839	0.829	0.834	0.892	0.917
AdaBoost	0.808	0.825	0.843	0.834	0.889	0.915
LightGBM	0.813	0.828	**0.850**	0.839	0.900	0.928
CatBoost	0.811	0.829	0.844	0.836	0.900	0.927
Hyperfast	0.821	0.838	0.831	0.835	0.896	0.924
Tabpfn	0.819	0.842	0.841	0.842	0.897	0.922
IDHPre	**0.822**	**0.852**	0.833	**0.843**	**0.902**	**0.929**

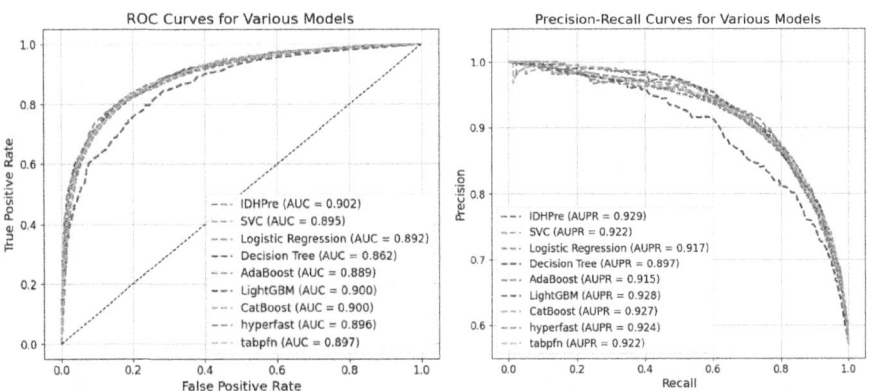

Fig. 2. ROC curve and PR curve comparing the IDHPre model with the baselines.

The confusion matrix of IDHPre is illustrated in Fig. 3, evaluates the classification performance of our model, where the rows represent the true classes and the columns represent the predicted classes.

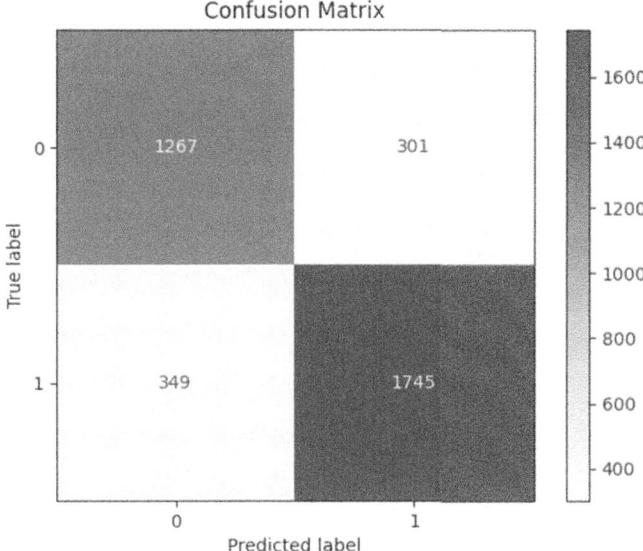

Fig. 3. Confusion matrix of IDHPre

4.3 Ablation Study

In this section, we will evaluate the performance of various parts of the model. Firstly, we removed the missing value imputation module and used only feature selection to generate classification prediction results. We compared our method with simple imputation methods (mean, median, KNN imputation). We also conducted experiments on the feature selection part. We also evaluated different filtering results obtained by controlling the parameters in the module to illustrate the impact of the feature selection module. According to the quantitative results shown in Table 2, it is evident that our proposed method have a positive impact on IDH prediction Table 3.

Table 2. The quantitative ablation study results for the missing value imputation module

Model	Acc	Precision	Recall	F1-score	AUC	AUPR
without imputation	0.770	**0.899**	0.683	0.773	0.880	0.912
imputation mean	0.797	0.851	0.781	0.815	0.874	0.904

(continued)

Table 2. (*continued*)

Model	Acc	Precision	Recall	F1-score	AUC	AUPR
imputation median	0.785	0.857	0.748	0.799	0.869	0.902
imputation KNN	0.794	0.851	0.775	0.811	0.871	0.906
IDHPre	**0.822**	0.852	**0.833**	**0.843**	**0.902**	**0.929**

Table 3. The quantitative ablation study results for the feature selection module

Model	Acc	Precision	Recall	F1-score	AUC	AUPR
Low-priority features	0.724	0.777	0.726	0.751	0.797	0.827
With Parameters α 1	0.811	0.856	0.804	0.829	0.887	0.913
With Parameters α 2	0.815	0.833	**0.846**	0.839	0.890	0.915
With Parameters α 3	0.815	**0.868**	0.798	0.832	0.896	0.922
IDHPre	**0.822**	0.852	0.833	**0.843**	**0.902**	**0.929**

5 Conclusion

In this paper, we first analyze the mechanism of hypotension during dialysis and propose an innovative machine learning prediction method based on it. To overcome the bias and information loss issues that may arise from traditional data imputation methods, we specifically adopt a covariance-based imputation technique. By introducing a shrinkage estimator in the imputation step in relation to Ridge regression, we reduce the variance of estimates, ensuring that the data are properly handled while maintaining their original correlations, interpretability, and confidence intervals. Additionally, we introduce a feature dimension reduction module adept at handling nonlinear features. By employing a weighted Lasso optimization criterion, this module selects crucial features for predicting outcomes swiftly, isolating a set of important features for further analysis. Such a design aims to streamline the model structure, reduce interference from redundant information, and thus enhance the stability and prediction accuracy of the model. Through this series of methods, we provide a more reliable and efficient solution to the prediction problem of hypotension during dialysis.

Acknowledgement. This work was supported in part by the Xinjiang Tianchi Talents Program (E33B9401), in part by the Natural Science Foundation of Xinjiang Uygur Autonomous Region under grant (2023D01E15) in part by the Youth Program of Natural Science Foundation of Xinjiang Uygur Autonomous Region (Grant No.2022D01B67), and in part by the Tianshan Talent Training Program (2023TSYCLJ0021).

References

1. Lu, Y., Du, X.: Personalized prevention and treatment strategies for hypotension patients during maintenance hemodialysis. MEDS Clin. Med. **4**(1), 59–64 (2023)
2. Santos, S.F., Peixoto, A.J., Perazella, M.A.: How should we manage adverse intradialytic blood pressure changes. Adv. Chronic Kidney Dis. **19**, 158–165 (2012)
3. Hussein, W.F., Schiller, B.: Dialysate sodium and intradialytic hypotension. Semin. Dial. Dial. **30**(6), 492–500 (2017). https://doi.org/10.1111/sdi.12634
4. Bradshaw, W., Bennett, P.N.: Asymptomatic intradialytic hypotension: the need for pre-emptive intervention. Nephrol. Nurs. J. **42**(5), 479–485 (2015)
5. Sornmo, L., Sandberg, F., Gil, E., Solem, K.: Noninvasive techniques for prevention of intradialytic hypotension. IEEE Rev. Biomed. Eng. **5**, 45–59 (2012)
6. Hamrahian, S.M.: Prevention of intradialytic hypotension in hemodialysis patients: current challenges and future prospects. Int. J. Nephrol. Renovascular Dis. **Volume 16**, 173–181 (2023). https://doi.org/10.2147/IJNRD.S245621
7. Rocha, A., Sousa, C., Teles, P., Coelho, A., Xavier, E.: Frequency of intradialytic hypotensive episodes: old problem, new insights. J. Am. Soc. Hypertens. **9**(10), 763–768 (2015)
8. Vito, D.: New clinical indexes for the automatic management of the dialysis treatment (2017)
9. Zhang, Y., Zhang, Z., Liu, X., Zha, L., Fengcong, Su, X., and Hu, P. (2023, July). A Deep Learning Approach Incorporating Data Missing Mechanism in Predicting Acute Kidney Injury in ICU. In: International Conference on Intelligent Computing, pp. 335-346. Singapore: Springer Nature Singapore.https://doi.org/10.1007/978-981-99-4749-2_29
10. Hu, P., Hu, L., Wang, F., Mei, J.: Computing and artificial intelligence in digital therapeutics. Front. Med. **10**, 1330686 (2024)
11. Tong, L., et al.: LiViT-Net: A U-Net-like, lightweight Transformer network for retinal vessel segmentation. Comput. Struct. Biotechnol. J. **24**, 213–224 (2024)
12. Li, J., Lin, C., Wang, Z., Song, H., Tan, F., Hu, L., & Hu, P. MLGL: Model-free Lesion Generation and Learning for Diabetic Retinopathy Diagnosis. In 2023 IEEE International Conference on Bioinformatics and Biomedicine (BIBM), pp. 1266–1271. IEEE (2023, December).
13. Li, J., et al.: Artificial intelligence accelerates multi-modal biomedical process: a survey. Neurocomputing **558**, 126720 (2023)
14. Xiang, Y., et al.: External validation of the prediction model of intradialytic hypotension: a multicenter prospective cohort study. Ren. Fail. **46**(1), 2322031 (2024)
15. Mendoza-Pittí, L., Gómez-Pulido, J.M., Vargas-Lombardo, M., Gómez-Pulido, J.A., Polo-Luque, M.L., Rodréguez-Puyol, D.: Machine-learning model to predict the intradialytic hypotension based on clinical-analytical data. IEEE Access **10**, 72065–72079 (2022)
16. L'Heureux, A., Grolinger, K., Elyamany, H.F., Capretz, M.A.: Machine learning with big data: Challenges and approaches. IEEE Access **5**, 7776–7797 (2017). https://doi.org/10.1109/ACCESS.2017.2696365
17. Zhang, S., Zhang, J., Zhu, X., Qin, Y., Zhang, C. (2008). Missing value imputation based on data clustering. In Transactions on computational science I, pp. 128–138. Berlin, Heidelberg: Springer Berlin Heidelberg https://doi.org/10.1007/978-3-540-79299-4_7
18. Mehala, B., Thangaiah, P.R.J., Vivekanandan, K.: Selecting scalable algorithms to deal with missing values. Int. J. Recent Trends Eng. **1**(2), 80 (2009)
19. Xu, X., Chong, W., Li, S., Arabo, A., Xiao, J.: MIAEC: Missing data imputation based on the evidence chain. IEEE Access **6**, 12983–12992 (2018)
20. Batista, G.E., Monard, M.C.: A study of K-nearest neighbour as an imputation method. His **87**(251–260), 48 (2002)
21. Wilson, D.R., Martinez, T.R.: Improved heterogeneous distance functions. J. Artif. Intell. Res. **6**, 1–34 (1997)

22. Folch-Fortuny, A., Arteaga, F., Ferrer, A.: Missing data imputation toolbox for MATLAB. Chemom. Intell. Lab. Syst. **154**, 93–100 (2016)
23. Cahan, E., Bai, J., Ng, S.: Factor-based imputation of missing values and covariances in panel data of large dimensions. J. Econometrics **233**(1), 113–131 (2023)
24. Bellazzi, R., Zupan, B.: Predictive data mining in clinical medicine: current issues and guidelines. Int. J. Med. Informatics **77**(2), 81–97 (2008)
25. Peng, M., Xiang, L.: Correlation-based joint feature screening for semi-competing risks outcomes with application to breast cancer data. Stat. Methods Med. Res. **30**(11), 2428–2446 (2021)
26. Sørlie, T., et al.: Gene expression patterns of breast carcinomas distinguish tumor subclasses with clinical implications. Proc. Natl. Acad. Sci. **98**(19), 10869–10874 (2001)
27. Singh, R.S., Gupta, B.P.: Genes and genomes and unnecessary complexity in precision medicine. NPJ Genom. Med. **5**(1), 21 (2020)
28. Nguyen, X. V., Chan, J., Romano, S., Bailey, J.: Effective global approaches for mutual information based feature selection. In: Proceedings of the 20th ACM SIGKDD International Conference on Knowledge Discovery and Data Mining, pp. 512–521 (2014, August)
29. Wang, X., Guo, B., Shen, Y., Zhou, C., Duan, X.: Input feature selection method based on feature set equivalence and mutual information gain maximization. IEEE Access **7**, 151525–151538 (2019)
30. Cule, E., De Iorio, M.: Ridge regression in prediction problems: automatic choice of the ridge parameter. Genet. Epidemiol. **37**(7), 704–714 (2013)
31. Ding, P.: On the conditional distribution of the multivariate t distribution. Am. Stat. **70**(3), 293–295 (2016)
32. Tibshirani, R.: Regression shrinkage and selection via the lasso. J. R. Stat. Soc. Ser. B Stat Methodol. **58**(1), 267–288 (1996)
33. Breiman, L.: Random forests. Mach. Learn. **45**(1), 5–32 (2001)
34. Paszke, A., et al.: Automatic differentiation in Pytorch (2017)
35. Freund, Y., Schapire, R.E.: A decision-theoretic generalization of on-line learning and an application to boosting. J. Comput. Syst. Sci. **55**(1), 119–139 (1997)
36. Ke, G., et al.: Lightgbm: a highly efficient gradient boosting decision tree. Adv. Neural Inf. Process. Syst. 30 (2017)
37. Prokhorenkova, L., Gusev, G., Vorobev, A., Dorogush, A. V., Gulin, A.: CatBoost: unbiased boosting with categorical features. Adv. Neural Inf. Process. Syst. 31 (2018)
38. Bonet, D., Montserrat, D. M., Giró-i-Nieto, X., Ioannidis, A.: HyperFast: instant classification for tabular data. In: NeurIPS 2023 Second Table Representation Learning Workshop **38**, (10), pp. 11114–11123 (2023, October)
39. Hollmann, N., Müller, S., Eggensperger, K., Hutter, F. TabPFN: A transformer that solves small tabular classification problems in a second. In: The Eleventh International Conference on Learning, p.01848 (2022, September)

Machine Learning Models for Improved Cell Screening

Jia-Song Liu, Zhi-Heng Yi, Bo Huang, Fan Wu, and Zu-Ping Zhang$^{(\boxtimes)}$

School of Computer Science and Engineering, Central South University,
Changsha 410083, China
234711109@csu.edu.cn

Abstract. In the cell engineering pharmaceutical industry, the quality of the cell screening process directly affects the quality of monoclonal cell lines, which in turn impacts the cost and efficiency of the entire production process. Current screening methods largely rely on time-consuming and costly manual operations, not only increasing the risk of subjective bias but also reducing the efficiency and quality of screening. To address this challenge, this paper proposes a method based on binary classification prediction, introducing two machine learning models: the Stacked Machine Learning Model (SMLM) and the Simple Linear Model (SLM). SMLM makes initial predictions using four different base learning models, which are then integrated through a shallow artificial neural network to produce a final prediction. In contrast, SLM employs a deeper artificial neural network for direct prediction. We analyzed these two models from the perspectives of generalizability and interpretability and conducted a quantitative evaluation using accuracy as the primary metric. The results show that while SMLM offers stronger interpretability, SLM performs better on multiple evaluation metrics and has a simpler structure. This research not only provides a new direction for the automation of cell screening but also opens up new pathways for the development of the cell engineering pharmaceutical industry.

Keywords: Cell Screening · Model Stacking · Artificial Neural Network · Machine Learning

1 Introduction

In recent years, the rapid advancements in medicine and biotechnology have catalyzed the rise of biotechnological pharmaceuticals as a pivotal new drug production modality [1]. These biopharmaceuticals not only facilitate the mass production of otherwise scarce natural drugs but also exhibit high efficacy and targeted specificity towards diseases. Their structural diversity allows selective binding to targets, presenting therapeutic benefits surpassing traditional chemical drugs [2, 3]. Cell engineering plays a crucial role in scaling production [4]. An essential initial phase in the cell engineering pharmaceutical process involves developing stable, high-yielding cell lines through steps like host cell line selection, expression vector engineering, transfection, cloning, cell expansion, and integrated screening phases [5]. Post-transfection, the brief process of cell screening is illustrated in Fig. 1. This screening aims to identify cells with optimal vitality, robust growth, and maximal levels of expression, stability, and quality, crucial for establishing monoclonal cell lines from vast populations [6]. The quality of selected cells significantly influences the monoclonal lines' quality, impacting the cost and efficiency of industrial-scale production in biopharmaceuticals [7]. Thus, the screening post-transfection is vitally important.

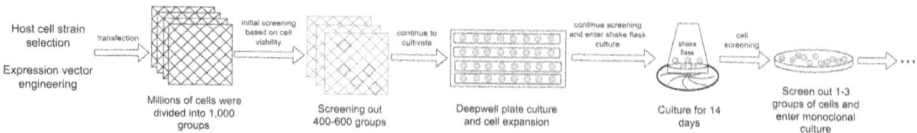

Fig. 1. Partial process schematic of cell culture and screening in cell engineering pharmaceutics.

In the shaker flask culture stage depicted in Fig. 1, to discern whether cells meet production requirements, a 14-day cultivation and observation period is necessary. During this process, cells that do not meet the criteria are removed from cultivation, completing the cell screening [8]. However, traditional cell screening methods rely on inefficient manual processing and judgment [6]. Throughout the cell culture process, data such as the live cells density, cell viability, and average diameter in the culture dish are measured daily to determine whether it is necessary to continue cultivation [9]. This process is not only labor-intensive but also requires operators to possess extensive professional knowledge and experience. Even so, judgments made by highly skilled operators often carry a strong subjective element. Due to the complexity, low efficiency, and lengthy duration of manual screening, prolonged exposure to the cell culture environment can easily contaminate it, compromising the precision of the entire cell culture process. The aforementioned factors all contribute to the high time and economic costs of manual screening methods, which in turn elevate the costs and prices of biopharmaceuticals, hindering the development of cell engineering pharmaceuticals. Therefore, inventing an efficient and accurate cell screening method that does not rely on human subjective experience, capable of improving the quality of cell screening and reducing its costs, holds significant commercial value and practical importance.

In this study, we frame cell screening as a binary classification challenge, aiming to predict whether cells should continue to be cultivated based on specific culture data to facilitate intelligent screening. This task encounters several obstacles. Firstly, the extended duration and high costs of cell engineering projects, coupled with the confidentiality of cell culture data, hinder extensive data collection, currently limited to two projects. Secondly, previous projects did not prioritize intelligent screening, resulting in a narrow range of data features available for model training and thereby limiting the feasibility of training sophisticated deep learning models. Thirdly, irregular and incomplete time series due to varied data collection intervals and discontinuation of non-compliant cell cultures complicate the use of time series models. We propose two predictive approaches: the Stacked Machine Learning Model (SMLM) and the Simple Linear Model (SLM). The SMLM integrates the initial predictions of four different base learners using a shallow neural network to produce the final prediction, aiming to enhance the accuracy and interpretability of predictions through ensemble learning. In contrast, the SLM employs a deeper neural network architecture, offering a simpler structure and higher operational efficiency, albeit at the cost of some interpretability.

Our research contributions include a comprehensive analysis of the cell screening process, the introduction of two novel predictive methods for intelligent cell screening, and a detailed discussion of their strengths, weaknesses, and applicable scenarios. Through testing on datasets from two different projects, we have demonstrated that these methods not only improve the efficiency and accuracy of cell screening but also possess strong generalizability. This opens up new directions for research and practical applications in the cell engineering pharmaceutical field.

2 Related Work

2.1 Mainstream Cell Line Screening Methods

The most common methods for stable cell line screening currently include limited dilution and flow cytometry-based cell sorting.

Limited dilution involves diluting cells and using microscopic visual screening to isolate individual cells [10], ideal for low integration transfection and monoclonal screening in suspension cultures, offering high uniformity in monoclonal antibodies and stable cell line selection. However, it suffers from low throughput, high contamination risk, and labor-intensive processes [11, 12]. This paper is dedicated to addressing some of the shortcomings of the limited dilution method in cell screening.

Flow cytometry-based cell sorting aligns fluorophore-labeled cells into droplets that are electrically charged and sorted into containers by high-voltage deflection, separating them effectively [13, 14]. This method achieves high purity and recovery of cells in a contamination-reduced, enclosed environment. However, it demands costly, complex equipment and skilled operators, and does not detect secreted proteins due to specific diameter requirements [15]. Moreover, it can cause irreversible cell damage due to high shear forces during sorting, compromising cell viability and integrity [16].

2.2 Model Stacking

We utilized model stacking to merge multiple models, aiming to reduce generalization error [17]. Model stacking, an ensemble learning strategy, enhances machine learning outcomes by amalgamating various base learners [18] and has proven superior to individual models across diverse applications [17–19]. This method includes two algorithmic layers: the base learners form the first layer, and a meta-learner comprises the second [20]. The process involves training base learners on the dataset to generate predictions, which then serve as input for the meta-learner that produces the final outcome [21]. Model stacking optimizes prediction accuracy by integrating strengths and minimizing weaknesses across different models' data analyses, improving the model's generalization and robustness [22]. To prevent overfitting, base learners are roughly adjusted, whereas the meta-learner is finely tuned.

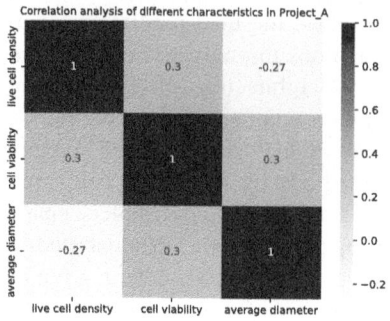 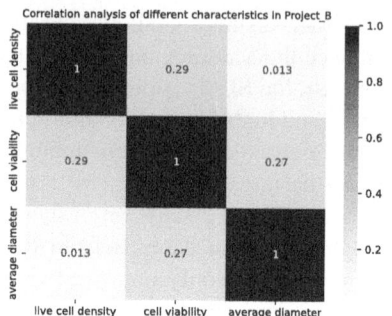

(a) Correlation analysis results of Project_A   (b) Correlation analysis results of Project_B

Fig. 2. Correlation analysis results of Project_A and Project_B

3 Dataset

To assess the potential for enhancing cell screening processes through machine learning, we gathered datasets from two real-world cell culture projects, designated Project_A and Project_B. Project_A contributed 1,446 entries, while Project_B provided 703 entries. The data, derived from stringent production protocols, focused on identifying cells with high and stable expression levels, ensuring a high-quality baseline for our research.

We applied the limiting dilution method for manual cell screening, utilizing this data to train, validate, and test our models and to benchmark their generalizability. For machine learning readiness, we partitioned the Project_A data into training, validation, and test sets using a 6:2:2 ratio, whereas the Project_B data served as an independent test set, allowing us to evaluate model performance on new, unseen data and simulate real-world application conditions.

Cell culture was capped at a maximum duration of 14 days, although cultivation could end earlier if cells did not meet production standards. Daily metrics collected included live cell density, cell viability, and average cell diameter, with labels assigned based on whether cells met the criteria to proceed to the next cultivation day. A label of '0' indicated failure to meet growth standards, halting cultivation, while a label of '1' signified adherence to growth criteria, permitting continued cultivation. The formatted data structure is presented in Table 1.

Table 1. Dataset format example.

Live cell density(cells/ml)	Cell viability(%)	Average diameter(μm)	Label
3.79E + 06	99.05	15.53	1
3.90E + 06	99.54	15.28	0
4.99E + 06	99.43	16.01	1
3.36E + 06	99	15.19	0

We conducted a correlation analysis on the data from Project_A and Project_B, with the results shown in Fig. 2. Different projects used different DNA fragments for cell transfection and produced different cell expression products, leading to some differences in the data. This also raised higher demands on the generalizability of the model.

4   Proposed Methods

The first method we propose is SMLM, which is based on four machine learning models and a shallow artificial neural network. The four machine learning models serve as base learners for initial predictions, and a shallow artificial neural network is used to integrate the predictions of the base learners to produce a final prediction. The structure of SMLM is shown in Fig. 3. The second method we propose is SLM, which solely utilizes an artificial neural network. The structure of SLM is shown in Fig. 4.

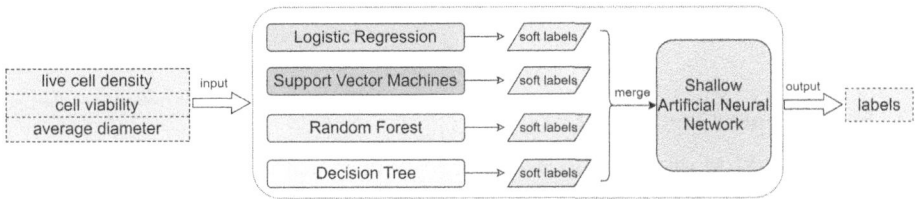

Fig. 3. Architecture of SMLM.

4.1   Stacked Machine Learning Method (SMLM)

SMLM employs four commonly used machine learning models as its base learners. These include the Decision Tree (DL) model, Logistic Regression (LR) model, Support Vector Machine (SVM) model, and Random Forest (RF) model.

These base learners are trained on input data featuring three dimensions. In traditional binary classification tasks, data typically feature "0" and "1" as hard labels. However, our approach utilizes the built-in functions of these models to produce soft labels, ranging from 0 to 1, thereby capturing more nuanced inter-class relationships [23, 24]. Each model outputs a tensor of size (batch_size, num_labels), and their combined outputs form a tensor of size (batch_size, 4, num_labels), enriching the label representation [25].

The training of these four distinct models on the same dataset often results in varied outcomes, complicating the selection of the most accurate prediction. To resolve this, we employ a shallow artificial neural network as a meta-learner. This network not only consolidates predictions from the base learners, enhancing decision-making, but also adapts to the distinct data characteristics presented by each learner, thus boosting overall predictive accuracy.

The meta-learner's architecture is a two-layer neural network designed to manage the complexity and prevent overfitting, crucial given the data's scale and feature number.

It processes the merged output tensor (batch_size, 4, num_labels) from the base learners. The tensor is first transformed to (batch_size, 8) by a linear layer and subsequently to (batch_size, 2) by another. The final predictions are derived using an argmax function.

Although we explored replacing the shallow network with a self-attention mechanism to enhance learning from the outputs of the machine learning models, this modification led to reduced performance, as detailed in Sect. 5.

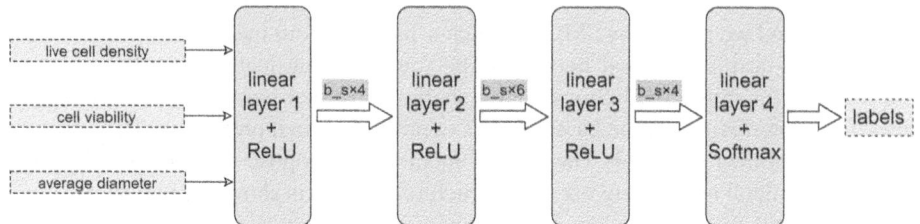

Fig. 4. Architecture of SLM. Batch_size is abbreviated as b_s.

4.2 Simple Linear Method (SLM)

Unlike SMLM, SLM directly utilizes a four-layer deep neural network to process the data. Through a series of experiments, we optimized the network structure, including the number of layers, the number of nodes in each layer, and the choice of activation functions. We use the ReLU function as the activation function for non-output layers and the Softmax function for the output layer, to enhance the model's nonlinear fitting capabilities and classification performance. The input data is the same as for SMLM, a tensor of size (batch_size, 3). After processing through the neural network, it outputs the prediction values.

4.3 Model Characteristics and Applicability Analysis

In this study, we introduced two machine learning-based cell screening models: the Stacked Machine Learning Model (SMLM) and the Simple Linear Model (SLM). The design of these models aims to address the challenges faced in the cell screening process within the cell engineering pharmaceutical industry, especially under conditions of limited data and feature dimensions.

Characteristics and Applicability of SMLM. The essence of SMLM lies in its use of ensemble learning from multiple base learners, enhancing prediction accuracy and interpretability by stacking different machine learning models. This method's advantage is its ability to integrate the strengths of various models, reduce the biases that might exist in a single model, and further improve predictive performance through a meta-learner. SMLM is particularly suited for scenarios requiring high interpretability, such as in the pharmaceutical industry, where understanding the model's decision-making process is crucial for ensuring drug safety and compliance. Additionally, SMLM performs excellently with limited feature dimensions, effectively extracting key features from sparse information.

Characteristics and Applicability of SLM. Unlike SMLM, SLM utilizes a deeper neural network structure for direct prediction. The advantages of SLM include its simplicity, high operational efficiency, and the ability to quickly process large volumes of data and provide predictions. This model is especially suitable for scenarios with high demands on prediction speed and efficiency. However, SLM has slightly less interpretability compared to SMLM, which may not be a critical issue in certain applications where model transparency is not a high priority.

Innovative Analysis. The innovation of this study lies in the first-time application of machine learning methods to the cell screening process in the cell engineering pharmaceutical industry. We not only proposed two novel predictive models but also validated their accuracy and generalizability through experiments. The design of SMLM and SLM fully considers the actual challenges in the cell screening process, overcoming these difficulties through innovative approaches. Moreover, our research showcases the powerful application prospects of machine learning models in the biotechnology field, paving new directions for future research and practical applications.

Through this innovative approach, we can enhance the automation level of cell screening, reduce production costs and improv the overall efficiency and quality of biopharmaceuticals. These achievements have significant theoretical and practical implications for advancing the development of the cell engineering pharmaceutical industry.

5   Experimental Results

5.1   Experimental Setup

The algorithms in this study were implemented based on the open-source pytorch version 2.0.1. For SMLM, the main parameter settings are as follows: four base learners were roughly tuned, and the shallow artificial neural network consisted of two linear layers. The model was trained for 35 epochs, with a batch_size of 4 and an initial learning rate set to 0.006. For SLM, the primary settings are: the number of linear layers was set to four, using ReLU as the activation function for intermediate layers and Softmax for the output layer. This model was trained for 40 epochs, with a batch size of 4 and an initial learning rate set to 0.008. In this experiment, the cross-entropy loss function was used throughout.

Ablation studies were conducted to compare model performance under different conditions. For SMLM, we varied the number of base learners and replaced the shallow neural network with a voting mechanism. For SLM, the number of linear layers and the activation function were altered.

We utilized accuracy, recall, and F1 score to evaluate our models, primarily focusing on accuracy as the primary performance metric.

5.2   Experimental Analysis

The experimental results are presented in Table 2 and Table 3.

Table 2. Results of the models on the project_A test set.[1]

Models	Accuracy(%)	Recall(%)	F1 score(%)
LR	84.8	81	88.9
SVM	87.5	83.3	90.9
RF	85.4	81	89.3
DT	85.1	80.2	89
SMLM	**89.9**	86.5	**92.8**
SMLM with Self-Attention	84.2	83.7	90.9
SLM	89	**87.7**	92.3

Table 3. Results of the models on the entire project_B dataset.

Models	Accuracy(%)	Recall(%)	F1 score(%)
LR	67.4	66.1	79.5
SVM	79.5	78.7	88
RF	73	72	83.6
DT	80.7	79.9	88.8
SMLM	83.5	83	90.6
SMLM with Self-Attention	81.3	80.8	89.2
SLM	**85.1**	**84.7**	**91.6**

Analysis was performed on the test set from Project_A. Within SMLM, Random Forest demonstrated the best performance, while Logistic Regression was the weakest, showing differences of 2.7%, 2.3%, and 2% in accuracy, recall, and F1 score, respectively. Random Forest excels at modeling complex data with high variance and low bias, which is characteristic of cell culture data, hence the effective performance. The average accuracy among the four base learners was 85.7%. SMLM improved accuracy by 5.1% compared to the weakest Logistic Regression, 2.4% over the best-performing Random Forest, and 4.2% over the average. SMLM's performance was notably superior to that of the base learners. SLM's accuracy was 0.9% lower than SMLM, but had a 1.2% higher recall, and both models exhibited identical F1 scores. Judging solely by accuracy, SMLM marginally outperforms SLM.

Analyzing the performance on the entire Project_B dataset, the gap between the base learners in SMLM was more significant on Project_B than on Project_A, with the best-performing Decision Tree outpacing the weakest Logistic Regression by 13.3% in accuracy. This indicates varying generalization capabilities and data fitting among

---

[1] LR, SVM, RF and DT are the abbreviations of Logistic Regression, Support Vector Machines, Random Forests and Decision Trees respectively.

the models. The average accuracy, recall, and F1 scores for the four base learners were 75.2%, 74.2%, and 85%, respectively. By stacking them with a shallow artificial neural network, SMLM enhanced these metrics by 8.2%, 8.8%, and 5.6% respectively compared to the average performance of the base learners. When compared to the top-performing base learner, the improvements were 2.7%, 3.1%, and 2.8%, respectively. On Project_B, SLM's accuracy, recall, and F1 scores surpassed SMLM's by 1.7%, 1.5%, and 1% respectively. In this scenario, SLM slightly outperformed SMLM.

We analyzed the results of using different meta-learners in SMLM. Employing shallow neural networks as meta-learners, compared to using self-attention, resulted in improvements of 5.7% in accuracy, 2.8% in recall rate, and 1.9% in F1 score on the test set of Project_A; on the Project_B dataset, accuracy, recall rate, and F1 score improved by 2.2%, 2.2%, and 1.4%, respectively. We believe the decreased performance when using self-attention is due to the fact that the four machine learning models have a total of eight output values, a relatively small data dimension, which prevents self-attention from achieving better results.

Table 4. The results of the ablation study on SMLM.

Changes to SMLM	Accuracy(%)	Recall(%)	F1 score(%)
Remove DT	81.4	80.8	89.2
Remove SVM and DT	77.8	76.9	86.9
Voting strategy	74.8	73.8	84.9

Table 5. The results of the ablation study on SLM.

Change to SLM	Accuracy(%)	Recall(%)	F1 score(%)
3 linear layers	80.4	79.8	88.6
5 linear layers	80.7	80.1	88.8
5 linear layers and use Sigmoid	83.9	83.5	90.9

Ablation Study. To explore the influence of different parts of the model, we conducted an ablation study and recorded the variations in evaluation metrics on the Project_B dataset. The results are presented in Table 4 and Table 5.

We altered the number of base learners to observe the performance changes of SMLM. Firstly, we removed the decision tree. Compared to the original model, the accuracy, recall, and F1 score of SMLM dropped by 2%, 2.2%, and 1.4%, respectively. Subsequently, upon removing the support vector machine, further deteriorations were observed, with declines of 5.6%, 6.1%, and 3.7% respectively. Each base learner focuses on distinct information of the data; reducing them inevitably leads to diminished performance due to a lesser breadth of learned knowledge.

We also removed the shallow artificial neural network from SMLM and let the base learners output hard labels. By employing a majority voting strategy for final predictions, we eliminated the decision tree to maintain an odd number of base learners, ensuring no tie situations. The model's performance significantly decreased compared to the scenario with the neural network, with drops of 6.6%, 7%, and 4.3% across the three metrics. This may be attributed to the base models making majority incorrect predictions for certain types of data, resulting in incorrect ensemble decisions. The shallow neural network in SMLM seems to have an adaptive capability to correct this, enhancing its performance. Majority voting is also limited as it requires an odd number of base learners.

For SLM, we varied its number of linear layers and analyzed the outcomes. Reducing the layers to three caused drops of 4.7%, 4.9%, and 3% across the metrics, suggesting underfitting. When increased to five layers, there were declines of 4.4%, 4.6%, and 2.8%, indicative of overfitting due to the model's excessive complexity relative to the data's dimensionality. Furthermore, replacing the ReLU activation function with Sigmoid led to decreased performance. ReLU, due to its ability to render some neurons inactive (output of 0), introduces sparsity into the network, mitigating overfitting issues. It also offers computational efficiency over Sigmoid.

Generalizability. Pearson correlation analyses for data from Project_A and Project_B are shown in Fig. 2. The Pearson coefficient between cell density and average diameter in Project_A was -0.27, while in Project_B it was 0.013. This stark contrast between the datasets underscores the need for robust generalizability in the models.

Comparing the performance of SMLM and SLM on the test set of Project_A and the entire dataset of Project_B, we noticed significant variations in base learner outcomes between the projects. The accuracy of Logistic Regression, Support Vector Machines, Random Forests, and Decision Trees dropped by 17.4%, 8%, 12.4%, and 4.4% respectively in Project_B compared to Project_A, with an average decline of 10.6%. Interestingly, despite its intrinsic weaker generalizability, the decision tree exhibited strong performance, possibly due to parameter tuning to prevent overfitting. SMLM's accuracy decreased by 6.4%, indicating the shallow neural network in SMLM mitigated the deterioration of base learners, enhancing the model's robustness. SLM's accuracy on Project_B declined by 3.9% compared to Project_A. Both SMLM and SLM managed to limit their accuracy drop to less than 7% on Project_B, achieving satisfactory outcomes with over 80% accuracy, showcasing their predictive capabilities and robust generalizability.

Table 6. Models prediction results and real labels.[2]

LR	SVM	RF	DT	SMLM	True label
0.61	0.8	0.76	0.88	1	1
0.89	0.11	0.05	0.94	0	0
0.21	0.93	0.84	0.22	1	1
0.11	0.85	0.72	0.69	0	0

---

[2] The data of the base learner represents the probability of predicting the label 1.

Interpretability. The outputs of SMLM are generated by the shallow artificial neural network, and each output can be traced back to individual base learners, as illustrated in Table 6. Contrarily, SLM's predictions lack such traceability. Therefore, compared to SLM, SMLM offers enhanced interpretability, a crucial aspect in the field of biopharmaceuticals.

6   Conclusion and Pen Question

In this study, we introduced two prediction methods, SMLM and SLM, aimed at enhancing the efficiency and quality and reducing the costs of cell screening processes. SMLM, a stacked predictive model, combines the initial predictions from four distinct machine learning models and utilizes a shallow artificial neural network to produce the final prediction. This approach leverages the strengths of different algorithms to provide a more robust and interpretable model. On the other hand, SLM directly employs a deeper artificial neural network for data processing, aiming to simplify the model structure while improving prediction accuracy. Our experimental results demonstrate that both methods exhibit good performance and strong generalizability in cell screening tasks, even with limited data volume and dimensions. Compared to the traditional limiting dilution method, both SMLM and SLM achieve comparable screening outcomes, which can help reduce the dependence on skilled operators and minimize the impact of human factors on screening results, thereby enhancing the consistency and reliability of the screening process. With significant advantages in improving screening efficiency and reducing costs, these methods hold promising application prospects and value. If the project emphasizes the interpretability and traceability of prediction results, SMLM would be the more suitable choice. For scenarios that prioritize higher prediction accuracy, SLM presents its unique applicability.

Given the current limitations in the volume and dimensions of data, we have not incorporated time series methods or more complex deep learning models. In the future, we plan to extensively collect various feature data during the cell culture process and increase the data volume from more projects. We will also explore the introduction of time series analysis and other advanced models to further enhance the predictive performance and generalizability of our models.

References

1. Kesik-Brodacka, M.: Progress in biopharmaceutical development. Biotechnol. Appl. Biochem. **65**(3), 306–322 (2018)
2. Deshaies, R.J.: Multispecific drugs herald a new era of biopharmaceutical Innovation. Nature **580**(7803), 329–338 (2020)
3. Homayun, B., Lin, X., Choi, H.J.: Challenges and recent progress in oral drug delivery systems for biopharmaceuticals. Pharmaceutics **11**(3), 129 (2019)
4. Madhavan, A., et al.: Customized yeast cell factories for biopharmaceuticals: from cell engineering to process scale-up. Microb. Cell Fact. **20**(1), 124 (2021)
5. Huang, R., et al.: Recent advances in CAR-T cell engineering. J. Hematol. Oncol. **13**(1), 1–19 (2020). https://doi.org/10.1186/s13045-020-00910-5

6. Greenfield, E.A.: Single-cell Cloning of hybridoma cells by limiting dilution. Cold Spring Harb. Protoc. **11**, pdb-prot103192 (2019)
7. Bashor, C.J., Hilton, I.B., Bandukwala, H., et al.: Engineering the next generation of cell-based therapeutics. Nature Rev. Drug Discov. **21**(9), 655–675 (2022)
8. Kaushik, N., Lamminmäki, U., Khanna, N., et al.: Enhanced cell density cultivation and rapid expression-screening of recombinant pichia pastoris clones in microscale. Sci. Rep. **10**(1), 7458 (2020)
9. Johnson, M.J., Laoharawee, K., Lahr, W.S., et al.: Engineering of primary human B cells with CRISPR/Cas9 Targeted Nuclease. Sci. Rep. **8**(1), 12144 (2018)
10. Fazekas de St Groth, S..: The evaluation of limiting dilution assays. J. Immunol. Methods **49**(2), R11-R23 (1982)
11. White, A.K., et al.: High-throughput microfluidic single-cell RT-qPCR. Proc. Natl. Acad. Sci. **108**(34), 13999–14004 (2011)
12. Ye, M., Wilhelm, M., Gentschev, I., et al.: A modified limiting dilution method for monoclonal stable cell line selection using a real-time fluorescence imaging system: a practical workflow and advanced applications. Methods Protoc. **4**(1), 16 (2021)
13. Cossarizza, A., et al.: Guidelines for the use of flow cytometry and cell sorting in immunological studies. Eur. J. Immunol. **51**(12), 2708–3145 (2021)
14. Vitelli, M., et al.: Applications of flow cytometry sorting in the pharmaceutical industry: a Review. Biotechnol. Prog. **37**(4), e3146 (2021)
15. Erdbrügger, U., et al.: Imaging flow cytometry elucidates limitations of microparticle analysis by conventional flow cytometry. Cytometry A **85**(9), 756–770 (2014)
16. Nolan, J.P.: Flow cytometry of extracellular vesicles: potential, pitfalls, and prospects. Curr. Protoc. Cytometry **73**(1), 13–14 (2015)
17. Palmal, S., Saha, S., Tripathy, S.: HIV-1 protease cleavage site prediction using stacked autoencoder with ensemble of classifiers. In: 2022 International Joint Conference on Neural Networks (IJCNN), pp. 1–8 (2022)
18. Zhang, H., Zhu, T.: Stacking model for photovoltaic-power-generation prediction. Sustainability **14**(9), 5669 (2022)
19. Fernandes, F., et al.: Long short-term memory stacking model to predict the number of cases and deaths caused by COVID-19. J. Intell. Fuzzy Syst. **42**(6), 6221–6234 (2022)
20. Mohammed, M., Mwambi, H., Mboya, I.B., et al.: A stacking ensemble deep learning approach to cancer type classification based on TCGA Data. Sci. Rep. **11**(1), 15626 (2021)
21. Martinez-Gil, J.: A comprehensive review of stacking methods for semantic similarity measurement. Mach. Learn. Appl. **10**, 100423 (2022)
22. Sun, S., Wang, S., Wei, Y.: A new ensemble deep learning approach for exchange rates forecasting and trading. Adv. Eng. Inf. **46**, 101160 (2020)
23. Collins, K.M., Bhatt, U., Weller, A.: Eliciting and learning with soft labels from every annotator. In: Proceedings of the AAAI Conference on Human Computation and Crowdsourcing, vol. 10, no. 1, pp. 40–52 (2022)
24. Hinton, G., Vinyals, O., Dean, J.: Distilling The Knowledge in A Neural Network. arXiv preprint arXiv:1503.02531 (2015)
25. Alshahrani, A., Ghaffari, M., Amirizirtol, K., et al.: Optimism/Pessimism Prediction of twitter messages and users using bert with soft label assignment. In: 2021 International Joint Conference on Neural Networks (IJCNN), pp. 1–8 (2021)

Prediction of Bladder Cancer Prognosis by Deep Cox Proportional Hazards Model Based on Adversarial Autoencoder

Jing Wu[1], Yanqiong Ren[1,2(✉)], Fei Han[1,2], and Xiang Bao[1]

[1] School of Computer Science and Communication Engineering, Jiangsu University, Zhenjiang 212013, Jiangsu, China
`yanqiongren@ujs.edu.cn`

[2] Jiangsu Key Laboratory of Security Technology for Industrial Cyberspace, Zhenjiang 212013, Jiangsu, China

Abstract. Due to the highly aggressive and heterogeneity of bladder cancer, it is important to predict bladder cancer prognosis accurately and identify the survival-related biomarkers. Recent studies have increasingly concentrated on exploring the influence of genes on cancer prognosis. Research indicates that how to deal with the obstacles caused by the redundant genes in genetic data for prognosis remains a challenge. To address this problem, in this study we propose a deep network (AAE-Cox) to accurately predict bladder cancer prognosis, which is designed by combining an improved adversarial autoencoder (AAE) with a deep Cox proportional network. Specifically, on the basis of normal AAE model, we maintain adversarial training while removing the decoder, to use the improved model to better gene extraction in the code space. The improved AAE model was then combined with a deep Cox network to achieve higher prognostic accuracy. Comprehensive experiments reveal the success of our approach in significantly enhancing the prognosis prediction accuracy for bladder cancer, and ablation experiment verifies the effectiveness of the improved model. Moreover, our method outperforms the existing advanced techniques in this domain.

Keywords: Survival analysis · Bladder cancer · Deep learning · Cancer prognosis · Cox proportional hazards

1 Introduction

Bladder cancer ranks among the top 10 cancers globally due to its high incidence and recurrence rates, and poses a significant health challenge [1]. Although early diagnosis and advances in robotic surgery have improved the survival rate, the cancer still threatens human life and health [2]. Therefore, accurate prognostic prediction is essential for the prevention and treatment [3]. Clinical studies have demonstrated that accurately distinguishing high-risk and low-risk patients and selecting appropriate symptomatic treatment can significantly enhance the treatment outcome for bladder cancer [4]. Various methods have been proposed for predicting cancer patient prognosis, with the Cox

D.-S. Huang et al. (Eds.): ICIC 2024, LNBI 14881, pp. 123–134, 2024.
https://doi.org/10.1007/978-981-97-5689-6_11

Proportional Hazards model (CPH) being one of the widely utilized approaches [5]. The CPH model estimates the relationship between patient features and the likelihood of event occurrence.

In modern medicine, accurate prognosis aims to provide customized treatment for patients. However, this approach heavily relies on a profound understanding of the patient's genetic makeup. The application of high-throughput technology facilitates the acquisition of large amounts of genetic data, laying the necessary foundation for accurate prognosis. On this basis, scholars have devoted efforts to exploring the relationship between genes and diseases, aiming for improved prognostic accuracy. However, the presence of redundant genes in genetic data brings noise that significantly hampers research progress. In particular, CPH model faces substantial challenges when handle the large gene inputs containing noise.

To tackle high-dimensional inputs, various enhanced Cox model methods have been proposed. Cox-LASSO [6] and PCA-Cox [7] employed LASSO regularization and principal component analysis respectively to select key genes from high-dimensional inputs, but ignore the correlation of genes. Cox-EN [8] employed elastic net which merged l_1 and l_2 norms to make up for the deficiency of LASSO. Tong et al. proposed HC-Cox [9] using unsupervised clustering for feature reduction. These models take into account the correlation of genes on a certain basis, but ignore the complex networks of genes.

With the rapid development of deep learning, it has been noticed that complex networks of genes are similar to neural networks. Katzman et al. first proposed the concept of deep Cox Proportional Hazards model and analyzed the relationship between patient genes and treatment effects (DeepSurv) [10]. AE-Cox [11] employed autoencoder to learn representative genes, enabling the differentiation of patient subgroups. However, traditional autoencoder is limited in their ability to deal with redundant genes [12]. Tong et al. enhanced AE-Cox by leveraging gene complementarity and consensus for improved gene extraction (CAE-Cox) [13]. Chai et al. applied transfer learning to AE-Cox to exploit inter-cancer correlations, enhancing the ability of autoencoder to identify effective genes (TCAP) [14].

Recently, experts have consistently sought to innovate upon the foundation of traditional autoencoder. Chai et al. used unsupervised denoising autoencoders (DAE) [15] to extract representative genes. However, the complex patterns and structures of noise make it challenging to accurately model and identify, potentially leading to the misidentification of noise as useful signals. Recently, the development of generative adversarial networks (GAN) [16] has attracted much attention. Ahmed et al. leveraged it to augment high-dimensional, small-sample genetic data, but the authenticity of the generated samples has been questioned [17]. Makhzani et al. proposed a novel approach (AAE) [18] that combines autoencoder with generative adversarial network and demonstrate its effectiveness in terms of feature dimensionality reduction.

Redundant features unrelated to cancer development bring significant noise, and the above methods still have certain limitations in dealing with noise problems, leading to mediocre model predictions. To address these problems and conduct a thorough analysis for predicting bladder cancer prognosis using genetic data, we propose a novel approach. On the basis of normal AAE and DeepSurv, we propose a model that combines adversarial training with deep Cox networks (AAE-Cox). This method can minimize the noise effect caused by redundant genes and extract effective genes from genetic data better. Comprehensive experimental results confirm that the method greatly improve the prognostic performance of bladder cancer. The main contributions of this study can be summarized as follows:

- A novel deep network for accurate prognostic prediction of bladder cancer is proposed, combining adversarial autoencoders with deep Cox proportional risk model for the first time. The model can extract key genes associated with disease to help guide precision treatment strategies.
- Compared with normal AAE model, the improved model pays more attention to the feature extraction process of the latent code space, so that the key genes obtained through adversarial training are closely related to bladder cancer, and ultimately greatly improve the accuracy of the prognosis.
- The comprehensive experimental results show that the proposed method is superior to the existing methods in improving the accuracy of prognosis, and ablation experiment verifies the effectiveness of the improved model.

2   Methods

2.1   The Framework of the Study

We propose a novel prognostic method for bladder cancer, Fig. 1 shows the framework of the study. Firstly, we preprocess the training dataset downloaded from TCGA (The Cancer Genome Atlas) [19]. Details of the preprocessing steps of different datasets are provided in Sect. 3.1. Following this, we employ the AAE-Cox model within the feature extraction module to reduce the dimensions of high-dimensional gene features, and subsequently obtain the prediction results. After that, we divide the samples into high-risk and low-risk groups according to the prediction results, and evaluate the survival difference discrimination ability of the model by using relevant evaluation indicators. Then, three datasets downloaded from GEO (Gene Expression Omnibus) [20] are used to test the model independently, and the generalization ability of AAE-Cox is verified. Finally, the difference analysis of risk subgroups identified by the model is carried out to find the genes that play a leading role in the development of bladder cancer, and their biological analysis is performed.

2.2   Adversarial Autoencoders

The normal AAE model combines the autoencoder with the adversarial component of the GAN. It learns a latent code representation by using the encoder and decoder of the autoencoder. In the AAE, we have an input data x and a latent code vector z, where the encoder defines the posterior distribution of the latent code given the input data, denoted as $q(z|x)$. At the same time, we want the latent code to follow a prior distribution, denoted as $p(z)$.

To achieve this, in Fig. 2, the AAE introduces a discriminator that distinguishes between the posterior distribution $q(z)$ and the prior distribution $p(z)$. Through adversarial training, the discriminator guides the encoder to generate latent codes similar to the prior distribution. The decoder learns the mapping from the latent code to the original data by using the reconstruction loss. By combining the autoencoder and adversarial training, the AAE can learn a latent representation that follows the prior distribution and generate realistic data samples.

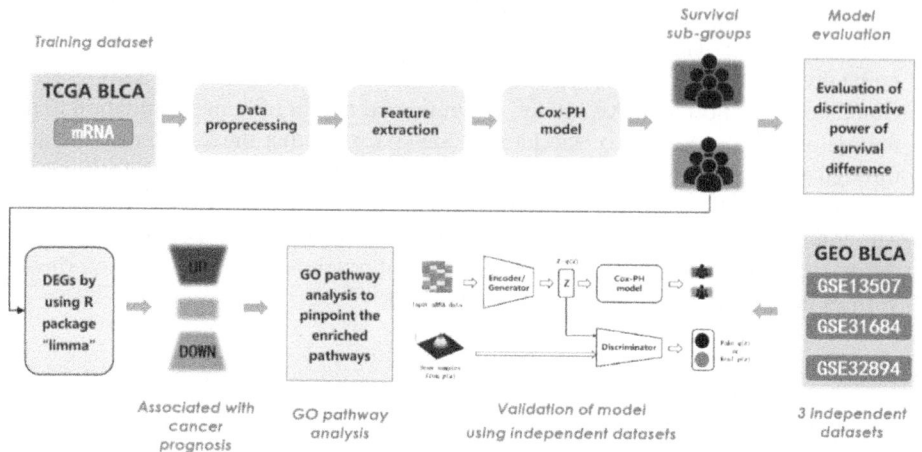

Fig. 1. The framework of the study

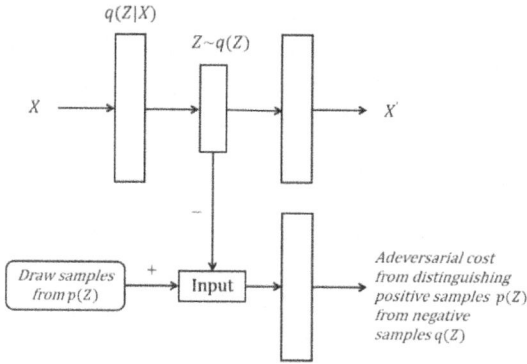

Fig. 2. An adversarial autoencoder consists of an autoencoder and a discriminator. The autoencoder learns to reconstruct input data from a latent code, while the discriminator distinguishes between hidden code samples and samples drawn from a specified prior distribution $p(z)$.

2.3 The Architecture of AAE-Cox

Based on the normal AAE, we are attempting to integrate a deep Cox network and remove the decoder part to obtain an improved model. Figure 3 shows the network structure AAE-Cox used for bladder cancer prediction in this study. The mRNA data downloaded from

TCGA were used in AAE-Cox to estimate the bladder cancer outcomes. By combining adversarial training and risk prediction, this network demonstrated enhanced performance in predicting the risk associated with the bladder cancer, and laid a foundation for the formulation of accurate treatment strategies.

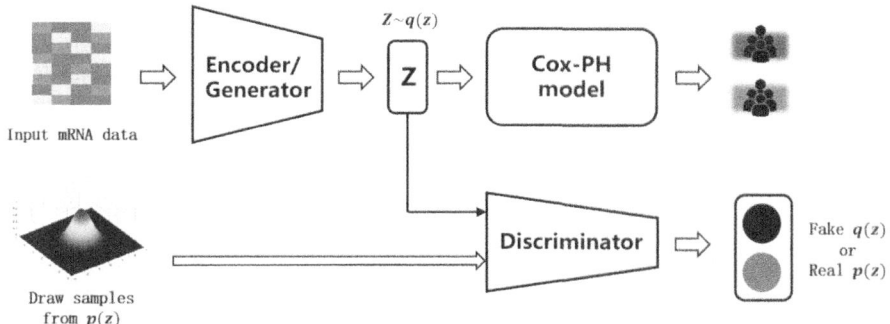

Fig. 3. Our proposed framework for bladder cancer prognosis prediction consists of a deep neural network in AAE-Cox. This network combines adversarial training and CPH loss for improved accuracy.

Adversarial Training of AAE-Cox. In our formulation, we define $X = (x\_1, x\_2, ..., x\_n)$ as the list of input features for the sample, and z represents the corresponding representation in the bottleneck layer, obtained through the encoder function (*i.e.*, $z = E(X)$).

In normal AAE, the main task is to enforce that the output of the encoder follows a given prior distribution (which can be a normal distribution, gamma distribution, etc.). We will use the encoder as our generator and the discriminator as a classifier to determine whether a sample comes from the prior distribution ($p(z)$) or from the encoder (z). In AAE, the generator plays the role of generating synthetic data, while the discriminator is trained to differentiate between real data from datasets and the artificially generated data by the generator. The training process continues until the generator becomes adept at producing synthetic data that becomes indistinguishable from real data, leading the discriminator to struggle in distinguishing between the two.

In this study, there is no longer a need for AAE to synthesize samples for data expansion. Conversely, the decoder part is removed, and the low-dimensional feature representation of the latent code space is obtained through adversarial training to achieve the extraction of high-dimensional gene features. Therefore, during the training phase, only the regularization loss is performed and remove the reconstruction loss. In this stage, we need to train both the discriminator and the generator (referring to the encoder).

Firstly, we train the discriminator to classify the outputs from the encoder (z) and some random inputs (z'). In this study, the random inputs fed into the discriminator are assumed to follow a Gaussian distribution. Therefore, if we pass random inputs with a Gaussian distribution (where $z' \sim p(z)$ is sampled from this distribution and defined as real data) into the discriminator, it should give us an output of 1 (real data). On the other hand, when we pass the output from the encoder (z) into the discriminator, it should give us an output of 0 (fake data). Hence, the output from the encoder and the input to the discriminator should have the same size.

The game rules of the confrontation training part can be expressed as following:

$$\min_{G} \max_{D} E_{z' \sim p_{(z)}} \left[\log D(z') \right] + E_X \left[\log(1 - D(G(X))) \right] \tag{1}$$

Discriminator: When training the discriminator, the output of the encoder is used as input. The corresponding labels for the encoder's output should be 0, indicating a fake sample. For the inputs sampled from a normal distribution, the corresponding labels should be 1, indicating a real sample. In the optimization process, in order to fit the gradient descent algorithm, the maximum discriminator loss is transformed into the minimization problem by minus sign. The loss of Discriminator is expressed as:

$$l_D = -\log\left(D(z')\right) + \log(1 - D(z)) \tag{2}$$

Generator (Encoder): In this context, the generator refers to the encoder. The loss function for the generator is defined as the negative of the discriminator's loss. When the encoder's output is utilized as input, the corresponding labels are set to 1 to indicate real samples. However, no loss is computed for inputs sampled from the normal distribution. The loss expression for the generator (encoder) can be represented as follows:

$$l_G = -\log(D(z)) \tag{3}$$

Cox Proportional Hazard Model for Patients' Risk Estimation. In this study, we employ the deep Cox proportional risk from DeepSurv. The encoder output z is fed into a fully-connected layer to predict bladder cancer outcomes. The survival function, which represents the probability of patient survival before time t, can be expressed as follows:

$$S(t) = Pr(T > t) \tag{4}$$

T is survival time, representing how long the patient survived before the end of the experiment or death. The risk function of patient death at time t is expressed as follows:

$$h(t) = \lim_{\Delta t \to 0} \frac{Pr(t \leq T \langle t + \Delta t | T \geq t)}{\nabla t} \tag{5}$$

The proportional hazard function is:

$$h(t, X) = h_0(t) exp^{(\beta X)} \tag{6}$$

In the expression, β represents the correlation coefficients of X, while $h_0(t)$ represents the basic risk function at time t. The maximum partial likelihood function can be formulated as:

$$L_c(\beta) = \prod_{i:E_i=1} \frac{exp^{h_\beta(x_i)}}{\sum_{j \in R(T_i)} exp^{h_\beta(x_i)}} \tag{7}$$

The deep neural network is updated based on the weight, and the computation of the risk prediction loss is performed as follows:

$$l_p = -\sum_{i:E_i=1} \left(h_\theta(x) - \log \sum_{j \in R(T_i)} exp^{h_\theta(x_j)} \right) \tag{8}$$

3 Results

3.1 Datasets

We utilize cancer datasets obtained from TCGA for training, while three additional datasets collected from GEO are employed to validate the robustness of our method.

TCGA: The mRNA data is downloaded by using the R package "TCGA assembler2", and the expression data is normalized through log transformation. Features with missing values exceeding 20% are removed, and for the left samples, missing values are imputed using the median values with the R package "imputeMissings".

GEO: Three cancer datasets obtained from GEO are used to independent tests. The GSE13507 dataset consists of RNA-seq data from 165 primary bladder cancer patients obtained from Chungbuk National University Hospital. GSE31684 provides data from 93 bladder cancer patients from the Dana-Farber Cancer Institute. GSE32894 contains information on 225 bladder cancer patients collected by the SCIBLU Genomics Centre at Lund University Sweden. To address potential batch effects, we utilize the "limma" R package to remove them from all the datasets.

3.2 Experiments

This study presents the evaluation of the method by showcasing the average C-index values obtained through 5-fold cross-validation. Furthermore, the robustness of the model is verified using three independent datasets sourced from GEO. Table 1 provides a comprehensive list of hyperparameters used in this study. These parameters are optimized step-by-step based on the performance on the test set during the experimental process. During the training phase, the mRNA data downloaded from TCGA is divided into a training set and a test set in according to 4:1, with a training epoch set to 3000. The *ReLu* function is employed as the nonlinear activation function in each layer.

Table 1. The parameters setting in the experiments

Encoder/Generator		Discriminator	
	(2000,1000)		(50,1000)
Three hidden layers	(1000,1000)	Tow hidden layers	(1000,1000)
	(1000,50)		
Dropout	0.25	Dropout	0.2
Learning risk	lr_en = 0.0001	Learning risk	lr_dis = 0.00005
	lr_gen = 0.00005		

3.3   Evaluations of Cancer Outcomes Prediction

To evaluate the prognostic effect of this model on bladder cancer, we utilize two evaluation metrics: the C-index and the log-rank p-value. C-index measures the proportion of correctly ordered pairs of samples based on predicted survival times, as defined by Harrell's C statistics. The higher C-index value is, the more accurate prediction performance we obtained. The p-value is applied to measure the different significance of the different risk subgroups divided by the different methods. In general, p-value less than 0.05 means there is a significant difference between the different risk subgroups.

3.4   Method Comparison

We evaluate the prediction accuracy of various traditional and deep learning methods using the BLCA (bladder cancer) dataset obtained from TCGA. The four traditional methods included in the evaluation are single regression (Cox), PCA-based compressed features (PCA-Cox), LASSO-constrained Cox (Cox-LASSO), and elastic net regularized Cox (Cox-EN). The five deep learning methods considered are deep networks (Deep-Surv), autoencoder-based Cox (AE-Cox), contact autoencoder-based compressed features (CAE-Cox), transfer learning strategy (TCAP), and denoising autoencoder-based Cox (DCAP).

In order to verify the effectiveness of the modification of the traditional AAE model, we conduct an ablation study. The AAE-Cox_D model shown in Table 2 is based on the AAE-Cox model, which restores the decoder part, and trains the model by using adversarial loss and reconstruction loss.

All investigated methods achieve satisfactory performance, and deep learning-based methods usually outperform traditional methods. Among the four traditional methods, a single Cox network (Cox) obtains the lowest C-index. In the same case of regularization, the Cox-EN using elastic network is better than Cox-LASSO using LASSO alone, because the elastic network combines the advantages l_1 and l_2 norm regularization.

Among methods utilizing deep learning, those employing autoencoder tend to yield higher C-index values compared to methods without autoencoder (DeepSurv). Notably, TCAP which utilizes an autoencoder and transfer learning achieves the fairly high C-index among all the methods evaluated. The results of DCAP using denoising autoencoder do not perform much better than other models using traditional autoencoder, reflecting that the principle of applying extra noise is not suitable for dimensionality reduction of genetic data with high noise. The AAE-Cox_D model of the recovery decoder obtained a higher C-index than other models, proving that the combination of adversarial autoencoders can obtain better prediction results. Furthermore, our method reaches the better C-index value and better than AAE-Cox_D, which is significantly superior to all the mentioned methods, which demonstrates that our method can accurately estimate the prognosis of bladder cancer and our improvement on the normal AAE is successful.

3.5   Independent Test

To validate the prognosis prediction models constructed by AAE-Cox, three independent datasets associated with bladder cancer are employed to be the independent tests:

GSE13507 (165), GSE32894 (225) and GSE31684 (93). The results presented in Fig. 4 demonstrate promising results for the evaluated method across three real datasets. The C-index values for all datasets exceeded 0.66, indicating a relatively high level of accuracy in predicting bladder cancer outcomes. Additionally, the survival curves of different cancers show significant differences exist between the different risk subgroups. The p-values were all less than 0.05, suggesting statistically significant differences in survival outcomes. These results indicate the effectiveness and robustness of AAE-Cox in predicting bladder cancer prognosis.

Table 2. The C-index values obtained by different methods

	C-index		C-index
Cox	0.572(±0.086)	DeepSurv	0.582(±0.081)
PCA-Cox	0.586(±0.074)	AE-Cox	0.598(±0.074)
Cox-LASSO	0.588(±0.032)	CAE-Cox	0.633(±0.064)
Cox-EN	0.602(±0.081)	TCAP	0.641(±0.064)
		DCAP	0.645(±0.047)
		AAE-Cox	**0.693(±0.056)**
		AAE-Cox_D	**0.662(±0.076)**

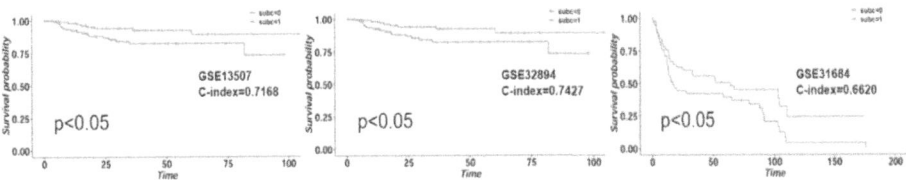

Fig. 4. The independent test results from three GEO datasets showed distinct survival patterns. The red lines represent high-risk patients, indicating a poorer prognosis, while the blue lines represent low-risk patients, indicating a better prognosis.

3.6   Identification of Cancer-Related Prognostic Markers and Pathways

The bladder cancer patients are divided into high-risk and low-risk clinical subgroups based on the median prognostic value. DEGs (Differentially Expressed Genes) [21] associated with bladder cancer risk subtypes are identified using the "limma" R package. In total, 299 genes (225 up-regulated and 74 down-regulated) are found to be significantly associated with bladder cancer prognosis (Fig. 5).

With these 299 DEGs, we conduct GO (Gene Ontology) [22] pathway analysis to pinpoint the enriched pathways. A total of 22 GO pathways associated with bladder cancer's prognosis are identified that exhibited corrected p-values below 0.05 and contained more than 20 genes. We show the number of genes in these 22 pathways in Fig. 6. We focus on *ECM* (collagen-containing extracellular matrix) [23] pathway which achieved the highest gene numbers in the enrichment of DEGs. *ECM* can promote the invasion and metastasis of tumor cells by providing suitable environment and signaling molecules. By interacting with the *ECM*, tumor cells can migrate from the primary tumor to other sites through the extracellular matrix by altering cell-*ECM* adhesion, movement, and migration. Collagen and other components in *ECM* can regulate signaling pathways in cells. These signaling pathways can affect the proliferation, survival, apoptosis, invasion and metastasis of tumor cells *ECM* can limit the spread and penetration of anticancer drugs in tumor tissues. The dense structure and protein composition of *ECM* can impede the delivery of drugs to tumor cells and reduce the effectiveness of drugs. In addition, cell signaling molecules in *ECM* and cell-*ECM* interactions can also lead to drug resistance in cancer cells.

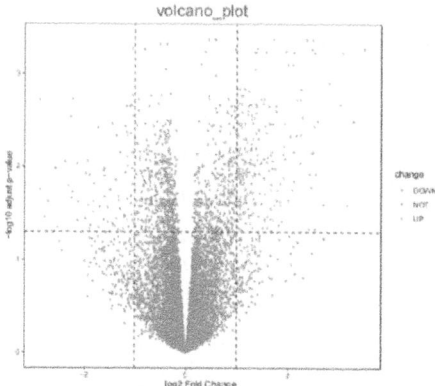

Fig. 5. Differentially expressed gene selection revealed distinct patterns: blue nodes represent down-regulated risk genes, red nodes represent up-regulated risk genes, and green nodes represent unselected genes.

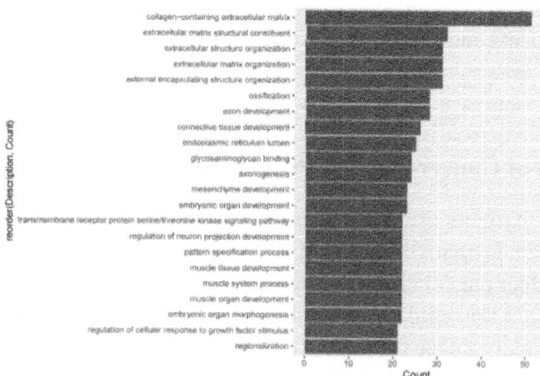

Fig. 6. The gene numbers in these 22 GO pathways related to cancer prognosis.

4   Conclusion and Discussion

Accurate prediction of bladder cancer outcomes is crucial for selecting appropriate treatment strategies. To address this, we propose a novel approach called AAE-Cox, an adversarial training based deep proportional hazards network. Instead of continuing the tradition of using AAE model to expand the unbalanced data, we remove the decoder part and use adversarial training combined with deep Cox model to extract high-dimensional gene features to improve the prediction effect of Cox model. Results indicate our method outperformed other traditional and deep learning-based methods in experiments. The constructed prediction model is validated in the independent datasets collected from GEO. By utilizing AAE-Cox to divide bladder cancer patients into risk subgroups, we successfully identify 225 up-regulated genes and 74 down-regulated genes that significantly impact bladder cancer survival, then identify 22 GO pathways associated with bladder cancer prognosis. The above results prove that our proposed bladder cancer prognosis prediction model combined with adversarial training is robust, accurate, and has biological significance. Additionally, this method can effectively help to formulate the accurate treatment strategies by enabling more targeted and individualized treatment approaches for bladder cancer patients.

In summary, our study has important implications for the prognosis of bladder cancer. However, we believe that certain problems remain. First of all, although we reduce the dimension of the genetic data through the model to better select effective features, the high deletion rate of gene data still limits the accuracy of bladder cancer prediction. Secondly, there is still a large amount of feature redundancy in gene data. In the future, we will consider combining clinical information, slide images and other multi-omics data to better predict the prognosis of bladder cancer.

Acknowledgments. This study was funded by the National Natural Science Foundation of China under Grant (grant number 61976108).

Code Availability.   https://github.com/MaiGavin/AAE-Cox

Disclosure of Interests.   The authors have no competing interests to declare that are relevant to the content of this article.

References

1. Sanli, O., Dobruch, J., Knowles, M.A., et al.: Bladder cancer. nature Rev. Dis. primers **3**(1), 1–19 (2017)
2. Sung, H., Ferlay, J., Siegel, R.L., et al.: Global cancer statistics 2020: GLOBOCAN estimates of incidence and mortality worldwide for 36 cancers in 185 countries. CA a cancer j. clin. **71**(3), 209–249 (2021)
3. Saginala, K., Barsouk, A., Aluru, J S., et al.: Epidemiology of bladder cancer. Med. Sci. **8** (1), 15 (2020)

4. Yeh, R.W., Secemsky, E.A., Kereiakes, D.J., et al.: Development and validation of a prediction rule for benefit and harm of dual antiplatelet therapy beyond 1 year after percutaneous coronary intervention. JAMA **315**(16), 1735–1749 (2016)
5. Cox, D.R.: Regression models and life-tables. J. Roy. Stat. Soc. Ser. B (Methodol.) **34**(2), 187–202 (1972)
6. Tibshirani, R.: The lasso method for variable selection in the cox model. Stat. Med. **16**(4), 385–395 (1997)
7. Bair, E., Tibshirani, R.: Semi-supervised methods to predict patient survival from gene expression data. PLoS Biol. **2**(4), e108 (2004)
8. Simon, N., Friedman, J., Hastie, T., et al.: Regularization paths for cox's proportional hazards model via coordinate descent. J. Stat. Softw. **39**(5), 1 (2011)
9. Tong, D., Tian, Y., Zhou, T., et al.: Improving prediction performance of colon cancer prognosis based on the integration of clinical and multi-omics data. BMC Med. Inform. Decis. Mak. **20**(1), 1–15 (2020)
10. Katzman, J.L., Shaham, U., Cloninger, A., et al.: DeepSurv: personalized treatment recommender system using a Cox proportional hazards deep neural network. BMC Med. Res. Methodol. **18**, 1–12 (2018)
11. Lee, T.Y., Huang, K.Y., Chuang, C.H., et al.: Incorporating deep learning and multi-omics autoencoding for analysis of lung adenocarcinoma prognostication. Comput. Biol. Chem. **87**, 107277 (2020)
12. Vincent, P., Larochelle, H., Bengio, Y., et al.: Extracting and composing robust features with denoising autoencoders. In: Proceedings of the 25th International Conference on Machine Learning, pp. 1096–1103 (2008)
13. Tong, L., Mitchel, J., Chatlin, K., et al.: Deep learning based feature-level integration of multi-omics data for breast cancer patients survival analysis. BMC Med. Inform. Decis. Mak. **20**, 1–12 (2020)
14. Chai, H., Zhang, Z., Wang, Y., et al.: Predicting bladder cancer prognosis by integrating multi-omics data through a transfer learning-based Cox proportional hazards network. CCF Trans. High Perform. Comput. **3**(3), 311–319 (2021)
15. Chai, H., Zhou, X., Zhang, Z., et al.: Integrating multi-omics data through deep learning for accurate cancer prognosis prediction. Comput. Biol. Med. **134**, 104481 (2021)
16. Goodfellow, I., Pouget-Abadie, J., Mirza, M., et al.: Generative adversarial nets. Adv. Neural Inf. Process. Syst. **27** (2014)
17. Ahmed, K.T., Sun, J., Cheng, S., et al.: Multi-omics data integration by generative adversarial network. Bioinformatics **38**(1), 179–186 (2022)
18. Makhzani, A., Shlens, J., Jaitly, N., et al.: Adversarial autoencoders. In: International Conference on Learning Representations (2016)
19. Tomczak, K., Czerwińska, P., Wiznerowicz, M.: Review the cancer genome atlas (TCGA): an immeasurable source of knowledge. Contemp. Oncol. Współczesna Onkologia **2015**(1), 68–77 (2015)
20. Clough, E., Barrett, T.: The gene expression omnibus database. In: Mathé, E., Davis, S. (eds.) Statistical Genomics. Methods in Molecular Biology, vol. 1418. Humana Press, New York, NY (2016). https://doi.org/10.1007/978-1-4939-3578-9_5
21. Clark, N.R., Hu, K.S., Feldmann, A.S., et al.: The characteristic direction: a geometrical approach to identify differentially expressed genes. BMC Bioinf. **15**, 1–16 (2014)
22. Yang, J., Chen, L., Kong, X., et al.: Analysis of tumor suppressor genes based on gene ontology and the KEGG pathway. PLoS ONE **9**(9), e107202 (2014)
23. Lee, Y.C., Lam, H.M., Rosser, C., et al.: The dynamic roles of the bladder tumour microenvironment. Nat. Rev. Urol. **19**(9), 515–533 (2022)

SGEGCAE: A Sparse Gating Enhanced Graph Convolutional Autoencoder for Multi-omics Data Integration and Classification

Junliang Shang[1], Limin Zhang[1], Linqian Zhao[1], Xin He[1], Yan Zhao[1], Daohui Ge[1], Jin-Xing Liu[2], and Feng Li[1(✉)]

[1] School of Computer Science, Qufu Normal University, Rizhao 276800, China
`lifeng_10_28@163.com`
[2] School of Health and Life Sciences, University of Health and Rehabilitation Sciences, Qingdao 266000, China

Abstract. Integration of multi-omics data is essential for obtaining comprehensive insights into molecular mechanisms of complex diseases. While several methods have been proposed for analyzing multi-omics data in various applications, challenges persist in effectively handling heterogeneous and rich multi-omics data. In this paper, a Sparse Gating Enhanced Graph Convolutional AutoEncoder, named SGEGCAE, is proposed for multi-omics data integration and classification. Specifically, an enhanced graph convolutional autoencoder is developed, which integrates a basic autoencoder with a sparse gating strategy, aiming to combine attribute information with topological structure information of the graph for obtaining more comprehensive feature representations. To address the inherent variability and fluctuations in different omics data quality among samples, true class probability is introduced into the SGEGCAE to acquire reliable classification confidence. Furthermore, a tensor fusion network is designed to explore both inter-omics and intra-omics relationships in the label space to achieve ultimately multi-omics integration and classification. Extensive biomedical classification experiments are carried out on four datasets. In these experiments, the superior performance of the SGEGCAE is clearly validated compared to some state-of-the-art integrative analysis methods, demonstrating that the SGEGCAE might serve as an alternative method for multi-omics data integration and classification. The code and datasets for the SGEGCAE are available online at https://github.com/CDMBlab/SGEGCAE.

Keywords: Multi-omics Integration · Graph Convolutional Autoencoder · Sparse Gating Strategy · True Class Probability · Tensor Fusion Network

1 Introduction

The rapid advancement of high-throughput sequencing technologies has significantly improved the efficiency and scale of acquiring multiple omics (multi-omics) data. Numerous studies indicate that a single type of omics (single-omics) data is insufficient for a thorough comprehension of biological complexities, lacking a holistic insight

© The Author(s), under exclusive license to Springer Nature Singapore Pte Ltd. 2024
D.-S. Huang et al. (Eds.): ICIC 2024, LNBI 14881, pp. 135–146, 2024.
https://doi.org/10.1007/978-981-97-5689-6_12

into interactions across molecular levels [1, 2]. In contrast, multi-omics data integrate information across multiple molecular levels, including genomics, transcriptomics, and epigenomics, providing a more holistic perspective for unraveling the molecular mechanisms underlying complex diseases [3]. Therefore, it is crucial to propose efficient integrative analysis methods to handle these heterogeneous and rich multi-omics data.

Many methods have been proposed for analyzing multi-omics data in diverse applications. Some traditional statistical and machine learning methods have been employed for integrating multi-omics data. Nonetheless, they often employ a simplistic splicing strategy for multi-omics data integration, which fail to effectively capture the complex relationships between different omics data types [4, 5]. Deep learning methods has provided new opportunities for the integration of multi-omics data, which exhibit stronger expressive power and better adaptability [6, 7]. For instance, Moon et al. [8] presented a multi-omics integration approach with an attention mechanism to enhance diagnostic performance. Han et al. [9] proposed a dynamical multimodal fusion strategy, which uses encoder networks to consider both the feature-level and modality-level dynamicities. Zhao et al. [10] proposed CLCLSA, a method for multi-omics integration utilizing contrastive learning and self-attention mechanism. Recently, graph convolutional networks (GCNs), as a crucial branch of deep learning, excel in modeling complex relationships and network structures [11, 12]. For example, Wang et al. [13] presented MOGONET, a supervised multi-omics integration method based on GCNs for biomedical classification. Li et al. [14] combined similarity network and GCN for cancer subtype classification with multi-omics data. Despite recent advancements, feature representations learned by these methods are still not sufficiently comprehensive. Furthermore, complex diseases exhibit biological heterogeneity between different samples, which increases the difficulty of extracting consistent information from multi-omics data. To address these challenges, there is a pressing need for innovative integrative analysis methods to capture the informativeness of each feature and type of omics data for different samples to enhance the learning performance.

There has been an increasing emphasis on integrating graph autoencoder (GAE) and autoencoder (AE) [15, 16]. On one hand, training an AE involves minimizing differences between inputs and reconstructions, and the learned low-dimensional representation preserves essential attribute information of features. On the other hand, GAE focus on processing graph-structured data by learning feature representations of nodes, taking into account connections between nodes. Therefore, integrating AE and GAE allows for a better balance between attribute information and topological structure of the graph, enabling the acquisition of a comprehensive feature representations.

In this study, a Sparse Gating Enhanced Graph Convolutional AutoEncoder, SGEG-CAE for short, is proposed for multi-omics data integration and classification. Specifically, SGEGCAE introduces an Enhanced Graph Convolutional AutoEncoder (EGCAE) that collaborates with an AE and a sparse gating strategy to obtain comprehensive feature representations. Furthermore, True Class Probability (TCP) allows for flexible handling of variability in omics data quality, enhancing classification robustness. Tensor Fusion Network (TFN) is designed to integrate different types of omics data in the label space, considering both inter-omics relationships and intra-omics variations. Extensive biomedical classification experiments are conducted on four datasets, results of which demonstrate that SGEGCAE offers a promising alternative for integrating and classifying multi-omics data.

2 Methods

2.1 Overview of SGEGCAE

The SGEGCAE, shown in Fig. 1, comprises five key components. Specifically, an AE is initially utilized to extract attribute information representation for each omics data type. Subsequently, an EGCAE integrates attribute information while capturing graph structural information, resulting in comprehensive feature representations. Additionally, a sparse gating strategy is introduced to enhance the focus on critical information while minimizing redundancy or unnecessary details. Furthermore, TCP is introduced to consider classification confidence across different omics data types, thereby reducing prediction errors. Finally, a TFN is designed for multi-omics data integration at the label space, taking into account both correlations between different omics data types and the preservation of information unique to each omics data type.

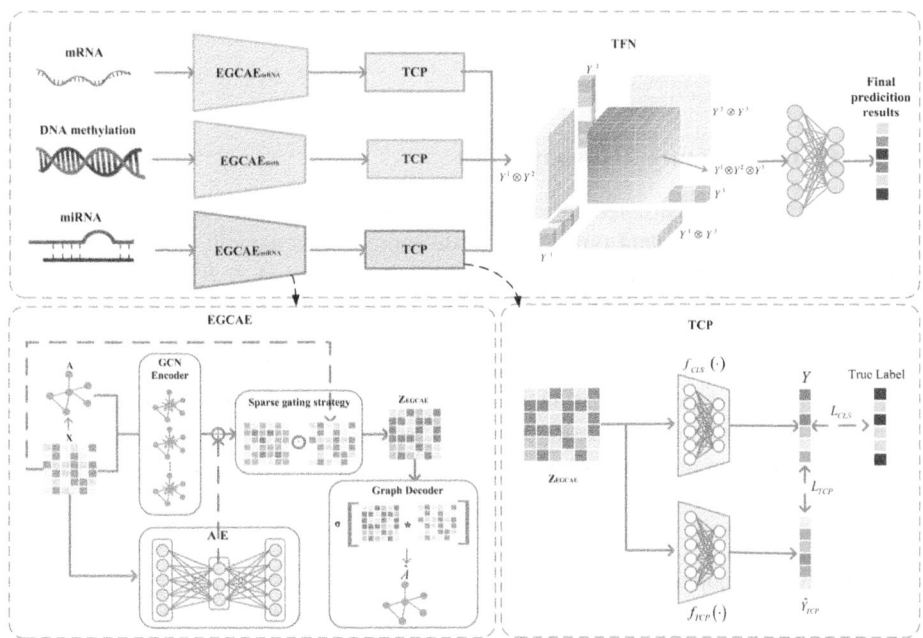

Fig. 1. The flowchart of SGEGCAE.

2.2 AE for Attribute Information Representation

Considering the m-th omics data type, its feature matrix can be denoted as $X^m \in R^{n \times d_m}$, where $m = 1, 2, 3$, n represents the number of samples and d_m denotes the number of input features. An AE is trained to extract meaningful attribute information representation. The encoder of the AE is defined as,

$$Z_{AE}^m = f_{enc}(X^m) \tag{1}$$

where $f_{enc}(\cdot)$ represents the encoder function, mapping the initial feature matrix $X^m \in R^{n \times d_m}$ into $Z_{AE}^m \in R^{n \times d_m'}$. Z_{AE}^m denotes the attribute information representation of the m-th omics data, and d_m' is its corresponding feature dimension.

The decoder of the AE approximately reconstructs the initial input feature matrix through the reverse operations of the encoder, which is defined as,

$$\hat{X}^m = f_{dec}(Z_{AE}^m) \tag{2}$$

where $f_{dec}(\cdot)$ represents the decoder function, $\hat{X}^m$ denotes the reconstructed feature matrix of the m-th omics data. The loss function of the AE is defined as,

$$L_{AE} = \sum_m \| X^m - \hat{X}^m \|_F^2 \tag{3}$$

2.3 EGCAE for Feature Representations

A sample graph is constructed to capture their neighborhood relationships between samples for each omics data type. Each node represents a sample, and each edge is measured by the cosine similarity between linking samples. $A \in R^{n \times n}$ represents the adjacency matrix corresponding to the sample graph. To obtain comprehensive feature representations for each omics data type, an EGCAE is presented. The graph encoder of the EGCAE is constructed by stacking multiple graph convolutional layers.

To integrate attribute information with topological structure information of the m-th omics data, the attribute information representation Z_{AE}^m from the AE and topological structure information representation H^m from the graph encoder, are combined as,

$$Z_I^m = (1 - \lambda)H^m + \lambda Z_{AE}^m \tag{4}$$

where the hyperparameter λ allows for a flexible balance between contributions of AE and GAE, and $0 \leq \lambda \leq 1$. $Z_I^m \in R^{n \times d_m'}$ is the integrated feature representations which ensures that the information learned by AE and GAE complements each other.

2.4 Sparse Gating Strategy for Enhanced Feature Representations

While Z_I^m incorporates both attribute information and topological structure information of the graph, it still suffer from redundancy and noise. Therefore, to enhance the focus on critical information while minimizing redundancy or unnecessary details, a sparse gating strategy is introduced, which is defined as,

$$Z_{EGCAE}^m = Z_I^m \odot X^m \tag{5}$$

where $\odot$ denotes the element-wise multiplication. X^m denotes the initial feature matrix that contains the raw information of the data. $Z_{EGCAE}^m \in R^{n \times d_m'}$ is the final feature representations learned by the EGCAE.

The graph decoder is designed to conduct a dot product with the low-dimensional representation generated by the EGCAE, thereby approximating the reconstruction of the original adjacency matrix. The reconstruction loss of the EGCAE is constructed using the mean square error loss function, which is defined as,

$$L_{EGCAE} = \sum_m \left\| \tilde{A}^m - \hat{A}^m \right\|_F^2 \tag{6}$$

where L_{EGCAE} represents the loss function of the EGCAE.

2.5 TCP for Omics Informativeness Estimation

In the context of multi-omics data classification tasks, it is critical to account for the inherent variability and fluctuations in the quality of different omics data across diverse samples, which is essential for developing a robust multi-omics data classification method. Considering classification confidence can help us assess the reliability of a model's classification results for different samples. When classification confidence is high, we can confidently trust the model's results; however, caution is warranted when confidence is low, as the model may lack certainty in sample classification. TCP utilizes the maximum softmax output for correctly classified samples, it tends to yield comparatively lower values in instances of misclassification, reflecting the model's tendency towards erroneous predictions [17]. Therefore, TCP is employed to address variability in informativeness and fluctuations in data quality for multi-omics data classification.

Specifically, a omics-classifier is employed for each omics data type to obtain predicted label probability distribution for each sample, which is defined as,

$$Y^m = f_{CLS}\left(Z_{EGCAE}^m\right) \tag{7}$$

$$L_{CLS} = -\sum_m \sum_{c=1}^{c} y_c \log p_c^m \tag{8}$$

where $f_{CLS}(\cdot)$ represents the omics-classifier, Y^m is the predicted label probability distribution for samples under the m-th omics data. L_{CLS} represents the loss function of omics-classifier. p_c^m indicates the probability that the sample belongs to class c under the m-th omics data. $p = [p_1, p_2, ..., p_c]$ is the one-hot representation of sample labels.

TCP provides a confidence measure for the classification decision of each sample. This adaptive capability enables SGEGCAE to flexibly handle variations in quality between omics data types.

$$Y_{TCP}^m = y \cdot p^m\left(y | Z_{EGCAE}^m\right) = \sum_{c=1}^{c} y_c p_c^m \tag{9}$$

where · denotes inner product. When misclassification occurs, the confidence is calculated using the softmax output probability corresponding to the true label, rather than the maximum softmax output probability.

Despite TCP offering more robust confidence estimates, it cannot be directly used during the testing phase due to the requirement for label information. Therefore, a confidence neural network is utilized to quantify the classification confidence of samples under different omics data types, which is defined as,

$$\hat{Y}_{TCP}^m = f_{TCP}\left(Z_{EGCAE}^m\right) \tag{10}$$

where $f_{TCP}(\cdot)$ represents the confidence neural network.

The loss function for omics informativeness estimation is defined as,

$$L_{TCP} = \sum_m \left(\hat{Y}_{TCP}^m - Y_{TCP}^m\right)^2 + L_{CLS} \tag{11}$$

Where L_{TCP} represents the loss function of omics informativeness estimation, Y_{TCP} can be approximated with the omics classifier and confidence neural network. Thus, through TCP, the classification uncertainty of SGEGCAE on several samples can be identified, reducing the likelihood of erroneous predictions.

2.6 TFN for Multi-omics Integration

The TFN can represent unimodal, bimodal, and trimodal interactions among behaviors [18]. In SGEGCAE, TFN is introduced to integrate predictions of class labels from different omics data types at the label space, considering both inter-omics and intra-omics aspects.

Specifically, matrix augmentation is performed on Y^1, Y^2, and Y^3 respectively, by concatenating a column filled with ones. Then, the Cartesian product operation is performed to generate feature representations that encompass all possible combinations among three omics data types. This process captures interactions and correlations between different omics data types. The specific calculation is defined as,

$$\hat{Y} = \begin{bmatrix} Y^1 \\ 1 \end{bmatrix} \otimes \begin{bmatrix} Y^2 \\ 1 \end{bmatrix} \otimes \begin{bmatrix} Y^3 \\ 1 \end{bmatrix} \tag{12}$$

where $\hat{Y}$ represents the feature representations fused from different omics data, simultaneously considering inter-omics correlations and preserving original features of each omics data type. Finally, the SGEGCAE inputs $\hat{Y}$ into TFN to obtain the ultimate classification results.

The loss function of the TFN is defined as,

$$L = \sum_i^n \left(\alpha L_{AE} + \beta L_{EGCAE} + L_{TCP} + L_{TFN}\right) \tag{13}$$

where α and β are hyperparameters used to balance the losses of AE and EGCAE, aiming to achieve precise classification predictions through continuous minimization of the loss function.

3 Experiments and Results

3.1 Datasets and Evaluation Metrics

To demonstrate the effectiveness of SGEGCAE, experiments were conducted on four biomedical datasets. The datasets used in this study are the same as those in previous studies [9, 13]. Classification tasks were conducted on the ROSMAP dataset to differentiate Alzheimer's disease patients (AD) from normal controls (NC), the LGG dataset for grading low-grade glioma, the BRCA dataset for classifying breast invasive carcinoma subtypes, and the KIPAN dataset for categorizing kidney cancer types. In these classification tasks, three types of omics data were utilized, that is, mRNA expression data (mRNA), DNA methylation data (meth), and miRNA expression data (miRNA). Detailed descriptions are provided in Table 1.

To quantitatively assess the classification performance of SGEGCAE, three evaluation metrics are applied to each dataset. Metrics like ACC, F1, and AUC assess binary classification tasks, while multi-class classification tasks utilize ACC, F1_weighted, and F1_macro. To ensure robust evaluations, each dataset is partitioned into training and testing sets, and ten experiments are conducted to facilitate comprehensive comparisons and enhance the statistical significance of the results.

Table 1. Datasets.

Dataset	Categories	mRNA	meth	miRNA
ROSMAP	NC: 169, AD: 182	200	200	200
LGG	Grade 2: 246, Grade 3: 264	2000	2000	548
BRCA	Normal-like: 115, Basal-like: 131, HER2- enriched: 46, Luminal A: 436, Luminal B: 147	1000	1000	503
KIPAN	KICH: 66, KIRC: 318, KIRP: 274	2000	2000	445

3.2 Analysis of Classification Results

To fully demonstrate the classification performance of SGEGCAE, seven supervised classification methods were selected for comparison: (1) KNN; (2) RF; (3) NN; (4) MOMA [8]; (5) MOGONET [13]; (6) MMdynamics [9]. (7) CLCLSA [17]. The comparative experimental results are reported in Fig. 2.

In summary, we have shown that SGEGCAE demonstrates its effectiveness and adaptability by outperforming methods on the ROSMAP, BRCA, and LGG datasets. It achieves superior performance across various evaluation metrics compared to state-of-the-art methods. On the KIPAN dataset, the classification performance of SGEGCAE is comparable to advanced algorithms such as CLCLSA, MMDynamics, and MOGONET, with all three evaluation metrics reaching 0.999, thereby validating its competitive classification performance.

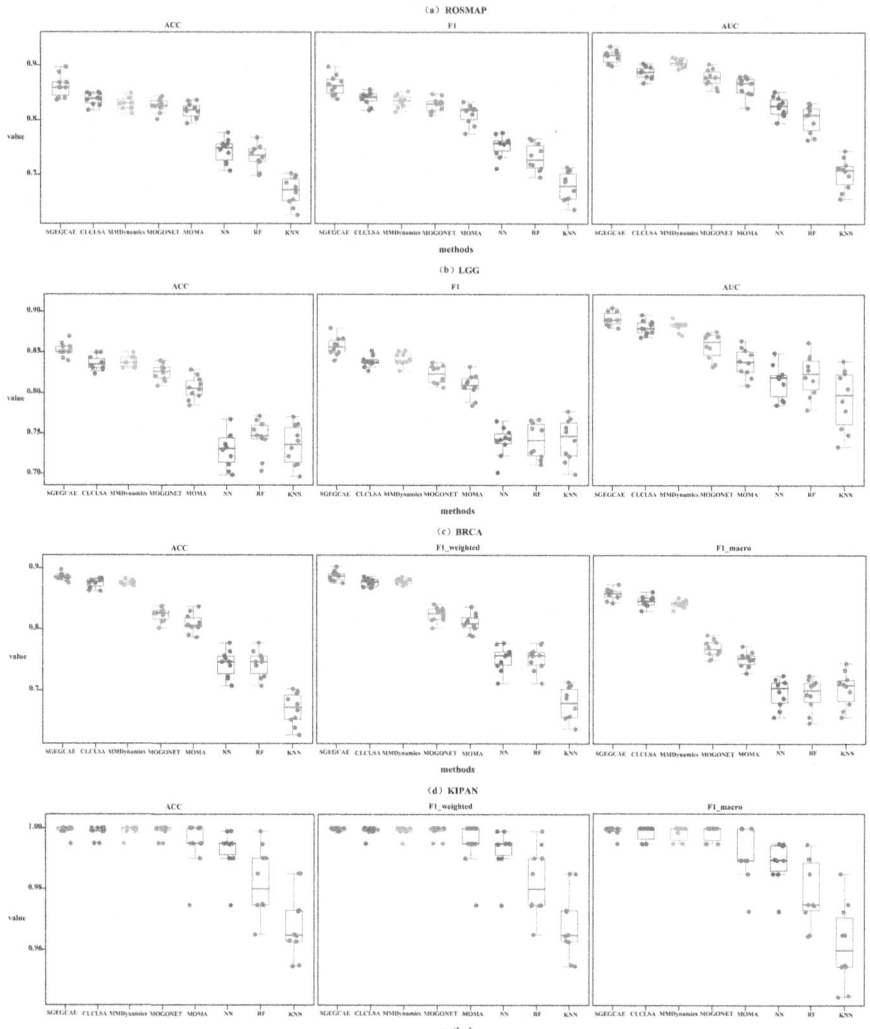

Fig. 2. Comparison of classification performance on different datasets.

3.3 Ablation Studies

In this study, ablation experiments were also conducted to assess the contributions of EGCAE, TCP, and TFN in the multi-omics data classification task. These three key components were selectively ablated to better understand their impact on classification performance. The experimental results are shown in Table 2 and Table 3, respectively.

In EGCAE ablation experiments, GAE was used to extract feature information from multi-omics data, while attribute information and sparse gating strategy were not considered. TFN was applied in TCP ablation experiments for multi-omics data integration by generating predicted label probabilities from classifiers without quantifying classification confidence of omics data types. In TFN ablation experiments, class label predictions

Table 2. Results of ablation experiments under the binary classification datasets.

Dataset	Method	ACC	F1	AUC
ROSMAP	SGEGCAE_EGCAE	0.835 ± 0.016	0.837 ± 0.014	0.905 ± 0.014
	SGEGCAE_TCP	0.839 ± 0.020	0.842 ± 0.021	0.895 ± 0.017
	SGEGCAE_TFN	0.844 ± 0.018	0.845 ± 0.016	0.907 ± 0.015
	SGEGCAE	**0.859 ± 0.014**	**0.860 ± 0.012**	**0.919 ± 0.011**
LGG	SGEGCAE_EGCAE	0.822 ± 0.015	0.819 ± 0.013	0.854 ± 0.013
	SGEGCAE_TCP	0.827 ± 0.021	0.828 ± 0.022	0.861 ± 0.019
	SGEGCAE_TFN	0.838 ± 0.017	0.836 ± 0.018	0.874 ± 0.013
	SGEGCAE	**0.845 ± 0.017**	**0.846 ± 0.015**	**0.892 ± 0.016**

Table 3. Results of ablation experiments under the multi-class classification datasets.

Dataset	Method	ACC	F1_weighted	F1_macro
BRCA	SGEGCAE_EGCAE	0.862 ± 0.015	0.864 ± 0.016	0.831 ± 0.016
	SGEGCAE_TCP	0.857 ± 0.020	0.860 ± 0.020	0.824 ± 0.018
	SGEGCAE_TFN	0.868 ± 0.014	0.874 ± 0.014	0.840 ± 0.013
	SGEGCAE	**0.881 ± 0.010**	**0.889 ± 0.011**	**0.856 ± 0.014**
KIPAN	SGEGCAE_EGCAE	0.994 ± 0.004	0.994 ± 0.004	0.994 ± 0.005
	SGEGCAE_TCP	0.994 ± 0.002	0.994 ± 0.002	0.994 ± 0.002
	SGEGCAE_TFN	0.998 ± 0.004	0.998 ± 0.004	0.996 ± 0.004
	SGEGCAE	**0.999 ± 0.002**	**0.999 ± 0.002**	**0.999 ± 0.002**

from TCP across omics data types were concatenated and fed into a fully connected neural network for classification. Overall, EGCAE, TCP, and TFN are crucial components in SGEGCAE, significantly improving accuracy in multi-omics data classification tasks.

3.4 Analysis of Hyper-parameter λ

Since specific task requirements vary across datasets, it is crucial to determine how to integrate and balance these two autoencoders. In SGEGCAE, a hyperparameter λ is introduced to balance their contributions, where $0 \leq \lambda \leq 1$. Experimental results for parameter selection of λ are shown in Fig. 3.

It is seen that optimal values in multiple evaluation metrics were achieved while $\lambda = 0.1$ across four datasets, implying that the structural information about the graph may play a crucial role in the performance of SGEGCAE for the classification task. When $\lambda = 0$, attribute information representations that the AE learned were excluded, and the SGEGCAE only focuses on the topological structure information of the graph, resulting in its classification performance significantly lower than that at $\lambda = 0.1$. This observation indicates that the attribute information from omics data can enhance the performance of SGEGCAE in classification tasks.

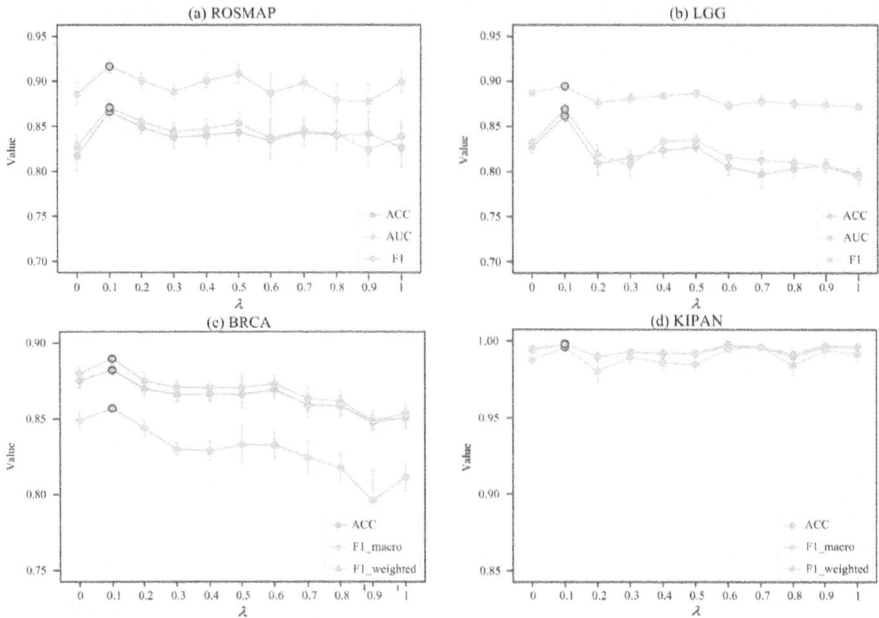

Fig. 3. Performance of SGEGCAE under different values of hyperparameter λ.

3.5   Analysis of Different Omics Data Types

To conduct performance evaluations of diverse omics data selections, we compared the classification predictions of single omics data and their combinations through experiments involving 7 different omics data type combinations. The results are detailed in Fig. 4. When combining mRNA, meth, and miRNA, SGEGCAE shows superior performance across various evaluation metrics. Additionally, it is evident that different types of omics data contribute differently to the classification task. The ROSMAP dataset shows superior classification performance with mRNA alone compared to meth alone. These findings illustrate that combining various omics data types enhances comprehension of biological processes and offers a more holistic approach to revealing disease mechanisms.

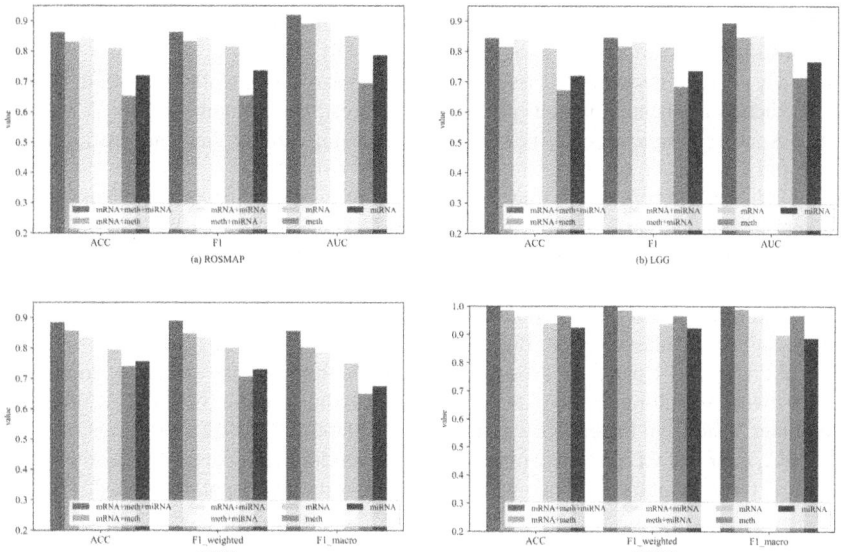

Fig. 4. Performance comparison of SGEGCAE under different omics data types.

4 Conclusion

The integration of multi-omics data is pivotal for comprehending complex disease mechanisms. This study presents SGEGCAE, a method for integrating and classifying multi-omics data, focusing on EGCAE, TCP, and TFN. Results from experiments on four biomedical datasets have shown SGEGCAE's superior performance in data classification. Ablation experiments further support the effectiveness of EGCAE, TCP, and TFN, suggesting SGEGCAE as a viable approach for multi-omics data integration and classification.

Future directions of this study may include further optimizing the performance of SGEGCAE, exploring alternative types of omics data fusion, and validating its applicability in broader biomedical contexts. Additionally, ensuring the interpretability of SGEGCAE results is crucial for facilitating practical applications in healthcare decision-making.

Acknowledgments. This study was funded by the National Natural Science Foundation of China (61972226 and 62172254).

References

1. Subramanian, I., Verma, S., Kumar, S., Jere, A., Anamika, K.: Multi-omics data integration, interpretation, and its application. Bioinform. Biol. Insights **14**, 1177932219899051 (2020)
2. Zhu, Y.T., Qiu, P., Ji, Y.: TCGA-assembler: open-source software for retrieving and processing TCGA data. Nat. Methods **11**(6), 599–600 (2014)

3. Chai, H., et al.: Integrating multi-omics data through deep learning for accurate cancer prognosis prediction. Comput. Biol. Med. **134**, 104481 (2021)
4. Acharjee, A., Kloosterman, B., Visser, R.G.F. Maliepaard, C.: Integration of multi-omics data for prediction of phenotypic traits using random forest. Bmc Bioinf. **17,** 363-373 (2016)
5. Ma, B.S., et al.: Diagnostic classification of cancers using extreme gradient boosting algorithm and multi-omics data. Comput. Biol. Med. **121**, 103761 (2020)
6. Leng, D.J., et al.: A benchmark study of deep learning-based multi-omics data fusion methods for cancer. Genome Biol. **23**(1), 171 (2022)
7. Yang, Y., Tian, S., Qiu, Y., Zhao, P. Zou, Q.: MDICC: novel method for multi-omics data integration and cancer subtype identification. Brief Bioinform **23**(3), bbac132 (2022)
8. Moon, S., Lee, H.: MOMA: a multi-task attention learning algorithm for multi-omics data interpretation and classification. Bioinformatics **38**(8), 2287–2296 (2022)
9. Han, Z., Yang, F., Huang, J., Zhang, C. Yao, J.: Multimodal dynamics: dynamical fusion for trustworthy multimodal classification. In: Proceedings of the IEEE/CVF Conference on Computer Vision and Pattern Recognition, pp. 20707–20717 (2022)
10. Zhao, C., et al.: CLCLSA: cross-omics linked embedding with contrastive learning and self attention for integration with incomplete multi-omics data. Comput. Biol. Med. **170**, 108058 (2024)
11. Zhang, S., Tong, H., Xu, J., Maciejewski, R.: Graph convolutional networks: a comprehensive review. Comput. Soc. Netw. **6**(1), 1–23 (2019)
12. Ouyang, D., et al.: Integration of multi-omics data using adaptive graph learning and attention mechanism for patient classification and biomarker identification. Comput. Biol. Med. **164**, 107303 (2023)
13. Wang, T.X. et al.: MOGONET integrates multi-omics data using graph convolutional networks allowing patient classification and biomarker identification. Nature Commun. **12**(1), 3445 (2021)
14. Li, X., et al.: MoGCN: a Multi-omics integration method based on graph convolutional network for cancer subtype analysis. Front. Genet. **13**(2), 806842 (2022)
15. Tu, W. et al.: Deep fusion clustering network. In: Proceedings of the AAAI Conference on Artificial Intelligence, pp. 9978–9987 (2021)
16. Gan, Y., Huang, X., Zou, G., Zhou, S. Guan, J.: Deep structural clustering for single-cell RNA-seq data jointly through autoencoder and graph neural network. Brief Bioinform **23**(2), bbac018 (2022)
17. Corbière, C., Thome, N., Bar-Hen, A., Cord, M., Pérez, P.: Addressing failure prediction by learning model confidence. Adv. Neural Inf. Process. Syst. **32** (2019)
18. Na, S.: Tensor fusion network for multimodal sentiment analysis. In: Proceedings of the 2017 Conference on Empirical Methods in Natural Language Processing, pp. 1103–1114 (2017)

Short-Term Blood Glucose Prediction Method Based on Signal Decomposition and Bidirectional Networks

Yili Zheng, Zhifang Liao, Jia Guo, and Song Yu[(✉)]

Central South University, Changsha 410083, China
ys@csu.edu.cn

Abstract. For diabetic patients, the monitoring and prediction of blood glucose concentration are crucial aspects of treatment. However, blood glucose concentration is a typical time series data with characteristics such as time variation, nonlinearity, and non-stationarity. In order to mitigate the impact of these characteristics on predictions and enhance the accuracy of blood glucose forecasting, a short-term blood glucose prediction method based on signal decomposition and bidirectional networks (SVBiGL) is proposed. This method employs the Variational Mode Decomposition (VMD) algorithm based on Sparrow Search Algorithm (SSA) to obtain the optimal decomposition of the patient's blood glucose time series. The decomposed sub-sequences are then predicted using a composite network (BiGL) consisting of Bidirectional Gated Recurrent Unit (BiGRU) and Bidirectional Long Short-Term Memory (BiLSTM) networks. Finally, the predicted sub-sequences are aggregated to obtain the final prediction results. Experimental results demonstrate that the proposed method outperforms BiGRU, BiGL, BiGRU with VMD, and BiGRU with combined SSA and VMD methods in blood glucose prediction at 30-min, 45-min, and 60-min time steps.

Keywords: Blood glucose prediction · Variational modal decomposition · SSA · BiLSTM · BiGRU

1 Introduction

Diabetes, marked by insufficient insulin production, necessitates real-time blood glucose monitoring to maintain balance [1]. Accurate prediction of glucose changes aids in diet control, exercise regulation, and insulin estimation.

Recent focus on blood glucose prediction models highlights their role in managing diabetes-related risks and treatments. Aliberti et al. [2] utilized deep learning, combining LSTM networks with multi-patient data to predict blood glucose concentrations effectively. BiLSTM networks, noted for capturing long-range dependencies, outperformed LSTM models. Siami-Namini et al. [3] found BiLSTM superior in financial time series prediction. Mhaskar et al. [4] demonstrated the efficacy of deep convolutional neural networks in blood glucose prediction. Zhu [5] introduced the FCNN framework for Type 1 Diabetes blood glucose prediction, demonstrating real-time alerts on smartphones for

© The Author(s), under exclusive license to Springer Nature Singapore Pte Ltd. 2024
D.-S. Huang et al. (Eds.): ICIC 2024, LNBI 14881, pp. 147–158, 2024.
https://doi.org/10.1007/978-981-97-5689-6_13

proactive control. Sun et al. [6] utilized BiLSTM for blood glucose prediction.Cho et al. [7] introduced GRU units as a variant of LSTM, with BiGRU extending GRU's bidirectional capability, improving model performance. Dudukcu [8] utilized LSTM, WaveNet, GRU, and their combinations for blood glucose prediction.

Due to blood glucose's nonlinear and non-stationary nature, direct prediction model application may compromise accuracy. Signal decomposition methods, however, effectively address these challenges [9]. Wang et al. [10] proposed a short-term prediction model integrating variational mode decomposition and LSTM networks, enhancing accuracy.Gao et al. [11] introduced a model combining EEMD and LS-SVM for precise short-term blood glucose predictions. Cai et al. [12] proposed an algorithm integrating CBCEEMD, adaptive noise, and BPNN to improve prediction accuracy.

Increasing network depth, rather than unit count per layer, enhances model generalization. Li et al. [13] proposed a convolutional recurrent neural network for blood glucose prediction, capturing features using CNN layers and analyzing sequence data with an improved RNN.Nemat et al. [14] employed meta-learning to enhance prediction accuracy, while D'Anton [15] optimized glucose prediction using edge computing boards.Shuvo [16] introduced a multi-task learning deep learning model for personalized blood glucose prediction, and also implemented a combined BiGRU and BiLSTM model for improved accuracy.

Increasing network depth improves blood glucose prediction. Composite networks leverage individual component advantages, enhancing accuracy further. Yet, signal decomposition method parameter determination, like decomposition mode numbers, remains challenging. Our proposed Short-Term Blood Glucose Prediction Method (SVBiGL) overcomes this by combining signal decomposition with bidirectional networks. The Sparrow Search Algorithm optimizes variational mode decomposition parameters, improving efficiency. Bidirectional composite networks predict blood glucose, with contributions including:

(1) Proposal of SVBiGL: Rapid optimization of VMD parameters enhances decomposition efficiency, mitigating non-stationarity and nonlinearity impacts on prediction. Fusion of bidirectional gated recurrent units and bidirectional LSTM networks elevates prediction accuracy.

(2) Application of SVBiGL to actual diabetic patient data: Experimental results confirm method effectiveness, achieving favorable predictions 30, 45, and 60 min in advance.

2 Short-Term Blood Glucose Prediction Method Based on Signal Decomposition and Bidirectional Networks

2.1 Overall Approach

Blood glucose time series exhibit strong nonlinearity and non-stationarity. To accurately predict blood glucose concentration, this paper proposes a short-term blood glucose prediction method based on signal decomposition and bidirectional networks (SVBiGL), as illustrated in Fig. 1. Building upon the reduction of the impact of nonlinearity and non-stationarity on predictions through signal decomposition, the method combines a bidirectional network composite model to leverage their respective advantages and further improve the accuracy of blood glucose prediction. The overall approach consists of three main parts:

(1) Particle Swarm Optimization for VMD Parameters: The first part utilizes the Sparrow Search Algorithm to iteratively update particle positions, rapidly obtaining the optimal penalty factor and the optimal number of decomposition modes for VMD.
(2) VMD Decomposition of Blood Glucose Time Series: The second part employs the optimal decomposition parameters to perform VMD decomposition on the blood glucose time series, resulting in Intrinsic Mode Functions (IMFs) representing glucose components in different frequency bands.
(3) Composite Bidirectional Networks: The third part involves the composition of BiGRU and BiLSTM networks. The decomposed blood glucose time series is input into the bidirectional composite network for prediction, generating predictions for each mode, and finally combined to obtain the final prediction.

The predictive values obtained using the SVBiGL method will be compared with actual values and predictions from other models. The accuracy of the proposed method will be validated through error metric analysis.

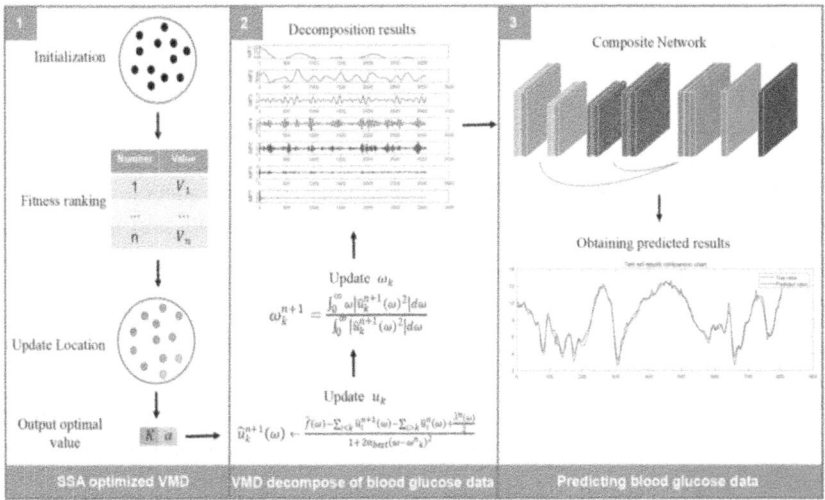

Fig. 1. Short-Term Blood Glucose Prediction Method Based on Signal Decomposition and Bidirectional Networks Illustration.

2.2 Variation Mode Decomposition Algorithm Based on Sparrow Search

The VMD method decomposes signals using parameters like K and penalty factor, impacting decomposition quality. Optimal parameters selection is crucial to avoid under or over-decomposition and bandwidth narrowing. To address this, a Sparrow Search-based VMD algorithm is proposed, categorizing the optimization population into discoverers, joiners, and perceivers of danger. Initialization involves setting parameters and defining fitness functions. Iterations update positionsbased on Eq. (1), checking stopping conditions for optimal parameters output.

$$X_{i,j}^{t+1} = \begin{cases} X_{i,j}^t \cdot \exp\left(-\frac{i}{\alpha \cdot \text{iter}_{max}}\right) & \text{if } R_2 < ST \\ X_{i,j}^t + Q \cdot L & \text{if } R_2 \geq ST \end{cases} \tag{1}$$

In this context, t is the current iteration, $j = 1,2,3,...d$. iter_{max} is the maximum iteration constant, $X_{i,j}$ denotes the positional information of parameters, $a \in (0, 1]$ is a random number, $R_2 (R_2 \in [0,1])$ and $ST (ST \in [0.5,1])$ represent warning and safety values, Q is a random number following a normal distribution, and L is a matrix of size $1 \times d$, with each element being 1.

Optimally obtained particles serve as decomposition parameters for Variational Mode Decomposition (VMD) in signal decomposition. VMD first applies the Hilbert transform to the original sequence, yielding analytical signals for K mode components, then converts them to one-sided spectra. Next, a central frequency exponent term is introduced to shift each mode's spectrum to its fundamental frequency band. The demodulated signals undergo Gaussian smoothing to estimate signal bandwidth, forming a variational constraint problem described by Eq. (2).

$$\begin{cases} \min_{\{u_k\},\{\omega_k\}} \left\{ \sum_k \left\| \partial_t \left[\left(\delta(t) + \frac{j}{\pi t} \right) * u_k(t) \right] e^{-j\omega_k t} \right\|_2^2 \right\} \\ s.t. \sum_k u_k = f \end{cases} \tag{2}$$

In the above equation, $\{v_k\} = \{v_1, ..., v_k\}$ and $\{\omega_k\} = \{\omega_1, ..., \omega_k\}$ respectively represent shorthand notations for all modes and their central frequencies.

Solve the above equation to obtain each decomposed mode, and the formula for solving mode u_k is Eq. (3):

$$\hat{u}_k^{n+1}(\omega) = \frac{\hat{f}(\omega) - \sum_{i \neq k} \hat{u}_i(\omega) + \frac{\hat{\lambda}(\omega)}{2}}{1 + 2\alpha(\omega - \omega_k)^2} \tag{3}$$

Finally, the optimal intrinsic mode functions of blood glucose levels in different frequency bands are obtained, facilitating subsequent input into the blood glucose prediction model for forecasting.

2.3 Composite Network of Bidirectional Gated Recurrent Unit (BiGRU) and Bidirectional Long Short-Term Memory (BiLSTM)

BiLSTM captures dependencies in sequences from both past and future data, but with increased computational complexity due to bidirectional processing. Conversely, BiGRU reduces model parameters, thus lowering training time and complexity. Combining their strengths, we propose a composite network, BiGL.

Initially, raw data is concatenated into input features and output labels based on specified time parameters. The dataset is divided, normalized to [0, 1], and formatted for neural network input.

The model consists of input, BiGRU, BiLSTM, and output layers. Each BiGRU layer includes two GRU networks followed by ReLU activation functions for non-linearity. BiLSTM layers incorporate dropout to prevent overfitting, and the second BiLSTM layer returns only the last time step's output. Fully connected layers connect previous outputs to the output layer for regression.

After establishing the model, training commences using training data and parameters. Simulation validation is performed on both training and testing sets, with results denormalized to original data range. Figure 2 illustrates the BiGL model's structure.

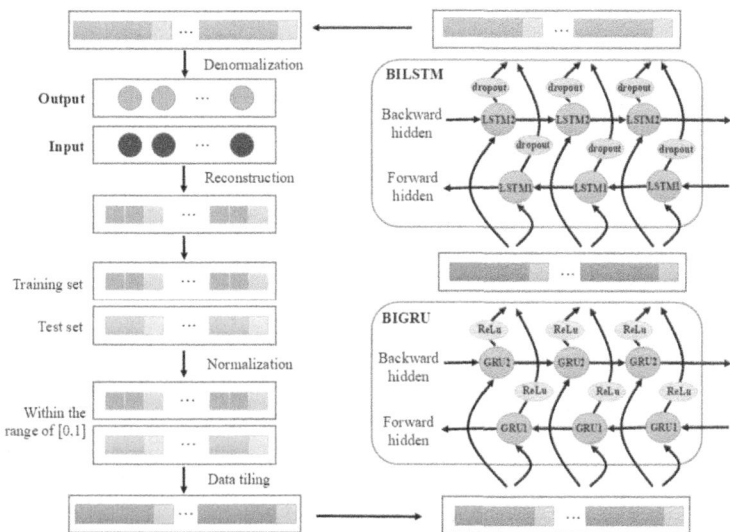

Fig. 2. BiGL Model Structure.

3   Results and Analysis

3.1   Experimental Environment and Parameter Settings

The study utilized data from Hunan Province Children's Hospital, consisting of CGM data sampled every 3 min. This dataset comprised blood glucose trends for 200 Type 1 diabetes patients. From this, blood glucose time series data from 20 patients were selected for experimentation. These data were randomly split into two subsets, approximately 70% and 30% of the initial samples, for training and testing, respectively. Table 1 shows the blood glucose data of one patient at 9 min.

In the experiment, VMD decomposed the blood sugar sequence, with penalty factor and number of decomposition modes as optimization parameters. The BiGL model comprised input, BiGRU, BiLSTM, and output layers. Trained for 500 epochs with a target of 0.0001, the model was implemented in MATLAB R2022b. Prediction horizons of 30, 45, and 60 min were chosen based on typical preventive measures for blood sugar events.

Table 1. CGM Blood Glucose Sample Data from Type 1 Diabetes Patients at the Children's Hospital of Hunan Province.

Number	Transmitter Serial Number	Patient Number	Collection time	Collection date	Blood Glucose Value
1	GN-D6M0137	118	2022/9/11 8:58	2022/9/11	11.3
2	GN-D6M0137	118	2022/9/11 8:55	2022/9/11	12.1
3	GN-D6M0137	118	2022/9/11 8:52	2022/9/11	12.0
4	GN-D6M0137	118	2022/9/11 8:49	2022/9/11	11.9

Research Question 1 (RQ1) aimed to identify the optimal blood sugar sequence decomposition method. Five experiments compared different methods using random parameters, with decomposed sequences input into the same prediction model for analysis.

Research Question 2 (RQ2) focused on determining the optimal fitness function for variational mode decomposition parameters. Four common fitness functions were compared, and the optimized parameters were used for blood sugar sequence decomposition.

Research Question 3 (RQ3) sought to evaluate the superiority of composite models over single models. Five experiments with random parameters compared different prediction models, including the proposed BiGL composite network.

Research Question 4 (RQ4) evaluated the proposed method's performance against BiGRU, BiGL, BiGRU with VMD, and BiGRU with SSA and VMD. All algorithms used the same training and testing sets for comparison.

3.2 Model Performance Evaluation Metrics

In order to evaluate the performance of the model, this study selected the following indicators as performance evaluation indicators Table 2.

R2 ranges from 0 to 1. A value of 0 suggests poor model fitting, while 1 indicates a perfect fit with no errors. Generally, a higher R2 value suggests a better model fit. Clark Error Grid Analysis (EGA) method employs Cartesian chart principles to assess blood glucose prediction accuracy based on predicted values falling into regions A to E.

3.3 Model Performance Evaluation Metrics

Comparison of Signal Decomposition Methods. Using nine signal decomposition methods, a 155 h blood glucose time series was obtained from specific patients (20 patients with a uniform distribution of age and gender). Feed these decomposed sequences into a blood glucose prediction model with a 30 min prediction range. Five experiments were conducted using random parameters for each method. Table 3 shows the average error and accuracy results, where CA represents the probability of the Clarke error grid falling into region A. We can see that the prediction after VMD decomposition

Table 2. Formula for each indicator.

Evaluation Metrics	Formula	Explaination
RMSE	$RMSE = \sqrt{\frac{1}{n}\sum_{i=1}^{N}(y_i - \hat{y}_i)^2}$	n: the sample size y_i: the actual value $\hat{y}_i$: the predicted value
MAPE	$MAPE = \frac{1}{n}\sum_{i=1}^{N}\frac{|\hat{y}_i - y_i|}{y_i} \times 100\%$	n: the sample size y_i: the actual value $\hat{y}_i$: the predicted value
MSE	$MSE = \frac{1}{m}\sum_{i=1}^{m}(y_i - \hat{y}_i)^2$	n: the sample size y_i: the actual value $\hat{y}_i$: the predicted value
R2	$R2 = 1 - \frac{\sum_i(\hat{y}_i - y_i)^2}{\sum_i(\bar{y}_i - y_i)^2}$	y_i: the actual observed values $\hat{y}_i$: the model's predicted values $\bar{y}_i$: the mean of the actual observed values

Table 3. Prediction Errors Using Different Decomposition Methods.

Decomposition Method	RMSE	MAE	MAPE(%)	R2	CA(%)
Wavelet Decom	1.5415	1.1196	10.9686	0.6240	74.66
Wavelet Packet Decom	1.5210	1.1067	10.8632	0.6329	76.35
EMD	0.8474	0.6262	6.1988	0.8862	87.75
EEMD	0.4668	0.3502	3.6544	0.9655	97.55
CEEMDAN	0.4329	0.3368	3.6029	0.9703	98.28
EWT	0.4185	0.3137	3.6365	0.9723	99.51
VMD	0.4146	0.3116	3.1738	0.9728	96.32

is the most accurate. Figure 3 shows the decomposition results achieved through VMD. VMD decomposition requires two parameters, and the next step is to optimize them.

Comparison of Different Fitness Functions in SSA-Optimized VMD. Using a 155-h blood glucose time series from a specific patient, VMD decomposition is applied. Initially, the SSA optimizes VMD parameters. Four fitness functions are considered. Each function undergoes 5 experiments, and results are averaged to obtain average error and accuracy, as depicted in Table 4. Here, α represents the penalty factor of VMD, and K represents the number of decomposition modes. Experimental results indicate that using the SSA to optimize VMD with Min. Permutation Entropy as the fitness function is the most suitable.

Comparison of Blood Glucose Prediction Models. Using 155 h of blood glucose time series from specific patients, various models are used to predict blood glucose with

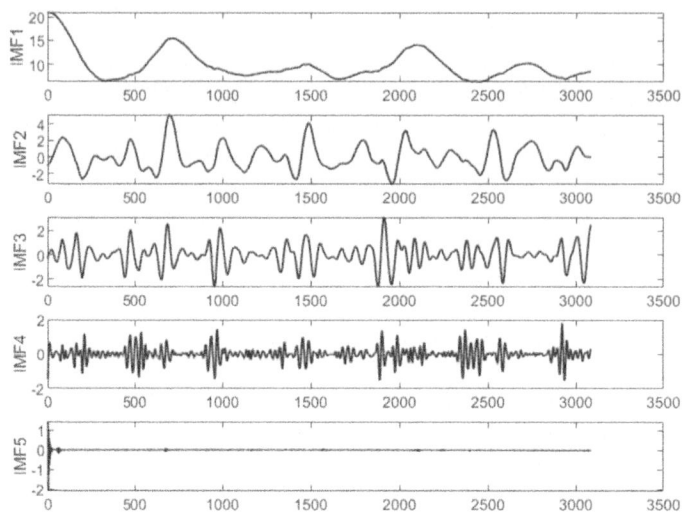

Fig. 3. VMD Decomposition Results.

Table 4. Optimization Results Using Different Fitness Functions in the Sparrow Search Algorithm.

SSA	α	K	RMSE	MAE	MAPE(%)	R2	CA(%)
No Optimization			0.4146	0.3116	3.1738	0.9728	96.32
Min. Information Entropy	2840	6	0.3152	0.2390	2.4523	0.9843	98.41
Min. Envelope Entropy	1287	4	0.5527	0.4199	4.2570	0.9516	95.59
Min. Sample Entropy	3000	10	0.2874	0.2228	2.3495	0.9869	98.77
Min. Permutation Entropy	2924	10	0.2801	0.2131	2.2159	0.9876	98.65

prediction intervals of 30, 45, and 60 min. As shown in Tables 5, 6, and 7, each model and interval under 5 experiments, resulting in average prediction error and accuracy We can see that prediction results of BiGRU and BiLSTM have small prediction errors and higher accuracy.

Sensitivity Analysis. After applying short-term blood glucose prediction methods, we compared the performance of various models in 60 min prediction. For the sake of article length, we have compared the results of each model in a single graph, as shown in Fig. 4: (1) Compared with the combination of VMD and BiGRU, BiGRU and BiGL are

Table 5. Experimental Results for Different Blood Glucose Prediction Models at 30-Min Intervals.

Prediction Model	RMSE	MAE	MAPE(%)	R2	CA(%)
RNN	1.3452	0.9744	9.5337	0.7133	79.17
CNN	1.4447	1.0737	18.4300	0.6696	76.69
DNN	1.8387	1.5045	22.4210	0.4648	63.80
LSTM	1.5453	1.1210	10.8045	0.6217	75.12
BiLSTM	1.3342	0.9632	9.4275	0.7180	79.29
GRU	1.5713	1.1146	10.7293	0.6088	74.88
BiGRU	1.3702	0.9721	9.4373	0.7026	78.43
BiGL	1.3000	0.9393	9.3225	0.7322	82.11

Table 6. Experimental Results for Different Blood Glucose Prediction Models at 45-Min Intervals.

Prediction Model	RMSE	MAE	MAPE(%)	R2	CA(%)
RNN	1.7389	1.2985	12.5527	0.5227	69.30
CNN	1.8642	1.4343	24.7030	0.4519	67.16
DNN	2.3839	1.8782	29.0340	0.1036	54.32
LSTM	1.9590	1.5410	15.7669	0.3943	63.87
BiLSTM	1.7414	1.3047	12.7604	0.5214	69.42
GRU	1.9781	1.5307	15.6764	0.3824	66.46
BiGRU	1.7416	1.2698	12.1975	0.5213	71.39
BiGL	1.6759	1.2498	12.4410	0.5567	73.74

insufficient. The latter effectively separates blood glucose components, reduces interference and non-stationary effects. (2) BiGRU with SSA and VMD is superior to VMD combined with BiGRU, as SSA optimization results in more accurate signal decomposition. (3) The performance of SVBiGL is superior to the combination of SSA, VMD, and BiGRU, as the BiGL composite network enhances the effectiveness of the model in capturing complex data relationships and improves prediction accuracy.

Figure 5 shows the Clark error grid for 60 min blood glucose prediction. Area A indicates that the deviation is less than 20%. The prediction rate of SVBiGL in Zone A exceeds 97%.

Clarke error grid analysis was conducted for the five prediction methods. Tables 8, 9, and 10 show the results. We can see that the SVBiGL predication model has a lower error value and higher R2 and Clarke error grid values.

Table 7. Experimental Results for Different Blood Glucose Prediction Models at 60-Min Intervals.

Prediction Model	RMSE	MAE	MAPE(%)	R2	CA(%)
RNN	2.2061	1.7223	16.7677	0.2342	59.18
CNN	2.0289	1.6196	25.4940	0.3525	58.51
DNN	2.4503	1.9273	27.5420	0.0557	51.18
LSTM	2.3691	1.8948	19.3650	0.1168	52.48
BiLSTM	1.9757	1.5290	15.3624	0.3858	64.02
GRU	2.1112	1.6466	16.3848	0.2987	61.41
BiGRU	1.9893	1.5013	14.5343	0.3774	62.78
BiGL	1.9420	1.4939	15.1573	0.4066	64.27

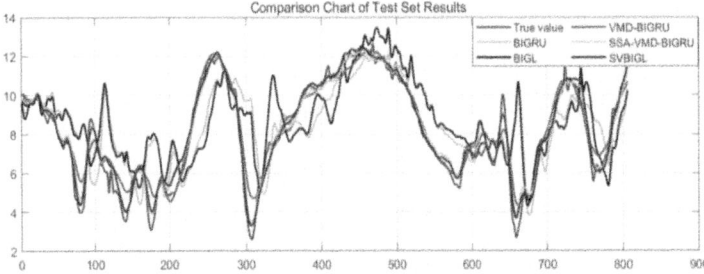

Fig. 4. Comparative Results of the Five Methods.

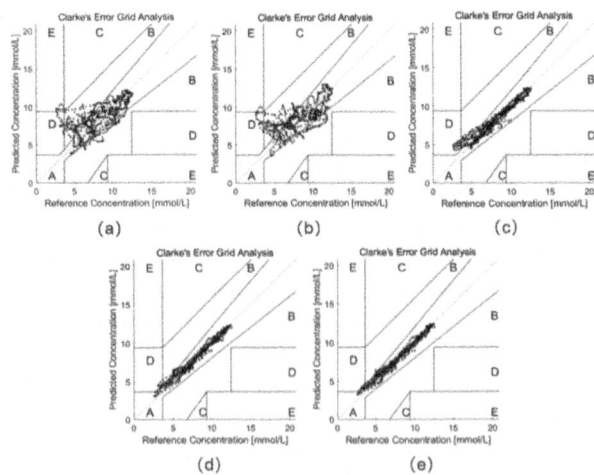

Fig. 5. Clarke Error Grids for the Five Methods (a) 60 min BiGRU; (b) 60 min BiGL; (c) 60 min BiGRU with VMD; (d) 60 min BiGRU with SSA and VMD; (e) 60 min SVBiGL.

Table 8. Experimental Results for Different Blood Glucose Prediction Methods at 30 min.

Prediction Model	RMSE	MAE	MAPE(%)	R2	CA(%)
BiGRU	1.3702	0.9721	9.4373	0.7026	78.43
BiGL	1.3000	0.9393	9.3225	0.7322	82.11
VMD-BiGRU	0.3941	0.2908	2.8959	0.9754	96.20
SSA-VMD-BiGRU	0.2791	0.2108	2.1193	0.9877	99.14
SVBiGL	0.2539	0.1906	1.9554	0.9898	99.39

Table 9. Experimental Results for Different Blood Glucose Prediction Models at 45-Min Intervals.

Prediction Model	RMSE	MAE	MAPE(%)	R2	CA(%)
BiGRU	1.7416	1.2698	12.1975	0.5213	71.39
BiGL	1.6759	1.2498	12.4410	0.5567	73.74
VMD-BiGRU	0.5041	0.3602	3.4743	0.9599	91.86
SSA-VMD-BiGRU	0.3546	0.2709	2.7188	0.9802	97.16
SVBiGL	0.3256	0.2437	2.4373	0.9833	97.90

Table 10. Experimental Results for Different Blood Glucose Prediction Models at 60-Min Intervals.

Prediction Model	RMSE	MAE	MAPE(%)	R2	CA(%)
BiGRU	1.9893	1.5013	14.5343	0.3774	62.78
BiGL	1.9420	1.4939	15.1573	0.4066	64.27
VMD-BiGRU	0.6672	0.4724	4.5314	0.9299	89.95
SSA-VMD-BiGRU	0.4328	0.3244	3.2639	0.9705	95.16
SVBiGL	0.3897	0.2985	3.0029	0.9761	97.15

4   Conclusion

In this study, SSA optimized the decomposition parameters of VMD to ensure accurate prediction results. The composite network BiGL that integrates BiGRU and BiLSTM further improves prediction accuracy.

The proposed method predicts blood glucose levels 30, 45, and 60 min in advance, which is superior to other methods. Future research should explore the impact of additional variables such as age and insulin on prediction accuracy, and use data fusion techniques to improve model performance. Integrating the influencing variables into the proposed model provides a promising avenue for further research.

Acknowledgments. This study was funded by Fund Project: Frontier Cross disciplinary Project of Central South University (2023QYJC041).

References

1. Yang, T., Yu, X., Ma, N., et al.: An autonomous channel deep learning framework for blood glucose prediction. Appl. Soft Comput. **120**, 108636 (2022)
2. Aliberti, A., et al.: A multi-patient data-driven approach to blood glucose prediction. IEEE Access **7**, 69311–69325 (2019)
3. Cao, Q., Zhang, H., Zhu, F., et al.: Multi-step-ahead flood forecasting using an improved BiLSTM-S2S model. J. Flood Risk Manag. **15**(4), e12827 (2022)
4. Mhaskar, H.N., Pereverzyev, S.V., van der Walt, M.D.: A deep learning approach to diabetic blood glucose prediction. Frontiers Appl. Math. Statist. **3**, 1–11 (2017)
5. Zhu, T., Li, K., Herrero, P., et al.: Personalized blood glucose prediction for type 1 diabetes using evidential deep learning and meta-learning. IEEE Trans. Biomed. Eng. **70**(1), 193–204 (2022)
6. Sun, Q., Jankovic M.V., Bally, L., et al.: Predicting blood glucose with an LSTM and Bi-LSTM based deep neural network. In: 2018 14th Symposium on Neural Networks and Applications (NEUREL), pp. 1–5. IEEE (2018)
7. Cho, K., Van Merriënboer, B., Gulcehre, C., Bahdanau, D., Bougares, F., Schwenk, H. and Bengio, Y.: Learning phrase representations using RNN encoder-decoder for statistical machine translation. arXiv preprint arXiv:1406.1078 (2014)
8. Dudukcu, H.V., Taskiran, M., Yildirim, T.: Blood glucose prediction with deep neural networks using weighted decision level fusion. Biocybernetics Biomed. Eng. **41**(3), 1208–1223 (2021)
9. Xue, J., Shen, B.: A novel swarm intelligence optimization approach: sparrow search algorithm. SystSciControl Eng. **8**(1), 22–34 (2020)
10. Wang, W., Tong, M., Yu, M.: Blood glucose prediction with VMD and LSTM optimized by improved particle swarm optimization. IEEE Access **8**, 217908–217916 (2020)
11. Gao, P., Lei, Y.: Prediction of blood glucose concentration based on CEEMD and improved particle swarm optimization LSSVM. Crit. Rev™. Biomed. Eng. **49**(2) (2021)
12. Cai, L., Ge, W., Liu, T.: Research of GPCEMBP glucose prediction algorithm based on continuous glucose monitoring. J. Sens. **2023**(1), 2579720 (2023)
13. Li, K., Daniels, J., Liu, C., et al.: Convolutional recurrent neural networks for glucose prediction. IEEE J. Biomed. Health Inf. **24**(2), 603−613(2019)
14. Nemat, H., Khadem, H., Eissa, M.R., et al.: Blood glucose level prediction: advanced deep-ensemble learning approach. IEEE J. Biomed. Health Inform. **26**(6), 2758–2769 (2022)
15. D'Antoni, F., Petrosino, L., Sgarro, F., et al.: Prediction of glucose concentration in children with type 1 diabetes using neural networks: an edge computing application. Bioengineering **9**(5), 183 (2022)
16. Shuvo, M.M.H., Islam, S.K.: Deep multitask learning by stacked long short-term memory for predicting personalized blood glucose concentration. IEEE J. Biomed. Health Inform. **27**(3), 1612–1623 (2023)

SLGNNCT: Synthetic Lethality Prediction Based on Knowledge Graph for Different Cancers Types

Jingru Chen[1], Jianyong Pan[2], Yan Zhu[1], and Junyi Li[1,3(✉)]

[1] School of Computer Science and Technology, Harbin Institute of Technology (Shenzhen), Shenzhen 518055, Guangdong, China
lijunyi@hit.edu.cn

[2] Shenzhen Third People's Hospital, The Second Affiliated Hospital, Southern University of Science and Technology, Shenzhen 518000, Guangdong, China

[3] Guangdong Provincial Key Laboratory of Novel Security Intelligence Technologies, Harbin Institute of Technology (Shenzhen), Shenzhen 518055, Guangdong, China

Abstract. Exploiting synthetic lethality could help researchers to expand the arsenal of potential cancer drug targets greatly and using computational methods to predict cancer synthetic lethal genes has become a highly regarded area of research. Although numerous methods are currently dedicated to the prediction of SL gene pairs, these methods generally adopt a hybrid prediction strategy that fails to adequately account for the heterogeneity of cancers. We propose a model named SLGNNCT, which is based on knowledge graph, to predict SL interactions categorized by cancer types. The model categorizes each SL gene pair by the types of human cancers it may cause, and divides the SL database into eight subsets for prediction separately. The main body of the SLGNNCT contains three parts: first, we consider the combinations of relationships in the gene-related knowledge graph as the key factors affecting SL interactions and model them; subsequently, we generate an initial embedding representation for genes through an explicit message aggregation mechanism; finally, we construct an SL graph based on known SL gene pairs; and finally, we construct an SL graph based on known SL genes. And model the key factors; subsequently, through the explicit message aggregation mechanism in the knowledge graph, we generate initial embedding representations for the genes; finally, we construct an SL graph based on known SL gene pairs and utilize a factor-based message aggregation approach, which leads to the derivation of the final gene embedding representations. By comparing the results of our model with those of the ELISL model for single-cancer prediction, SLGNNCT performs much better on three cancer predictions and is surprisingly prominent in colon cancer (COAD) prediction.

Keywords: Synthetic lethality · Graph neural network · Knowledge graph · Cancer type association prediction · Link prediction

1 Introduction

Synthetic lethality (SL) is a genetic principle that characterizes the occurrence where simultaneous inactivation of two or more genes leads to cell death, whereas the individual inactivation of any of these genes alone does not exert a fatal impact on the cell [1]. Currently, identification of SL gene pairs typically relies on two primary approaches: experimental screening and computational prediction. High-throughput wet-lab screening techniques, such as RNA interference (RNAi) [2] or clustered regularly interspaced short palindromes (CRISPR) [3], are commonly utilized in experimental screening. However, due to the limitations of laboratory screening in terms of high cost, relatively long study time, and susceptibility to off-target effects [4], researchers are focusing on developing computational methods that can effectively complement experimental screening approaches.

Computational methods used to predict SL interactions can be broadly categorized into two main groups: those based on domain knowledge and those based on machine learning [5]. Domain knowledge-based methods predict SL gene pairs using known biological knowledge or networks, relying on researchers' understanding of relevant domain knowledge. As an instance, DAISY [6] provides a method that employs hypothesis testing and data mining to identify SL interactions in the cancer genome based on co-expression rather than co-mutation of gene pairs. A synthetic lethal prediction model was developed by Jacunski et al. [7], utilizing connectivity homology. Sinha et al. [8] introduced the MiSL approach, aiming to identify mutation-specific synthetic lethal partners for specific cancer types. These methods, due to their reliance on existing knowledge, don't fully utilize valuable insights from known SL gene pairs, making them unsuitable for predicting novel pairs.

Machine learning-based approaches, unlike domain knowledge-driven methods, excel in integrating diverse data sources and conducting gene feature learning through parameter optimization, thereby enhancing information comprehensiveness for predicting synthetic lethality. These methods can be classified into two main types: traditional machine learning techniques and deep learning approaches. Liu et al. [9] developed SL2MF, a matrix decomposition model, to obtain gene features from dissimilarity matrices of various genes, predicting potential synthetic lethal interactions. However, machine learning techniques necessitate manual intervention for feature engineering and selection [10]. Currently, researchers have limited understanding of synthetic lethality's mechanism, facing challenges in constructing suitable gene features for prediction. Noise introduced during feature engineering can also impact model accuracy.

Deep network architectures, unlike classical methods, excel in capturing intricate, nonlinear relationships between inputs and outputs, effectively discerning complex data patterns. Graph neural networks (GNN) have gained widespread adoption for SL prediction tasks. Three main types of graph neural networks utilized in synthetic lethality are graph convolutional networks [11] (GCN), graph attention networks [12] (GAT), and graph autoencoders [13] (GAE). DDGCN [14] pioneered applying GNN to SL prediction, using a strategy of GCN with a double dropout mechanism to address data sparsity and prevent overfitting effectively. Introducing the KG4SL model [15], Wang et al. devised a novel approach based on knowledge graphs (KG), wherein a graph neural network is integrated. This model samples subgraphs surrounding individual genes within the KG, facilitating the learning of gene representations from these gene-centric subgraphs. Zhu et al. [16] introduced the SLGNN model, which aims to capture gene preferences across various relationships within the knowledge graph. Based on these preferences, they designed a GNN model to effectively learn gene representations.

Given the diversity of cancers, each with unique pathogenesis, molecular features, and therapeutic responses, accurate SL classification significantly influences the development and screening of tailored drugs for specific cancer types. The ELISL [17] uses multiple features such as gene expression data, mutation data, and clinical data to characterize the interactions between gene pairs for different types of cancers, and also uses a random forest (RF) model to train the SL prediction models for individual cancer types. Although ELISL classifies cancer genes into eight categories for prediction, its method still relies on a large amount of a priori systems biology knowledge and is based on traditional machine learning methods for prediction.

Inspired by the ELISL study, we present SLGNNCT, an enhanced approach based on SLGNN for predicting synthetic lethality. By categorizing and predicting cancer types associated with synthetic lethal interactions, this method sheds light on biological mechanisms, aiding drug development. Our model comprises four phases: initially categorizing SL pairs and KGs for specific cancer types, analyzing gene relationships in knowledge graphs, aggregating gene information, and constructing the SL graph to refine gene embeddings using a factor-based mechanism for improved accuracy and reliability.

2 Dataset

We use datasets from SynLethDB [1] and ELISL [17]. Specific information on the database can be found in the supplementary material.

Classification of SL Gene Pairs By Cancer Type. The ELISL dataset provides a dataset containing SL gene pairs related to eight types of cancers, namely breast cancer (BRCA), cervical cancer (CESC), colon cancer (COAD), Kidney renal clear cell carcinoma (KIRC), leukemia (LAML), lung cancer (LUAD), ovarian cancer (OV), and skin cancer (SKCM), in which each of the involved genes has a unique and fixed gene name. Due to the different data collection methods of the two databases we selected, some of the genes in SynLethDB do not exist in ELISL, so it is temporarily not possible to categorize the SL pairs containing these genes by the types of cancers they may cause. Therefore, to guarantee the precision of predicting each cancer type, this part of the genes and SL gene pairs were excluded during data processing. We also divided the SL gene pair dataset into 8 subsets based on cancer types. Table 1 presents the data regarding the count of SL gene pairs within each subset post-classification processing.

Division of Knowledge Graph Subgraphs By Categorized SL Gene Pairs. After performing the delineation of the SL gene pairs, the types of genes associated with each cancer type are not identical, and there exist certain genes that are only associated with specific cancers or multiple cancers, rather than all types of cancers. This indicates that certain genes within the subset of SL gene pairs for the eight cancer types may not be present in SynLethKG. Failure to address this issue could potentially compromise the overall performance of the model, as it may encounter difficulties in generating appropriate representations for these genes. Therefore, to ensure the validity of the experiment, we excluded these genes along with their associated relationships during the data processing phase. Table 2 displays the count of relationships within the subgraphs of the knowledge graphs corresponding to each of the eight cancer types.

Table 1. Number of SL pairs associated with 8 cancers.

Types of Cancer	Number of SL pairs
BRCA	1312
CESC	145
COAD	1566
KIRC	60
LAML	1147
LUAD	529
OV	255
SKCM	95
Total	5109

Table 2. Number of KG relationships associated with 8 cancers.

Types of Cancer	Number of KG relationships
BRCA	169957
CESC	73938
COAD	249218
KIRC	40169
LAML	96190
LUAD	136242
OV	34375
SKCM	54131
Total	468619

In order to map the entities and relations in the knowledge graph from the semantic level to the real vector space, we used the word vector approach to generate a unique embedding representation for each entity and relation. In the initialization phase, we randomly assigned values to these embeddings using a standard normal distribution to obtain the initial embedding representations of entities and relations in the knowledge graph.

Obtaining Synthetic Lethal Negative Samples. In the SynLethDB database, there are some non-synthetic lethal pairs, also known as negative SL gene pairs, although their quantity is significantly lower compared to the known synthetic lethal gene pairs. To achieve a balanced dataset comprising both positive and negative samples, we adopted the approach utilized in KG4SL [15] to randomly generate unknown pairs as negative counterparts, thereby ensuring an equal number of positive and negative SL pairs.

3   Method

The objective of the prediction model is to generate gene embeddings automatically via the knowledge graph. Subsequently, the model conducts link prediction within the SL graph to uncover potential synthetic lethal pairs. The model is mainly composed of two important graph convolutional networks: the relationship-aware knowledge graph convolutional neural network and the factor-based awareness graph convolutional network, and Fig. 1 illustrates the structure of the model. First, the hypothetical synthetic lethal correlates are modeled; second, an explicit message aggregation process is employed to generate the knowledge graph-level representation of gene entity embeddings. The final embeddings of genes are obtained through a message aggregation process that incorporates factor-awareness.

The data structures used by SLGNNCT are described in the supplementary material.

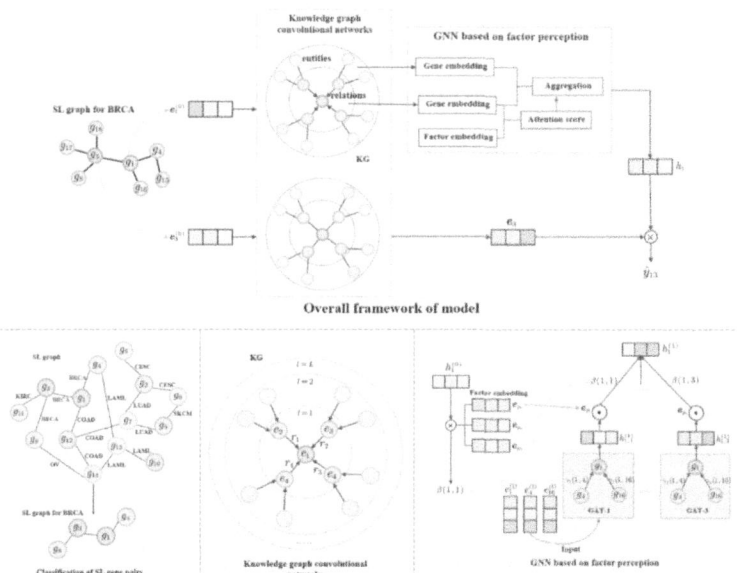

Fig. 1. Overall framework of the model SLGNNCT. (High-resolution figure is available at https://github.com/Lydia0228/SLGNNCT/tree/main/figures)

3.1 Knowledge Graph Level Gene Embedding Generation

In the knowledge graph level embedding stage of obtaining genes, we choose to use knowledge graph convolutional networks (KGCN) based on message passing mechanism to automatically generate embeddings of entities in the knowledge graph, which are then integrated to obtain gene embeddings. This method circumvents the need for manual gene feature design by leveraging the abundant information contained within the knowledge graph to generate these features.

In KGCN, the relationships of the knowledge graph play a key role, and they enable different entities to relate to each other. It is worth noting that the same entities may be connected to other entities through diverse relationships, and these different relationships correspond to unique biological processes, respectively. Therefore, a clear distinction between the various types of relationships is particularly important in the message delivery process. The process of generating knowledge graph-level gene embeddings is described in detail below.

Firstly, from Eq. 1, the embedding representation of entity i in the first layer of the network is obtained by aggregating the information of all its neighboring entities.

$$e_i^{(1)} = \frac{1}{|\mathcal{N}_i|} \sum_{(r,j) \in \mathcal{N}_i} e_r \odot e_j^{(0)} \tag{1}$$

where $e_i^{(1)}$ represents the embedding representation of entity i in the first layer of the network; $\mathcal{N}_i$ represents the set of neighboring entities and relationships of entity i; j represents the neighboring entities of entity i; r represents the relationship between entity i and entity j; e_r represents the embedding representation of relationship r; and $e_j^{(0)}$ represents the embedding representation of entity j in the zeroth (original) layer.

After obtaining one-layer embeddings for all entities, we set up the KGCN as a multilayer network in order to obtain messages from higher-order neighbors. In a multilayer convolutional network, the above process can be iterated many times, and each layer further aggregates the neighbor messages to obtain a higher-level embedding representation. Equation 2 is the iterative representation of the high-level entity embedding:

$$e_i^{(l+1)} = \frac{1}{|\mathcal{N}_i|} \sum_{(r,j) \in \mathcal{N}_i} e_r \odot e_j^{(l)} \tag{2}$$

where l represents the number of network layers for entity iteration.

We obtain the final gene embeddings by summing the entity embeddings at each layer (Eq. 3). Lower-level embeddings may contain more local features about the entity, while higher level embeddings may contain more global features about the entity. By summing up the information from these levels, the final embedding representation provides a more comprehensive description of the entity, including both local and global information.

$$e_i = e_i^{(0)} + e_i^{(1)} + \cdots + e_i^{(L)} \tag{3}$$

where L denotes the total amount of layers of the convolutional network.

3.2 Message Aggregation Based on Factors

After obtaining the gene embedding at the knowledge graph level using knowledge graph information aggregation, we will use it as an input to derive the gene embedding in the SL graph by introducing factors of different importance (see Supplementary Material for the modeling of synthetic lethality correlates).

Given that various factors exert differing levels of influence on the synthetic lethal interactions associated with a particular gene, we employed the attention mechanism to assess the significance of individual factors for each gene. For gene $i \in V$, its attention score is generated from Eq. 4:

$$\beta(i, p) = \frac{exp\left(e_p^T h_i^{(l)}\right)}{\sum_{p' \in P} exp\left(e_{p'}^T h_i^{(l)}\right)} \tag{4}$$

where $\beta(i, p)$ represents the attention score of gene i with respect to factor p; e_p represents the embedded representation of factor p; and $h_i^{(l)}$ represents the embedded representation of gene i in layer l.

Concurrently, prior to factor-based message aggregation, it's imperative that gene i generates distinct embeddings for each factor. To achieve this, we employ graph attention networks (GAT) to dynamically produce embedding representations tailored to different genes. We input the gene embeddings from the layer $l+1$ of the KGCN into $|P|$ different GATs (hereafter referred to as GAT-P), yielding an embedding representation $h_{i,p}^{(l+1)}$ corresponding to the layer $l + 1$ (Eq. 5).

$$h_{i,p}^{(l+1)} = \sum_{j \in \mathcal{N}_i} \gamma_p(i, j) e_j^{(l+1)} \tag{5}$$

where $h_{i,p}^{(l+1)}$ represents the embedded representation generated by gene i for factor p; j denotes the neighbor genes of gene i in the SL graph; $\mathcal{N}_i$ represents the set of neighbors of gene i in the SL graph; $\gamma_p(i, j)$ represents the attention score of GAT-p; and $e_j^{(l+1)}$ represents the embedded representation of gene j in the layer $l + 1$ of KGCN.

The attention score $\gamma_p(i, j)$ of GAT-p can be calculated from Eq. 6:

$$\gamma_p(i, j) = \frac{exp\left(\sigma\left(a_p^T\left[W_p e_i^{(l+1)} \| W_p e_j^{(l+1)}\right]\right)\right)}{\sum_{k \in \mathcal{N}_i} exp\left(\sigma\left(a_p^T\left[W_p e_i^{(l+1)} \| W_p e_k^{(l+1)}\right]\right)\right)} \tag{6}$$

where σ denotes the function LeakyReLU for activation; a_p represents the weight vector of factor p; and W_p represents the projection matrix associated with factor p that projects the embedding representation into a shared space.

Finally, we further summarized these gene embeddings generated by GAT-p based on the attention scores of the genes about the factor (Eq. 7).

$$h_i^{(l+1)} = \sum_{p \in P} \beta(i, p) e_p \odot h_{i,p}^{(l+1)} \tag{7}$$

where $h_i^{(l+1)}$ represents the embedded representation of gene i in the layer $l + 1$.

Adopting a similar approach to factorial modeling, we endeavored to ensure that the gene embeddings generated by distinct GATs remain mutually independent to convey a broader range of information. This is depicted in Eq. 8, which illustrates the regularization function.

$$\mathcal{L}_{GAT} = \sum_{p,p' \in P, p \neq p'} dCor\left(h_p^{avg}, h_{p'}^{avg}\right) \tag{8}$$

where h_p^{avg} denotes the mean embedding across all genes obtained through GAT-p; $h_{p'}^{avg}$ denotes the mean embedding across all genes obtained through GAT-p'; $dCor\left(h_p^{avg}, h_{p'}^{avg}\right)$ represents the distance correlation measure between gene embeddings h_p^{avg} and $h_{p'}^{avg}$; and $\mathcal{L}_{GAT}$ represents the regularized loss function.

Similar to knowledge graph convolutional networks, we derive the ultimate gene embeddings by summing each layer of gene embeddings (Eq. 9).

$$h_i = h_i^{(0)} + h_i^{(1)} + \cdots + h_i^{(L)} \tag{9}$$

where L represents the total number of layers of the convolutional network.

3.3 Calculation of Synthetic Lethal Interaction Probabilities

Link prediction is an important network analysis task aimed at determining whether a connection exists between two nodes in a network. We predicted the probability of the existence of SL interactions between i and j by performing dot product on the gene embeddings h_i and e_j, and mapping the result to the range [0, 1] via a Sigmoid function (Eq. 10).

$$\hat{y}_{ij} = \sigma\left((h_i)^T e_j\right) \tag{10}$$

where $\hat{y}_{ij}$ represents the likelihood of an SL interaction occurring between genes i and j; h_i represents the final gene embedding of gene i; and e_j represents the gene embedding of gene j at the knowledge map level.

We assessed the effectiveness of the reconstruction of the synthetic lethal network by employing a binary cross-entropy loss function to calculate the primary loss of the model (Eq. 11):

$$\mathcal{L}_{BCE} = -\frac{1}{|\mathcal{E}_t|} \sum_{i,j \in \mathcal{E}_t} y_{ij} \cdot \ln\left(\hat{y}_{ij}\right) + \left(1 - y_{ij}\right) \cdot \ln\left(1 - \hat{y}_{ij}\right) \tag{11}$$

where $\mathcal{E}_t$ represents the set of SL genes in the training set.

Our model incorporates four types of losses: independence loss for factor modeling $\mathcal{L}_{IND}$, independence loss for graph attention networks $\mathcal{L}_{GAT}$, regularization loss to prevent overfitting $\mathcal{L}_{NORM}$, and reconstruction loss for synthetic lethal networks $\mathcal{L}_{BCE}$. We introduced two weighting parameters λ_1 and λ_2 to modulate the importance of these loss functions as a way to obtain the overall loss of the model (Eq. 12):

$$\mathcal{L} = \mathcal{L}_{BCE} + \lambda_1(\mathcal{L}_{IND} + \mathcal{L}_{GAT}) + \lambda_2 \mathcal{L}_{NORM} \tag{12}$$

4   Experiment

4.1   Baselines

We contrast the experimental outcomes of the SL prediction model with those obtained by ELISL [17]. ELISL leverages a diverse array of features encompassing gene expression, mutation, and clinical data across various cancer types to delineate the interactions between gene pairs. Additionally, it employs a Random Forest model to train the SL prediction model tailored to individual cancer types.

4.2   Model Evaluation

SLGNNCT uses Python 3.9.18 and PyTorch 1.10.0 as the main tools. For the division of the dataset, we randomly divided the data into training set, validation set and test set according to the ratio of 8:1:1. Meanwhile, we used a 5-fold cross-validation method for each cancer type and selected the average of the five results as the final result.

Table 3. Comparison of AUC and AUPR metrics between the two models

Types of Cancer	SLGNNCT (AUC)	ELISL (AUC)	SLGNNCT (AUPR)	ELISL (AUPR)
BRCA	0.884	**0.948**	0.882	**0.951**
CESC	0.669	**0.837**	0.654	**0.847**
COAD	**0.951**	0.699	**0.964**	0.682
KIRC	0.667	**0.746**	**0.764**	0.733
LAML	**0.693**	0.667	**0.682**	0.653
LUAD	0.848	**0.874**	**0.901**	0.889
OV	0.755	**0.908**	0.761	**0.921**
SKCM	0.787	**0.854**	0.767	**0.837**
Total	**0.892**	0.805	**0.918**	0.805

In order to be able to make a more accurate judgment of the model performance, we use AUC and AUPR as the evaluation indexes of the model. The hyperparameters and model parameters specified in the original paper are employed for the baseline model. The comparison of our model with ELISL is shown in Table 3.

Based on the above comparison results, it can be found that SLGNNCT outperforms ELISL in colon cancer (COAD), leukemia (LAML), lung cancer (LUAD), and overall prediction, and slightly underperforms ELISL in breast cancer (BRCA), and renal clear cell carcinoma (KIRC). Overall, the prediction of SLGNNCT on cancer types with more data volume effect will be more excellent. Given that deep learning models typically necessitate a substantial volume of data for effective training to learn features and patterns with strong generalization ability, overfitting, underfitting, etc. may occur on small

Table 4. Results of ablation experiments

Types of Cancer	Model	AUC	AUPR	F1-score
BRCA	SLGNNCT	0.884	0.882	0.761
	SLGNNCT-D	0.758	0.760	0.665
CESC	SLGNNCT	0.669	0.654	0.769
	SLGNNCT-D	0.566	0.584	0.515
COAD	SLGNNCT	0.951	0.964	0.865
	SLGNNCT-D	0.884	0.908	0.816
KIRC	SLGNNCT	0.667	0.764	0.579
	SLGNNCT-D	0.533	0.614	0.408
LAML	SLGNNCT	0.693	0.682	0.532
	SLGNNCT-D	0.562	0.574	0.516
LUAD	SLGNNCT	0.848	0.901	0.760
	SLGNNCT-D	0.732	0.769	0.641
OV	SLGNNCT	0.755	0.761	0.720
	SLGNNCT-D	0.744	0.729	0.681
SKCM	SLGNNCT	0.787	0.767	0.670
	SLGNNCT-D	0.706	0.617	0.517
Total	SLGNNCT	0.892	0.918	0.823
	SLGNNCT-D	0.737	0.757	0.689

datasets. Cervical cancer (CESC) and skin cancer (SKCM) have significantly worse prediction results than ELISL due to fewer relevant SL pairs.

However, surprisingly, SLGNNCT performed exceptionally well in the prediction of colon cancer (COAD), with AUC and AUPR metrics improving by 40.2% and 44.5%, respectively, compared to ELISL, suggesting that the use of SLGNNCT could provide a strong support for the advancement of tailored pharmaceuticals for this type of cancer in the future.

4.3 Results and Analysis of Ablation Experiments

From the overall performance analysis, SLGNNCT showed a significant performance improvement over ELISL in the prediction of synthetic lethality for multiple cancers. In this section, we conduct experiments by removing a portion of the knowledge graph subgraph division component in the data processing part of the model to evaluate its contribution to the overall performance of the model. For this component, we design a corresponding sub-model SLGNNCT-D. The results of the ablation experiments are shown in Table 4.

The above experimental results show the difference in performance between the model and the sub-model, and the results show that the removal of the sub-graph delineation of the knowledge graph resulted in a significant decrease in the performance of SLGNNCT-D, which demonstrates the effectiveness of removing redundant relationships from the knowledge graph in generating appropriate representations of the genes, and indirectly demonstrates the contribution of the incorporation of the knowledge graph to the SLGNNCT.

5 Conclusion

We present the SLGNNCT synthetic lethality prediction model, leveraging factor-aware network representations. Recognizing the diverse pathogenesis, molecular characteristics, and treatment responses across various cancer types, our model categorizes predictions based on SL-associated cancer types. This approach aims to facilitate the development and screening of novel drugs tailored to specific cancer types. Initially, we analyze various relationships within the KG to model the factors influencing SL interactions, employing weighted summation and ensuring their independence by minimizing the distance correlation between them. Subsequently, we employ a GNN-based message aggregation mechanism to effectively derive the initial embedding representations of genes within the KG. Finally, leveraging a factor-based message aggregation approach, we optimize embeddings further using the SL graph and introduced GAT to enhance model performance.

We also evaluate this model in comparison with other model for SL prediction, and the outcomes of the experiments showed that our proposed model has superior results in the prediction of colon cancer (COAD), leukemia (LAML), lung cancer (LUAD), and overall, and showed unexpectedly high performance in the prediction of synthetic lethal gene pairs for colon cancer, which implies that the use of this model can provide a powerful support for the future development of such cancers for the This implies that the use of this model can provide strong support for the prospective advancement of precision medicines for this type of cancer.

Acknowledgements. This work was supported by the grants from the National Key R&D Program of China (2021YFA0910700), Shenzhen Science and Technology Program (JCYJ20200109113201726), Guangdong Basic and Applied Basic Research Foundation (2021A1515012461 and 2021A1515220115), Guangdong Provincial Key Laboratory of Novel Security Intelligence Technologies (2022B1212010005).

References

1. Guo, J., Liu, H., Zheng, J.: SynLethDB: synthetic lethality database toward discovery of selective and sensitive anticancer drug targets. Nucleic Acids Res. **44**(D1), D1011–D1017 (2016)
2. Luo, J., Emanuele, J., Li, D., et al.: A genome-wide RNAi screen identifies multiple synthetic lethal interactions with the Ras oncogene. Cell **137**(5), 835–848 (2009)

3. Du, D., Roguev, A., Gordon, E., et al.: Genetic interaction mapping in mammalian cells using CRISPR interference. Nat. Methods **14**(6), 577–580 (2017)
4. Topatana, W., Juengpanich, S., Li, S., et al.: Advances in synthetic lethality for cancer therapy: cellular mechanism and clinical translation. J. Hematol. Oncol. **13**, 1–22 (2020)
5. Long, Y., Wu, M., Liu, Y., et al.: Graph contextualized attention network for predicting synthetic lethality in human cancers. Bioinformatics **37**(16), 2432–2440 (2021)
6. Jerby-Arnon, L., Pfetzer, N., Waldman, Y., et al.: Predicting cancer-specific vulnerability via data-driven detection of synthetic lethality. Cell **158**(5), 1199–1209 (2014)
7. Jacunski, A., Dixon, S.J., Tatonetti, P.: Connectivity homology enables inter-species network models of synthetic lethality. PLoS Comput. Biol. **11**(10), e1004506 (2015)
8. Sinha, S., Thomas, D., Chan, S., et al.: Systematic discovery of mutation-specific synthetic lethals by mining pan-cancer human primary tumor data. Nat. Commun. **8**(1), 15580 (2017)
9. Liu, Y., Wu, M., Liu, C., et al.: SL 2 MF: Predicting synthetic lethality in human cancers via logistic matrix factorization. IEEE/ACM Trans. Comput. Biol. Bioinf. **17**(3), 748–757 (2019)
10. Zhang, K., Wu, M., Liu, Y., et al.: KR4SL: knowledge graph reasoning for explainable prediction of synthetic lethality. Bioinformatics **39**(Supplement_1), i158–i167 (2023)
11. Kipf, T.N., Welling, M.: Semi-supervised classification with graph convolutional networks. arXiv preprint arXiv:1609.02907 (2016)
12. Velickovic, P., Cucurull, G., Casanova, A., et al.: Graph attention networks. stat **1050**(20), 10–48550 (2017)
13. Kipf, N., Welling, M.: Variational graph auto-encoders. arXiv preprint arXiv:1611.07308 (2016)
14. Cai, R., Chen, X., Fang, Y., et al.: Dual-dropout graph convolutional network for predicting synthetic lethality in human cancers. Bioinformatics **36**(16), 4458–4465 (2020)
15. Wang, S., Xu, F., Li, Y., et al.: KG4SL: knowledge graph neural network for synthetic lethality prediction in human cancers. Bioinformatics **37**(Supplement_1), i418–i425 (2021)
16. Zhu, Y., Zhou, Y., Liu, Y., et al.: SLGNN: synthetic lethality prediction in human cancers based on factor-aware knowledge graph neural network. Bioinformatics **39**(2), btad015 (2023)
17. Tepeli, I., Seale, C., Gonçalves, P.: ELISL: early–late integrated synthetic lethality prediction in cancer. Bioinformatics **40**(1), btad764 (2024)

TransPBMIL: Transformer-Based Weakly Supervised Prognostic Prediction in Ovarian Cancer with Pseudo-Bag Strategy

Yongxin Mao[1,2], Ziwei Hu[1,2], Xinlin Zhang[1,2(✉)], and Tong Tong[1,2,3(✉)]

[1] College of Physics and Information Engineering, Fuzhou University, Fuzhou 350108, Fujian, China
xinlin@fzu.edu.cn, ttraveltong@gmail.com
[2] Fujian Key Lab of Medical Instrumentation and Pharmaceutical Technology, Fuzhou University, Fuzhou 350108, Fujian, China
[3] Imperial Vision Technology, Fuzhou 350002, China

Abstract. Ovarian cancer presents a notable health concern characterized by its unfavorable prognosis and elevated mortality rates in the female population. Accurate prognostic assessment is essential for tailoring treatment strategies and improving patient outcomes. Analysis of histopathological whole-slide images is the gold standard for pathological diagnosis, which contains rich phenotypic and molecular information. Multiple instance learning methods have been the dominant approach for processing megapixel whole slide images. However, the methods adopt the one image as a bag strategy, which will contain many noisy tiles leading to model overfitting during training. To mitigate the above situation, we propose a transformer-based multi-instance learning framework with a pseudo-bag strategy (TransPBMIL) for predicting overall survival within 3 years of ovarian cancer patients using pathological images. Extensive studies on multiple cancer prognostic datasets demonstrate the superiority of TransPBMIL.

Keywords: Deep Learning · Prognosis Prediction · Tissue Pathology Analysis · Ovarian Cancer

1 Introduction

Cancer is one of the leading causes of death globally [1], with ovarian cancer ranking among the most prevalent gynecological cancers, following cervical and uterine cancers as the third most common [2]. Ovarian cancer carries the poorest prognosis and highest mortality rate [3]. Due to its asymptomatic nature and inadequate screening, ovarian cancer often remains undetected until advanced stages, contributing to dismal outcomes and earning it the moniker 'silent killer' [4, 5]. Despite ovarian cancer's lower incidence compared to breast cancer, its mortality rate is estimated to be three times higher [4]. The 5-year survival rate post-diagnosis for ovarian cancer stands at a mere 48%. Over the past few decades, ovarian cancer survival rates have remained largely unchanged, even in resource-rich countries [6]. Accurate prediction of overall survival (OS) can provide reliable guidance for treatment planning and clinical practice.

D.-S. Huang et al. (Eds.): ICIC 2024, LNBI 14881, pp. 171–180, 2024.
https://doi.org/10.1007/978-981-97-5689-6_15

Pathological examination remains the gold standard for diagnosing ovarian cancer, offering opportunities for biomarker detection and prognostic stratification [7, 8]. However, it is labor-intensive, cumbersome and time-consuming, and has a negative impact on the accuracy of diagnosis [9]. With the advent of digital pathology, Whole Slide Imaging (WSI) provides a means of obtaining high-resolution images, aiding pathologists in identifying atypical microlesions. Yang et al. introduced a graph-based deep learning model to develop the Ovarian Cancer Digital Pathology Index (OCDPI) to predict prognosis and response to adjuvant therapy. Experimental results demonstrate that patients with low OCDPI have better survival benefits and lower relapse rates [10]. Yokomizo et al. used machine learning to learn image features using histopathological images as input, and finally predicted the overall survival rate of ovarian clear cell carcinoma patients with an accuracy of 0.9 [11]. Wei et al. proposed the MultiDeepCox-SC model, and the prognostic accuracy of MultiDeepCox-SC (C-index = 0.744) exceeded results based on histopathological images (C-index = 0.660) [12].

In this study, we included Hematoxylin and Eosin (H&E) stained histopathological images from 105 ovarian cancer patients with known OS information from The Cancer Genome Atlas (TCGA) public dataset. We developed a deep learning framework TransPBMIL based on pathological images to predict the 3-year OS of ovarian cancer patients accurately and directly by analyzing pathological images. Additionally, to assess the robustness of our framework in OS prediction tasks, we also included pathological images of 411 gastric cancer patients from TCGA as an additional dataset. The results indicate that TransPBMIL also demonstrates outstanding performance on external datasets.

2 Materials and Methods

2.1 Participants and Dataset Generation

This study used two histopathological image data sets: ovarian cancer histopathological images and gastric cancer histopathological images from TCGA (retrieved from https://portal.gdc.cancer.gov/). Initially, a total of 108 whole-slide tissue histopathological images were downloaded from 107 individual patients. After excluding 2 patients with missing survival data, 106 complete slides from 105 independent patients were analyzed.

In order to test the robustness of our framework on the OS prediction tasks, we also included 442 H&E-stained gastric cancer tissue histopathological images of 416 gastric cancer patients from TCGA as an additional dataset. After excluding 4 unlabeled slices and 1 slice of poor quality containing a large amount of handwriting, we finally included 437 gastric cancer tissue pathology images from 411 patients in the study. Table 1 lists the details of the two datasets.

2.2 TransPBMIL Framework

Initially, scanned H&E-stained pathology images are used to automatically identify tissue areas, and the pre-trained Resnet50 residual network [13] is used to extract image

Table 1. Training dataset for algorithm development and external dataset for validating model robustness

Dataset		Survived within 3 years	Death within 3 years
TCGA-OV			
	Case	56(53.3%)	49(46.7%)
	Slide	56(52.8%)	50(47.2%)
TCGA-STAD			
	Case	252(61.3%)	159(38.7%)
	Slide	272(62.2%)	165(37.8%)

features in the entire tissue area. Based on these features, TransPBMIL trains the network model by fusing transfer learning and weakly supervised learning strategies, fully exploring the morphological and spatial information among different instances. Finally, the information is aggregated to the classification layer to calculate the final score of the entire pathological image.

Weakly Supervised Training Methods and Transfer Learning to Extract Features. In this approach, each H&E-stained image is treated as a collection of small regions, and information is learned from the entire image without the need for manual annotation of regions of interest. All image patches cropped from pathological images are treated as instances, and their collection is used as a bag of pathological images. In the model establishment process, we only need coarse-grained labels of the bags, which are WSI-level labels.

The CLAM library [14] was used to automatically perform tissue region segmentation within the entire H&E-stained image. At a magnification of 40x, the segmented tissues were cropped into 256×256 image blocks. The specific implementation process is illustrated in Fig. 1A. The ResNet50 model pretrained on ImageNet is used as the feature extractor in this framework. After the third residual block of ResNet50, adaptive mean spatial pooling technique was applied to extract 1024-dimensional features of each patch. For each WSI at a magnification of 40x, we generated a low-dimensional feature set {N x 1024}, where N represents the number of patches contained in the WSI.

Furthermore, to minimize overfitting and improve the generalization ability of the model, we employed data augmentation techniques during the feature extraction process to increase the diversity of the dataset. This included horizontal flipping, vertical flipping, 45-degree rotation, as well as adjustments to brightness, contrast, saturation, and hue in the color space transformation. These augmentations enable the model to learn image representations that are invariant to rotation and color changes.

Self-attention Module and Recurrent Criss-Cross Attention. To effectively process image features comprehensively, aggregate morphological information, and enhance contextual relationships between features for improved inputs in subsequent tasks, we devised the attention module within our framework. This module draws inspiration from the Criss-Cross Attention (CCA) module [15] and Transformer layers commonly

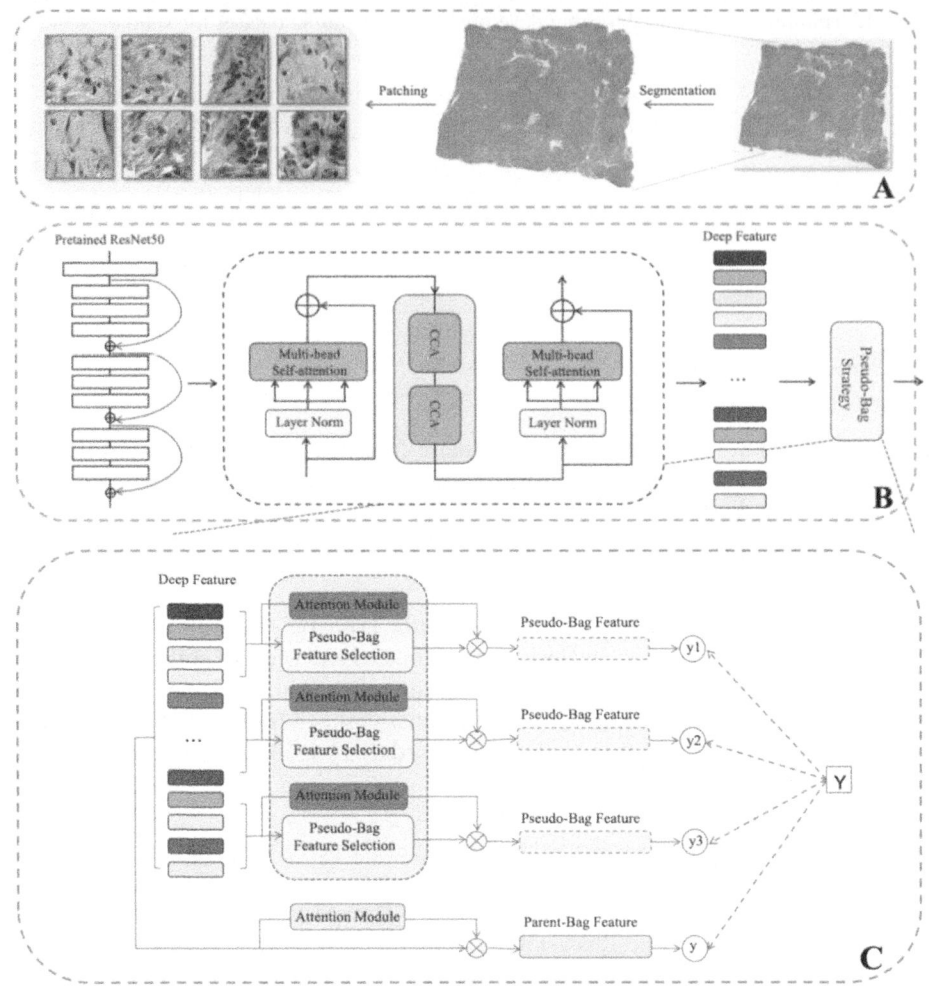

Fig. 1. Overview of the proposed TransPBMIL

employed in multi-instance learning [16]. It consists of two Transformer layers and a Recurrent Criss-Cross Attention (RCCA) module. The specific workflow for this segment is depicted in Fig. 1B.

Currently, Transformer is widely used in many vision tasks due to its powerful ability in describing the correlation between different fragments in a sequence and modeling long-range information [16]. In our network, sequences are first derived from feature embeddings of tissue region parts extracted from each WSI. The Transformer layer approximately calculates the softmax of the attention weight through the Nystrom method to solve the problem of excessive memory and time complexity of the standard self-attention mechanism under long input sequences [16].

The RCCA module serves as an extension of the cross-attention module. In the first iteration, the input feature mappings from the Transformer layer are fed into the cross-attention module, which generates new output feature mappings that aggregate long-range contextual information from each pixel in a cross-wise manner. The generated output feature mappings are fed back into the cross-attention module, and through multiple iterations, the module captures dependencies at the pixel level and gradually generates new feature mappings with dense and rich context information.

Pseudo-Bag Strategy. In order to mitigate the impact of overfitting, the concept of pseudo-bags was used in this study. We divide the instances of a bag into several smaller pseudo-bags b according to the corresponding instance selection strategy. These smaller bags are pseudo-bags, with labels identical to the original bag (parent-bag), as illustrated in Fig. 2. The pseudo-bag strategy increases the number of bags, and consequently reduces the number of instances within each bag. The instance features are weighted by attention scores. The weighted instance features can be considered as the refined representative features of the pseudo-bags, which are then inputted into the classifier for bag prediction. The parent-bag is also weighted by the attention network to generate a feature vector, which is fed into the network together with the pseudo-bags features for learning. The specific implementation process is illustrated in Fig. 1C.

The three instance selection strategies are as follows:

LPS (Larger Probability Subset) - Selects a larger probability subset: Extracts features from the top 20% of instances ranked by positive probability in the pseudo-bag.

MAS (Maximum Probability Selection) - Selects the maximum probability: Extracts features from the instances with the highest positive probability in the pseudo-bag.

AFS (Aggregate Feature Selection) - Aggregates features from all instances in the pseudo-bag.

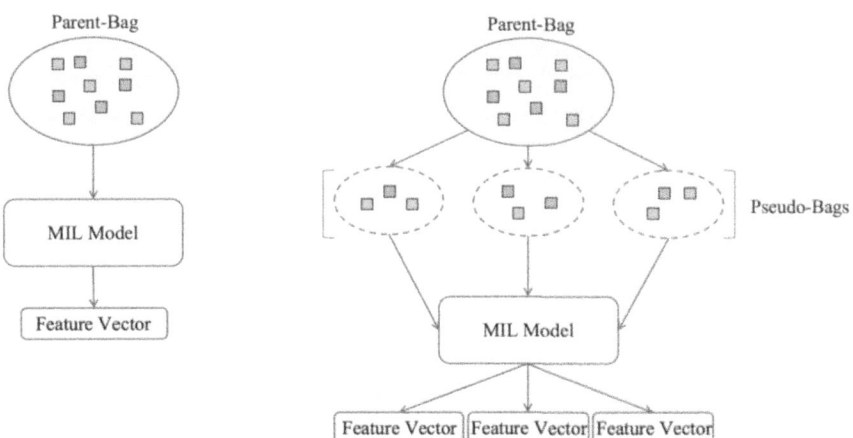

Fig. 2. Conventional MIL model and MIL model with Pseudo-bag Strategy

3 Result

3.1 Comparison with Existing Weakly Supervised Works

We present experimental results of the proposed method using three pseudo-bags and compare it with other existing weakly supervised methods. The baseline models we selected include Mean-Pooling and Max-Pooling; the classical attention-based multiple instance learning method ABMIL [17]; the non-local attention DSMIL [18], the single attention branch CLAM-SB [14], and the multi-attention branch CLAM-MB [14]; the Transformer-based multiple instance learning TransMIL [16]; and the Double-Tier Feature Distillation Multiple Instance Learning DTFD-MIL [19]. Table 2 presents the results of the 3-year OS prediction task on the TCGA-OV dataset, while Table 3 presents the results on the TCGA-STAD dataset.

In the TCGA-OV dataset, among TransPBMIL using different instance feature extraction methods, MAXS exhibited the worst performance, yet still outperformed other existing weakly supervised methods. Notably, the model outperformed all other competing methods when employing the best-performing LPS instance selection strategy, with AUC, Acc, and F1 score higher than the second-ranked method by 4.2%, 5.6%, and 3.4%, respectively.

In the TCGA-STAD dataset, some slides exhibited extensive ink stains and lacked complete tissue structures, which could potentially impact the experimental results to some extent. Among the other instance selection strategies used in the TransPBMIL model, the best-performing AUC was 2.9% higher than TransMIL, and the F1 score was 4.4% higher. Overall, our framework achieved superior results in the 3-year OS prediction task across both datasets.

Table 2. Results with the corresponding 95% confidence intervals on TCGA-OV set.

Methods	AUC	Acc	F1
Mean-Pooling	0.750 ± 0.126	0.764 ± 0.117	0.753 ± 0.120
Max-pooling	0.815 ± 0.125	0.848 ± 0.066	0.827 ± 0.082
AB-MIL	0.797 ± 0.107	0.804 ± 0.088	0.811 ± 0.081
DSMIL	0.827 ± 0.106	0.822 ± 0.089	0.830 ± 0.082
CLAM-SB	0.797 ± 0.128	0.812 ± 0.120	0.813 ± 0.104
CLAM-MB	0.825 ± 0.136	0.850 ± 0.098	0.828 ± 0.124
TransMIL	0.816 ± 0.139	0.859 ± 0.084	0.833 ± 0.114
DTFD-MIL	0.785 ± 0.113	0.700 ± 0.105	0.694 ± 0.097
TransPBMIL	0.835 ± 0.136	0.869 ± 0.105	0.855 ± 0.112
TransPBMIL(LPS)	**0.869 ± 0.101**	0.878 ± 0.092	0.864 ± 0.124
TransPBMIL(MAXS)	0.843 ± 0.108	0.859 ± 0.075	0.846 ± 0.084
TransPBMIL(AFS)	0.866 ± 0.088	**0.878 ± 0.072**	**0.869 ± 0.086**

Table 3. Results with the corresponding 95% confidence intervals on TCGA-STAD set.

Methods	AUC	Acc	F1
Mean-Pooling	0.572 ± 0.041	0.554 ± 0.041	0.562 ± 0.031
Max-pooling	0.590 ± 0.042	0.609 ± 0.056	0.528 ± 0.064
AB-MIL	0.603 ± 0.027	0.609 ± 0.049	0.579 ± 0.035
DSMIL	0.551 ± 0.063	0.563 ± 0.063	0.564 ± 0.034
CLAM-SB	0.580 ± 0.047	0.604 ± 0.037	0.502 ± 0.095
CLAM-MB	0.575 ± 0.041	0.613 ± 0.011	0.545 ± 0.044
TransMIL	0.635 ± 0.012	0.661 ± 0.018	0.567 ± 0.027
DTFD-MIL	0.578 ± 0.023	0.599 ± 0.056	0.505 ± 0.050
TransPBMIL	0.630 ± 0.040	0.609 ± 0.068	0.595 ± 0.033
TransPBMIL(LPS)	0.638 ± 0.032	**0.664 ± 0.032**	0.557 ± 0.067
TransPBMIL(MAXS)	0.663 ± 0.020	0.661 ± 0.027	0.573 ± 0.027
TransPBMIL(AFS)	**0.664 ± 0.029**	0.654 ± 0.043	**0.613 ± 0.048**

3.2 The Performance Improvement Brought by the Pseudo-Bag Strategy.

In Table 2 and Table 3, we present the results of our model applied to the TCGA-OV and TCGA-STAD datasets using different pseudo-bag strategies and without using a pseudo-bag strategy. After incorporating pseudo-bag strategies, TransPBMIL achieves significant performance improvement, breaking through the performance bottleneck. It is worth mentioning that we validated our framework across three pseudo-bag strategies, all of which outperform existing MIL methods.

Figure 3 and Fig. 4 respectively illustrate the AUC, Acc, and F1 score scores for different numbers of pseudo-bags proposed on the TCGA-OV and TCGA-STAD datasets. We can observe that TransPBMIL is more sensitive to the number of pseudo-bags in the TCGA-OV dataset compared to the TCGA-STAD dataset. One potential reason for this scenario is that when the OS is closely linked to tumor-related information, a small tumor area on the slice coupled with an excessive number of pseudo-bags may lead to insufficient allocation of tumor information, consequently diminishing the model's effectiveness. Conversely, when the tumor area on the slice is larger, the influence of the number of pseudo packets on the results might be less pronounced.

3.3 Visualization of Detection Results

To gain deeper insights into the predictive patterns of TransPBMIL, we generated WSI heatmaps to further interpret the model's explain ability (Fig. 5). These attention heatmaps visually illustrate the regions within the image that the model focuses on when making predictions, highlighting the relative importance of different areas. Importantly, these heatmaps are generated from WSIs where tumor regions are not annotated. For the high-risk group, the heatmaps reveal the model's ability to concentrate on tumor

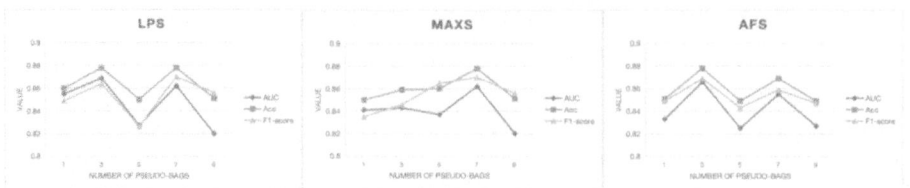

Fig. 3. The Performance of Three Feature Selection Strategies on the TCGA-OV Dataset under Different Numbers of Pseudo-bags

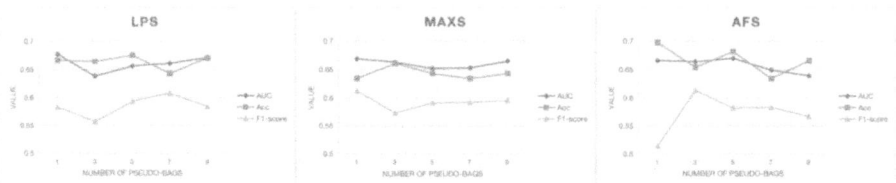

Fig. 4. The Performance of Three Feature Selection Strategies on the TCGA-STAD Dataset under Different Numbers of Pseudo-bags

regions. Despite the absence of pixel-level or patch-level annotations during training, the areas attended to by the network closely align with tumor regions. This suggests that tumor regions provide valuable cues for the network to predict the overall survival status of patients.

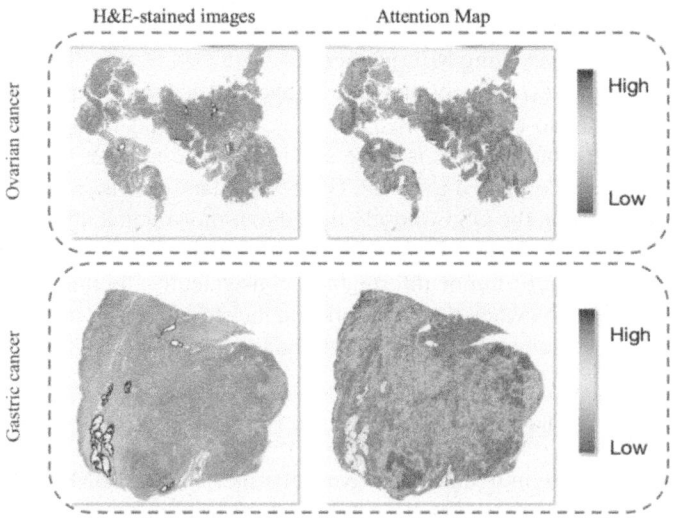

Fig. 5. Attention visualization on TCGA-OV and TCGA-STAD

4   Conclusion

The alarming mortality associated with ovarian cancer is well known. Accurate prediction of OS in ovarian cancer patients can provide reliable guidance for treatment design and clinical practice. Accurate prediction of OS in ovarian cancer patients remains a huge challenge for pathologists due to tumor heterogeneity. Using artificial intelligence to develop an effective and labor-saving method to predict the prognosis of ovarian cancer and other cancer types is novel and meaningful work. In this study, we propose a novel deep learning model named TransPBMIL for the tumor prognosis prediction task. We evaluated our model using data from the TCGA-OV ovarian cancer dataset and the TCGA-STAD gastric cancer dataset. We compared our model with other existing weakly supervised learning models and found that TransPBIL showed good performance on both datasets. We believe that through further refinement and optimization of the model, combined with the requirements of clinical practice, our research will have broad application prospects in tumor prognosis prediction tasks. This can provide important auxiliary support for medical diagnosis and treatment.

References

1. Momenimovahed, Z., Ghoncheh, M., Pakzad, R., et al.: Incidence and mortality of uterine cancer and relationship with human development Index in the world. Cukurova Med J **42**(2), 233–240 (2017)
2. Bray, F., Ferlay, J., Soerjomataram, I., et al.: Global cancer statistics 2018: GLOBOCAN estimates of incidence and mortality worldwide for 36 cancers in 185 countries. CA a cancer J. clin. **68**(6), 394–424 (2018)
3. Coburn, S.B., Bray, F., Sherman, M.E., et al.: International patterns and trends in ovarian cancer incidence, overall and by histologic subtype. Int. J. Cancer **140**(11), 2451–2460 (2017)
4. Yoneda, A., Lendorf, M.E., Couchman, J.R., Multhaupt, H.A.: Breast and ovarian cancers: a survey and possible roles for the cell surface heparan sulfate proteoglycans. J. Histochem. Cytochem. **60**(1), 9–21 (2012). https://doi.org/10.1369/0022155411428469.PMID: 22205677;PMCID:PMC3283135
5. Badgwell, D., Bast, R.C., Jr.: Early detection of ovarian cancer. Dis. Markers **23**(5–6), 397–410 (2007)
6. Guo, Y., Lu, Y., Jin, H.: Appraising the role of circulating concentrations of micro-nutrients in epithelial ovarian cancer risk: a mendelian randomization analysis. Sci. Rep. **10**(1), 7356 (2020)
7. Desbois, M., Udyavar, A.R., Ryner, L., et al.: Integrated digital pathology and transcriptome analysis identifies molecular mediators of T-cell exclusion in ovarian cancer. Nat. Commun. **11**(1), 5583 (2020)
8. Saillard, C., Schmauch, B., Laifa, O., et al.: Predicting survival after hepatocellular carcinoma resection using deep learning on histological slides. Hepatology **72**(6), 2000–2013 (2020)
9. Metter, D.M., Colgan, T.J., Leung, S.T., Timmons, C.F., Park, J.Y.: Trends in the US and Canadian pathologist workforces from 2007 to 2017. JAMA Netw. Open **2**(5), e194337V (2019). https://doi.org/10.1001/jamanetworkopen.2019.4337.PMID:31150073; PMCID:PMC6547243
10. Yang, Z., Zhang, Y., Zhuo, L., et al.: Prediction of prognosis and treatment response in ovarian cancer patients from histopathology images using graph deep learning: a multicenter retrospective study. Eur. J. Cancer **199**, 113532 (2024)

11. Yokomizo, R., Lopes, T.J.S., Takashima, N., et al.: O3c glass-class: a machine-learning frame-work for prognostic prediction of ovarian clear-cell carcinoma. Bioinform. Biol. Insights **16**, 11779322221134312 (2022)
12. Wei, T., Yuan, X., Gao, R., et al.: Survival prediction of stomach cancer using expression data and deep learning models with histopathological images. Cancer Sci. **114**(2), 690–701 (2023)
13. He, K., Zhang, X., Ren, S., et al.: Deep residual learning for image recognition. In: Proceedings of the IEEE Conference on Computer Vision and Pattern Recognition, pp. 770–778 (2016)
14. Lu, M.Y., Williamson, D.F.K., Chen, T.Y., et al.: Data-efficient and weakly supervised computational pathology on whole-slide images. Nature biomed. Eng. **5**(6), 555–570 (2021)
15. Huannlg, Z., Wang, X., Huang, L., et al.: CCNet: criss-cross attention for semantic segmentation. In: Proceedings of the IEEE/CVF International Conference on Computer Vision, pp. 603–612 (2019)
16. Shao, Z., Bian, H., Chen, Y., et al.: Transmil: transformer based correlated multiple instance learning for whole slide image classification. Adv. Neural. Inf. Process. Syst. **34**, 2136–2147 (2021)
17. Ilse, M., Tomczak, J., Welling, M.: Attention-based deep multiple instance learning. In: International Conference on Machine Learning, pp. 2127–2136. PMLR (2018)
18. Li, B., Li, Y., Eliceiri, K.W.: Dual-stream multiple instance learning network for whole slide image classification with self-supervised contrastive learning. In: Proceedings of the IEEE/CVF Conference on Computer Vision and Pattern Recognition, pp. 14318–14328 (2021)
19. Zhang, H., Meng, Y., Zhao, Y., et al.: DTFD-MIL: double-tier feature distillation multiple instance learning for histopathology whole slide image classification. In: Proceedings of the IEEE/CVF Conference on Computer Vision and Pattern Recognition, pp. 18802–18812 (2022)

Biomedical Informatics Theory and Methods

A Heterogeneous Cross Contrastive Learning Method for Drug-Target Interaction Prediction

Qi Wang, Jiachang Gu, Jiahao Zhang, Mingming Liu$^{(\boxtimes)}$, Xu Jin, and Maoqiang Xie

College of Software, Nankai University, Tianjin 300350, China
liumingming@nankai.edu.cn

Abstract. Identification of interactions between drugs and target proteins plays an important role in drug developments. In recently years, deep biological learning methods including graph neural networks, contrastive learning and random walk have demonstrated notable advantages in DTI prediction. However, most of the current contrastive learning-based models perform contrast between the original graph and the graph of diffusion or augment, which is limited by the source of available interaction information from heterogeneous graph. The relationships from DDI and PPI are not involved in contrastive learning. Therefore, we propose a heterogeneous cross contrastive learning method (HCCL-DTI) for DTI prediction. It demonstrates a novel approach for heterogeneous graph contrastive learning, which uses two modes including self-contrastive learning and cross-contrastive learning to obtain more expressive node embeddings. In the method, another view of DDI and PPI is extracted from the heterogeneous graph, which is complementary to the original view of DDI and PPI. Self-contrast is performed in the two views respectively, while cross-contrast is performed between the two views in a supervised way. After that, an attention module is introduced to selectively fuse the features of the two views. HCCL-DTI is fully compared with state-of-the-art baselines on three heterogeneous DTI datasets, experimental results demonstrate that HCCL-DTI outperforms advanced baselines and achieves improved performance in DTI prediction.

Keywords: Drug-target Interaction Prediction · Heterogeneous Networks · Contrastive Learning

1 Introduction

Finding the associations between potential drugs and target proteins plays an important role in drug development and repurposing. Determining the drug-target interaction (DTI) relationship can help drug developers screen compounds that interact with specific target proteins. However, it usually takes more than 2 years to discover and identify new drug-protein associations through biochemical experiments, and requires plenty of time, experiments and other costs [1]. Recently, it has become possible to identify DTI using machine learning, which significantly reduced the development time and experimental costs for all parties.

© The Author(s), under exclusive license to Springer Nature Singapore Pte Ltd. 2024
D.-S. Huang et al. (Eds.): ICIC 2024, LNBI 14881, pp. 183–194, 2024.
https://doi.org/10.1007/978-981-97-5689-6_16

The earliest methods based on machine learning that used to discover potential DTI were similarity-based methods, which mainly focused on pharmacological similarities between drugs, while genomic similarities [2] or topological similarities [3] are defined between target proteins. The disadvantage is that they require sufficient known DTIs to be effective, which is not possible in real sparse datasets. In addition, there are methods based on matrix factorization [4, 5], which mainly learn the interpretable potential features of proteins and drugs by factorizing the drug-target protein interaction matrix, but with high computational costs in large datasets.

With the application of graph neural networks in bioinformatic research, methods for aggregating semantic information and topological structure features from adjacent nodes have emerged. GraphDTA [6] is a representative GNN-based method which views drug molecules as graph structures and uses multiple GNNs to obtain feature representations of molecules to conduct drug-target affinity prediction. Manoochehri et al. proposed a GNN-based model [7] that is integrated with encoder-decoder technique to learn more informative and interpretable embeddings for drug and protein in the graph. ELPH [8] is a novel full-graph GNN that approximates the SGNN mechanism by passing messages of subgraph sketches without explicit subgraph construction.

Recently, network-based methods have acquired outstanding achievements in link prediction. NRWRH [9] provides a method using random walk with restart mechanism in heterogeneous network, and calculates the jump probability from the drug node to the protein node to predict DTIs. DTI-HeNE [10] is another DTI prediction method based on heterogeneous graph, it splits the heterogeneous graph into some sub-graphs, and use a random forest to conduct the prediction. Tang et al. [11] provided a model converting the predicted DTI into an edge denoising model of heterogeneous networks, which solves the false positive problem caused by the sparseness of the heterogeneous network. DTINet [12] collects heterogeneous network to acquire the low-dimensional embedding for nodes, it finds the best projection from drug to target to perform the DTI prediction. Another solution is the construction of line graphs [13], GCN-DTI [14] constructs drug-protein pair (DPP) and uses graph convolutional networks [15] to generate embedding for all of DPPs, then uses deep neural networks to get the prediction scores.

With the rapid development of deep learning, contrastive learning approach has been applied in the field of bioinformatics. CML [16] provides a contrastive learning method to research relationships between drug and target, it uses task inconsistency loss to measure the differences of the task. SGCL-DTI [17] learns graph node representations in a supervised way, the contrastive learning is performed from semantic and topological perspectives. SHGCL-DTI [18] is a recent research work that predicts potential DTI by reconstructing the original heterogeneous network, which maximizes the positive pairwise similarity between neighbor views and meta-path views. Petar et al. proposed a general method by constructing a corrupted graph, it uses mutual information to maximize the mutual information between the patch embedding and graph embedding [19]. Some other contrastive learning methods are mainly based on the graph diffusion or graph augmentation. MVGRL [20] obtains a diffusion graph by diffusing the original graph, and simultaneously samples subgraphs on the original graph and the diffusion graph. GCA provides an adaptive augmented graph contrastive learning method [21], it designs an adaptive Edge Dropping mechanism to generate two augmented graphs, which aims to identify whether a pair of features come from the same node in two views. Yang et al. [22] add a graph learning layer after each graph convolution layer, it calculates each interaction feature in the original and augmented graph, the objective is to maximize the mutual information between the augmented graph and the original graph.

Despite the success of aforementioned methods in DTI prediction, there are some limitations. The prediction quality of the network-based methods is limited by the available correlation information, and the prediction for new drugs and targets with rare correlation does not perform well. In addition, when perform contrastive learning between the original graph and the graph of diffusion or augment, the source interaction only comes from the regular sample graph of the heterogeneous graph, and the relationships from DDI and PPI are not involved in contrastive learning. The result is the embeddings source of drugs and proteins used in DTI prediction are relatively single. Furthermore, there are no contrastive learning methods that perform contrast between the different drug-drug interaction views and the target-target interaction views, which contain effective information that contributes to DTI prediction. However, this information, which is not necessarily included in the known DTIs, has never been explored in previous works.

To improves on this basis, a heterogeneous cross contrastive learning method (HCCL-DTI) is proposed for DTI prediction. The contributions are as follows:

- HCCL-DTI provides a novel idea for heterogeneous graph contrastive learning, which uses two modes including self-contrastive learning and cross-contrastive learning to obtain more expressive node embeddings. Self-contrast is performed in two views respectively and cross-contrast is performed between two views.
- Another DDI and PPI information contained in the heterogeneous graph is extracted, which provides another complementary view of the drug-drug interactions and protein-protein interactions, it is contrasted with the original view of DDI and PPI to improve prediction accuracy effectively.
- HCCL-DTI is extensively compared with state-of-the-art baselines on three datasets. The experimental performance illustrate that HCCL-DTI outperforms advanced baselines and achieves improved results in DTI prediction.

2 Method

The HCCL-DTI model takes the 3 networks R, D and T as inputs, and outputs the interaction score $S(d_i, t_j)$ for each candidate drug d_i and target protein t_j. HCCL-DTI consists of four modules: the graph embedding module, the self-contrast module, the cross-contrast module, and the pairwise judgment module. The whole model framework is shown in Fig. 1. Next, we will introduce these four parts in detail one by one.

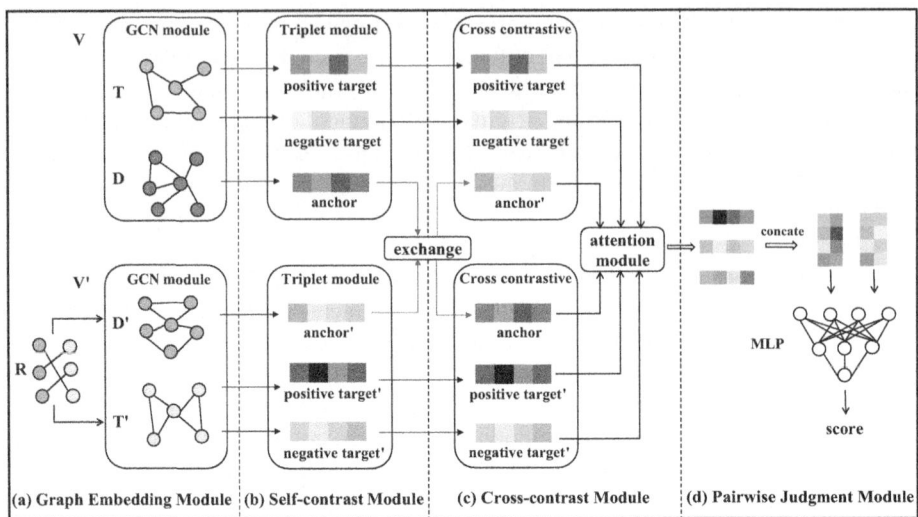

(a) Graph Embedding Module | (b) Self-contrast Module | (c) Cross-contrast Module | (d) Pairwise Judgment Module

Fig. 1. Flow chart of HCCL-DTI for DTI prediction, the model consists of four modules: (a) Graph Embedding Module: in this module, another view V' is generated by R, which is complementary to the initial view, then multiple graph encoders is utilized to learn various representation from view V and V'. (b) Self-contrast Module: triplet contrastive learning is conducted between anchor, positive sample, and negative sample in two views, respectively. (c) Cross-contrast Module: The anchor objects in the two views cross-contrast with positive and negative samples from each other, which extracts the implicit information through a supervised way, and an attention mechanism is introduced to fuse the embeddings of drug and target from two views. (d) Pairwise Judgment Module: the final embeddings of the drug and protein are concatenated to represent the DTI, finally a multi-layer perceptron is used to generate prediction score for DTIs.

2.1 Graph Embedding Module

In this module, a model of two-layer GCNs are used to generate the embedding for the drugs and targets of current dataset based on the existing DDI and PPI, and use one-hot encoding for the initial features of drugs and proteins. The mode of propagation is expressed as follows:

$$z_d^{(l)} = ReLU\left(D_d^{-\frac{1}{2}}\tilde{A}_d D_d^{-\frac{1}{2}} z_d^{(l-1)} W_d^{(l)}\right) \tag{1}$$

where $\tilde{A}_d = A_d + I$, A_d is the adjacency matrix of DDI, D_d denotes the diagonal degree matrix of $\tilde{A}_d$, $z_d^{(l)}$ is the l − th layer embedding matrix. The $W_d^{(l)}$ is the matrix with parameters of the l-th layer that needs to be learned. The initial embeddings inputted into the module are denoted by $z_d^{(0)} \in \{0, 1\}^{m*m}$. The same operation is performed on PPI:

$$z_t^{(l)} = ReLU\left(D_t^{-\frac{1}{2}}\tilde{A}_t D_t^{-\frac{1}{2}} z_t^{(l-1)} W_t^{(l)}\right) \tag{2}$$

In order to effectively utilize implicit drug-drug associations and protein-protein associations from known DTIs, considering that when two drugs are associated with the same known protein, we consider that the two drugs may have chemical structural or pharmacological similarities in some degree. Likewise, two proteins related to a known drug may share primary sequences, have genomic similarity or topological similarity. Based on this assumption, we calculated the basic drug-drug interaction network D' from the known DTI associations:

$$D'_{ij} = \begin{cases} 1, \ \exists k : R_{ik} = 1 \wedge R_{jk} = 1 \\ 0, \ \forall k : R_{ik} = 0 \vee R_{jk} = 0 \end{cases} \tag{3}$$

where R_{ik} denotes the association between drug i and protein j. Likewise, the protein-protein interaction network T' is obtained. After that, another GCN module is performed on D' and T' to generate the embeddings of drugs and proteins of view V'. Finally the output of the module is the representations for all of the drugs and targets from the two views.

2.2 Self-contrast Module

In this part, a new self-supervised method is proposed to pull the drug closer to proteins which are related to it, and to maximize the distance between the drug and proteins which are not related. This objective operates on triplets in the two views, respectively.

Weinberger proposed the triplet loss firstly [23]. The innovation in our method is that d and p are not samples with the same category, but the drug and the protein associated with it in the same view. Likewise, d and n form a pair of negative samples with different node types, there is no observed link between d and n. Our loss function forces the Euclidean distance of representations of d and p in the feature space to get closer, and makes the distance of the representations of d and n get farther in the feature space. This loss function is defined as follows:

$$L_{tcl}^{V} = \sum_{i=1}^{N} ReLU \left(||z_{di} - z_{ti}^{+}||_2 - ||z_{di} - z_{ti}^{-}||_2 + m \right) \tag{4}$$

where m is the margin, z_{di} is the embedding of drug i, which is the anchor. z_{ti}^{+} And z_{ti}^{-} denote the target embeddings of positive and negative targets, respectively. N is the number of all triplets selected. The training process will keep optimizing the feature embedding until the distance between the positive sample and anchor is short enough and the distance between the negative sample and anchor is large enough. Similar operation is used in view V' to get $L_{tcl}^{V'}$:

$$L_{tcl} = \alpha L_{tcl}^{V} + (1 - \alpha) L_{tcl}^{V'} \tag{5}$$

where L_{tcl} is the triplet contrastive learning loss of the model.

2.3 Cross-Contrast Module

Supervised contrastive learning is conducted between the two views, the objective of cross-contrast is to ensure that the drug is close to the protein which it is related, and far away from the protein which it is not related, similar to the self-supervised graph contrast in SGL [24]. The difference is that we do not use graph augmentations to generate different node views, and we do not treat the node of another view which from the same node as positive sample. Our anchor and positive samples (negative samples) are not augmented representations from the same (different) nodes, but from the exchange of different views. In addition, the input triplet contains data types with different meanings, which including (drug, positive target, negative target).

The contrastive objective uses the selected drug under V as the anchor, selects protein under V' which has a known association with the anchor as the positive sample, the negative sample comes from randomly selected proteins under V' which have no association or have not yet confirmed association with the anchor. The supervised cross-contrastive learning objective in view V can be defined as follows:

$$L_{ccl}^V = -\sum_{i=1}^{N_d} log \frac{exp(s(z_{di}, z_{tj}')/t_p)}{\sum_{k \in N_i} exp(s(z_{di}, z_{tk}')/t_n)} \qquad (6)$$

where the z_{di} represents the embedding of drug i in view V, z_{tj}' is the embedding of target j in view V' which has an association with i, z_{tk}' is the embedding of target k in view V' without a known association with i. The function $s(\cdot)$ is cosine function to calculate the similarity of embeddings. The temperature coefficients t_p and t_n control the convergence speed. The loss encourages agreement between z_{di} and z_{tj}' which form a positive sample pair, while minimizing the consistency between z_{di} and z_{tk}'.

Similarly, we use the drug selected under V' as anchor, proteins under V are selected as positive or negative samples. Loss of the view V' can be computed similarly, which denotes by $L_{ccl}^{V'}$. Then we obtain the cross-contrastive learning objective defined as:

$$L_{ccl} = \beta L_{ccl}^V + (1 - \beta) L_{ccl}^{V'} \qquad (7)$$

Then an attention module is used to fuse the representations of drug and target in view V and V'. For a drug node i, let z_{di} and z_{di}' denote the representations of i in two views respectively, firstly a weight score w_{di} is computed for z_{di}:

$$w_{di} = p^T \cdot tanh\left(W \cdot (z_{di})^T + b\right) \qquad (8)$$

where $W \in R^{d_l \times d_h}$ is a parameters matrix, $b \in R^{d_l \times 1}$ is a vector of bias and $p \in R^{d_l \times 1}$ is a shared attention vector. d_l is the attention layer dimension and d_h is the drug embedding dimension. The score w_{di}' of drug i can also be obtained for z_{di}' in view V'. Then we use $softmax$ function which normalizes the weight scores to obtain α_{di}:

$$\alpha_{di} = softmax(w_{di}) = \frac{exp(w_{di})}{exp(w_{di}) + exp(w_{di}')} \qquad (9)$$

where α_{di} is the attention score for z_{di}, and $\alpha_{di}' = softmax(w_{di}')$ can be calculated for z_{di}'. Then the final representation of drug i can be obtained by weighted summation of z_{di} and z_{di}' from two views:

$$Z_{di} = \alpha_{di} \cdot z_{di} + \alpha_{di}' \cdot z_{di}' \qquad (10)$$

Through the same operation, the final representation Z_{tj} can be obtained for target j.

2.4 Pairwise Judgment Module

In the previous sections, an attention module is used to get the final embeddings of the drug and target from V and V'. Afterwards, we concatenate the final representation of the drug and protein to represent DTI representations:

$$Z_{ij} = Z_{di}||Z_{tj}, i \in N_d, j \in N_t \tag{11}$$

where Z_{ij} denotes the DTI representation between drug i and target j. Finally, Z_{ij} is input into a multi-layer perceptron for judgment, the output S_{ij} is the prediction score for the existence of the DTI:

$$S_{ij} = MLP(Z_{ij}) \tag{12}$$

We use the loss function of mean squared error to obtain the objective:

$$L_{pairwise} = ||S_{DT} - R||_F^2 \tag{13}$$

where S_{DT} represents calculated matrix that including all of prediction score S_{ij} in the pairwise judgment module. The goal of the objective is to maximize the consistency between the heterogeneous network R and the calculated score matrix S_{DT}.

The final loss of HCCL- DTI is a weighted combination of the self-contrastive loss, cross-contrastive loss, and the DTI prediction objective. We derive the final formulation of the complete loss for HCCL-DTI model as follows:

$$L = \lambda \cdot L_{pairwise} + (1 - \lambda) \cdot L_{tcl} + \gamma \cdot L_{ccl} \tag{14}$$

where the λ and the γ are the weight parameters in the experiment.

3 Experiments

3.1 Datasets

We select three heterogeneous networks as datasets. The dataset DrugBank&HPRD was collected from the latest versions of DrugBank[25] and HPRD[26]. The datasets Luo's dataset [12] and Zheng's dataset [27] were obtained from the literature. A summary is shown in Table 1. The DTI Sparsity refers to the proportion of known DTIs to all possible DTIs, which can be calculated by $Sparsity = \frac{Interactions}{(Drug\#)*(Target\#)}$.

Table 1. The details of experiment datasets.

Dataset	Drug#	Target#	Interactions#	Sparsity
DrugBank&HPRD	1331	2738	6088	0.17%
Luo's dataset	708	1512	1923	0.17%
Zheng's dataset	1094	1596	11819	0.68%

3.2 Experimental Settings

Parameter Settings. The known positive DTIs are divided into training set, validation set and testing set in a ratio of 8:1:1 in the experiments, an equal quantity of negative samples is then randomly sampled from unobserved interactions. For HCCL-DTI, the dimension of drug and protein representation is 128. For methods that need iterative optimization, the value of epochs is set to 200 and we use algorithm of Adam to optimize the loss functions. The hyper parameters of baselines are tuned according to their papers, all methods are implemented with pytorch 1.12.0 and executed on GPUs. AUC, AUPR and F1 score are selected as evaluation metrics.

Baselines. We select eight competing methods as baselines, including methods based on GNNs, contrastive learning, matrix factorization, random walk and network integration.

- **GCN** [15]: The most original spectral GCN has been improved with first-order approximation. Currently, most spectral GCNs are based on this paper.
- **DGI** [19]: DGI first learns a patch representation that contains subgraph information, then uses mutual information to optimize the objective.
- **SimGCL** [28]: The model provides a simple contrastive learning method which do not use the graph augmentations, but add noises to generate another view of the node representations.
- **MVGRL** [20]: MVGRL obtains a diffusion graph by diffusing the original graph, then perform contrastive learning on the feature representation of the two subgraphs.
- **GCA** [21]: The model generates two graph views through adaptive graph structure and random augmentation. Contrastive learning is performed by shrinking the same node representations and larging the different node representations in the two views.
- **EGLN** [22]: It obtains the graph features from the original graph and samples the negative relation from the augmented graph. The contrast learning is performed between original graph and augmented graph.
- **NIMCGCN** [29]: The model uses the inductive matrix completion method to generate correlation matrix completion to obtain accurate DTI prediction performance.
- **DTI-HeNE** [10]: DTI-HeNE learns high-quality DTI embeddings from bipartite DTI networks, which are then fed into a RF model to perform predictions.
- **DTINet** [12]: This method acquires low-dimensional embeddings of nodes through random walks with restarting. It utilizes the best projection space from the drug to target to conduct the prediction for DTIs.

3.3 Experimental Results.

Performance Comparison. HCCL-DTI is compared with the baselines selected in three datasets, the AUC, AUPR and F1 results are shown in Table 2.

From the above results, we can draw the conclusions as follows:

- Contrastive learning-based methods including DGI [19], MVGRL [20] and EGLN [22] significantly outperform the matrix factorization-based method NIMCGCN [29], which indicates the significance of contrastive learning in graph learning.

Table 2. Performance of HCCL-DTI and other baselines in three datasets.

Methods	DrugBank&HPRD			Luo's DTI			Zheng's DTI		
	AUC	AUPR	F1	AUC	AUPR	F1	AUC	AUPR	F1
GCN	0.8352	0.8647	0.7525	0.7802	0.8291	0.6788	0.8109	0.8350	0.7346
DGI	0.8907	0.9039	0.8179	0.8530	0.8898	0.7668	0.9111	0.9227	0.8411
SimGCL	0.8394	0.8097	0.8066	0.8527	0.8459	0.8031	0.9141	0.9150	0.8707
MVGRL	0.8458	0.8468	0.7810	0.8790	0.9145	0.8031	0.9208	0.9083	0.8436
GCA	0.8514	0.8537	0.7946	0.8260	0.8264	0.7617	0.8061	0.8454	0.7498
EGLN	0.8218	0.8663	0.7492	0.8622	0.8752	0.7927	0.9224	0.9265	0.8504
NIMCGCN	0.8236	0.8908	0.8111	0.8585	0.8794	0.7565	0.8015	0.8191	0.7219
DTI-HeNE	0.8856	0.9057	0.0704	0.9242	0.9265	0.0144	0.8855	0.8868	0.1972
DTINet	0.8362	0.8808	0.7734	0.8798	0.9063	0.8394	0.9054	0.9206	0.8718
HCCL-DTI	0.9033	0.9281	0.8311	0.9229	0.9366	0.8549	0.9553	0.9584	0.8884

- Compared with network integration-based models DTI-HeNE [10] and DTINet [12], we achieve improved performance by fusing heterogenous information of two views.
- HCCL-DTI achieves state-of-the-art performance on the three datasets, particularly Zheng's DTI gets an improvement of at least 3% in both AUC and AUPR. The possible reason is that Zheng's DTI has the highest sparsity, which shows that it achieves improved performance on heterogeneous networks with more DTIs.

Ablation Experiment. As a comparison, the node embedding is acquired by taking the average of the embeddings from V and V' to test the effectiveness of attention module:

$$Z_{di}' = \frac{z_{di} + z_{di}'}{2} \tag{15}$$

where Z_{di}' is the final representation of drug i, z_{di} and z_{di}' are the embeddings of drug i in view V and V' respectively. Similarly, Z_{tj}' for target j is calculated.

We can obtain the following information from Table 3:

- Either without self-contrast, cross-contrast or attention module, the performance decrease. HCCL-DTI with three modules has the best experimental performance.
- Self-contrast module is the most critical part of the model. Both of the AUC and AUPR drop by about 3% without it, especially in Luo's DTI, they declined by about 5% respectively.
- The result of the attention module is better than the average of the two representations, which demonstrates that the features learned from the two views are complementary to each other, and selectively focusing on the important features through a weighted approach can improve the prediction results.

Table 3. The AUC and AUPR of ablation from HCCL-DTI

Ablation method	DrugBank&HPRD		Luo's dataset		Zheng's dataset	
	AUC	AUPR	AUC	AUPR	AUC	AUPR
Without self-contrast module	0.8701	0.8926	0.8732	0.8900	0.9325	0.9248
Without cross-contrast module	0.8773	0.9031	0.9222	0.9298	0.9380	0.9387
Without attention module	0.8888	0.9184	0.9160	0.9348	0.9422	0.9430
HCCL-DTI full model	**0.9033**	**0.9281**	**0.9229**	**0.9366**	**0.9553**	**0.9584**

3.4 Parameter Sensitivity Analysis.

In this work, drug and protein representations learning is mainly determined by GCN, which includes the amount of graph encoder layers and the dimensionality of the embeddings. To evaluate the impact of these two properties in GCN, we performed parameter sensitivity experiments to test the robustness of HCCL-DTI.

Number of GCN Layers. The results are shown in Fig. 2, it is evident that the optimal number of the layers is negatively related with the sparsity of the datasets. HCCL-DTI performs best with 2 layers of GCN on DrugBank&HPRD and Luo's DTI, while it is enough to use 1 layer of GCN on Zheng's DTI. When the number of layers increases, the performance declines because this will cause the representation of each node to be closer. Considering the potential issue of over-smoothing in the approximation of convolutional operations, it is not suitable to use too many layers of GCN in the experimental process.

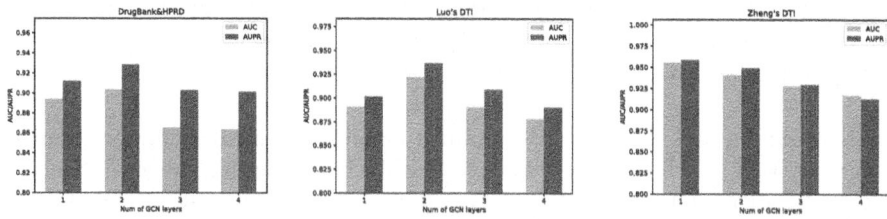

Fig. 2. The impact of number of GCN layers for DTI prediction performance on datasets.

Dimensions of Embeddings. The experiments were conducted with dimensions of 32, 64, 128, 200 and 300 for datasets, which are shown in Fig. 3. We can obtain that the best dimensionality of the embeddings is positively correlated with the scale of the datasets. HCCL-DTI performs best with a 200-dimensional embedding in DrugBank&HPRD and

Zheng's DTI, while a 128-dimensional embedding is most suitable for Luo's DTI. Setting a dimension that is too large in small datasets may introduce redundant information unrelated to DTI prediction, resulting in a decline in performance.

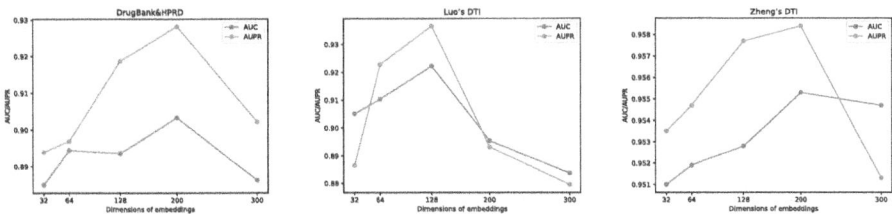

Fig. 3. The impact of dimensions of embeddings for DTI prediction performance on datasets.

4   Conclusion

An end-to-end method HCCL-DTI is proposed for DTI prediction. We first use the interactions in the known DTI to generate another view, which is complementary to the initial view. Then the self-contrast is conducted inside the two views and cross-contrast is conducted between the two views. Pairwise judgment module is finally used to predict the score for each drug-protein pair. Extensive experiments illustrate that HCCL-DTI outperforms state-of-the-art baselines and demonstrates its practical significance.

Acknowledgements. This work is supported by the Tianjin Natural Science Foundation under Grant No. 22JCYBJC01020.

References

1. Kapetanovic, I.M.: Computer-aided drug discovery and development (CADDD): in silico-chemico-biological approach. Chem. Biol. Interact. **171**(2), 165–176 (2008)
2. He, Z., Zhang, J., Shi, X.H., et al.: Predicting drug-target interaction networks based on functional groups and biological features. PLoS ONE **5**(3), e9603 (2010)
3. Mei, J.P., Kwoh, C.K., Yang, P., et al.: Drug–target interaction prediction by learning from local information and neighbors. Bioinformatics **29**(2), 238–245 (2013)
4. Cobanoglu, M.C., Liu, C., Hu, F., et al.: Predicting drug–target interactions using probabilistic matrix factorization. J. Chem. Inf. Model. **53**(12), 3399–3409 (2013)
5. Liu, Y., Wu, M., Miao, C., et al.: Neighborhood regularized logistic matrix factorization for drug-target interaction prediction. PLoS Comput. Biol. **12**(2), e1004760 (2016)
6. Nguyen, T., Le, H., Quinn, T.P., et al.: GraphDTA: predicting drug–target binding affinity with graph neural networks. Bioinformatics **37**(8), 1140–1147 (2021)
7. Manoochehri, H.E., Pillai, A., Nourani, M.: Graph convolutional networks for predicting drug-protein interactions. In: 2019 IEEE International Conference on Bioinformatics and Biomedicine (BIBM), pp. 1223–1225. IEEE, San Diego (2019)

8. Chamberlain, B.P., Shirobokov, S., Rossi, E., et al.: Graph neural networks for link prediction with subgraph sketching. In: The Eleventh International Conference on Learning Representations (2023)
9. Chen, X., Liu, M.X., Yan, G.Y.: Drug–target interaction prediction by random walk on the heterogeneous network. Mol. BioSyst. **8**(7), 1970–1978 (2012)
10. Yue, Y., He, S.: DTI-HeNE: a novel method for drug-target interaction prediction based on heterogeneous network embedding. BMC Bioinf. **22**, 1–20 (2021)
11. Tang, C., Zhong, C., Chen, D., et al.: Drug-target interactions prediction using marginalized denoising model on heterogeneous networks. BMC Bioinf. **21**, 1–29 (2020)
12. Luo, Y., Zhao, X., Zhou, J., et al.: A network integration approach for drug-target interaction prediction and computational drug repositioning from heterogeneous information. Nat. Commun. **8**(1), 573 (2017)
13. Cai, L., Li, J., Wang, J., et al.: Line graph neural networks for link prediction. IEEE Trans. Pattern Anal. Mach. Intell. **44**(9), 5103–5113 (2021)
14. Zhao, T., Hu, Y., et al.: Identifying drug–target interactions based on graph convolutional network and deep neural network. Brief. Bioinform. **22**(2), 2141–2150 (2021)
15. Kipf, T.N., Welling, M.: Semi-supervised classification with graph convolutional networks. arXiv preprint arXiv:1609.02907 (2016)
16. Li, M., Xu, S., Cai, X., Zhang, Z., Ji, H.: Contrastive meta-learning for drug-target binding affinity prediction. In: 2022 IEEE International Conference on Bioinformatics and Biomedicine (BIBM), pp. 464–470. IEEE, Las Vegas (2022)
17. Li, Y., Qiao, G., Gao, X., et al.: Supervised graph co-contrastive learning for drug–target interaction prediction. Bioinformatics **38**(10), 2847–2854 (2022)
18. Yao, K., Wang, X., Li, W., et al.: Semi-supervised heterogeneous graph contrastive learning for drug–target interaction prediction. Comput. Biol. Med. **163**, 107199 (2023)
19. Velickovic, P., Fedus, W., et al.: Deep graph infomax. ICLR (Poster) **2**(3), 4 (2019)
20. Hassani, K., Khasahmadi, A.H.: Contrastive multi-view representation learning on graphs. In: International Conference on Machine Learning, pp. 4116–4126. PMLR, (2020)
21. Zhu, Y., Xu, Y., Yu, F., et al.: Graph contrastive learning with adaptive augmentation. Proc. Web Conf. **2021**, 2069–2080 (2021)
22. Yang, Y., Wu, L., et al.: Enhanced graph learning for collaborative filtering via mutual information maximization.In: Proceedings of the 44th International ACM SIGIR Conference on Research and Development in Information Retrieval, pp. 71–80 (2021)
23. Weinberger, K.Q., Saul, L.K.: Distance metric learning for large margin nearest neighbor classification. J. Mach. Learn. Res. **10**(2), (2009)
24. Wu, J., Wang, X., Feng, F., et al.: Self-supervised graph learning for recommendation. In: Proceedings of the 44th International ACM SIGIR Conference on Research and Development in Information Retrieval, pp. 726–735 (2021)
25. Wishart, D.S., Feunang, Y.D., Guo, A.C., et a.l: DrugBank 5.0: a major update to the DrugBank database for 2018. Nucleic Acids Res. **46**(D1), D1074–D1082 (2018)
26. Keshava Prasad, T.S., Goel, R., Kandasamy, K., et al.: Human protein reference database—2009 update. Nucleic Acids Res. **37**(suppl_1), D767–D772 (2009)
27. Zheng, Y., Peng, H., Zhang, X., Gao, X., Li, J.: Predicting drug targets from heterogeneous spaces using anchor graph hashing and ensemble learning. In: 2018 International Joint Conference on Neural Networks (IJCNN), pp. 1–7. IEEE, Rio de Janeiro (2018)
28. Yu, J., Yin, H., et al.: Are graph augmentations necessary? Simple graph contrastive learning for recommendation. In: Proceedings of the 45th International ACM SIGIR Conference on Research and Development in Information Retrieval, pp. 1294–1303. ACM Madrid (2022)
29. Li, J., Zhang, S., Liu, T., et al.: Neural inductive matrix completion with graph convolutional networks for miRNA-disease association prediction. Bioinformatics **36**(8), 2538–2546 (2020)

A Retrieval-Based Molecular Style Transformation Optimization Model

Cheng Wang[1], Ya-Jie Zhang[1], Xin Xia[2], Yan-sen Su[2,4], Chun-hou Zheng[4], and Qing-Wen Wu[3(✉)]

[1] School of Computer Science and Technology, Anhui University, Hefei 230039, China
[2] School of Artificial Intelligence, Anhui University, Hefei 230039, China
[3] Affiliated Hospital of Jining Medical University, Jining 272000, China
wqwcyf@126.com
[4] Information Materials and Intelligent Sensing Laboratory of Anhui Province, Anhui University, Hefei 230039, China

Abstract. Molecular optimization endeavors to enhance molecular properties through the alteration of intricate molecular structures, has important roles in drug discovery. In the past years, a lot of computational models have been proposed for molecular optimization. However, there is still great room for existing models to improve their performance, which is due to the following two facts. First, the majority of existing models are based on basic molecular descriptors, such as molecular weights and property values. Second, most of them are lack of the guidance from high-quality molecules. To address this issue, in this paper we propose a retrieval-based molecular style transformation model, which is terms as RMST. In the model, distinctive attribute features of high-property molecules are integrated with SMILES sequence data to navigate the optimization pathway. Besides, the model uses an attention-driven retrieval module to extract molecular attributes, which are fused with the SMILES sequence information to enrich the representation of molecules. Experimental results on two representative optimization tasks show that the proposed model is superior to three baseline models.

Keywords: drug discovery · molecular optimization · style transformation

1 Introduction

According to statistics, it takes approximately 13 years and nearly one billion dollars to develop a new drug [1]. Early molecular design and screening methods, such as Quantitative Structure-Activity Relationship (QSAR) [2], involve extracting structural descriptors, selecting variables, constructing complex empirical equations, and establishing relationships between molecular structures and corresponding physicochemical properties, which are heavily dependent on expert knowledge and manual rules. The molecule optimization task is highly challenging since there is a vast number of chemically viable molecules with their number being estimated to be between 10^{23} and 10^{60} [3]. Although with the advancement of high-throughput screening (HTS) technologies, the daily experimental detection of candidate molecules can reach nearly hundreds of thousands, there are still challenges such as low hit rates and high costs [4]. Recent years have shown that machine learning, through the fusion of domain expertise and data-driven learning, is a promising approach for molecular optimization, enabling efficient discoveries [5–7]. Molecular optimization refines input molecules to create new molecular structures with desired properties, fulfilling application requirements.

The related work in machine learning for molecular optimization can be classified into two categories, namely the search-based and transformation-based methods. The search-based methods can be further divided into two subcategories, genetic algorithms and random search methods. Genetic algorithms search the molecular space and improve the properties of molecules through gene encoding, genetic operations, and fitness evaluation [8–10]. These algorithms typically conduct searches in the form of populations and utilize operations like selection, crossover, and mutation to generate new molecules. Random search-based methods involve searching the molecular space by randomly sampling [11]. However, these methods lack explicit optimization and discover good molecules by randomly selecting molecules and evaluating their performance.

Transformation-based methods can also be categorized into two categories, molecular graphs-based molecular optimization methods [12–14] and those based on SMILES sequences [15, 16]. For instance, Mol-CG [17] employs a CycleGAN loss to generate molecules with desired properties from given molecules. UGMMT designs a transformer to facilitate conversions between different domains, thus achieving molecular optimization. These works focus on molecular optimization based on the SMILES representation of molecules. SMILES is preferred over molecular graphs due to its superior uniformity and completeness in covering chemical space [7]. Transformation-based methods transfer the molecular generation as a sequence-to-sequence conversion problem [18–20]. Compared to search-based methods, transformation-based approaches require additional information about paired sequences, which needs to learn the way of converting a molecule with low properties to an improved one. However, existing methods do not fully utilize information of high-property molecules.

To make better use of the information from high-quality molecules, The molecule optimization problem is considered as a molecule style transfer task, analogous to the text style transfer task [21]. The text style transfer task aims to change the style of a text sequence with its content being preserved. Text style transfer combines text content with a specific style to generate text with a desired style based on contextual grammar rules. As for molecule style transfer task, molecular property optimization combines molecular sequence information with property features to generate molecules with improved properties based on molecular structure rules. The aforementioned two tasks share the commonality that they both rely on contextual rules and involve a transformation between two domains. A retrieval-based molecular style transformation model (RMST) is proposed. RMST utilizes the property features of high-quality molecules and combines them with the molecular SMILES sequence information to guide molecular optimization. The main contributions of this paper are as follows:

- Considering the similarity between text-style transformation and molecular optimization tasks, this study applies the concept of text-style transformation to molecular optimization. It guides the optimization process by combining the attribute features of high-quality molecules with molecular SMILES sequence information.

● To better utilize the attribute features of high-quality molecules, a retrieval module is designed based on the attention mechanism. This module integrates molecular sequence information with attribute features to perform molecular optimization.

2 Methods

The attribute optimization of molecules combines the molecular sequence information with attribute features to generate molecules with higher attributes. The main idea behind the retrieval-based molecular style transformation model (RMST) is to utilize the attribute features of high-quality molecules to guide molecular optimization, effectively improving the molecular attributes while maintaining a high degree of similarity. The concept of style transformation permeates the entire RMST algorithm: the Maximum Inner Product Search (MIPS) [22] algorithm is employed to retrieve the top K molecules with the highest scores from the high-quality molecule library, and then use a cross-attention mechanism to fuse the attribute features of the K molecules into a latent embedding. This embedding is combined with the embedding of the molecule to be optimized to guide the molecular optimization process. Specifically, our proposed Retrieval-based Molecular Style Transformation Optimization Model (RMST) consists of three modules: an encoder, a retriever, and an information fusion and decoder, as shown in Fig. 1.

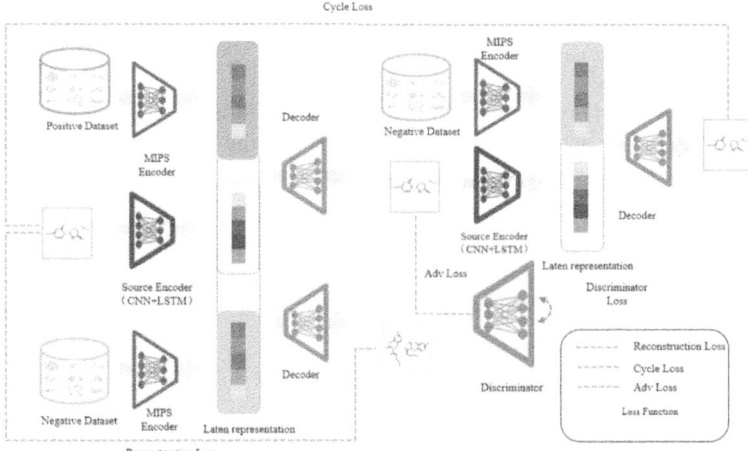

Fig. 1. RMST Model Architecture.

2.1 Overview

The RMST model is a molecular optimization model based on molecular SMILES sequences. The objective of the encoder is to encode the input molecule into an embedding vector, which is then used in subsequent retrieval and optimization processes. Given an input molecule $X = (x_1, x_2, \ldots, x_n)$, the output of the encoder is the hidden state of the molecule,

$$E^{enc} = \text{Encoder}(X), \tag{1}$$

where $E^{enc} = [h_1^{enc}, h_2^{enc}, \ldots, h_n^{enc} \in R^{n \times d}]^{\mathrm{T}}$ and d is the dimensionality of the latent states. It is worth noting that our content encoder and style encoder utilize different encoders. To be specific, the style encoder employs a bidirectional LSTM [23]. As for the content encoder, it uses a combination of CNN and bidirectional LSTM.

2.2 Molecular Retriever

Considering the rapid expansion of molecular library data over the past few decades, an efficient retrieval tool has been utilized to identify the best-suited high-quality molecules for guiding the optimization of molecules. Lately, dense retrieval methods, along with their efficient implementation of Maximum Inner Product Search (MIPS), have been suggested for capturing semantics. It is utilized to capture the style information of molecules. For the molecular retriever, the embedding vector of molecules is crucial, as it is relied upon to find similar samples in different attribute subsets. Figure 1 illustrates the main framework of RMST. First, a set of molecules that meet certain constraints, such as X, is constructed. Specifically, for the lead molecules, Maximum Inner Product Search (MIPS) is utilized to retrieve from the high-quality molecule library, selecting the top K molecules with the best attribute scores, as follows:

$$s(x) = Softmax([\alpha(x_1), \ldots, \alpha(x_n)]) \cdot E^{enc}, \tag{2}$$

The parameter $\alpha(x_i)$ represents the weight of each atom x_i. They are initialized as $\alpha(x_1) = 1 - \sum_j |f_{s_j}(x_i) - \frac{1}{M}|$, where $f_{s_j}(x_i)$ denotes the frequency of atom x_i under style s_j, $\frac{1}{M}$ represents the ideal uniform distribution, meaning that the probability of each atom appearing in every style is the same. Retrieve the top K similar sentences from the target-style training subset using the MIPS retrieval method, and measure similarity using the cosine similarity function. A bidirectional LSTM molecular style encoder is used to obtain the latent representations $E_{\text{top\_k}} = [h_1, h_2, \ldots, h_k]^T$ of the top K molecule.

2.3 Information Fusion Module and Decoder

This module allows retrieved high-quality molecules to adjust the molecules being optimized according to the target design criteria. It achieves this by merging the embedding of the input molecule with those of the retrieved high-quality molecules, a mechanism similar to the attention mechanism in Irwin [24]. Specifically, the fused embedding e is given by the following formula:

$$v_t^h = Attn(h_{t-1}^{dec}, E^{enc}), \tag{3}$$

$$v_t^r = Attn(h_{t-1}^{dec} W_h, E_{\text{top\_k}} W_z), \tag{4}$$

where $W_h \in R^{d_{dec} \times d_{dec}}$ and $W_z \in R^{d \times d}$ with d_{dec} being the dimension of the decoder's hidden state.

$$Attn(h, E) = \frac{\sum_{j=1}^{n} \exp(e_{ij}) E_j}{\sum_{k=1}^{n} \exp(e_{ik})}, \tag{5}$$

$$e_{ij} = v^T tanh(W_d h + W_e E_j), \tag{6}$$

where v, W_e, W_d are three parameters.

The decoder generates molecules atom by atom. At each step, its input comprises three components: 1) The output of the previous step $\hat{y}_{t-1}$, 2) The latent embedding of x, i.e v_t^h, and 3) The latent embedding of the retrieved molecule v_t^r. The generated text $y = (\hat{y}_1, \hat{y}_2, \ldots, \hat{y}_m)$ is obtained from the following formula:

$$h_t = Decoder\left(\hat{y}_{t-1}, v_t^h, v_t^r\right), \tag{7}$$

where v_t^h and v_t^r represent the attended input sentence and retrieved molecule, respectively. This step can also employ various decoding techniques, and in this model, LSTM is used.

2.4   Retrieval-Based Molecular Style Transformation Generative Network

The objective of the optimization model is to generate new effective molecules with the similarity being maintained to a certain level, which requires the model to understand chemical principles. The total loss of the optimization model consists of four losses. In addition to the three typical losses utilized in prior research on style transformation, including the reconstruction loss, cycle loss, and adversarial style loss, we introduce a new loss related to the retriever, termed the retrieval loss. Therefore, given an input molecule x, its style s_i, and the target style s_j, the optimization objective can be represented as follows:

$$L = \lambda_1 L_{rec} + \lambda_2 L_{cyc} + \lambda_3 L_{adv} + \lambda_4 L_{ret}, \tag{8}$$

where L_{rec}, L_{cyc}, L_{adv} and L_{ret} represent the reconstruction loss, cycle reconstruction loss, adversarial style loss, and retrieval loss, respectively.

Reconstruction loss: this loss as shown in (9) is used to capture the information features for self-reconstruction.

$$L_{rec} = -logP(x|x, si), \tag{9}$$

Cycle reconstruction loss: for the output molecule y, the transformed output back to the original style should be as consistent as possible with x.

$$L_{cyc} = -logP(x|\hat{y}, si), \quad \hat{y} = G(x, sj), \tag{10}$$

where G refers to the generator of RMST.

Adversarial loss: Adversarial training is adopted to establish style supervision.

$$L_{adv} = -logP_D(j|\hat{y}), \tag{11}$$

where D refers to the discriminator.

Retrieval loss: the design of the retrieval loss as shown in (12) is aimed at capturing the similarity of embedding vectors that are independent of the attribute features between the input molecule and the corresponding transformed molecule.

$$L_{ret} = 1 - \cos(s(x), s(\hat{y})), \tag{12}$$

According to Algorithm 1, the detailed procedure for optimizing the loss function during the training process is as follows.

Algorithm 1: Training and Optimization of the SMOM Model

Input: Molecule to be optimized: x; High and low attribute style libraries: s_i and s_j; Number of iterations: $N = 500$

Output: Trained model; Objective loss function L

1   for $n = 1$; $n \leq N$; n++ do

2 | Samples from x, s_i and s_j

3 | $G_x(x|x, s_i)$ ←Generator G_x transforms feature x

4 | $\hat{y} = G(x, s_j)$ ← $\hat{y}$ is generated from the feature x and s_j

5 | $G_{\hat{y} \to x}(x|\hat{y}, s_i)$ ←Generator $G_{\hat{y} \to x}$ performs style transfer based on the
     | feature $\hat{y}$ and s_i

6 | $D_{\hat{y}}^C(G(x, s_j))$ ←Discriminator $D_{\hat{y}}^C$ classifies the fake molecule

7 | L_{ret} ←Calculate the retrieval loss during the generation process

8   Optimize the generator and discriminator L during the iteration process

9   end

---

3 Results

3.1 Datasets and Performance Metrics

In the experiments, the performance of the proposed model is compared with three baseline models, namely the JTVAE, Mol-CG, and UGMMT, where the quantitative estimation of drug-likeness (QED) and penalized logP (PlogP) optimization tasks are used as measurement criteria.

To ensure the fairness of performance comparison, the scores of QED and PlogP are calculated using the RDkit library [25], following the approach adopted by Fu [26]. For the QED optimization task, the datasets in Barshatski [15]. Are used. For the PlogP optimization task, the datasets in Jin [27] are used.

In the experiments, the experimental settings of Barshatski have been followed. For each molecule in the test set, N output molecules are generated, where N is set to 20.

In the following, some performance metrics are introduced for performance comparisons. To be specific, for the QED optimization tasks, there are totally five performance metrics that are used for performance assessments, including the property value, success rate, similarity, novelty, and diversity.

- Property is the average score of optimized molecules. For the calculation, the attribute values of the optimized molecules exceeding the threshold are first summed up and then the average value is calculated.
- Success is usually defined by comparing the ratio of the number of successfully optimized molecules to the total number of optimized molecules.
- Similarity denotes the average resemblance between the optimized molecules and those in the test set, computed through Morgan fingerprints.
- Novelty is an important evaluation metric used to assess whether the optimized molecules are unknown or unrecorded in the existing dataset, meaning they do not appear in the drug molecules of the training and test sets.
- Diversity is used to assess how different the optimized molecular populations are from each other.

As for the PlogP optimization task, two performance metrics are used for performance comparisons, namely the property improvement and success rate. It is worth noting that, on the QED optimization task, the property and similarity thresholds are set to 0.8 and 0.4, respectively. On the PlogP optimization task, three different similarity thresholds are used, which include 0.2, 0.4, and 0.6.

3.2 Results on the QED and PlogP Tasks

Table 1 shows the statistical results obtained for the proposed model RMST and three baseline models on the QED optimization task. It can be seen that the proposed model exhibits the best overall performance, having obtained the best optimization results in terms of property value, success rate, and similarity. In addition, the optimization results of the proposed model have the advantage of competing with the comparative baseline model in terms of novelty and diversity.

The optimization results of the proposed model and three baseline models on the PlogP optimization task are presented in Table 2. As shown in the table, under the three different similarity thresholds, the proposed model RMST obtains the largest property improvement and success rate, which are obviously better than the three baseline models.

Table 1. The comparative results of RMST and other baseline models on the QED task.

Method	Success rate	Novelty	Similarity	Property	Validity	Diversity
JTVAE	0.126	0.979	0.304	0.878	**1.000**	1.000
Mol-CG	0.081	0.961	0.302	0.857	0.988	1.000
UGMMT	0.489	0.998	0.427	0.879	0.971	1.000
RMST	**0.635**	**1.000**	**0.530**	**0.879**	0.955	**1.000**

Table 2. The comparative results of RMST and other baseline models on the PlogP task.

Method	$\delta = 0.2$		$\delta = 0.4$		$\delta = 0.6$	
	impro	success	impro	success	impro	success
JTVAE	1.74 ± 1.81	49.88%	0.13 ± 0.68	5.3%	0.2 ± 0.17	1.00%
Mol-CG	0.97 ± 1.35	34.34%	0.12 ± 0.60	8.6%	0.02 ± 0.14	2.8%
UGMMT	3.11 ± 2.76	60.2%	0.80 ± 1.22	6.4%	0.27 ± 0.26	3.2%
RMST	$\mathbf{3.21 \pm 1.72}$	**60.6%**	$\mathbf{1.62 \pm 1.94}$	**35.9%**	$\mathbf{0.51 \pm 1.17}$	**13.9%**

3.3 Ablation Experiments

In this subsection, a number of ablation experiments are conducted to confirm the validity of each key component of the proposed model. For this purpose, three different variant of the model are designed. Specifically, the first variant model (RMST1) is obtained by removing the CNN in the content encoder, the second variant model (RMST2) is designed by deleting the retrieval module, and the third variant model (RMST3) is obtained by removing the attention mechanism in the feature fusion module.

Table 3 gives the optimization results obtained for the proposed RMST and three variant models on the QED optimization task. The table clearly demonstrates that RMST outperforms the three variants in terms of success rate, similarity, and validity. This highlights a significant decrease in performance for the proposed model when the three key components are excluded. The optimization results obtained by the proposed model and the three variant models on the PlogP optimization task are presented in Table 4. Similarly, when the three components are removed from the proposed model, its performance deteriorates.

Table 3. The comparative results of RMST.

Method	Success rate	Novelty	Similarity	Validity	Diversity
RMST	**0.635**	**1.000**	**0.530**	**0.955**	**1.000**
RMST[1]	0.081	1.000	0.521	0.955	0.995
RMST[2]	0.489	1.000	0.462	0.946	1.000
RMST[3]	0.135	1.000	0.273	0.842	1.000

In short, based on the experimental results as mentioned above, it can be concluded that the CNN in the content encoder, the retrieval module, and the attention mechanism in the feature fusion module have important effects on the performance of the proposed model.

Table 4. The comparative results of RMST and other baseline models on the PlogP task.

Method	$\delta = 0.2$		$\delta = 0.4$		$\delta = 0.6$	
	impro	success	impro	success	impro	success
RMST	**3.21 ± 1.72**	**60.6%**	**1.62 ± 1.94**	**35.9%**	**0.51 ± 1.17**	**13.9%**
RMST[1]	3.15 ± 2.73	37.1%	0.78 ± 1.83	9.8%	0.02 ± 0.289	0.4%
RMST[2]	3.18 ± 2.66	38.2%	0.799 ± 1.83	10.4%	0.05 ± 0.404	0.9%
RMST[3]	3.46 ± 2.71	40.6%	0.950 ± 1.96	15.0%	0.08 ± 0.57	1.2%

3.4 Visualized Optimization Results

Figure 2 displays the distribution of successfully optimized molecules for JTVAE, Mol-CG, and UGMMT, and the proposed RMST on the QED molecular optimization task. On the one hand, compared with JTVAE and Mol-CG, the UGMMT and RMST obtain more molecules that are successfully optimized. On the other hand, when the molecules obtained by UGMMT concentrate on the bottom-left corner, the molecules obtained by the proposed RMST have better performance metrics, showing a better distribution.

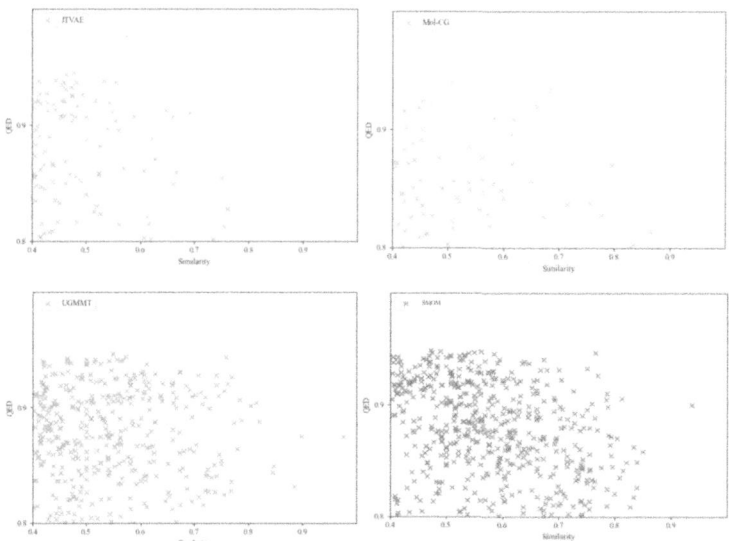

Fig. 2. Distribution of successfully optimized molecules of JTVAE, Mol-CG, UGMMT, and the proposed RMST on the QED molecular optimization task.

In this subsection, the molecular structures have also been compared. An effective optimization model should not only produce molecules with the desired properties, but also molecules with structures similar to the originals. As shown in Fig. 3, several molecules have been selected from the generated results to observe the changes in the optimized molecules, where the first row and the second row in each sub-figure denote the lead and optimized molecules, respectively. From the figure, it can be seen that after optimization, the QED values of the molecules have significantly improved. Moreover, by observing the structure of the optimized molecules, It is clear that the proposed model achieves significant attribute value enhancement by modifying only a small portion of the molecule, ensuring that good similarity is maintained between the input and output molecules.

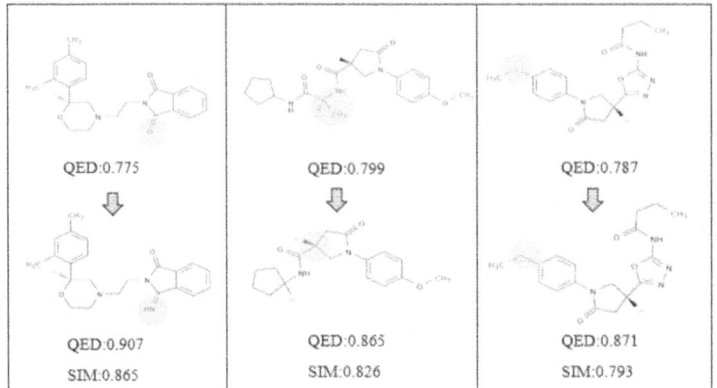

Fig. 3. Molecular structures of the input and optimized molecules.

3.5 Parameter Analysis

In this subsection, the effect of the parameter K on the model performance will be investigated. To this aim, five different parameter settings are tested, namely K = 1, 5, 10, 15, 20. For the parameter analysis experiments, similarity thresholds of 0.4 and 0.2 are applied to the QED and PlogP tasks, respectively. From the statistical results as shown in Fig. 4, when K is set to 10, the proposed model obtains the best performance.

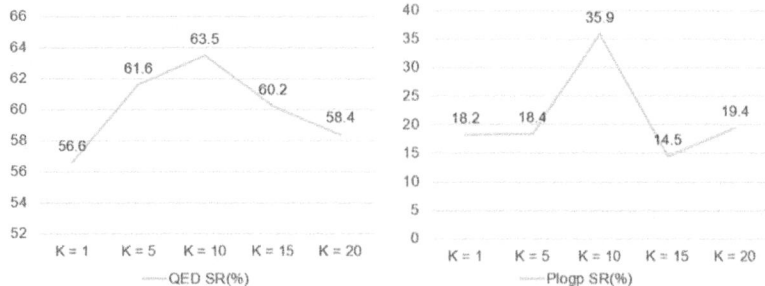

Fig. 4. Success rate obtained by the proposed model with different parameter settings on the QED and PlogP optimization tasks.

4 Conclusion

In the paper, we have proposed a retrieval-based molecular style transformation model, namely RMST. Given the SMILES sequence of a lead molecule, the proposed RMST used a CNN+biLSTM encoder to obtain the latent embedding vector. At the same time, the model used a biLSTM encoder to obtain the embedding vectors of molecules that were to be retrieved by a MIPS retriever. Afterwards, an attention mechanism was used to fuse the latent embedding vectors of the input and retrieved molecules, guiding the model to obtain the target molecule with improved properties. In the experiments, the performance of the proposed model has been verified by comparing it with three state-of-the-art baseline models. Experimental results on the QED and PlogP tasks demonstrated that the proposed model achieved superior performance in both improving properties and maintaining similarity.

Funding. This work has been supported by the National Key Research and Development Program of China (Grant No. 2021YFE0102100), the University Synergy Innovation Program of Anhui Province (Grant No. 2021-039), Anhui University outstanding youth research project(Grant No. 2022AH020010), National Natural Science Foundation of China (Grant Nos. 62172002, 62122025, 61872309, U19A2064) and Anhui Provincial Natural Science Foundation (Grant No. 2108085QF267).

References

1. Paul, S.M., Mytelka, D.S., Dunwiddie, C.T., et al.: How to improve R&D productivity: the pharmaceutical industry's grand challenge. Nat. Rev. Drug Discovery **9**(3), 203–214 (2010)
2. Lien, E.J., Ren, S., Bui, H.H., et al.: Quantitative structure-activity relationship analysis of phenolic antioxidants. Free Radical Biol. Med. **26**(3–4), 285–294 (1999)
3. Polishchuk, P.G., Madzhidov, T.I., Varnek, A.: Estimation of the size of drug-like chemical space based on GDB-17 data. J. Comput. Aided Mol. Des. **27**, 675–679 (2013)
4. Phatak, S.S., Stephan, C.C., Cavasotto, C.N.: High-throughput and in silico screenings in drug discovery. Expert Opin. Drug Discov. **4**(9), 947–959 (2009)
5. Bartók, A.P., De, S., Poelking, C., et al.: Machine learning unifies the modeling of materials and molecules. Sci. Adv. **3**(12), e1701816 (2017)
6. Button, A., Merk, D., Hiss, J.A., et al.: Automated de novo molecular design by hybrid machine intelligence and rule-driven chemical synthesis. Nat. Mach. Intell. **1**(7), 307–315 (2019)
7. Kotsias, P.C., Arús-Pous, J., Chen, H., et al.: Direct steering of de novo molecular generation with descriptor conditional recurrent neural networks. Nat. Mach. Intell. **2**(5), 254–265 (2020)
8. Ahn, S., Kim, J., Lee, H., et al.: Guiding deep molecular optimization with genetic exploration. In: Advances in Neural Information Processing Systems, vol. 33, pp. 12008–12021 (2020)
9. Nigam, A.K., Friederich, P., Krenn, M., et al.: Augmenting genetic algorithms with deep neural networks for exploring the chemical space. arXiv preprint arXiv:1909.11655 (2019)
10. Gómez-Bombarelli, R., Wei, J.N., Duvenaud, D., et al.: Automatic chemical design using a data-driven continuous representation of molecules. ACS Cent. Sci. **4**(2), 268–276 (2018)
11. Hoffman, S.C., Chenthamarakshan, V., Wadhawan, K., et al.: Optimizing molecules using efficient queries from property evaluations. Nat. Mach. Intell. **4**(1), 21–31 (2022)
12. Zhou, Z., Kearnes, S., Li, L., et al.: Optimization of molecules via deep reinforcement learning. Sci. Rep. **9**(1), 10752 (2019)

13. Sun, M., Xing, J., Meng, H., et al.: Molsearch: search-based multi-objective molecular generation and property optimization. In: Proceedings of the 28th ACM SIGKDD Conference on Knowledge Discovery and Data Mining, pp. 4724–4732 (2022)
14. Shin, B., Park, S., Bak, J.Y., et al. Controlled molecule generator for optimizing multiple chemical properties. In: Proceedings of the Conference on Health, Inference, and Learning, pp. 146–153 (2021)
15. Barshatski, G., Radinsky, K.: Unpaired generative molecule-to-molecule translation for lead optimization. In: Proceedings of the 27th ACM SIGKDD Conference on Knowledge Discovery & Data Mining, pp. 2554–2564 (2021)
16. Barshatski, G., Nordon, G., Radinsky, K.: Multi-property molecular optimization using an integrated poly-cycle architecture. In: Proceedings of the 30th ACM International Conference on Information & Knowledge Management, pp. 3727–3736 (2021)
17. Maziarka, Ł, Pocha, A., Kaczmarczyk, J., et al.: Mol-CycleGAN: a generative model for molecular optimization. J. Cheminform. 12(1), 1–18 (2020)
18. Griffen, E., Leach, A.G., Robb, G.R., et al.: Matched molecular pairs as a medicinal chemistry tool: miniperspective. J. Med. Chem. 54(22), 7739–7750 (2011)
19. Dalke, A., Hert, J., Kramer, C.: Mmpdb: An open-source matched molecular pair platform for large multiproperty data sets. J. Chem. Inf. Model. 58(5), 902–910 (2018)
20. Bahdanau, D., Cho, K., Bengio, Y.: Neural machine translation by jointly learning to align and translate. arXiv preprint arXiv:1409.0473 (2014)
21. Yi, X., Liu, Z., Li, W., et al.: Text style transfer via learning style instance supported latent space. In: Proceedings of the Twenty-Ninth International Conference on International Joint Conferences on Artificial Intelligence, pp. 3801–3807 (2021)
22. Cai, D., Wang, Y., Li, H., et al.: Neural machine translation with monolingual translation memory. arXiv preprint arXiv:2105.11269 (2021)
23. Hochreiter, S., Schmidhuber, J.: Long short-term memory. Neural Comput. 9(8), 1735–1780 (1997)
24. Irwin, R., Dimitriadis, S., He, J., et al.: Chemformer: a pre-trained transformer for computational chemistry. Mach. Learn. Sci. Technol. 3(1), 015022 (2022)
25. Landrum, G., et al.: RDKIT: open-source cheminformatics software (2016)
26. Fu, T., Xiao, C., Sun, J.: Core: automatic molecule optimization using copy & refine strategy. In: Proceedings of the AAAI Conference on Artificial Intelligence, vol. 34, no. 01, pp. 638–645 (2020)
27. Jin, W., Yang, K., Barzilay, R., Jaakkola, T.: Learning multimodal graph-to-graph translation for molecular optimization. In: 7th International Conference on Learning Representations, ICLR (2019)

Aggregation Strategy with Gradient Projection for Federated Learning in Diagnosis

Huiyan Lin[1,2], Yunshu Gao[1,2], Heng Li[1(✉)], Xiaotian Zhang[1,2], Xiangyang Yu[1,2], Jianwen Chen[3], and Jiang Liu[1,2]

[1] Research Institute of Trustworthy Autonomous Systems, Southern University of Science and Technology, Shenzhen 518000, China
lih3@sustech.edu.cn

[2] Department of Computer Science and Engineering, Southern University of Science and Technology, Shenzhen 518000, China

[3] Department of Orthopedics Medicine, Southern University of Science and Technology Hospital, Shenzhen 518000, Guangdong, China

Abstract. Federated learning aims to address privacy and data security concerns associated with distributed data resources. However, data across different clients typically is not independently and identically distributed, resulting in different local optimal objectives. This disparity will hinder the convergence and performance of global models. Moreover, the presence of noisy labels in client data further complicates matters, making it harder to efficiently deploying global models on a single client. To overcome these issues, we propose a novel algorithm called Federated Learning Aggregation Strategy with Gradient Projection Memory (FedGPM), which leverages gradient projection to refine the model aggregation process. FedGPM reduces the impact of data heterogeneity by projecting gradients into orthogonal directions to remove inconsistent gradient components. Based on the gradient projection memory, the server maintains a federated projection matrix for each client, accurately quantifying the distribution difference between that client's data and the rest. Adaptive update strategy is employed for each layer during local model training, based on the consistency of local and others' gradient directions, ensuring positive contributions to global model progress. Experimental results conducted on disease diagnosis tasks using the OCT dataset, with varying levels of data heterogeneity and noise label ratios, demonstrate the superior performance of our algorithm over state-of-the-art methods.

Keywords: Federated Learning · Gradient Projection · Disease Diagnosis · Data Heterogeneity · Model Aggregation

1 Introduction

With advancements in deep learning, computer-assisted diagnosis [1, 2] has made significant strides, increasing the demand for centralized platforms offering remote automated diagnostic services. However, data privacy conflicts among data centers pose a challenge

D.-S. Huang et al. (Eds.): ICIC 2024, LNBI 14881, pp. 207–218, 2024.
https://doi.org/10.1007/978-981-97-5689-6_18

to the development of algorithm [3–6]. Deploying algorithms individually at each hospital is costly, and the variability in data distribution [7, 27] between centers may render a single algorithm unsuitable for all hospitals. Federated learning (FL) [8, 9] addresses these challenges by providing a general model and enhancing client performance without compromising client privacy.

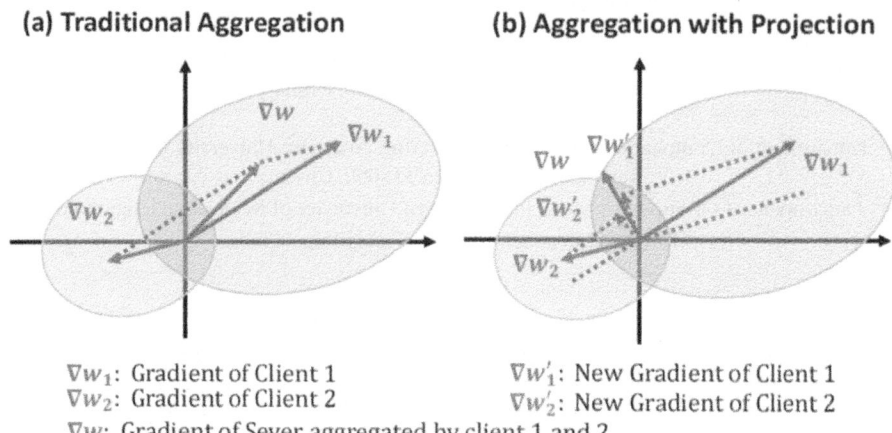

(a) Traditional Aggregation   **(b) Aggregation with Projection**

∇w_1: Gradient of Client 1 $\nabla w_1'$: New Gradient of Client 1
∇w_2: Gradient of Client 2 $\nabla w_2'$: New Gradient of Client 2
∇w: Gradient of Sever aggregated by client 1 and 2

Fig. 1. Illustration of traditional aggregation and GPM aggregation when client update directions are inconsistent. (a) Directly sum ∇w_1 and ∇w_2 to get ∇w, which has the opposite direction to ∇w_2. (b) Project ∇w_1 to the direction of ∇w_2 to obtain $\nabla w_1 /$ that is orthogonal to ∇w_2. In this way, ∇w_1 is consistent with the direction of both clients.

Recently, FL has made significant progress in medical applications. Guo et al. [10] introduced a method utilizing MRI from multiple institutions to train a deep learning model for magnetic resonance imaging (MRI) reconstruction. Bercea et al. [11] presented a FL algorithm that utilizes MRI from four different clients to unsupervisedly train a convolutional autoencoder for abnormal region segmentation. Moreover, Liu et al. [12] proposed a methodology combining FL and domain generalization to develop a robust medical image segmentation model. Despite the successful integration of FL into medical settings by these methods, the coordination of weight updates between the central general model parameters and the client model parameters remains an open question.

Some methods have explored the update strategy of the central general model parameters and the client model parameters. FedAvg, proposed in [13], involves each client updating the model locally using stochastic gradient descent, followed by model averaging by the server to achieve global performance improvement. However, FedAvg faces challenges with heterogeneous data (i.e., non-independent and non-identically distributed data across the network), resulting in client drift. The heterogeneity of data results in variations in the update directions of client models, making it challenging to achieve balance among these models. Later studies prioritized mitigating the impact of data heterogeneity. In [14], the issue of client drift is addressed by introducing control variables, prioritizing the reduction of communication costs. In [15], FedAvg is reparameterized and added a proximal term to prevent significant deviation of the local model from the global model, albeit at the expense of increased computational and memory overhead. In [16], an adaptive weight clipping mechanism is introduced to reduce weight differences between client updates, thereby enhancing convergence speed and communication efficiency. However, these existing methods have limitations, such as overlooking heterogeneous client gradient interaction and disregarding potentially valuable client data, and cannot handle out-of-distribution data. Hence, it is imperative to devise strategies that better leverage information from all clients to enhance efficiency and convergence in FL.

Inspired by [17], we introduce the concept of Core Gradient Space (CGS) into federated learning. The CGS analyzes gradient updates during model training to derive a basis for the model's gradient space, capturing the principal directions of model updates. By analyzing the CGS of each client, we can quantify the differences in update directions between the local models of clients and adjust them accordingly to mitigate the impact of data heterogeneity on the global model. We proposed a novel aggregation approach called Federated with Gradient Projection Memory (FedGPM). With FedGPM, our goal is to mitigate direction inconsistencies during model aggregation in FL, thereby enhancing the stability and efficiency of model updates. Specifically, we compute the Core Gradient Space (CGS) within client's data domain to extract key domain knowledge. Alongside uploading gradients, the client also uploads spatial information related to the CGS. On the server side, a federal projection matrix is maintained for each client, which integrates gradient space information from other clients. Unlike conventional aggregation methods, FedGPM eliminates gradient parts that hinder the training of other clients during local parameter updates by projecting gradients.

Due to the influence of data heterogeneity, inconsistencies in update directions between clients may emerge. Figure 1 visually illustrates this phenomenon: (a) showcases outcomes from traditional aggregation methods where the gradient direction of the aggregated global model conflicts with ∇w_2, , whereas (b) demonstrates outcomes achieved using gradient projection. Here, each client's gradient is projected onto the update directions of others, resulting in orthogonal gradients, thereby ensuring their update directions remain independent. Subsequent aggregation maintains the performance of all clients without degradation. Our contributions can be summarized as follows:

- We propose a novel method for aggregating weights (FedGPM) in FL, which effectively addresses the challenge of inconsistent client updates.
- The federated projection matrix is introduced to record the gradient information of other clients and quantify the disparities between local and other clients data.
- Leveraging gradient projections to update local models alleviates data heterogeneity issues, specifically by eliminating portions of the gradients located in the CGS that are exclusive of other clients.
- The effectiveness of FedGPM is demonstrated through diagnosis experiments under different heterogeneity levels and noise label conditions, where FedGPM achieves superior performance compared to state-of-the-art FL algorithms.

2 Method

In this paper, we tackle the issue of inconsistent client update directions in FL. To address this challenge, we propose a novel algorithm called FedGPM, which leverages gradient projection memory. Our approach, illustrated in Fig. 2, involves clients uploading their gradient projection matrices to the server and receiving a federal projection matrix that represents the update directions of other clients. During local training, we selectively discard gradient parts that oppose the update directions captured by the federal projection matrix, utilizing directional consistency as a criterion. This ensures that the uploaded model gradients do not negatively affect other clients' training, leading to efficient aggregation and improved model convergence.

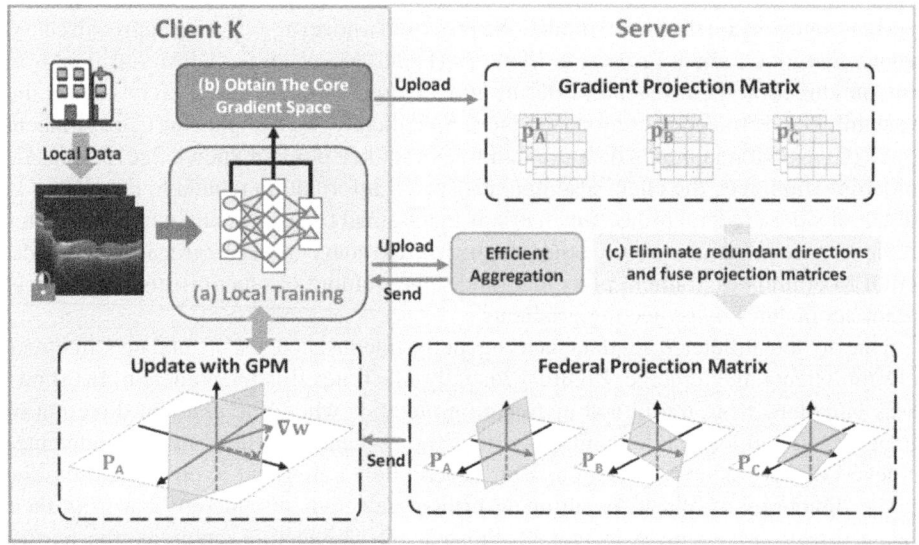

Fig. 2. Overview of FedGPM. (a) Client k uses local data to train the model and updates the model with GPM. (b) Extract the input from each layer of the model to obtain the information of CGS, which is depicted using a gradient projection matrix and uploaded to the server. (c) The server aggregates the gradient projection matrix from all clients, eliminating redundant directions to generate federal projection matrix. This matrix is then sent back to the respective client to facilitate further model updates.

2.1 Problem Definition

Assuming there are k distinct clients, and the dataset of each client is denoted by $\{D_1, D_2, \ldots, D_k\}$. The data distribution across clients is non-iid due to factors such as varying medical devices and the diversity of patients' races. Usually, a shared general model is established, and the server instructs each client to train the model locally. Subsequently, the server evenly aggregates the model gradients uploaded by each client as following:

$$w^{t+1} = \sum_{i=0}^{k} \frac{n_i}{\sum_{j=0}^{k} n_j} w_i^t, \tag{1}$$

where t is the training epoch, w is the model weight, and n is the number of training samples.

In the given case, it overlooks inconsistencies in the direction of gradient updates during training across different clients. This simplistic aggregation approach can negatively impact the updates of non-uniformly gradients. The aggregated global gradient tends to be biased towards the majority gradient direction, disregarding the contribution of minority gradients. To address this aggregation problem caused by inconsistent gradient directions, we introduce gradient projection into FL.

2.2 Federal Projection Matrix

Inspired by the concept of gradient projection memory introduced in [18, 19] to preserve historical learning content and prevent forgetting, we extend its application to FL. Our objective is to ensure that the training of a client's local model does not interfere with that of other clients. Building on insights from [20, 21], we recognize that during the weight update process using stochastic gradient descent, these weights exist within the parameter space defined by the input data D_k. This underscores the importance of weights in the space defined by D_k for each local client. Accordingly, our goal is to identify the CGS that encompasses these critical parameters by exploring the input space of D_k, aiming to mitigate adverse effects stemming from the interplay among the weights of each client.

Singular value decomposition (SVD) is an efficient way to obtain the CGS. The model we used consists of convolutional layers and fully connected layers. For each layer, denoting the input of layer l of client i as R_i^l, we perform SVD on $R_i^l = U_i^l \Sigma_i^l (V_i^l)^{\mathsf{T}}$. Then, we identify its k-rank approximation, which encompasses all directions with the highest singular values in the representation for obtaining the CGS. This process can be expressed as:

$$\left\| \left(R_i^l \right)_k \right\|_F^2 \geq \epsilon_i^l \left\| R_i^l \right\|_F^2, \tag{2}$$

where $\| \cdot \|_F$ is the Fibonacci norm and ϵ_i^l is the threshold hyperparameter. For layer l of client i, the CGS can be given by a projection matrix $p_i^l = [u_{i,1}^l, u_{i,2}^l, \ldots, u_{i,k}^l]$, comprising the first k vectors of U_i^l.

The server maintains a federated projection matrix for each client, used to store the essential vectors constituting other clients' CGS. After completing local model training, each client computes its CGS and upload it to update the federated projection matrices of other clients. Assuming the current client is denoted as C_i, and the federated projection matrix of the other clients is denoted as $\{P_j^l, j \neq i\}_{j=1}^k$. When updating the blank matrix P_j^l, we directly incorporate the CGS basis vectors into $P_{j,}^l$. However, if certain CGS directions of a client already exist in P_j^l, we first remove these redundant directions, denoted as $R^l_{i,proj}$, from R^l_i before performing SVD and k-rank approximation. This process ensures the non-redundancy of P^l_j. The representation is obtained as follows:

$$\hat{R}^l_i = R^l_i - R^l_{i,proj} = R^l_i - P^l_j \left(P^l_j\right)^T R^l_i. \tag{3}$$

Then perform SVD on $\hat{R}^l_i = \hat{U}^l_i \hat{\Sigma}^l_i (\hat{V}^l_i)^T$, and select the least k new orthogonal bases that satisfy the threshold according to the following formula:

$$\left\| \left(R^l_i\right)_k \right\|^2_F + \left\| \hat{R}^l_{i,proj} \right\|^2_F \geq \epsilon^l_i \left\| R^l_i \right\|^2_F. \tag{4}$$

Finally, the federated projection matrix of the l-th layer of C_j is given by $P^l_j = \left[P^l_j, \hat{u}^l_{i,1}, \ldots, \hat{u}^l_{i,k} \right]$.

2.3 Local Training with GPM

The server produces a federal projection matrix P^l_j, which captures the gradient update directions of other clients and sends it to C_i. To achieve efficient federated aggregation, we employ gradient projection during local training. If the update direction of local training is inconsistent with that of other clients, it becomes necessary to eliminate the part on the CGS of other clients before back propagation, retaining only the orthogonal gradients. However, if the update direction aligns with the others update direction, gradient projection is unnecessary. Here, we utilize cosine similarity to quantify the consistency of gradient directions, expressed by the following formula:

$$S^l_i = \frac{p^l_i \cdot P^l_i}{||p^l_i|| \cdot ||P^l_i||}. \tag{5}$$

If $S^l_i < 0$, it indicates that the gradient of local training solely benefits the current data without considering the data distribution of other clients. In such cases, gradient projection helps to mitigate this situation. Conversely, if $S^l_i \geq 0$, the gradient contributes positively to the overall performance. Therefore, the final update gradient δ is given by:

$$\nabla \hat{w}^l_i = \begin{cases} \nabla w^l_i - P^l_i (P^l_i)^T \nabla w^l_i \; if \; S^l_i < 0 \\ \nabla w^l_i \qquad\qquad\qquad if \; S^l_i \geq 0 \end{cases} \tag{6}$$

The adaptive update described above ensures that FedGPM mitigates the adverse effects on model aggregation resulting from the client's local update phase in FL.

For the local loss function, we employ cross-entropy loss, which is widely used in disease diagnosis tasks. It effectively quantifies the disparity between the predicted output and the true label.

3 Experiment

3.1 Datasets and Experiment Settings

To assess the effectiveness of FedGPM, our experiments are conducted on OCT2017 dataset [22], which is a widely employed benchmark dataset in the field of medical image analysis. It consists of 84,484 high-resolution optical coherence tomography (OCT) images captured from patients with various retinal diseases, including Choroidal Neo-vascularization (CNV), Diabetic Macular Edema (DME), Drusen, and Normal. This dataset provides a diverse range of images capturing different pathological conditions, retinal layers, and disease severity levels, enabling comprehensive research and evaluation of automated diagnostic algorithms. For data preprocessing, each image is resized to 256 × 256 pixels and normalized. To simulate real-world FL scenarios, we partition the dataset into four clients using Dirichlet distribution [3], with smaller α indicating higher data heterogeneity. The impact of different α on the statistical heterogeneity of the OCT2017 dataset is visualized in Fig. 2. When $\alpha =0.1$, the discrepancies in both the types and numbers of diseases across each client become even more pronounced.

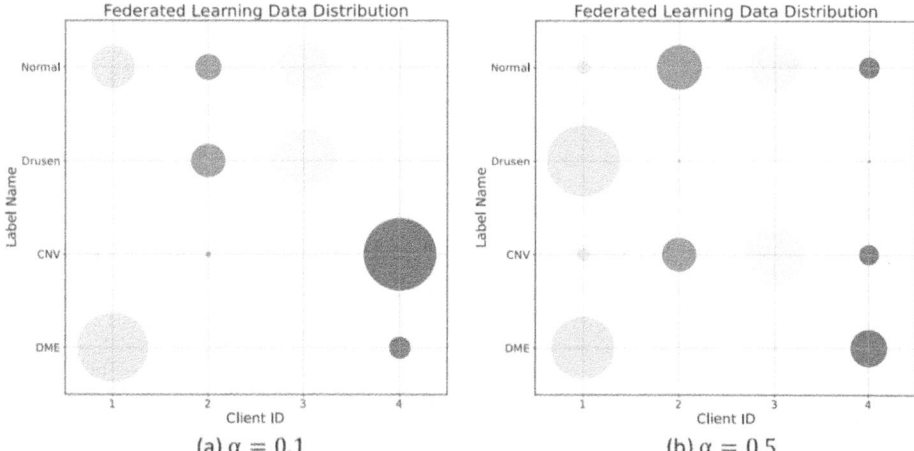

(a) $\alpha = 0.1$ (b) $\alpha = 0.5$

Fig. 3. Visualization illustrating the statistical heterogeneity among clients in the OCT2017 dataset, with the x-axis denoting the client ID, the y-axis indicating the disease label, and the size of the scatter points representing the number of training samples.

3.2 Implementation Details

Our proposed method and the compared method employ the same convolutional neural network with 5 convolutional layers followed by 2 linear layers. During training, SGD is utilized as the optimizer, with a learning rate set to 0.001 and a batch size of 16. Each communication round involves local training for 1 epoch, followed by training the global model for 100 rounds, with the best-performing model on the validation set selected as the final model. The entire framework is implemented using PyTorch 1.9.0 and trained on an NVIDIA GeForce RTX 2080 Ti GPU.

3.3 Evaluation and Discussion

The traditional Fl algorithm FedAvg [13] is used as the baseline. SCAFFOLD [14] employs control variables to address client drift during local updates. FedProx [15] integrates proximal terms into optimization objectives to prompt clients to update local models towards global optimality. MOON [4] incorporates contrastive learning to handle non-iid challenges in FL. FedNTD [5] introduces knowledge distillation, assuming that mitigating the forgetting problem will alleviate data heterogeneity issues. The comparative experiments are quantified using accuracy and F1 score metrics. Experiments are conducted under two different FL scenarios, each with different heterogeneity levels based on different Dirichlet distributions $\alpha = 0.1$ and $\alpha = 0.5$. Additionally, to showcase the robustness of our approach against noise interference, various levels of noise labels are introduced to the clients.

Table 1. Performance results of global models for different methods

Metrics	ACC			F1		
Noise Rate	0%	10%	30%	0%	10%	30%
Dirichlet Parameter	$\alpha = 0.1$					
FedAvg [13]	65.80	65.50	61.36	73.67	72.86	68.45
SCAFFOLD [4]	66.60	59.08	56.52	72.81	61.36	57.19
FedProx [5]	66.15	65.67	61.51	73.85	72.99	68.44
MOON [4]	64.59	65.12	59.49	72.89	72.86	65.74
FedNTD [5]	65.93	66.20	61.44	72.44	73.26	66.78
FedGPM (Ours)	**66.73**	**66.52**	**61.64**	**73.87**	**74.49**	**69.11**
Dirichlet Parameter	$\alpha = 0.5$					
FedAvg [13]	58.66	58.08	57.40	67.91	67.50	66.14
SCAFFOLD [4]	**59.85**	57.30	56.40	65.51	64.48	64.12
FedProx [5]	58.66	57.50	57.01	68.14	67.11	66.32
MOON [4]	58.94	59.11	56.98	67.57	68.06	64.46
FedNTD [5]	56.75	57.04	56.21	62.77	64.20	61.60
FedGPM (Ours)	59.43	**59.29**	**57.59**	**68.59**	**68.54**	**66.92**

The experimental results are shown in Table 1, highlighting the superior results in bold. Our FedGPM consistently outperforms the baseline in all settings, achieving state-of-the-art results in most cases. Conversely, suboptimal methods exhibit varying performance under different conditions, indicating instability in the presence of noisy labels. In the case of high heterogeneity, FedAvg [13], SCAFFOLD [4], and FedProx [5] experience declining performance with increasing noise ratios. Despite enhancements over FedAvg [13], SCAFFOLD [4], and FedProx [5] remain constrained by its framework. Introducing contrastive learning and distillation learning through MOON [24] and

FedNTD [25] respectively shows promise, especially with minimal noise. However, due to complexity and tuning challenges, their performance trails behind SCAFFOLD [14] or FedProx [15]. In the case of low heterogeneity, although SCAFFOLD [14] exhibits superior accuracy compared to FedGPM, its F1 score notably lags behind. MOON [24] demonstrates impressive performance with minimal noise. But when the noise ratio further increases, FedAvg [13] and FedProx [15] exhibit greater robustness.

Compared with these methods, our FedGPM shows superior performance, particularly in terms of F1 score. FedGPM demonstrates minimal impact when incorporating minor noise. The consistency-based gradient direction projection effectively filters out adverse information during model updates. However, extensive noise may still affect the model to some extent. As our approach is rooted in supervised learning, we plan to further refine and enhance it in future iterations.

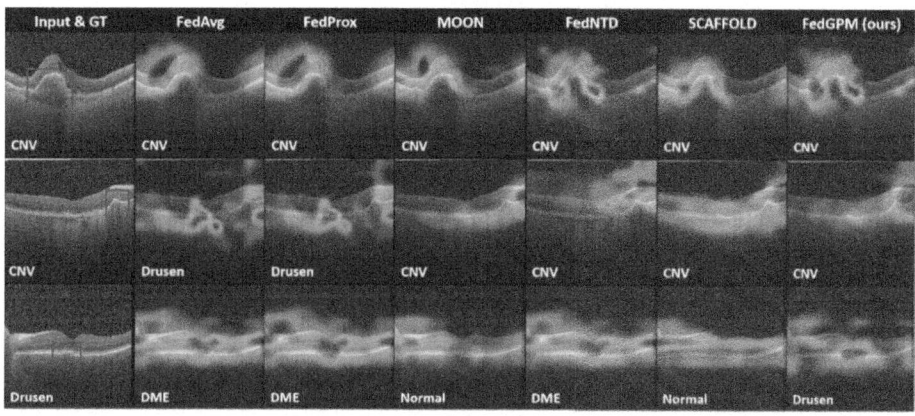

Fig. 4. Visualization of the models by CAM [6]. The leftmost image depicts the original input image alongside its ground truth. The prediction results obtained from comparison methods and our proposed method are presented on the right. The predicted category is indicated in the lower left corner of each image. The regions corresponding to the lesions are highlighted by a red frame.

The prediction heat maps of different models drawn by CAM are shown in the Fig. 4, with lesion areas marked in red boxes. For samples correctly classified by all models, the area of concern typically aligns closely with the lesion, with FedGPM exhibiting higher accuracy. In the case of partially correct samples, FedAvg [13] and FedProx [5] concentrate on other areas, leading to misclassifications. FedNTD [25] and SCAFFOLD [14] show overly broad areas of concern, while MOON's [24] area closely matches the lesion's size but lacks the precision of FedGPM's location. In instances where only FedGPM achieves correct classification, it demonstrates superior capability in identifying subtle lesions compared to other methods. In summary, FedGPM, built upon FedAvg [13], maintains stability and achieves optimal performance across diverse settings. The gradient backpropagation method, rooted in GPM, effectively filters information, aiding in the efficient generation of the optimal global model.

3.4 Ablation Studies

Ablation studies conducted under various Dirichlet distributions aim to validate the efficacy of the proposed gradient projection and consistency constraints, as outlined in the Table 2. Results indicate a decline in performance when solely employing gradient projection. This decline is attributed to the indiscriminate projection of gradients into directions orthogonal to other clients, leading to the loss of pertinent information when their distributions are similar. Hence, the inclusion of consistency constraints becomes imperative. In scenarios where client distributions are consistent, requiring identical update directions, additional operations are unnecessary. Conversely, in cases of notable distribution disparities, gradient projection onto orthogonal directions becomes essential to mitigate the adverse impact of uploaded gradients on other clients.

Table 2. Ablation experiment

Baseline	Projection	Consistency	$\alpha = 0.1$		$\alpha = 0.5$	
			ACC	F1	ACC	F1
✓			65.80	73.67	58.66	67.91
✓	✓		63.37	72.33	52.45	62.22
✓	✓	✓	**66.73**	**73.87**	**59.43**	**68.59**

4 Conclusion

Federated learning offers a promising avenue for resolving privacy concerns and enhancing data security in medical data through decentralized learning. However, challenges arise from data heterogeneity and noisy data, which complicate aggregation. To tackle this hurdle, we propose FedGPM, a gradient projection algorithm tailored for efficient federated learning aggregation. FedGPM leverages gradient consistency constraints to gauge data distribution disparities among clients, facilitating precise updates of local models and streamlined aggregation of global models. Experimental findings demonstrate that FedGPM outperforms existing methods in non-IID scenarios and exhibits resilience against noise interference to a considerable degree.

Acknowledgments. This work was supported in part by General Program of National Natural Science Foundation of China (Grant No. 82272086), Shenzhen Science and Technology Program (JCYJ20210324103800001, JCYJ20220530112609022), Guangdong Basic and Applied Basic Research Fund (2022A1515010487), Nanshan District Healthcare Program (NSZD2023058) and SUSTech Undergraduate Innovation and Entrepreneurship (S202314325016). We appreciate Shenzhen DE Sci&Tech Company for their data support in this study. School Level Projects of the 2024 "Climbing Program" Special Funds (Y01141820).

Competing Interests. The authors declare that they have no competing interests.

References

1. Shanmugavadivel, K., Sathishkumar, V., Cho, J., Subramanian, M.: Advancements in computer-assisted diagnosis of Alzheimer's disease: a comprehensive survey of neuroimaging methods and AI techniques for early detection. Ageing Res. Rev. **91**, 102072 (2023)
2. Juan, J., et al.: Computer-assisted diagnosis for an early identification of lung cancer in chest X rays. Sci. Rep. **13**, 7720 (2023)
3. Lu, Z., Pan, H., Dai, Y., Si, X., Zhang, Y.: Federated learning with non-IID data: a survey. IEEE Internet Things J. (2024)
4. Li, H., et al.: A generic fundus image enhancement network boosted by frequency self-supervised representation learning. Med. Image Anal. **90**, 102945 (2023)
5. Li, H., et al.: Enhancing and adapting in the clinic: Source-free unsupervised domain adaptation for medical image enhancement. IEEE Trans. Med. Imaging (2024)
6. Li, H., et al.: Frequency-mixed single-source domain generalization for medical image segmentation. In: Medical Image Computing and Computer Assisted Intervention, MICCAI (2023)
7. Li, H., et al.: An annotation-free restoration network for cataractous fundus images. IEEE Trans. Med. Imaging **41**, 1699–1710 (2022)
8. Li, Q., Diao, Y., Chen, Q., He, B.: Federated learning on non-IID data silos: an experimental study. In: International Conference on Data Engineering, ICDE (2022)
9. Mahon, P., et al.: A federated learning system for precision oncology in Europe: DigiONE. Nat. Med. **30**, 334–337 (2024)
10. Guo, P., Wang, P., Zhou, J., Jiang, S., Patel, V.M.: Multi-institutional collaborations for improving deep learning-based magnetic resonance image reconstruction using federated learning. In: Conference on Computer Vision and Pattern Recognition, CVPR (2021)
11. Bercea, C.I., Wiestler, B., Rueckert, D., Albarqouni, S.: Federated disentangled representation learning for unsupervised brain anomaly detection. Nature Machine Intelligence (2022)
12. Liu, Q., Chen, C., Qin, J., Dou, Q., Heng, P.: FedDG: federated domain generalization on medical image segmentation via episodic learning in continuous frequency space. In: Conference on Computer Vision and Pattern Recognition, CVPR (2021)
13. McMahan, B., Moore, E., Ramage, D., Hampson, S., y Arcas, B.A.: Communication-efficient learning of deep networks from decentralized data. In: International Conference on Artificial Intelligence and Statistics, AISTATS (2017)
14. Karimireddy, S.P., Kale, S., Mohri, M., Reddi, S.J., Stich, S.U., Suresh, A.T.: SCAFFOLD: stochastic controlled averaging for federated learning. In: International Conference on Machine Learning, ICML (2020)
15. Li, T., Sahu, A.K., Zaheer, M., Sanjabi, M., Talwalkar, A., Smith, V.: Federated optimization in heterogeneous networks. In: Machine Learning and Systems, MLSys (2020)
16. Wang, J., Liu, Q., Liang, H., Joshi, G., Poor, H.V.: Tackling the objective inconsistency problem in heterogeneous federated optimization. In: Annual Conference on Neural Information Processing Systems, NeurIPS (2020)
17. Shu, K., et al.: Replay-oriented gradient projection memory for continual learning in medical scenarios. In: IEEE International Conference on Bioinformatics and Biomedicine, BIBM (2022)
18. Luo, K., Li, X., Lan, Y., Gao, M.: GradMA: a gradient-memory-based accelerated federated learning with alleviated catastrophic forgetting. In: Conference on Computer Vision and Pattern Recognition, CVPR (2023)
19. Saha, G., Garg, I., Roy, K.: Gradient projection memory for continual learning. In: International Conference on Learning Representations, ICLR (2021)

20. Saha, G., Roy, K.: Continual learning with scaled gradient projection. In: Conference on Artificial Intelligence, AAAI (2023)
21. Zhang, C., Bengio, S., Hardt, M., Recht, B., Vinyals, O.: Understanding deep learning (still) requires rethinking generalization. Commun. ACM **64**, 107–115 (2021)
22. Kermany, D., Zhang, K., Goldbaum, M.: Labeled optical coherence tomography (Oct) and chest X-ray images for classification. Mendeley Data (2018)
23. Zhu, Z., Hong, J., Zhou, J.: Data-free knowledge distillation for heterogeneous federated learning. In: International Conference on Machine Learning, ICML (2021)
24. Li, Q., He, B., Song, D.: Model-contrastive federated learning. In: Conference on Computer Vision and Pattern Recognition, CVPR (2021)
25. Lee, G., Jeong, M., Shin, Y., Bae, S., Yun, S.: Preservation of the global knowledge by not-true distillation in federated learning. In: Annual Conference on Neural Information Processing Systems, NeurIPS (2022)
26. Fernandez, F.G.: TorchCAM: class activation explorer (March 2020). https://github.com/frgfm/torch-cam
27. Li, H., et al.: RaffeSDG: random frequency filtering enabled single-source domain generalization for medical image segmentation. arXiv preprint arXiv:2405.01228 (2024)

Coronary Artery 3D/2D Registration Based on Particle Swarm Optimization of Contextual Morphological Features

Yibo Liu and Ting Ke[✉]

College of Artificial Intelligence, Tianjin University of Science and Technology, Tianjin 300222, China

keting@tust.edu.cn

Abstract. This paper proposes a novel method for coronary artery registration of computed tomography angiography and digital subtraction angiography (DSA) images. Firstly, the vascular centerlines of DSA are extracted, the sub-pixel corner points of the vessel centerline are identified by Shi-Tomas corner detection algorithm and gray central moment model. Then, the 2D vessel intersections are extracted by combining the skeleton erosion algorithm and the contextual morphological feature descriptors. For the extraction of 3D vessel intersections, we extract the 3D vessel centerline through the axial section centroid fitting algorithm, the 3D vessel intersections are extracted according to the centerline intersecting model. Finally, the pose of the 3D vessel model is searched by the particle swarm optimization algorithm based on the acquired 2D and 3D vessel intersections. The proposed method was validated on 10 clinical coronary cases, the accuracy of the proposed method is 0.25 ± 0.09 mm for the corner detection of 2D vessel centerlines, and 0.56 ± 0.11 mm for the extraction of 3D vessel centerline. The final 3D and 2D vessel registration accuracy is 2.28 ± 0.21 mm, which is better than the comparison methods.

Keywords: Coronary artery 3D/2D registration · vascular intersection extraction · contextual morphological feature · particle swarm optimization

1 Introduction

Registration of computed tomography angiography (CTA) three-dimensional images of coronary arteries with digital subtraction angiography (DSA) images can effectively expand the surgical field of view for physicians, compensating for spatial topological structural information. Methods for 3D-2D registration of blood vessels can be categorized into intensity-based, gradient-based, feature-based, and deep learning-based approaches. Kerrien et al. [1] proposed an image translation method based on maximizing correlation and improved optical flow techniques to recover residual rigid motion. This method achieves comparison and registration between vascular imaging images and maximum intensity projection of vascular imaging volume. Hipwell et al. [2] reconstructed MRI images into digitally reconstructed radiograph (DRR) images, setting the

© The Author(s), under exclusive license to Springer Nature Singapore Pte Ltd. 2024
D.-S. Huang et al. (Eds.): ICIC 2024, LNBI 14881, pp. 219–228, 2024.
https://doi.org/10.1007/978-981-97-5689-6_19

intensity of the MRI image corresponding side to zero according to the DSA image, to enhance the similarity between DSA and DRR images. Nevertheless, this method exhibits lower robustness when the initial position is far from the registration position.

Livyatan et al. [3] perform initial pose estimation and final fine matching by maximizing projection gradients and extracting edge pixels from X-ray fluoroscopic images. Markelj et al. [4] introduced a gradient-matching criterion function to establish the registration relationship with the three-dimensional gradient vector field by reconstructing gradients from two-dimensional images. Subsequently, the optimal correspondence is iteratively searched based on anatomical structures, employing a robust random sample consensus algorithm to ensure the robustness of the matching process and the accuracy of the matching results. In recent years, deep learning-based methods have been increasingly employed for 3D-2D vascular registration. Miao et al. [5] demonstrated the highly nonlinear mapping from residuals to transformation parameters. Utilizing convolutional neural networks, they extract deep features from residual image sub-blocks to achieve 3D/2D registration of blood vessels. However, this method entails high training workload, memory usage, and demands a large amount of labeled high-quality vascular data, limiting its clinical applicability in vascular matching.

To address the challenges posed by the aforementioned methods, this paper proposes a coronary artery 3D/2D registration method based on particle swarm optimization (PSO) of contextual morphological features. By rapidly extracting contextual morphological features from CTA and DSA coronary artery images to construct feature mapping pairs in 3D/2D space, the spatial pose of CTA vessels is searched using the PSO algorithm to register 3D/2D morphological features, thereby achieving accurate matching of CTA and DSA coronary arteries.

2 Proposed Method

The flowchart of the proposed coronary artery 3D-2D registration method is depicted in Fig. 1. Firstly, the Mask-RCNN network [6] is initially employed to extract the 2D vessel structures from DSA images. Subsequently, the Steger centerline extraction algorithm [7] is utilized to obtain the vessel centerlines, followed by combining the Shi-Tomas corner detection algorithm [8] and the gray-level centroid model to extract corner points of vessel centerlines. A contextual feature descriptor is then proposed to filter 2D vessel intersection points from the corner points of centerlines. Secondly, the vessel's three-dimensional model is first acquired using a dynamic threshold segmentation algorithm. Then, an axial slice centroid fitting method is proposed to extract the 3D vessel centerline, and the intersection method based on centerlines is utilized to obtain 3D vessel intersection points. Finally, the similarity measure between the projected vessel intersection points and the 2D vessel intersection points in DSA images is calculated. Based on the PSO algorithm [9], the optimal pose of the 3D vessel model is searched to maximize the similarity measure, thereby achieving accurate matching of 3D-2D vessels.

2.1 DSA Vessel Intersection Extraction

The registration of coronary artery vessels in 3D/2D relies on corresponding vessel intersection points. To extract vessel intersection points from DSA images, we extract

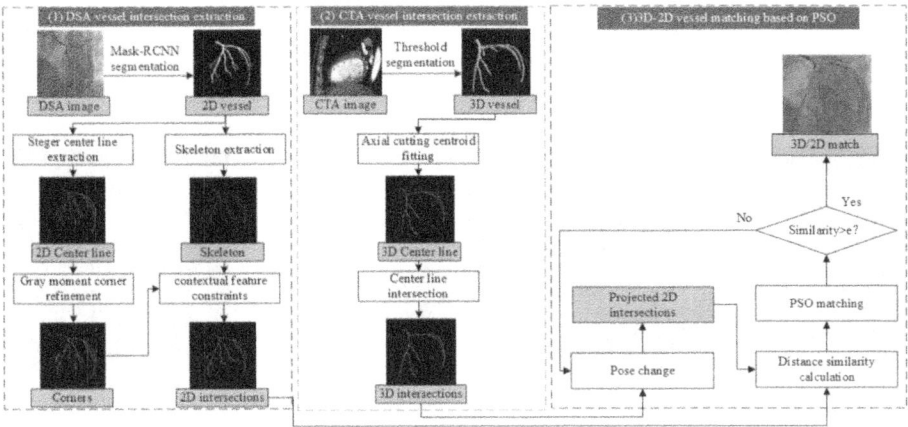

Fig.1. Flowchart of the proposed method

corner points from the vessel centerline and further optimize the selection to obtain the final vessel intersection points. In the process of extracting vessel corner points, we first preliminarily extract the coordinates of corner points of the vessel centerline using the Shi-Tomasi corner detection algorithm. Subsequently, with these corner point coordinates as initial values, we refine the initial corner points using the second-order gray-level moment model. In this study, the second-order central moment of neighboring pixels in the k neighborhood of the initial corner point is calculated using the following formula:

$$m_{ab} = \sum_{x=0}^{w-1} \sum_{y=0}^{h-1} x^a y^b T(x, y) \tag{1}$$

where $a, b = 0,1, 2$, w and h are the width and height of the neighborhood pixel block, respectively, and $T(x, y)$ is the gray value of the 2D vascular centerline coordinate point (x, y). From the above formula, the origin moment and first moment of the pixel block in the k neighborhood can be obtained as follows:

$$\begin{cases} m_{00} = \sum_x \sum_y f(x, y) dx dy \\ m_{10} = \sum_x \sum_y xf(x, y) \\ m_{01} = \sum_x \sum_y yf(x, y) \end{cases} \tag{2}$$

The origin moment m_{00} describes the area of the target region, and the first order moment m_{10} and m_{01} describe the centroid of the target region. The centroid coordinates of the pixel block are calculated as follows:

$$\bar{x} = \frac{m_{10}}{m_{00}}, \bar{y} = \frac{m_{01}}{m_{00}} \tag{3}$$

The above centroid coordinates are the corner coordinates of the blood vessel center line, as shown in Fig. 2. Figure 2 (a) is the 2D segmentation result of coronary vessels, Fig. 2 (b) is the centreline extraction result, Fig. 2 (c) is the centerline displayed on the DSA image, and Fig. 2 (d) is the centreline corner detection result. As can be seen from Fig. 2 (d), corner points are located at the intersection and endpoint of blood vessel branches.

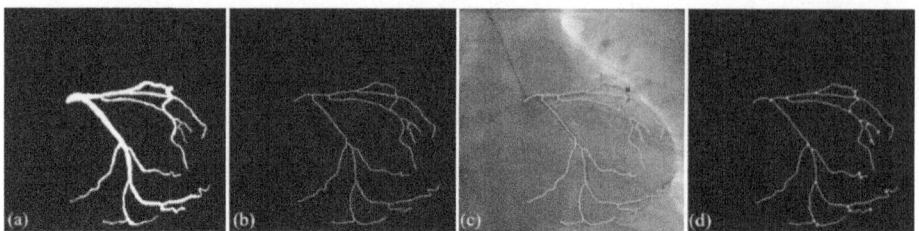

Fig. 2. Extraction results of vascular center lines and corners. (a) Results of vascular segmentation; (b) centerline extraction results; (c) centerline displayed on the DSA image; (d) Center line corner detection results

To obtain the intersection points of the 2D vascular centerline, we filter the corner points of the DSA vascular centerline to remove non-intersection points. To accurately eliminate non-intersection corner points, we first utilize the Zhang-Suen thinning algorithm [10] to extract the vascular skeleton information. Subsequently, we propose a vascular intersection point extraction algorithm based on contextual morphological feature constraints, leveraging both the vascular skeleton and corner points of the vascular centerline. A circle is drawn within the r neighborhood of the candidate point P, and the r neighborhood is divided into six equal parts denoted as S_0 to S_5. Let N_0 to N_5 represent the number of neighboring pixels around P in each of the S_0 to S_5 regions. Among N_0 to N_5, let N_{max}, N_{max-1}, and N_{max-2} denote the sets of points with the highest three counts. The percentage of N_{max}, N_{max-1}, and N_{max-2} relative to the total number of neighboring pixels is calculated as follows:

$$\begin{cases} P_{N_{max}} = \frac{N_{max}}{\sum_{i=0}^{5} N_i} \times 100\% \\ P_{N_{max-1}} = \frac{N_{max-1}}{\sum_{i=0}^{5} N_i} \times 100\% \\ P_{N_{max-2}} = \frac{N_{max-2}}{\sum_{i=0}^{5} N_i} \times 100\% \end{cases} \quad (4)$$

The contextual morphological constraints of neighboring pixels around the corner point can be utilized to determine the number of vascular segments in the vicinity. If the number of vascular segments in the neighborhood is greater than or equal to 3, then the corner point is considered a vascular intersection point. The specific judging formula is defined by the following equation:

$$\begin{cases} \left| P_{N_{max}} - P_{N_{max-1}} \right| < 15\% \\ \left| P_{N_{max-1}} - P_{N_{max-2}} \right| < 15\% \\ \left| P_{N_{max}} - P_{N_{max-2}} \right| < 15\% \end{cases} \quad (5)$$

If all the conditions described in the above equation are satisfied, then the corner point is considered a vascular intersection point.

2.2 CTA Vessel Intersection Extraction

For the extraction of the 3D vascular centerline, this study proposes an axial slice centroid fitting algorithm. Firstly, two endpoints are selected from each vascular branch of the 3D vascular model, and they are connected to form a straight line, which serves as the axial guiding line for the current vascular branch. Secondly, virtual slicing planes perpendicular to the axial guiding lines are constructed, and the vascular branches are segmented based on these axial guiding lines, with the slicing planes perpendicular to the direction of the axial guiding lines. By uniformly slicing each vascular branch into n intervals, $n+1$ slicing cross-sections are obtained. Finally, the centroid points of the $n+1$ cross-sections are extracted using a connected component centroid extraction algorithm. Based on the coordinates of the centroid points and utilizing a curve fitting algorithm, the 3D centerline of the vascular branch is obtained. Subsequently, by utilizing the intersecting relationship of the centerlines of vascular branches, the intersection points of the 3D vasculature can be determined.

2.3 3D-2D Vessel Matching Based on PSO

Initially, multiple angular transformations are applied to the 3D vascular model to construct the 3D pose search space. Subsequently, the PSO algorithm is employed to perform rapid global optimization of the 3D pose search space, thereby obtaining accurate 3D vascular poses. During the angular transformation process of the 3D vasculature, the 3D vascular model is projected onto the DSA 2D plane each time to compute the vascular matching quality of the current pose. To measure the matching quality of CTA and DSA images after projection transformation, the following distance similarity measure is introduced:

$$C\left(A\left(x_A^{2D}\right), B\left(x_B^{2D}\right)\right) = \frac{1}{\frac{1}{N}\sum_{j=1}^{N} D\left(x_{Aj}^{2D}, x_{Bj}^{2D}\right) + 1} \tag{6}$$

where x_{Aj}^{2D} and x_{Bj}^{2D} respectively represent the coordinates of the projected 2D intersection point and the DSA image 2D intersection point, while $\left(x_{Aj}^{2D}, x_{Bj}^{2D}\right)$ denotes their Euclidean distance. The parameter C ranges from (0,1], where a higher value indicates better 3D/2D matching performance, and conversely for lower values. If C is greater than the threshold value e, the search ends and the final matching result is obtained.

3 Experiments and Results

The proposed algorithm is implemented in the C++ programming language, and experiments are conducted on the Windows platform. Performance validation is carried out on 10 clinical coronary artery datasets. Performance metrics include errors in DSA vessel crossing point detection, CTA vessel centerline and crossing point detection, and CTA/DSA vessel matching. The dimensional information of CTA images is $512 \times 512 \times 291$, with a slice thickness of 0.375 mm, while the resolution of DSA images is 512×512. Each patient's imaging data includes the left and right coronary arteries. The parameters of the proposed algorithm are set as follows: In the Steger algorithm, the σ value for Gaussian filtering is 3, and the similarity threshold for coronary 3D/2D matching is set to 0.9. In the PSO algorithm, the search dimension is set to 6, representing the dimensions of vessel pose freedom. To balance the search speed and matching accuracy, 300 particles are used for optimization, the maximum number of iterations is set to 1000, the update step is 1, the acceleration factor is 1.49, and the starting and ending weight coefficients are 0.9 and 0.4, respectively.

3.1 DSA Vessel Intersection Extraction Results

The segmented DSA coronary artery images were subjected to skeleton erosion, resulting in vessel skeletons composed of a series of discrete points. The first and third columns of Fig. 3 show the corner points of the left and right coronary DSA images before filtering, while the second and fourth columns display the vascular crossing points of the left and right coronary DSA images after filtering. In this regard, there are more vascular crossing points in the left coronary DSA images, while relatively fewer in the right coronary ones. This difference stems from the physiological fact that the left coronary artery typically has a greater number of branches than the right coronary artery. In the experiments conducted in this study, the correct rate of vascular crossing point selection for the 10 clinical coronary artery datasets reached 100%. Therefore, the proposed algorithm can accurately select crossing points from 2D vascular corner points using context morphological feature descriptors.

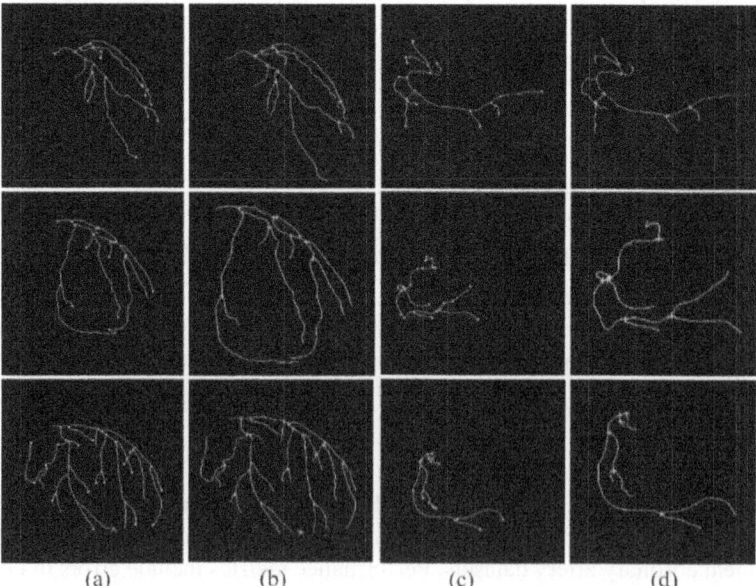

(a)   (b)   (c)   (d)

Fig. 3. Extraction results of vascular crossing points in DSA images. (a) DSA corners of the left coronary artery; (b) DSA intersections of the left coronary artery; (c) DSA corners of the right coronary artery; (d) DSA intersections of the right coronary artery

3.2 Results of CTA Vascular Center Line and Intersection

The Mimics dynamic threshold segmentation algorithm was employed for 3D vessel segmentation of CTA images. The segmentation results of the CTA images are shown in the second and fourth columns of Fig. 4. The 3D vessel centerlines obtained using the axial cutting centroid fitting algorithm proposed in this paper are depicted in the third and fifth columns of Fig. 4. The crossing points at the intersections of the left and right coronary artery centerlines are marked with green dots in Fig. 4.

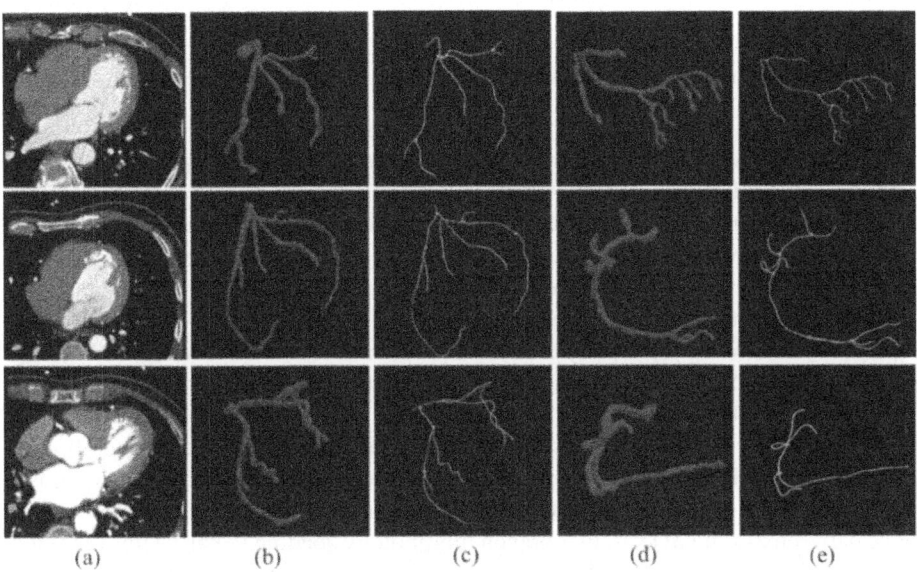

(a)   (b)   (c)   (d)   (e)

Fig. 4. Extraction results of vascular centerlines and intersections in CTA images. (a) CTA image; (b) Left coronary artery CTA segmentation results; (c) Left coronary artery center lines and intersections; (d) Right coronary artery CTA segmentation results; (e) Right coronary artery center lines and intersections

3.3 Results of Vascular Matching Between CTA and DSA

To determine the matching accuracy of 3D/2D vessels, we calculate the average Euclidean distance between the 3D vessel crossing points projected onto 2D points in the DSA images and the original vessel crossing points in the DSA images. This error metric is referred to as the Mean Projection Distance (MPD). Figure 5 shows the fusion results of the registration between clinical coronary CTA images and DSA images. It can be observed that the eight depicted 3D/2D matching schematic diagrams all exhibit relatively accurate matching results. However, the accuracy of vessel matching varies across different experimental subjects. The partial misalignment between some 2D vessel branches and 3D vessel branches is attributed to certain non-rigid deformations present in coronary vessels during the acquisition of CTA and DSA images.

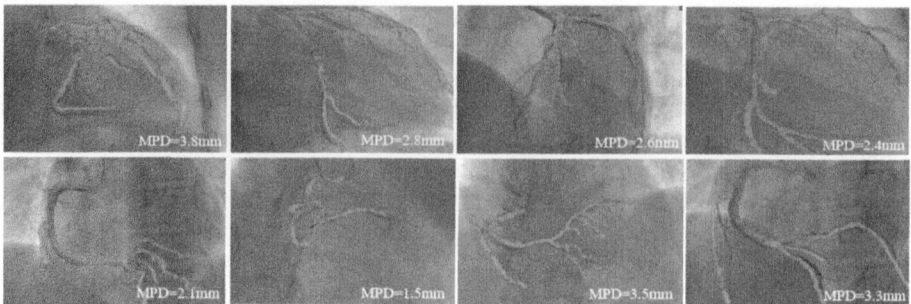

Fig. 5. Coronary 3D/2D registration results. The first act is the left coronary result, and the second act is the right coronary registration result

Comparing the proposed method with four advanced 3D/2D vascular registration methods, namely ICP-BP [11], ICP-PnP [11], ICC [12], and OGMM [13], the registration errors between CTA and DSA vessels on 10 clinical coronary datasets are presented in Table 1. As shown in the table, the proposed method achieves an average registration error of 2.41 mm on the left coronary artery. Compared to the other four comparison methods, the registration accuracy is improved by 8.37%, 22.76%, 10.41%, and 65.22%, respectively. On the right coronary artery, the proposed method achieves an average registration error of 2.23 mm. Compared to other methods, the registration accuracy is improved by 56.44%, 60.04%, 82.13%, and 55.67%, respectively. These results indicate that the proposed method achieves the best registration performance among all comparison methods. However, as observed in Fig. 5, the registration error of the proposed method fluctuates significantly. The main reason for this is the incomplete correspondence between 2D and 3D vascular branch morphology, leading to partial misalignment of 3D-2D crossing points. Additionally, during the acquisition of DSA images, the lack of spatial depth information often results in visual errors in stacked vascular branches, causing the number of vascular crossing points in DSA images to be higher than that in CTA images, thereby affecting the accuracy of 3D/2D crossing point matching.

Table 1. The vascular 3D/2D registration error results of the proposed method and 4 comparison methods in 10 cases of coronary clinical data.

Methods	Left coronary artery (mm)	Right coronary artery (mm)
ICP-BP [11]	2.63 ± 0.13	5.12 ± 0.22
ICP-PnP [11]	3.12 ± 0.21	5.58 ± 0.36
ICC [12]	2.69 ± 0.15	12.48 ± 2.35
OGMM [13]	6.93 ± 0.35	5.03 ± 0.25
Proposed	**2.41 ± 0.32**	**2.23 ± 0.26**

4   Conclusions

The proposed vascular 3D/2D registration method based on context-aware feature particle swarm optimization provides three-dimensional visualization assistance for physicians to identify vascular structures intraoperatively. The proposed method accurately extracts vascular crossing points from CTA and DSA images and completes vascular matching based on corresponding crossing points. As the information from CTA and DSA images complements each other for vascular intervention navigation, vascular registration is widely used in clinical diagnosis and treatment, assisting in procedures such as stent implantation, real-time 3D tracking of guidewires, and 3D-2D fusion visualization of interventional surgery. For instance, S Goreczny et al. [14] use a 3D/2D vascular registration system, the Philips Vessel Navigator, to guide complex cardiac catheterization and interventions in coronary atherosclerotic heart disease. Sailer A M et al. [15] apply 3D/2D vascular registration to guide endovascular repair of complex aortic aneurysms.

Acknowledgments. This research is supported by the project of Tianjin science and technology plant (grant number: 23YDTPJC00470), and the graduated research innovation project of Tianjin (grant number: 2022SKY124).

Disclosure of Interests. The authors declare that there are no conflicts of interest related to this article.

References

1. Kerrien, E., Berger, M. -O., Maurincomme, E., Launay, L., Vaillant, R., Picard, L.: Fully automatic 3D/2D subtracted angiography registration. In: Taylor, C., Colchester, A. (eds.) MICCAI 1999. LNCS, vol. 1679, pp. 664–671. Springer, Heidelberg (1999). https://doi.org/10.1007/10704282_72
2. Hipwell, J.H.: Intensity-based 2-D-3-D registration of cerebral angiograms. IEEE Trans. Med. Imaging. **22**, 1417–1426 (2003)
3. Livyatan, H., Yaniv, Z., Joskowicz, L.: Gradient-based 2-D/3-D rigid registration of fluoroscopic X-ray to CT. IEEE Trans. Med. Imaging. **22**, 1395–1406 (2003)
4. Markelj, P., Tomazevic, D., Pernus, F., Likar, B.: Robust gradient-based 3-D/2-D registration of CT and MR to X-ray images. IEEE Trans. Med. Imaging. **27**, 1704–1714 (2008)
5. Miao, S., Wang, Z.J., Liao, R.: A CNN regression approach for real-time 2D/3D registration. IEEE T Med Imaging. **35**, 1352–1363 (2016)
6. He, K., Gkioxari, G., Dollár, P., Girshick, R.: Mask R-CNN. In: Proceedings of the IEEE International Conference on Computer Vision, pp. 2961–2969 (2017)
7. Steger, C.: An unbiased detector of curvilinear structures. IEEE Trans. Pattern Anal. **20**, 113–125 (1998)
8. Bansal, M., Kumar, M., Kumar, M., Kumar, K.: An efficient technique for object recognition using Shi-Tomasi corner detection algorithm. Soft. Comput. **25**, 4423–4432 (2021)
9. Shi, Y., Eberhart, R.: A modified particle swarm optimizer. In: 1998 IEEE International Conference on Evolutionary Computation Proceedings. IEEE World Congress on Computational Intelligence (Cat. No. 98TH8360), pp. 69–73. IEEE (1998)

10. Zhang, T.Y., Suen, C.Y.: A fast parallel algorithm for thinning digital patterns. Commun. ACM **27**, 236–239 (1984)
11. Besl, P.J., McKay, N.D.: A method for registration of 3-d shapes. IEEE Trans. Pattern Anal. **14**, 239–256 (1992)
12. Benseghir, T., Malandain, G., Vaillant, R.: Iterative closest curve: a framework for curvi-linear structure registration application to 2D/3D coronary arteries registration. In: Mori, K., Sakuma, I., Sato, Y., Barillot, C., Navab, N. (eds.) MICCAI 2013. LNCS, vol. 8149, pp. 179–186. Springer, Heidelberg (2013). https://doi.org/10.1007/978-3-642-40811-3_23
13. Baka, N., Metz, C.T., Schultz, C.J., van Geuns, R.J., Niessen, W.J., van Walsum, T.: Oriented Gaussian mixture models for nonrigid 2D/3D coronary artery registration. IEEE Trans. Med. Imaging **33**, 1023–1034 (2014)
14. Goreczny, S., Dryzek, P., Morgan, G.J., Lukaszewski, M., Moll, J.A., Moszura, T.: Novel three-dimensional image fusion software to facilitate guidance of complex cardiac catheter-ization: 3D image fusion for interventions in CHD. Pediatr. Cardiol. **38**, 1133–1142 (2017)
15. Sailer, A.M., De Haan, M.W., Peppelenbosch, A.G., Jacobs, M.J., Wildberger, J.E., Schurink, G.W.H.: CTA with fluoroscopy image fusion guidance in endovascular complex aortic aneurysm repair. Eur. J. Vasc. Endovasc. **47**, 349–356 (2014)

Enhancing Drug-Drug Interaction Predictions in Biomedical Knowledge Graphs Through Integration of Householder Projections and Capsule Network Techniques

Xia Li[1], Sensen Zhang[2], Yang Liu[1(✉)], Peng Bi[1], and Tiangui Hu[1]

[1] Yangtze River Delta Research Institute (Quzhou),, University of Electronic Science and Technology of China, Quzhou 324003, China
18810928931@163.com
[2] School of Information Technology, Renmin University of China, Beijing 100872, China

Abstract. Investigating Drug-Drug Interactions (DDIs) is crucial for optimizing drug treatments, improving therapeutic outcomes, and reducing adverse reactions, aiding in safer drug therapy strategies. Knowledge Graph Embedding (KGE) models, vital in DDI research, map drugs and their interactions in the Biomedical Knowledge Graph (BioKG), revealing potential and unreported drug interactions. However, due to diverse drug relationship complexities and significant Relationship Mapping Properties (RMPs), existing models are limited in capturing complete relationship patterns, struggling with prevalent RMPs. This study introduces HPCap, combining Householder projections with capsule networks to enhance modeling of complex drug interactions in BioKG. Householder projections first address complex RMPs, followed by embedding entities and their relationships into a projection space for dynamic interaction modeling. A capsule network with reverse dot-product dynamic routing captures in-depth information, enhancing entity representation. Experimental results show HPCap's superior performance over contemporary methods on three BioKG datasets.

Keywords: DDIs · BioKG · Householder projections · capsule network · projection space

1 Introduction

Healthcare professionals frequently encounter the imperative to prescribe combinations of pharmacotherapies to optimize therapeutic outcomes. However, the concurrent administration of multiple medications often precipitates Drug-Drug Interactions (DDIs). DDIs transpire when one pharmaceutical agent interacts with another, resulting in diverse adverse drug reactions. These reactions span a spectrum from mild side effects to severe injuries or, in extreme cases, fatalities. Empirical studies have demonstrated that the probability of DDIs significantly escalates with the augmentation of the number of concurrently administered drugs, rising from 6% with two drugs to 50% with five drugs,

D.-S. Huang et al. (Eds.): ICIC 2024, LNBI 14881, pp. 229–240, 2024.
https://doi.org/10.1007/978-981-97-5689-6_20

and reaching a zenith of 100% with ten drugs [1]. Consequently, healthcare practitioners must exercise prudence and diligence in considering the existing medication regimens of their patients to avert adverse reactions and potential treatment failures. The identification of DDIs represents a formidable challenge in pharmaceutical management and development, and the precise anticipation of such interactions assumes paramount importance in safeguarding patient health and welfare.

In the initial stages of DDIs research, conventional methodologies predominantly relied on traditional feature-based machine learning techniques. Chowdhury et al. [2] employed kernel methods as part of their approach to extract DDIs, while Thomas et al. [3] utilized majority voting methods for DDI extraction. Subsequent investigations sought to enhance results by incorporating diverse feature types, encompassing relative position features, syntactic structure features, and phrasal auxiliary verb features [4–6]. Nevertheless, these methodologies heavily leaned on manual feature engineering and selection, inherently constraining their efficacy and potentially overlooking essential patterns. In recent years, deep neural network-based models have gained extensive adoption, leveraging their capacity to glean knowledge from high-dimensional spaces. These models can be broadly categorized into three groups: (I) Matrix Factorization (MF)-based models [7, 8], (II) Random Walk (RW)-based models [9, 10], and (III) Neural Network (NN)-based models [11, 12]. Notably, these NN-based models have demonstrated superior performance compared to traditional machine learning-based approaches.

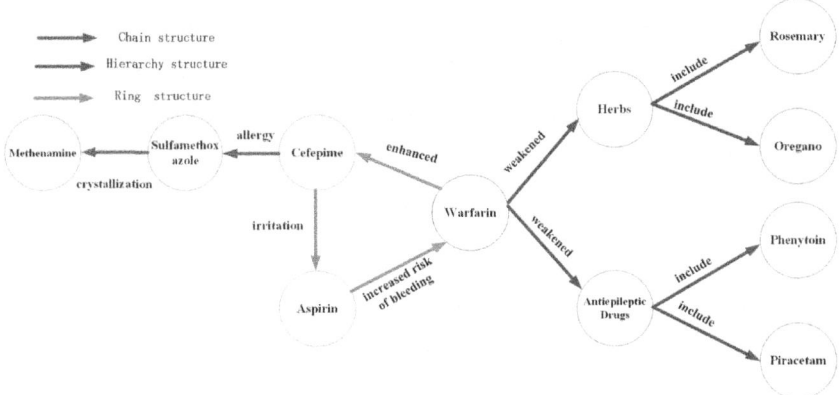

Fig. 1. The illustration of the Warfarin drug knowledge graph structure is depicted, wherein redarrows denote the Chain structure, green arrows signify the Hierarchy structure, and purple arrowsrepresent the Ring structure.

In spite of recent strides in KG-based models for predicting DDIs, certain pivotal facets remain inadequately addressed. Firstly, within the realm of the Biomedical Knowledge Graph (BioKG), drug entities manifest intricate relationship patterns characterized by Relation Mapping Properties (RMPs), encompassing 1-to-1, 1-to-N, N-to-1, and N-to-N relationships, along with a hierarchical structure. Despite the notable exception of KG2ECapsule [13], which successfully infers these relationship mappings through the integration of the Bernoulli distribution with capsule networks, the efficacy of models employing Bernoulli distribution sampling remains suboptimal. The current landscape of knowledge graph-based models falls short in the ability to reason about such nuanced relationship patterns. Furthermore, as shown in Fig. 1, the patterns of interaction between drugs include not only common chain structures but also hierarchy structures and ring structures. Unfortunately, existing models overlook these pivotal features, creating a deficiency in their capacity to comprehensively capture the spectrum of information associated with drug entities. Consequently, there is an imperative demand for a more sophisticated model that can adeptly reason about intricate relationship patterns and RMPs within the BioKG.

To address the aforementioned challenges, we introduce, for the first time, Household projections into the transformation of the projection space and integrate them with capsule networks to construct HPCap. Specifically, we implement Household projections for the head and tail entities of BioKG triplets in the projection space. Subsequently, the transformed head entity is subjected to Möbius Representation transformation based on the triplet's relationship parameters in projection space. The training is conducted using the similarity of the transformed entity and the triplet's scoring function. Finally, the trained triplets serve as input to the capsule network, where different dimensions of triplet information are enriched through one convolutional layer and two capsule layers, facilitating deep-level information exchange and enhancing the representation of triplets in the projection space during training.

2 Preliminaries

2.1 Projective Space

A projective transformation is a mapping from the Riemann Sphere to itself. The Riemann Sphere is an extension of the complex plane, incorporating a point at infinity to represent points at an infinite distance, effectively transforming the entire complex plane into a sphere.

Projective transformations are commonly employed in the theory of complex functions, where they belong to a special class known as projective or fractional linear transformations. These transformations can be expressed by a fractional expression:

$$f(z) = \frac{az + b}{cz + d},\tag{1}$$

where z is a complex number, and a, b, c and d are complex constants with $ad\text{-}bc \neq 0$. This transformation maps a point z on the Riemann Sphere to another point $f(z)$. Points on the Riemann Sphere can be seen as points on the complex plane augmented by a point at infinity on the surface of the sphere. The projective transformation effectively maps this sphere back onto itself while preserving certain properties, such as the properties of circles and straight lines.

2.2 Advanced Formulation of Householder Projections

Consider a unit vector $\mathbf{p}^k$ and a scalar τ. The $k \times k$ Householder matrix, hereafter referred to as the modified Householder matrix and denoted by $\mathbf{M}(\mathbf{p}, \tau)$, is formally defined as: $\mathbf{M}(\mathbf{p}, \tau) = \mathbf{I} - \tau\mathbf{p}\mathbf{p}^{\mathrm{T}}$, where $\mathbf{I}$ is the $k \times k$ identity matrix, and $\|\mathbf{p}\|_2 = 1$ ensuring normalization. This matrix is characterized by $k-1$ eigenvalues of unity and a single distinct eigenvalue of $1-\tau$, thereby preserving its invertibility, except in the special case when $\tau = 1$. Extending this concept, for a set of real scalars $T = \{\tau_c\}_{c=1}^m$ and corresponding unit vectors $P = \{\mathbf{p}_c\}_{c=1}^m$ in ^{k}, with m being a positive integer, we define a composite mapping as follows:

$$\mathbf{P}(P, T) = \prod_{c=1}^{m} \mathbf{M}(\mathbf{p}_c, \tau_c), \tag{2}$$

where $\mathbf{P}(P, T)$ invariably results in an invertible matrix. This property stems from the fact that the product of invertible matrices retains its invertibility.

The mechanism encapsulated by $\mathbf{P}(P, T)$, termed the **Householder Projection**, represents a series of m modified Householder reflections. Notably, these projections are renowned for their distance-preserving characteristics. However, they uniquely facilitate systematic and reversible modifications of relative distances between point pairs. This attribute renders Householder projections exceptionally suitable for modeling complex relational properties, enabling the representation of intricate relational patterns while maintaining structural integrity.

3   Model

The work presented in this article encompasses three primary components. Initially, we implement Householder projections for the head and tail entities of BioKG triples within a projective space. Following this, the transformed head entities in the projection space are subjected to a Möbius representation transformation, which is parameterized by the relationships within the triples. This step is then combined with the similarity assessment of the triples to formulate a scoring function for training objectives. Finally, the trained triples are fed into a capsule network, where

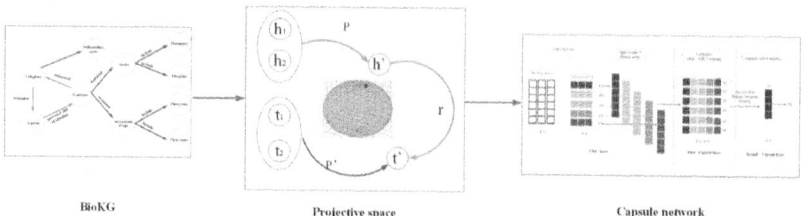

Fig. 2. Overall Process Diagram: BioKG triples undergo Household projection within the projective space, followed by embedding the trained triples into the capsule network.

They pass through a convolutional layer followed by two capsule layers. This architecture facilitates deep inter-dimensional information exchange among the triples, thereby enriching their representation within the projective space. The entire process is illustrated in Fig. 2, delineating the sophisticated training regimen and the underlying architectural framework.

3.1 Relational Householder Projections

Consider a BioKG $\mathcal{G}$, represented as a set of triple facts in the form of medicine-relation-medicine, denoted by $\{(h, r, t)|h, t \in \varepsilon, r \in \delta\}$. Here, ε represents the set of medicine entities, and δ denotes the set of relations. Each head (h) and tail (t) entity is embedded into a d-dimensional complex projective line, expressed as $\mathbf{h}, \mathbf{t}^d$.

The Householder projection in the relational space is defined by axes $\mathbf{p}_r \in \mathbb{C}^{d \times m}$ and scalars $\tau_r \in^{d \times m}$, where d is the dimensionality of the entity embeddings and m is the count of modified Householder matrices. For a given triple (h, r, t), the transformation of the head and tail entities through r_1-specific Householder projections is formalized as follows:

$$\mathbf{h}' = \mathbf{P}(\mathbf{p}_r, \tau_r)\mathbf{h} = \prod_{r=1}^{m} \mathbf{M}(\mathbf{p}_r[i], \tau_r[i])\mathbf{h}, \mathbf{t}' = \mathbf{p}'(\mathbf{p}_r, \tau_r)\mathbf{h} = \prod_{r=1}^{m} \mathbf{M}'(\mathbf{p}_r[i], \tau_r[i])\mathbf{t} \quad (3)$$

3.2 Möbius Representation Transformation

In the proposed model, a relation r_2 is embedded into a d-dimensional matrix representation $\mathbf{r}$, defined as:

$$\mathbf{r} = \begin{bmatrix} \mathbf{r}_a & \mathbf{r}_b \\ \mathbf{r}_c & \mathbf{r}_d \end{bmatrix}, \quad (4)$$

where $\mathbf{r}_a, \mathbf{r}_b, \mathbf{r}_c, \mathbf{r}_d \in \mathbb{C}^d$. This formulation facilitates the application of a relation-specific Möbius transformation to project the head entity $\mathbf{h}'$ from the source complex plane to a target complex plane. This transformation leverages the principles of stereographic projection and is defined with respect to the Riemann Sphere as follows:

$$\mathbf{h}'_r = \frac{\mathbf{r}_a \mathbf{h}' + \mathbf{r}_b}{\mathbf{r}_c \mathbf{h}' + \mathbf{r}_d}, \quad (5)$$

where the condition $\mathbf{r}_a \mathbf{r}_d - \mathbf{r}_b \mathbf{r}_c \neq 0$ ensures the transformation's validity. The model assesses the validity of triples within a KG based on the similarity between the relation-specific transformed head $\mathbf{h}'_r$ and the tail entity $\mathbf{t}'$. The scoring function for the model is expressed as:

$$f(h, r, t) = Re\left(\langle \mathbf{h}'_r, \overline{\mathbf{t}'} \rangle\right), \quad (6)$$

where $Re(.)$ signifies the operation of extracting the real part. Model optimization is carried out using the Adam optimizer, following the methodology outlined in Kingma and Ba [14], as we minimize the loss function defined in [15]:

$$L = \sum_{(h,r,t) \in T} \sum_{o'} log(1 + exp(y_o f(h, r, t))). \quad (7)$$

Here, T represents the set of valid triplets, T' denotes the set of corrupted triplets, and $y_o = 1$ if $(h, r, t) \in T$, otherwise, $y_o = 0$.

3.3 Capsule Network Layer

We utilize the embeddings of trained entities and relationships within drug pairs as inputs for the capsule network layer. These embeddings are defined as a matrix $\mathbf{B} = [\mathbf{h}_e, \mathbf{r}_e, \mathbf{t}_e] \in \mathbb{CP}^{k \times 3}$. The i-th row of $\mathbf{B}$ is denoted as $\mathbf{B}_i$, and a filter ω operates repeatedly over each row of $\mathbf{B}$ to generate a feature map $\mathbf{q} = [q_1, q_2, q_3, \cdots, q_k]$, as expressed by the following equation:

$$q_i = g(\omega \cdot \mathbf{B}_i + b_c), \qquad (8)$$

where, the filter $\omega \in \mathbb{CP}^{1 \times 3}$, $(\cdot)$ denotes the dot product, b_c represents a bias term, and g denotes a non-linear activation function such as ReLU or Sigmoid.

In the convolution layer, Eq. 8 is used to compute the values of k capsules. We introduce N as the number of filters, resulting in N k-dimensional feature maps. Each feature map is designed to capture a distinct characteristic among entries within the same dimension. The initial capsule layer consists of k capsules, each with a vector output $\mathbf{Q}_i \in \mathbb{CP}^{N \times 1}$. These vector outputs $\mathbf{Q}_i$ are then multiplied by weight matrices $\omega_i \in \mathbb{CP}^{d \times N}$ to generate vectors $\mathbf{v}_i \in \mathbb{CP}^{d \times 1}$. These vectors are subsequently summed to yield a vector input $\mathbf{s} \in \mathbb{CP}^{d \times 1}$ for the capsules in the second layer. Subsequently, the capsules produce vector outputs $\mathbf{e} \in \mathbb{CP}^{d \times 1}$ following a normalization operation. Finally, these vector outputs $\mathbf{e}$ are multiplied by the weight matrix $\mathbf{W} \in \mathbb{CP}^{d \times 1}$ to generate a score used for assessing the accuracy of a given quadruple. The detailed process is described by the following equations:

$$\mathbf{v}_i = \omega_i \cdot \mathbf{Q}_i, \mathbf{s} = \sum_i r_i v_i, \mathbf{e} = layerNorm(\mathbf{s}), f = \mathbf{eW}. \qquad (9)$$

We employ Layer Normalization[16] as the normalization technique, which we have empirically observed to enhance convergence in the context of routing. The coupling coefficients, denoted as r_i, are determined during this process.

Our proposed model is depicted in Fig. 3. In this illustration, the embedding size (k) is set to 6, the number of filters (N) is 5, the number of neurons within the capsules in the first capsule layer is equal to N, and the number of neurons within the capsules in the second capsule layer is denoted as d, which is set to 4. In summary, we define the score function as follows:

$$f(h, r, t) = caps(g([\mathbf{h}_e, \mathbf{r}_e, \mathbf{t}_e] * \varphi)) \cdot \mathbf{W}, \qquad (10)$$

where, the set of filters φ and $\mathbf{W}$ are shared hyper-parameters within the convolution layer, the $*$ symbol represents the convolution operator, and the capsule network operator is denoted as caps. The activation function used is ReLU. Negative sampling, which has been proven effective in learning knowledge graph embeddings [16] and word embeddings [17], is employed in our model optimization. The loss function is described as follows:

$$L = -log\sigma(\gamma - f(h, r, t)) - \frac{1}{k} \sum_{i=1}^{n} log\sigma(f(h'_i, r, t'_i) - \gamma) \qquad (11)$$

In this equation, γ represents a fixed margin hyper-parameter, σ is the sigmoid function, and (h'_i, r, t'_i) refers to the i-th negative triple.

4 Experiment and Analysis

4.1 Datasets, Benchmark, and Evaluation Criteria

To evaluate our model, we perform link prediction task on three commonly used BioKG benchmark datasets, namely OGB-Biokg [18], DrugBank [19] and KEGG [20].

In our study, we've undertaken an extensive comparative analysis of our models against a variety of standard methods. For assessing the efficacy of conventional network representation learning techniques, we included these baseline models: MF-based (Laplacian [8]), RW-based (DeepWalk [10]), and NN-based (LINE [12]). Furthermore, KG-based models were also evaluated, specifically KGN [21], KGAT [22], R-GCN [23], BERTKG [24], Xin [6], and KG2ECapsule [13]. In addition, to ascertain the utility of Householder projections, this research has developed an Pcap model that omits Householder projections, employed as a baseline in our experiments.

The evaluation of the model's performance involved six metrics, specifically Accuracy (Acc.), Precision (Pre.), Recall (Rec.), F1 Score (F1), Area Under the ROC Curve (AUC), and Area Under the Precision-Recall Curve (AUPR).The model's evaluation metrics were systematically selected to provide insights into its predictive accuracy and robustness.

4.2 Results and Analysis

The analysis of experimental outcomes on three distinguished BioKG datasets, as delineated in Table 1, highlights the exceptional effectiveness of the model wepropose. Exhibiting superior performance across a broad range of evaluative metrics, our model distinctly positions itself as a leading contender. Notably, the discernible preeminence of the biomedical model, predicated on the KG framework, is evident in these results. Additionally, our research sheds light on the notable efficacy of the PCap model. Although its performance metrics show a slight underperformance in comparison to those predominantly based on the Bernoulli distribution, as exemplified in the KG2ECapsule [13], and our newly introduced model, it is significant to note that the approach focusing on projection for articulating inter-entity RMPs also demonstrates appreciable results. Nevertheless, while these findings are promising, they exhibit potential for enhancement when contrasted with outcomes derived from both the Bernoulli distribution and Householder projection methodologies.

Table 1. Experimental Results of HPCap and Baseline Models on Three Datasets.

Datasets	Methods	ACC	Pre	Rec	F1	Auc	AUPR
OGB-Biokg	Laplacian	.5710	.5296	.5934	.5597	.5692	.5861
	DeepWalk	.5681	.5473	.5223	.5345	.5419	.5325
	LINE	.5786	.5534	.5386	.5459	.5418	.5374
	KGNN	.7389	.7541	.7245	.7390	.7849	.7378
	KGAT	.7489	.7559	.7191	.7370	.7962	.8011
	RGCN	.8467	.8773	.8063	.8403	.9172	.9268
	BERTKG	.8326	.8835	.8243	.8529	.8967	.9167
	Xin.et al	.8627	.9105	.8467	.8774	.9276	.9341
	KG2ECapsule	.9078	.9219	.8914	.9064	.9656	.9672
	PCap	.8921	.9178	.8836	.8999	.9521	.9549
	HPCap	.9151	.9291	.9011	.9149	.9728	.9754
DrugBank	Laplacian	.5923	.4455	.3372	.3838	.6724	.4782
	DeepWalk	.6163	.6059	.5904	.5980	.6501	.4782
	LINE	.6374	.6283	.6189	.6236	.6926	.4923
	KGNN	.7947	.7959	.7931	.7945	.8602	.8587
	BERTKG	.8469	.8524	.5681	.6817	.8925	.8726
	Xin.et al	.87364	.8672	.8620	.8646	.9224	.9341
	KG2ECapsule	.9078	.9219	.8914	.9064	.9656	.9672
	PCap	.8882	.9113	.8737	.8921	.9518	.9616
	HPCap	.9132	.9271	.8974	.9120	.9734	.9751
KEGG	Laplacian	.5694	.3683	.3781	.3731	.5608	.2916
	DeepWalk	.5800	.3801	.3762	.3781	.5751	.3005
	LINE	.5528	.3546	.3390	.3466	.5462	.2810
	KGNN	.7282	.4790	.4237	.4497	.8314	.4484
	KGAT	.7798	.5340	.4185	.4692	.8202	.5382
	RGCN	.8330	.4969	.4392	.4663	.8358	.4590
	BERTKG	.8216	.5773	.4587	.5112	.8267	.4937
	Xin.et al	.8367	.5837	.4592	.5140	.8426	.5887
	KG2ECapsule	.8348	.6278	.4794	.5131	.8505	.6644
	PCap	.8317	.6251	.4706	.5370	.8468	.6541
	HPCap	.8472	.6414	.4826	.5508	.8591	.6738

4.3 Ablation Experiment

Pojection Space. To explore the advantages of projection space modeling, we embedded triples into complex vector space, quaternion space, dual quaternion space, biquaternion space, and hybrid vector space, in accordance with RotatE [25], QuatE [26], DualE [27], BiQUE [28], and HousE [29]. This process resulted in the construction of the RPCap, QPCap, DPCap, BPCap, and HouPCap models. These models were then tested on the OGB-Biokg and DrugBank datasets. The experimental results, as depicted in Fig. 3, indicate that as the dimensionality of the space increases, there is a corresponding enhancement in model performance (as measured by ACC. and F1 scores), peaking in the projection space. This outcome indirectly demonstrates that the modeling capability of these models is correlated with the degree of rotational freedom in the respective spaces.

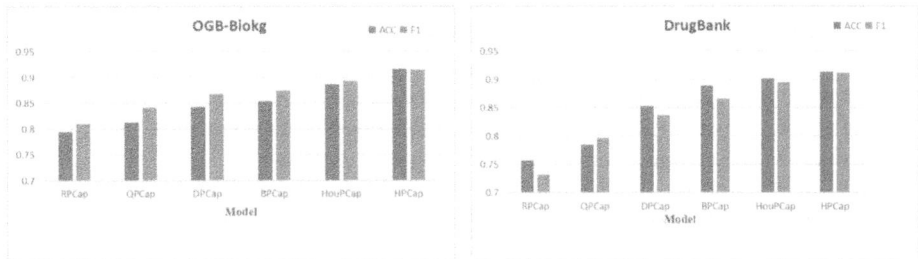

Fig. 3. The constructed model was evaluated on DrugBank, and the trend changes in the metrics ACC and F1 were graphically represented.

The Effect of Capsule Network. In our experimental design, we replaced Eq. 10 with the scoring mechanisms from notable models such as CovE [30], ConvKB [31], and CapsE [32]. This led to the development of variant models, designated as HPCovE, HPConvKB, and HPCapsE. The primary objective of this strategic replacement was to empirically assess and confirm their operational efficacy. These modified models were rigorously tested within the context of the OGB-Biokg and DrugBank datasets. The results, as delineated in Table 2, reveal that our HPCap model outperforms its comparable counterparts, thereby evidencing its superior experimental performance and efficacy in this context.

Table 2. The pattern modeling and inference abilities of several models.

Models	ACC	Pre	Rec	F1	Auc	AUPR
HPCovE	.8764	.8953	.8724	.8837	.9415	.9512
HPConvKB	.8962	.9026	.8757	.8889	.9509	.9551
HPCapsE	.9024	.9078	.8813	.8944	.9521	.9613
HPCap	.9132	.9271	.8974	.9120	.9734	.9751

5   Conclusion

To unearth potential and implicit drug interactions, including those not yet explicitly reported, and to address the limitations in the modeling capabilities of existing models, this study pioneers the application of household projection to transformations in the projective space. Utilizing the inverse dot-product dynamic routing technology of capsule networks, we capture deep-level information between different dimensions of drug triples, enriching the representation of entities in the projective space. Experiments conducted on three prevalent BioKG datasets demonstrate that the performance of HPcap significantly surpasses that of existing state-of-the-art methods.

Acknowledgments. This work was supported by Artificial intelligence vision chip development(2022D022). Xia Li and Sensen Zhang are co-first authors of the article. Yang Liu is the corresponding author of this paper.

Disclosure of Interests.   The authors declare that they have no known competing financial interests or personal relationships that could have appeared to influence the work reported in this paper.

References

1. Finkel, R., Clark, M.A., Cubeddu, L.X.: Pharmacology. Lippincott Williams & Wilkins (2009)
2. Chowdhury, M.F.M., Lavelli, A.: Title of a proceedings paper. In: Proceedings of the Naacl-Hlt Conference, Atlanta, Georgia, USA, pp. 351–355. The Association for Computer Linguistics (2013)
3. Thomas, P., Neves, M.L., Rocktäschel, T., Leser, U.: WBI-DDI: drug-drug interaction extraction using majority voting. In: Proceedings of the NAACL-HLT Conference, Atlanta, Georgia, USA, pp. 628–635. The Association for Computer Linguistics (2013)
4. Bokharaeian, B., Díaz, A.: Nil_ucm: extracting drug-drug interactions from text through combination of sequence and tree kernels. In: Proceedings of the 7th International Workshop on Semantic Evaluation, SemEval@NAACL-HLT 2013, Atlanta, Georgia, USA, pp. 644–650. The Association for Computer Linguistics (2013)
5. Kim, S., Liu, H., Yeganova, L., Wilbur, W.J.: Extracting drug-drug interactions from literature using a rich feature-based linear kernel approach. J. Biomed. Informatics **55**, 23–30 (2015)
6. Jin, X., Sun, X., Chen, J., Sutcliffe, R.F.E.: Extracting drug-drug interactions from biomedical texts using knowledge graph embeddings and multi-focal loss. In: Proceedings of the 31st ACM International Conference on Information & Knowledge Management, Atlanta, GA, USA, pp. 884–893. ACM (2022)

7. Belkin, M., Niyogi, P.: Laplacian eigenmaps for dimensionality reduction and data representation. Neural Comput. **15**(6), 1373–1396 (2003)
8. Cao, S., Lu, W., Xu, Q.: Grarep: Learning graph representations with global structural information. In: Proceedings of the CIKM 2015, Melbourne, VIC, Australia, pp. 891–900. ACM (2015)
9. Perozzi, B., Al-Rfou, R., Skiena, S.: Deepwalk: online learning of social representations. In: Proceedings of the KDD 2014, New York, NY, USA, pp. 701–710. ACM (2014)
10. Grover, A., Leskovec, J.: node2vec: scalable feature learning for networks. In: Proceedings of the SIGKDD 2016, CA, USA, pp. 855–864. ACM (2016)
11. Tang, J., Qu, M., Wang, M., Zhang, M., Yan, J., Mei, Q.: LINE: large-scale information network embedding. In: Proceedings of the WWW 2015, Florence, Italy, pp. 1067–1077. ACM (2015)
12. Wang, D., Cui, P., Zhu, W.: Structural deep network embedding. In: Proceedings of the SIGKDD 2016, CA, USA, pp. 1225–1234. ACM (2016)
13. Su, X., et al.: Biomedical knowledge graph embedding with capsule network for multi-label drug-drug interaction prediction. IEEE Trans. Knowl. Data Eng. **35**(6), 5640–5651 (2023)
14. Kingma, D.P., Ba, J.: Adam: a method for stochastic optimization. In: Bengio, Y., LeCun, Y. (eds.) Proceedings of the 3rd International Conference on Learning Representations, ICLR 2015, Conference Track Proceedings, San Diego, CA, USA (2015)
15. Trouillon, T., Welbl, J., Riedel, S., Gaussier, É., Bouchard, G.: Complex embeddings for simple link prediction. In: Proceedings of the ICML 2016, vol. 48, pp. 2071–2080. JMLR.org, New York City, NY, USA (2016)
16. Ba, L.J., Kiros, J.R., Hinton, G.E.: Layer normalization. CoRR abs/1607.06450 (2016)
17. Mikolov, T., Sutskever, I., Chen, K., Corrado, G.S., Dean, J.: Distributed representations of words and phrases and their compositionality. In: Advances in Neural Information Processing Systems 26: Proceedings of the 27th Annual Conference on Neural Information Processing Systems, pp. 3111–3119. Lake Tahoe, Nevada, USA (2013)
18. Hu, W., et al.: Open graph benchmark: datasets for machine learning on graphs. In: NeurIPS 2020, Virtual Conference, December 6–12 (2020)
19. Wishart, D.S., Feunang, Y.D., Guo, A.C.: Drugbank 5.0: a major update to the drugbank database for 2018. Nucleic Acids Res. **46**(Database-Issue), D1074–D1082 (2018)
20. Kanehisa, M., Furumichi, M., Tanabe, M., Sato, Y., Morishima, K.: KEGG: new perspectives on genomes, pathways, diseases and drugs. Nucleic Acids Res. **45**(Database-Issue), D353–D361 (2017)
21. Lin, X., Quan, Z., Wang, Z., Ma, T., Zeng, X.: KGNN: knowledge graph neural network for drug-drug interaction prediction. In: Proceedings of the Twenty-Ninth International Joint Conference on Artificial Intelligence, IJCAI 2020, pp. 2739–2745. ijcai.org (2020)
22. Wang, X., He, X., Cao, Y., Liu, M., Chua, T.: KGAT: knowledge graph attention network for recommendation. In: Proceedings of the KDD 2019, Anchorage, AK, USA, pp. 950–958. ACM (2019)
23. Schlichtkrull, M.S., Kipf, T.N., Bloem, P., van den Berg, R., Titov, I., Welling, M.: Modeling relational data with graph convolutional networks. In: Gangemi, A., et al. (eds.) ESWC 2018. LNCS, vol. 10843, pp. 593–607. Springer, Cham (2018). https://doi.org/10.1007/978-3-319-93417-4_38
24. Mondal, I.: BERTKG-DDI: towards incorporating entity-specific knowledge graph information in predicting drug-drug interactions. In: Proceedings of the AAAI 2021, vol. 2831 of CEUR Workshop Proceedings. CEUR-WS.org, Virtual Event (2021)
25. Sun, Z., Deng, Z., Nie, J., Tang, J.: Rotate: knowledge graph embedding by relational rotation in complex space. In: Proceedings of the ICLR 2019. OpenReview.net, New Orleans, LA, USA (2019)

26. Zhang, S., Tay, Y., Yao, L., Liu, Q.: Quaternion knowledge graph embeddings. In: Wallach, H.M., Larochelle, H., Beygelzimer, A., d'Alché-Buc, F., Fox, E.B., Garnett, R. (eds.) Advances in Neural Information Processing Systems 32: Annual Conference on Neural Information Processing Systems 2019, pp. 2731–2741. NeurIPS 2019, Vancouver, BC, Canada (2019)

27. Cao, Z., Xu, Q., Yang, Z., Cao, X., Huang, Q.: Dual quaternion knowledge graph embeddings. In: Proceedings of the Thirty-Fifth AAAI Conference on Artificial Intelligence, AAAI 2021, pp. 6894–6902. AAAI Press, Virtual Event (2021)

28. Guo, J., Kok, S.: Bique: Biquaternionic embeddings of knowledge graphs. arXiv preprint arXiv:2109.14401 (2021)

29. Li, R., et al.: House: Knowledge graph embedding with householder parameterization. In: Chaudhuri, K., Jegelka, S., Song, L., Szepesvári, C., Niu, G., Sabato, S. (eds.) Proceedings of the International Conference on Machine Learning, ICML 2022. PMLR, Baltimore, Maryland, USA, vol. 162, pp. 13209–13224 (2022)

30. Dettmers, T., Minervini, P., Stenetorp, P., Riedel, S.: Convolutional 2d knowledge graph embeddings. In: Proceedings of the AAAI 2018, pp. 1811–1818. AAAI Press (2018)

31. Jiang, X., Wang, Q., Wang, B.: Adaptive convolution for multi-relational learning. In: Proceedings of the NAACL-HLT 2019, pp. 978–987. Association for Computational Linguistics (2019)

32. Nguyen, D.Q., Vu, T., Nguyen, T.D.: A capsule network-based embedding model for knowledge graph completion and search personalization. In: Proceedings of the NAACL-HLT 2019, pp. 2180–2189. Association for Computational Linguistics (2019)

Feature Extraction Approach for Predicting Protein-DNA Binding Residues Using Transformer Encoder-Decoder Architecture

Yi Qiu, Long Cheng, Man Xu, Jing Chen$^{(\boxtimes)}$, and Hongjie Wu$^{(\boxtimes)}$

School of Electronic and Information Engineering, Suzhou University of Science and Technology, Suzhou 215009, Jiangsu, China
jingchen@usts.edu.cn, Hongjie.wu@qq.com

Abstract. In the realm of biology, the effects of protein binding with other molecules are of paramount importance, especially in the context of DNA binding. Precisely identifying the residues implicated in protein-DNA binding is crucial for gaining a more profound insight into the mechanisms governing protein-DNA interactions. The majority of existing methods presently utilize a two-step approach, which is plagued by drawbacks including low prediction efficiency and poor usability, thereby constraining their practical applicability. In the present study, we propose a novel method grounded in sequence-to-sequence (seq2seq) models. This model has the capability to accept variable-length complete protein sequences as input and employs Transformer encoder blocks along with feature extraction blocks for hierarchical feature extraction. Through this approach, our objective is to augment the identification capability of protein-DNA binding residues. We conducted comparative experiments on the benchmark datasets, with the results demonstrating the remarkable effectiveness of our proposed method in identifying protein-DNA binding residues. This approach presents a promising new avenue for tackling research on protein-DNA interactions.

Keywords: Protein-DNA binding · Sequence-to-sequence models · Residue identification · Transformer encoder blocks · Feature extraction

1 Introduction

Proteins fulfill multifaceted and indispensable roles within organisms, engaging in diverse functions including interaction with biomolecules such as Deoxyribonucleic Acid (DNA), Ribonucleic Acid (RNA), nucleotides, and a spectrum of metal ions such as Mn^{2+}, Zn^{2+}, Fe^{3+}, Ca^{2+}, Na^{1+}, among others [1–3]. Notably, the interaction between proteins and DNA stands as an indispensable process in biology, intricately entwined with pivotal biological processes like gene regulation, DNA replication, and transcriptional control. Consequently, conducting a comprehensive exploration of protein-DNA binding residues and elucidating their interaction mechanisms emerges as imperative. The investigation of protein-DNA binding residues assumes profound significance in elucidating the intricacies of protein-DNA interactions, thereby opening novel avenues and methodologies for the management and mitigation of associated diseases.

© The Author(s), under exclusive license to Springer Nature Singapore Pte Ltd. 2024
D.-S. Huang et al. (Eds.): ICIC 2024, LNBI 14881, pp. 241–250, 2024.
https://doi.org/10.1007/978-981-97-5689-6_21

Wet laboratory methods are commonly used for identifying protein-DNA binding residues. These methods encompass X-ray crystallography, Fast Chromatin Immuno-precipitation (Fast ChIP), and Electrophoretic Mobility Shift Assay (EMSA). Through these approaches, researchers can discern the binding locations and residues between proteins and DNA, thereby elucidating details of their interactions. However, wet labo-ratory methods possess certain drawbacks. Firstly, they are frequently costly and labor-intensive, necessitating considerable time and resources. Secondly, with the accelerating pace of protein sequence growth, these methods have struggled to keep pace with the rapid expansion of protein sequences, thereby impeding research progress. In recent years, with the advancement of computer theory, numerous computational methods have surfaced to tackle this issue. These methods encompass sequence-based approaches, structure-based approaches, and hybrid methods.

In the preceding decade, a plethora of sequence-based methodologies has surfaced, encompassing BindN [4], DP-Bind [4], BindN+ [5], MetaDBSite [6], TargetDNA [7], and DNAPred [8]. These methodologies prognosticate the binding residues between pro-teins and DNA by leveraging diverse features extracted from protein sequences alongside machine learning methodologies [23].

Techniques for protein structure prediction, exemplified by DBD-Hunter [10], DNABINDPROT [11], DR_bind [12], and PreDs [13], are dedicated to harnessing three-dimensional structural data of proteins in inferring the positions of residues involved in DNA binding. Nevertheless, they tend to disregard the potential information encoded within the protein sequence itself. In contrast, hybrid methodologies amalgamate both protein sequence and structural data, culminating in enhanced predictive capabilities. Examples of these hybrid methodologies encompass TargetATP [14], TargetS [15], SVMPred [16], and NsitePred [4], among others.

A pivotal pursuit centers on techniques grounded in computation and sequences, directed towards the prediction of protein-DNA binding residues. Past studies fre-quently employed sliding window methodologies to process features for individual residues. Drawing inspiration from DeepCSeqSite [17], we present a pioneering app-roach that constructs an encoder-decoder model proficient in predicting the entirety of the protein sequence. Experimental findings underscore the competitiveness of this new-found method across the PDNA-543 and PDNA-41 datasets, exhibiting superiority over alternative state-of-the-art methodologies.

2   Material and Method

2.1   Data Set

This investigation employed the PDNA-543 and PDNA-41 datasets as the foundation for both training and evaluation purposes. These datasets were curated by Hu et al. [8]. Initially, the researchers compiled 7186 meticulously annotated DNA-binding proteins from the Protein Data Bank (PDB). Subsequently, redundant sequences were eliminated through the utilization of CD-hit software [18], guaranteeing a homology threshold of under 30% among the retained protein sequences, there by yielding 584 sequences fulfilling the stipulated criteria. The aforementioned 584 sequences were partitioned into a training cohort (comprising 543 sequences) and a testing cohort (comprising 41 sequences), labeled as the PDNA-543 and PDNA-41 datasets, respectively.

The PDNA-543 dataset encompasses a total of 543 protein sequences possessing DNA-binding capabilities. Nevertheless, it is imperative to acknowledge that only a minor fraction of residues within these sequences participate in DNA binding. Furthermore, it is noteworthy that the homology between any pair of protein sequences in the dataset remains below 30%. Conversely, the PDNA-41 dataset exhibits resemblances to PDNA-543 but is of a reduced scale, encompassing solely 41 protein sequences (Table 1).

Table 1. Detailed Information of the PDNA-543 and PDNA-41 Datasets.

Dataset	Number				R_{DNABR}[1] (%)
	Proteins	Residues			
		DNA Binding	Non-DNA Binding	Total	
PDNA-543	543	9549	134995	144544	7.07
PDNA-41	41	734	14021	14755	5.24

[1]R_{DNABR}: ratio of DNA binding residues.

2.2 Feature Representation

The characteristics of input data are pivotal in determining the model's performance. To more effectively depict each residue within the protein, we have opted for two distinct features: Position-Specific Scoring Matrix (PSSM) [9] and Predicted Secondary Structure (PSS).

2.2.1 PSSM and PSS

The PSSM encapsulates evolutionary insights of the query protein, a factor that has been proven to exert a beneficial influence on numerous bioinformatics endeavors. The generation of the PSSM feature involved the utilization of the PSI-BLAST multiple sequence alignment tool, performing three iterations of searches against the Uniprot database [19], employing an e-value threshold of 10^{-3}. This process culminated in the acquisition of the PSSM feature for every residue. To guarantee uniform scaling of these features, we employed a normalization formula to rescale the values of PSSM within the range (0, 1), thereby ensuring standardization of units across disparate features. The normalization formula for PSSM is expressed as follows [20]:

$$y = \frac{1}{1+e^{-x}} \qquad (1)$$

where x represents each raw score in the PSSM, and y represents the normalized score. For a protein sequence of length L, the dimensionality of the PSSM feature is L × 20.

Within the realm of protein structure, secondary structure comprises three primary categories: helical, α-helical, and β-folded. Frequently employed tools such as PSIPRED and PSSpred [21] are conventionally utilized for the prediction of protein secondary structure. These tools produce a collection of three-dimensional features for each residue, with each value falling within the range of 0 to 1. In this study, the PSIPRED tool was selected for generating predicted secondary structure for the target protein.For a protein sequence of length L, the predicted secondary structure feature will possess a dimensionality of L × 3. In this context, every set of three values denotes the likelihood of a residue being attributed to each of the three secondary structure classifications (helical, α-helical, and β-folded).This elucidates the utilization of secondary structure prediction tools and the informational content encapsulated within the features they yield.

2.3 Model

Drawing inspiration from the seq2seq model [21, 22], this study introduces an encoder-decoder model that creatively circumvents task segmentation. The fundamental concept is to treat the entire sequence holistically, rather than partitioning it into individual segments for processing. The schematic representation of this innovative approach is depicted in Fig. 1.

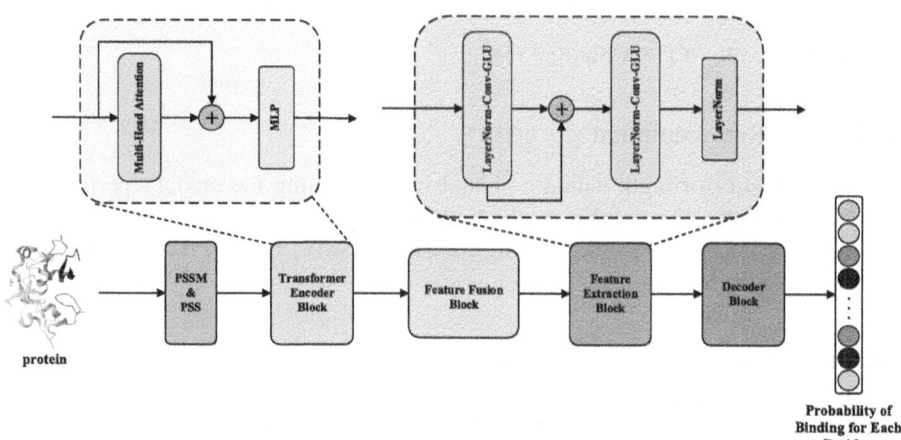

Fig. 1. The overarching architectural framework of the model. The model architecture comprises positional encoding, hierarchical feature extraction, and decoder components, designed to fulfill the task of predicting protein-DNA binding residues.

2.4 Transformer Encoder Block

In our study, the Transformer Encoder Block utilized is founded upon the self-attention mechanism [21], fostering direct interconnections among diverse positions, thereby empowering it to comprehensively capture vital information across the entire text. Subsequent to encoding protein sequence features with positional encoding, they are fed into the Transformer Encoder Block. The structure of the Transformer Encoder Block is delineated in Fig. 1. Initially, the embedded sequence traverses through the multi-head self-attention mechanism. The self-attention mechanism computes attention weights for all residues (including itself) concerning each residue in the protein sequence. The computation proceeds as follows [24]:

$$\text{Attention}(Q, K, V) = \text{softmax}\left(\frac{QK^T}{\sqrt{d_k}}\right)V \tag{2}$$

Within the multi-head attention mechanism, Q, K, and V symbolize the matrices of Query, Key, and Value, respectively, mirroring distinct representations of the input sequence. This equation delineates the procedure for computing weights predicated on the resemblance between queries and keys, followed by utilizing these weights to scale and aggregate the values, thereby acquiring the resultant representation under the attention mechanism. The computation proceeds as follows [24]:

$$\text{MultiHead}(Q, K, V) = \text{Concat}(\text{head}1, \ldots, \text{head}_h)W^0 \tag{3}$$

Specifically, for each attention head i (where i $= 1,2,\ldots,$h), the queries Q, keys K, and values V undergo linear transformations. The output outcomes from each attention head are concatenated, yielding the ultimate multi-head attention output. This process can be depicted as follows [24]:

$$\text{head}_i = \text{Attention}\left(QW_i^Q, KW_i^K, VW_i^V\right) \tag{4}$$

Here, W represents a spatial mapping function of identical dimensionality employed in this study.

Besides the multi-head attention mechanism, there is typically a Multi-Layer Perceptron (MLP) module. The MLP module comprises two fully connected layers and a non-linear activation function, facilitating additional non-linear transformation and feature extraction of the multi-head attention output.

2.5 Feature Fusion Block

The dependencies inherent in various sequence contexts can profoundly influence the prediction process of protein-DNA binding residues. In order to gain a deeper insight into the contextual relationships within sequence data, Bidirectional Gated Recurrent Units (BiGRU) were employed. Each GRU unit comprises a reset gate r_t, an update gate z_t, a candidate memory cell $\tilde{h}_t$, and a hidden state h_t utilized for capturing sequence features. The forward computation procedure of the bidirectional GRU proceeds as follows: (1) Compute the reset gate and update gate; (2) Determine the candidate memory cell for the current time step; (3) Update the hidden state for the current time step.

$$r_t = \sigma(W_{r_f} \cdot [h_{t-1}, x_t]) \tag{5}$$

$$z_t = \sigma(W_{z_f} \cdot [h_{t-1}, x_t]) \tag{6}$$

$$\tilde{h}_t = \tanh(W_f \cdot [r_t \odot h_{t-1}, x_t]) \tag{7}$$

$$h_t = (1 - z_t) \odot h_{t-1} + z_t \odot \tilde{h}_t \qquad (8)$$

Here, x_t denotes the input at the current time step, h_{t-1} signifies the hidden state from the previous time step, σ denotes the sigmoid function, and $\odot$ denotes element-wise multiplication, wherein W_{r_f}, W_{z_f}, and W_f represent the weight parameters of the forward GRU.

The backward computation process of the bidirectional GRU mirrors the forward computation process, albeit utilizing backward inputs and weight parameters instead.

2.6 Feature Extraction Block

To accomplish the acquisition of local features, the Feature Extraction Block embraces convolutional neural networks. Convolutional operations aggregate features through the traversal of convolutional kernels across feature maps, thereby facilitating improved extraction of local residue features and bolstering the model's comprehension of sequence features, consequently enhancing prediction accuracy.

The Gate Linear Unit (GLU) introduces a gating mechanism employing a sigmoid activation function. The feature extraction block comprises two LayerNorm-Conv-GLU blocks. The disparity between the two LayerNorm-Conv-GLU blocks lies in the inclusion of a residual connection following the first block, which is absent in the second block.

2.7 Decoder Block

The primary objective of the decoder is to produce results matching the length of the protein, discerning the nature of each residue, namely, whether it constitutes a protein-DNA binding residue. To accomplish this objective, a decoder architecture founded upon multiple layers of CNNs can be employed. Within this framework, it is imperative that the output channels of the ultimate convolutional layer in the decoder amount to 2, serving the purpose of identifying the residue type. Within the decoder, the input data initially undergoes a sequence of convolution operations. Subsequent to this, the resultant values traverse through the ReLU (Rectified Linear Unit) activation function to introduce nonlinearity, thus acquiring the capability to learn and depict intricate function relationships. The ReLU activation function is as follows [25]:

$$f(x) = max(0, x) \qquad (9)$$

Subsequently, the data proceeds to a dropout layer, which randomly omits a fraction of the neuron outputs. This simplification aids in diminishing the computational and storage demands of the model and safeguards the neural network against excessive dependence on particular neurons or features, thereby mitigating the risk of overfitting. With this architectural design, the decoder is capable of proficiently producing results commensurate with the protein's length while accurately discerning the type of each residue.

2.8   Model Evaluation

Prediction of protein-DNA binding residues constitutes a binary classification task, wherein the model endeavors to ascertain whether each residue manifests as a protein-DNA binding residue. Assessment of prediction performance commonly incorporates metrics encompassing sensitivity (SN), specificity (SP), accuracy (ACC), and Matthews correlation coefficient (MCC). Owing to the frequently encountered highly imbalanced distribution of positive and negative samples in protein-DNA binding residue prediction, MCC factors in the ratios of true positives, true negatives, false positives, and false negatives, furnishing a more accurate depiction of the model's efficacy on imbalanced datasets. Therefore, MCC is typically considered a more appropriate metric for evaluating the performance of this method.

$$SN = \frac{TP}{TP + FN} \tag{10}$$

$$SP = \frac{TN}{TN + FP} \tag{11}$$

$$ACC = \frac{TP + TN}{TP + TN + FP + FN} \tag{12}$$

$$MCC = \frac{TP \times TN - FP \times FN}{\sqrt{(TP + FP)(TP + FN)(TN + FP)(TN + FN)}} \tag{13}$$

where TP represents the predicted DNA binding residues that are indeed DNA binding residues (true positive). TN denotes the predicted non-DNA binding residues that indeed are non-DNA binding residues (true negative). FP signifies the predicted non-DNA binding residues that are in actuality DNA binding residues (false positive). FN indicates the predicted DNA binding residues that are essentially non-DNA binding residues (false negative). Indeed, the greater the magnitudes of these four metrics, the more superior the model's performance.

3   Result and Discussion

3.1   Setting of Hyperparameters

The selection of model hyperparameters profoundly influences the ultimate prediction outcomes. In this section, we employ ten-fold cross-validation on the PDNA-543 dataset to juxtapose the experimental outcomes stemming from diverse model hyperparameter selections.

The model encompasses a plethora of hyperparameters, rendering each experiment potentially time-consuming, often spanning days for completion. Such protracted timelines are impractical for swift iteration and model optimization endeavors. To tackle this challenge, we adopted an empirical strategy, setting the batch size to 1 and employing a reduced learning rate. By setting the batch size to 1, the model gains the capability to handle lengthier protein sequences sans the requirement for supplementary padding operations. This approach is geared towards stabilizing model training, alleviating the risk of overfitting, and amplifying overall computational efficiency (Table 2).

Table 2. Hyperparameter Values.

Hyperparameter	Values
Optimizer	Adam
Learning rate	0.00005
Loss function	CrossEntropyLoss
Batch size	1
Epoch	1000

3.2 Performance Comparison with Other Predictors

Independent testing was conducted on the PDNA-41 dataset to substantiate the efficacy of our experimental methodology. The comparative analysis in Table 3 juxtaposes our approach with established methodologies including BindN [4], DP-Bind [4], BindN+ [5], MataDBSite [6] and TargetDNA [7]. Notably, the MCC, SP, SN, and ACC stand at 0.348, 96.38%, 47.17%, and 94.85%. The experimental outcomes unmistakably illustrate that our methodology attains commendable performance relative to prior approaches.

Table 3. A Comparative Analysis of Various Classifiers on the PDNA-41 Dataset through Independent Evaluation.

Method	MCC	SP (%)	SN (%)	ACC (%)
BindN	0.143	80.90	45.64	79.15
BindN + (SP ≈ 95%)	0.178	95.11	24.11	91.58
BindN + (SP ≈ 85%)	0.213	85.41	50.81	83.69
MetaDBSite	0.221	93.35	34.20	90.41
DP-Bind	0.241	82.43	61.72	81.40
TargetDNA (SN ≈ SP)	0.269	85.79	60.22	84.52
TargetDNA (SN ≈ 95%)	0.300	93.27	45.50	90.89
Our Method	0.348	96.38	47.17	94.85

4 Conclusion

In this study, we present an encoder-decoder model meticulously crafted for the prediction of protein-DNA binding sites. The distinctiveness of this model resides in its utilization of two sequence-based features, namely, the evolutionary feature PSSM and the predicted secondary structure, for representing protein sequences. Through this methodology, we can hierarchically extract features from protein sequences while simultaneously acquiring global and local feature representations. On the PDNA-41 test set, our model exhibits promising performance, achieving an MCC of 0.348, SP of 96.38%, SN of 47.17%, and ACC of 94.85%. Presently, our model is restricted to processing one protein sequence at a time, somewhat constraining its versatility. We're exploring two strategies: using graph neural networks to represent protein sequences for identifying DNA binding residues, and integrating forecasted protein structural data to improve prediction of protein-DNA binding sites. Both approaches aim to enhance the model's performance and predictive ability.

Acknowledgments. This paper is supported by the National Natural Science Foundation of China (62073231, 62372318, 62176175), National Research Project (2020YFC2006602), Provincial Key Laboratory for Computer Information Processing Technology, Soochow University (KJS2166). Opening Topic Fund of Big Data Intelligent Engineering Laboratory of Jiangsu Province (SDGC2157).

References

1. Wu, H., Ling, H., Gao, L., et al.: Empirical potential energy function toward ab initio folding G protein-coupled receptors. IEEE/ACM Trans. Comput. Biol. Bioinf. **18**(5), 1752–1762 (2020)
2. Wu, H., Wang, K., Lu, L., et al.: Deep conditional random field approach to transmembrane topology prediction and application to GPCR three-dimensional structure modeling[J]. IEEE/ACM Trans. Comput. Biol. Bioinf. **14**(5), 1106–1114 (2016)
3. Lu, Y., Zhang, R., Jiang, T., et al.: TrGPCR: GPCR-ligand binding affinity predicting based on dynamic deep transfer learning. IEEE J. Biomed. Health Inform. (2023). https://doi.org/10.1109/JBHI.2023.3307928
4. Zhao, J., Cao, Y., Zhang, L.: Exploring the computational methods for protein-ligand binding site prediction. Comput. Struct. Biotechnol. J. **18**, 417–426 (2020)
5. Guan, S., Zou, Q., Wu, H., et al.: Protein-DNA binding residues prediction using a deep learning model with hierarchical feature extraction. IEEE/ACM Trans. Comput. Biol. Bioinform. (2022)
6. Si, J., Zhang, Z., Lin, B., Schroeder, M., Huang, B.: Metadbsite: a meta approach to improve protein DNA-binding sites prediction. BMC Syst. Biol. **5**(1), 1–7 (2011)
7. Lagunas-Rangel, F.A., Bermúdez-Cruz, R.M.: Natural compounds that target DNA repair pathways and their therapeutic potential to counteract cancer cells. Front. Oncol. **10**, 2567 (2020)
8. Zhu, Y.-H., Hu, J., Song, X.-N., Yu, D.-J.: Dnapred: accurate identification of DNA-binding sites from protein sequence by ensembled hyperplane-distance-based support vector machines. J. Chem. Inf. Modeling **59**(6), 3057–3071 (2019)
9. Kha, Q.H., Ho, Q.T., Le, N.Q.K.: Identifying SNARE proteins using an alignment-free method based on multiscan convolutional neural network and PSSM profiles. J. Chem. Inf. Model. **62**(19), 4820–4826 (2022)
10. Qian, Y., Meng, H., Lu, W., Liao, Z., Ding, Y., Wu, H.: Identification of DNA-binding proteins via hypergraph based laplacian support vector machine. Curr. Bioinform. **17**(1), 108–117 (2022)
11. Ozbek, P., Soner, S., Erman, B., Haliloglu, T.: Dnabindprot: fluctuation-based predictor of DNA-binding residues within a net work of interacting residues. Nucleic Acids Res. **38**(suppl 2), W417–W423 (2010)

12. Chen, Y.C., Wright, J.D., Lim, C.: DR bind: a web server for predicting DNA-binding residues from the protein structure based on electrostatics, evolution and geometry. Nucleic Acids Res. **40**(W1), W249–W256 (2012)

13. Jaspard, M., et al.: Development of the PREDS score to predict in-hospital mortality of patients with Ebola virus disease under advanced supportive care: Results from the EVISTA cohort in the Democratic Republic of the Congo. EClinicalMedicine, 54

14. Haas, M.J., Mooradian, A.D.: Potential therapeutic agents that target ATP binding cassette A1 (ABCA1) gene expression. Drugs **82**(10), 1055–1075 (2022)

15. Yu, D.-J., Hu, J., Yang, J., Shen, H.-B., Tang, J., Yang, J.-Y.: Designing template-free predictor for targeting protein-ligand binding sites with classifier ensemble and spatial clustering. IEEE/ACM Trans. Comput. Biol. Bioinf. **10**(4), 994–1008 (2013)

16. Yu, D.J., Hu, J., Li, Q.M., Tang, Z.M., Yang, J.Y., Shen, H.B.: Constructing query-driven dynamic machine learning model with application to protein-ligand binding sites prediction. IEEE Trans. Nanobiosci. **14**(1), 45–58 (2015)

17. Cui, Y., Dong, Q., Hong, D., Wang, X.: Predicting protein-ligand binding residues with deep convolutional neural networks. BMC Bioinform. **20**, 1–12 (2019)

18. Hauser, M., Steinegger, M., Söding, J.: MMseqs software suite for fast and deep clustering and searching of large protein sequence sets. Bioinformatics **32**(9), 1323–1330 (2016)

19. UniProt Consortium: UniProt: a worldwide hub of protein knowledge. Nucleic Acids Res. **47**(D1), D506–D515 (2019)

20. Wang, J., et al.: POSSUM: a bioinformatics toolkit for generating numerical sequence feature descriptors based on PSSM profiles. Bioinformatics **33**(17), 2756–2758 (2017)

21. Wu, H., Liu, J., Jiang, T., et al.: AttentionMGT-DTA: a multi-modal drug-target affinity prediction using graph transformer and attention mechanism. Neural Netw. **169**, 623–636 (2023)

22. Liu, Y., Guan, S.X., Jiang, T.S., et al.: DNA protein binding recognition based on lifelong learning. Comput. Biol. Med. **164**, 107094 (2023)

23. Wu, H., Lv, Q., Wu, J., et al.: A parallel ant colony method for De Novo prediction of protein backbone and its application in CASP8/9. Sci. China Inf. Sci. **42**(8), 1034–1048 (2012)

24. Jaegle, A., Gimeno, F., Brock, A., Vinyals, O., Zisserman, A., Carreira, J.: Perceiver: General Perception with Iterative Attention. In: International Conference on Machine Learning, pp. 4651–4664. PMLR (2021)

25. Banerjee, C., Mukherjee, T., Pasiliao Jr., E.: An empirical study on generalizations of the ReLU activation function. In: Proceedings of the 2019 ACM Southeast Conference, pp. 164–167 (2019)

GCEA: Contrastive-Enhanced Autoencoders with Adaptive Completion for Partial Multi-omics Integration in Cancer Subtyping

Weiting Yu, Zhimin Li, and Cheng Liang[✉]

Shandong Normal University, Jinan 250358, China
ALCS417@sdnu.edu.cn

Abstract. Cancer subtype identification represents a crucial research domain in bioinformatics. The precise differentiation of patients into distinct cancer subtypes by leveraging information among multi-omics data represents a major challenge in the current research landscape. However, the presence of missing multi-omics data further complicates the precise identification of cancer subtypes. To conquer the limitation, this paper proposes a novel cancer subtype identification model, named GCEA, capable of addressing scenarios involving missing multi-omics data. Specifically, by leveraging adversarial generative strategies, the model adaptively generates missing omics data, thereby enabling it to flexibly handle various types of missing multi-omics data. Additionally, the model enhances its learning capability by employing contrastive learning to acquire informative and distinct latent representations of samples, facilitating precise identification of cancer subtypes in patients. Extensive experimental results on ten cancer multi-omics datasets demonstrate that the proposed model exhibits competitive performance compared with existing methods.

Keywords: Cancer subtype identification · Contrastive learning · Partial multi-omics data

1 Introduction

Cancer is a heterogeneous disease which can vary widely between individuals and even within the same tumor. Accurate identification of cancer subtypes enables clinicians to develop more tailored treatment strategies for patients belonging to different subtypes, thereby improving patient survival and quality of life [1–4]. With the rapid development of modern high-throughput technologies, a large amount of multi-omics data for various cancer types has been deposited in public databases. For instance, the Cancer Genome Atlas (TCGA) [5] collects a wealth of multi-omics data, which allows researchers to explore cancer subtypes more efficiently by analyzing these data in silico [6–8].

Currently, several multi-omics clustering methods have been developed for cancer subtype identification. The Similarity Network Fusion (SNF) [9] approach generates similarity networks for each omics data type and subsequently performs network fusion on multiple similarity networks to obtain the final clustering results. Subtype-WESLR

D.-S. Huang et al. (Eds.): ICIC 2024, LNBI 14881, pp. 251–262, 2024.
https://doi.org/10.1007/978-981-97-5689-6_22

[10] fuses base clustering obtained by different methods through a weighted ensemble strategy, and projects the representation of each sample into a common subspace that can maintain local structure for clustering. Despite its power, the final clustering results obtained by Subtype-WESLR might be unstable due to the quality of the base clustering obtained by different methods. Subtype-GAN [12] utilizes the generative adversarial networks to learn latent representations and the identification of cancer subtypes is done through consensus clustering and the Gaussian Mixture model. DSML [13] learns its corresponding subspaces separately from available and overall omics data. Then perform clustering in the subspace. DSIR [14] employs similarity network fusion technique to capture local structures and utilizes an encoder-decoder framework to extract global structures. Subsequently, clustering analysis is performed utilizing the extracted global and local structural information.

The methods introduced above generally assume that the multi-omics data for samples is complete. However, in real life scenarios, data collection is often prone to various factors that lead to data incompleteness. Consequently, the identification of cancer subtypes in the presence of partial multi-omics data is a pressing challenge that researchers are actively endeavoring to tackle. Several methods have been proposed to tackle the problem of cancer subtype identification in the presence of partial data. Among them, MSNE [11] and NEMO [15] are two representative methods. MSNE embeds multiple similarity networks into a common vector space. Then, the low-dimensional representation of the samples is obtained using an embedding method and the final clustering results are obtained by K-means. NEMO first builds an inter-patient similarity matrix for each omics. Then, it integrates the omics-specific matrices into matrix and perform clustering for subtype identification. Although existing methods have achieved competitive results in the partial multi-omics data integration tasks, they still have some limitations. For example, MSNE only retains the similarity information of the k nearest neighbor samples, which might lead to underexplored extraction of information between the data. NEMO relies on strong assumptions about multi-omics data. It strictly requires that all samples have representations in at least one view, which limits the applicability of the NEMO method.

To address the challenges in cancer subtype identification with partial multi-omics data, we propose a contrastive enhanced autoencoders with adaptive interpolation for cancer subtype identification(GCEA). Our method consists of two modules: an autoencoder module with contrastive enhancement and adversarial generation interpolation(CAG-AE) and a contrastive clustering module(CEC). Specifically, we use an autoencoder to learn a latent representation for each omic and perform clustering directly on the latent representation, incorporating both instance-level contrastive loss and clustering-level contrastive loss. To handle missing omics, we employ the adversarial generation-based interpolation for imputation, which effectively captures the underlying distribution of multi-omics data and accommodates various types of missing data scenarios. We validate our method on ten cancer datasets and compare it with several representative incomplete multi-view clustering algorithms. The results demonstrate our method exhibits superior performance across diverse datasets and experimental conditions.

The contributions of this paper are as follows:

- We integrate adversarial generative strategies into the feature extraction module of the model, enabling different types of missing multi-omics data to be learned under a unified framework without the need to design separate models for each missing type.
- We further enhance the capacity of our model to obtain distinct latent representations for samples by designing a dual contrastive learning scheme and the experimental results validate the superiority of our method over existing alternatives.

2 Proposed Method

2.1 Overview

Figure 1 illustrates the overall architecture of the GCEA model proposed for cancer subtype identification. The model consists of two modules, namely the CAG-AE module and the CEC module. Firstly, the CAG-AE module takes the multi-omics data as input and encodes them into latent presentations. The decoder reconstructs the original data using the encoded features. During this process, the decoder automatically generates missing omics data based on the observed data distribution. The model then incorporates both the original input data and the reconstructed representation from the decoder into the discriminator. The discriminator discerns whether the input data is generated by the generator. The CEC module is responsible for clustering the data based on the encoded features from the encoder to obtain the final cancer subtype identification results.

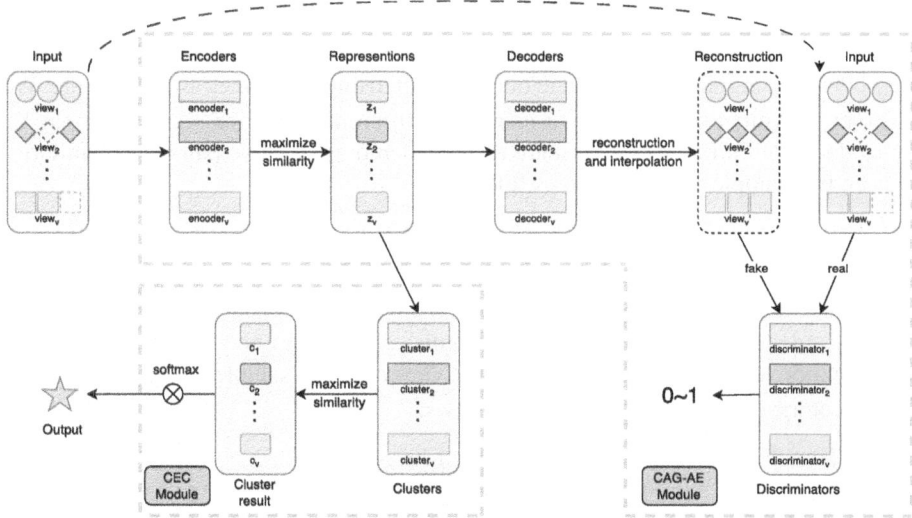

Fig. 1. The overall architecture of model GCEA. The latent representation of each omics is learned by the specific omics encoder, and then clustering is performed using these latent representations. Features of missing omics data are generated by the decoder with adversarial generative learning.

2.2 CAG-AE Module

Autoencoder is a commonly used deep learning model for feature extraction [16]. However, autoencoders in the past could only capture the global structure of the data, and such a representation does not have sufficient expressive power. To address this, we use contrastive learning to enhance the representation capacity of our model.

To this end, we employ three different loss functions to constrain the learning process: reconstruction loss, instance-level contrastive loss, and cluster-level contrastive loss. Specifically, we first learn a representation using these loss functions, and then use the resulting representation for clustering.

For each view, we use an encoder to learn the latent representation $Z^v = f_v(X^v)$, where $f(\cdot)$ is the encoder of v_{th} view. Then we use the decoder to get the reconstructed data $X^{v'} = g_v(Z^v)$, where Z^v is the output of encoder for the v_{th} view. We use the MSE loss as the reconstruction loss:

$$L_{rec} = \frac{1}{N} \sum_{i=1}^{v} \sum_{j=1}^{N} M_i^j \left(X_j^i - X_j^{i'} \right)^2 \tag{1}$$

where v represents the number of omics and N is the number of samples, M is the indicator matrix denoting the indices of available samples in the i_{th} view. The indicator matrix M is used to mitigate the effect of missing samples. The element M_i^j is defined as follows:

$$M_i^j = \begin{cases} 1, & \textit{if the } j_{th} \textit{ instance has the } i_{th} \textit{ view} \\ 0, & \textit{otherwise} \end{cases} \tag{2}$$

For the instance-level contrastive loss, we use the cosine distance to calculate the similarity between samples. We illustrate our approach using two omics as an example. We construct a positive pair $\{X_i^{k_1}, X_i^{k_2}\}$ using two different omics for a given sample X_i. Omics data from different samples are all considered as negative pairs. The similarity between two samples is calculated as follows:

$$s\left(Z_i^{k_1}, Z_j^{k_2} \right) = \frac{\left(Z_i^{k_1} \right)\left(Z_j^{k_2} \right)^T}{\| Z_i^{k_1} \| \| Z_j^{k_2} \|} \tag{3}$$

where $k_1, k_2 \in [1, v]$ means two different omics of sample i and sample j. $Z_i^{k_1}$ is the representation learned by encoder for the k_{1th} view of the i_{th} sample. For a given sample $X_i^{k_1}$, its contrastive loss is calculated as follows:

$$l^{k_1} = -\log \frac{\exp\left(s\left(Z_i^{k_1}, Z_i^{k_2} \right)/\tau_I \right)}{\sum_{j=1}^{N} \left[\exp\left(s\left(Z_i^{k_1}, Z_j^{k_2} \right)/\tau_I \right) + \exp\left(s\left(Z_j^{k_1}, Z_j^{k_2} \right)/\tau_I \right) \right]} \tag{4}$$

where τ_I is a hyperparameter controlling the softness of instance-level contrastive loss. Equation (5) demonstrates the total instance-level contrastive loss:

$$L_{ins} = \sum_{i=1}^{N} \left(l^{k_1} + l^{k_2} \right) \tag{5}$$

For incomplete data, the absence of certain omics in the samples can impact the clustering performance. To address this issue, we introduce an adversarial generation strategy in the

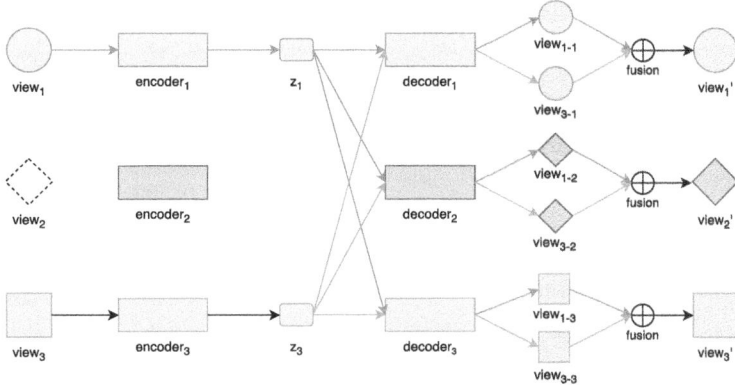

Fig. 2. Adaptive imputation illustration. The missing omics data are generated from existing counterparts. We use three omics for illustration purpose.

decoders, where we leverage samples from existing omics to adaptively generate missing omics. Figure 2 depicts the process of generating missing omics data from existing ones.

In our settings of incomplete data, the number of missing omics per sample is uncertain, that is the number of omics that exists for each sample is uncertain. Therefore, we design an adaptive fusion layer based on generative adversarial strategy to generate other missing omics based on existing omics. Assuming a sample possesses two extant omics, the process of adaptively generating an additional view follows the formula below:

$$X_i^{v_1^g} = \alpha_1 X_i^{v_m^g} + \alpha_2 X_i^{v_n^g} + \cdots + \alpha_k X_i^{v_k^g}$$
$$s.t. \quad \alpha_1 + \alpha_2 + \cdots + \alpha_k = 1 \tag{6}$$

where $X_i^{v_1^g}$ is a generated representation from the existing omics of sample i. After adaptively fusing $X_i^{v_1^g}$ and $X_i^{v_2^g}$, we obtain the representation reconstructed by the adversarial generative strategy. We combine the reconstruction loss and traditional GAN loss in this learning process. The reconstruction loss is defined as follows:

$$L_{g-rec} = \frac{1}{N} \sum_{i=1}^{v} \sum_{j=1}^{N} \left(X_j^i - X_j^{i'} \right)^2 \tag{7}$$

and the GAN loss with reconstructed samples is:

$$L_{GAN} = L_G + L_D \tag{8}$$

where L_G is the loss for generators:

$$L_G = E_{x \sim p_{data}(x)} \left\{ \sum_{\substack{v_i = 1, \\ v_m, v_n \neq v_i}}^{v} \log\left(1 - D_{v_i}\left(G_{v_i}(X^{v_m}, X^{v_n})\right)\right) \right\} \tag{9}$$

and the L_D is the loss for discriminators:

$$L_D = E_{y \sim p_{data}(y)} \left\{ \sum_{v_i=1}^v log D_{v_i} \left(X^{v_i^g} \right) \right\} \tag{10}$$

D_v and G_v are the view-specific generator and discriminator respectively. For full data, the total GAN Interpolation Module loss is:

$$L_{GIM} = L_{g-rec} + L_{GAN} \tag{11}$$

The overall loss function of CAG-AE module is:

$$L_{CAG-AE} = L_{rec} + L_{ins} + L_{GIM} \tag{12}$$

2.3 CEC Module

After we get the latent representations, we apply clustering directly to the latent representations. According to the idea of "label as representation" [17], we apply contrastive learning on the clustering representation. Similarly, cosine distance is used to measure the similarity of the cluster representations, which is defined as follows:

$$s\left(y_i^{k_1}, y_i^{k_2} \right) = \frac{\left(y_i^{k_1} \right)^T \left(y_j^{k_2} \right)^T}{\| y_i^{k_1} \| \| y_i^{k_2} \|} \tag{13}$$

where $k_1, k_2 \in [1, v]$ means two different omics of cluster representations, while $i, j \in [1, m]$ means two samples of dataset. For the distinction of different cluster representations, we calculate the cluster-specific contrastive loss as below:

$$l_i^{k_1} = -log \frac{\exp\left(s\left(y_i^{k_1}, y_j^{k_2} \right) / \tau_C \right)}{\sum_{j=1}^M \left[\exp\left(s\left(y_i^{k_1}, y_j^{k_2} \right) / \tau_C \right) + \exp\left(s\left(y_i^{k_1}, y_j^{k_2} \right) / \tau_C \right) \right]} \tag{14}$$

Similarly, τ_C is also a hyperparameter to control the softness. The total cluster-level contrastive loss is as follows:

$$L_{clu} = \frac{1}{M} \sum_{i=1}^M \left(l_i^{k_1} + l_i^{k_2} \right) - H(Y) \tag{15}$$

where the term $H(Y)$ is used to avoid trivial solutions and is calculated as: $H(Y) = -\sum_{i=1}^M \left[P\left(y_i^{k_1} \right) log P\left(y_i^{k_1} \right) + P\left(y_i^{k_2} \right) log P\left(y_i^{k_2} \right) \right]$ with $P(y_i^k)$ is $p(y_i^k) = \sum_{t=1}^N Y_{t_i}^k / \| Y^k \|, k \in \{k_1, k_2\}$.

The overall loss of GCEA is as follows:

$$L = L_{CAGAE} + L_{clu} \tag{16}$$

3   Experiments and Analysis

3.1   Dataset

We use ten cancer datasets in our experiments, and the detailed information for each dataset is shown in Table 1. The information of ten cancer datasets.

We set different missing rates η for the datasets, defined as follows: $\eta = (n - m)/n$ is the total number of samples and m is the number of complete samples. Please note that for the second and third type of missing data scenario, there is no missing rate, or in other words, the missing rate is 1, meaning that each sample will have some omics data missing. For noisy data, the variable η represents the proportion of data set to 0 across all omics data for each sample.

Table 1. The information of ten cancer datasets.

Dataset	AML	BIC	COAD	GBM	KIRC	LIHC	LUSC	SKCM	OV	SARC
No. of omics	3	3	3	3	3	3	3	3	3	3
No. of samples	170	621	220	274	183	167	341	448	287	257
No. of exp	2000	2000	2000	2000	2000	2000	2000	2000	2000	2000
No. of mirna	558	891	613	534	796	852	878	901	616	838
No. of methy	2000	2000	2000	2000	2000	2000	2000	2000	2000	2000

3.2   Experimental Settings

The number of enriched clinical labels and the p-value of the log-rank test are the most commonly used evaluation metrics for cancer subtype identification tasks. Therefore, we also adopt these two measurements to assess the performance of our model.

Moreover, to comprehensively evaluate the performance of all methods, we construct three types of incomplete datasets as depicted in Fig. 3. For the first type, as depicted in subfigure (a) of Fig. 3, there exists samples that are complete in all omics, while the rest of the samples are only available in certain omics. For the second type, as depicted in subfigure (b) of Fig. 3, all samples are partially missing in certain omics, i.e., none of the samples are complete across all omics. For the third type, as depicted in subfigure (c) of Fig. 3, at least one of the omics contains all samples, that is, samples are not missing in a specific omics, while in other omics, sample are missing randomly. In addition to the aforementioned three types of incomplete data, we also evaluate robustness of each method against random noise as depicted in subfigure (d) of Fig. 3.

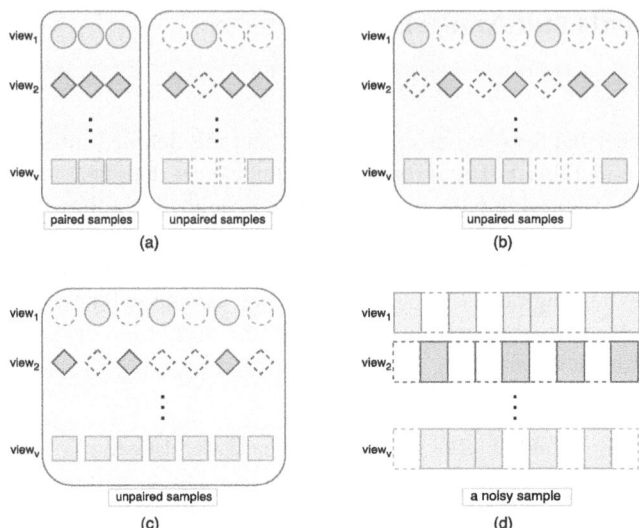

Fig. 3. Four schematic diagrams of the missing scenarios of multi-omics data. Subfigures (a)–(c) depict three types of missing multi-omics data, while subfigure (d) represents samples with various levels of noise.

We implement our code with PyTorch 1.5 as the core library and CUDA version 10.2. The trade-off hyperparameters for both instance-level contrastive loss and clustering-level contrastive loss are set to 0.5. In our experiments, we employ the Adam optimizer with a learning rate of 0.001.

3.3   Results Analysis

We compare GCEA with six representative multi-omics data integration methods, i.e. DLSF [22], MSNE [11], SGC-IMVC [18], NEMO [15], MDICC [23] and SNF [10]. The experimental results are demonstrated in Fig. 4.

As shown in the Fig. 4, it can be observed that, GCEA achieves comparable or superior performance compared to previously proposed methods in terms of the enriched clinical numbers and the survival p-values. Notably, in takes involving missing type 1 and type 2, GCEA attains the overall best results. It demonstrates superior performance in log-rank test indicators, signifying statistically significant differences among cancer subtypes derived through GCEA clustering. Additionally, GCEA outperforms other comparison methods comprehensively in clinical enrichment indicators, indicating that the subtype results obtained by GCEA align more closely with clinical reality. Similar observations are also obtained in the constructed noisy data. We find that GCEA did not achieve the best evaluation results in the third type of missing data scenario, this might because the information extracted from other missing omics data misleads the information extracted from the complete omics data, leading to a decrease in the learning effectiveness of GCEA. Interestingly, GCEA could still achieve the second best with different omics available in this missing scenario. Taken together, the extensive experimental results clearly confirm the superiority of GCEA in handling partial multi-omics data integration task.

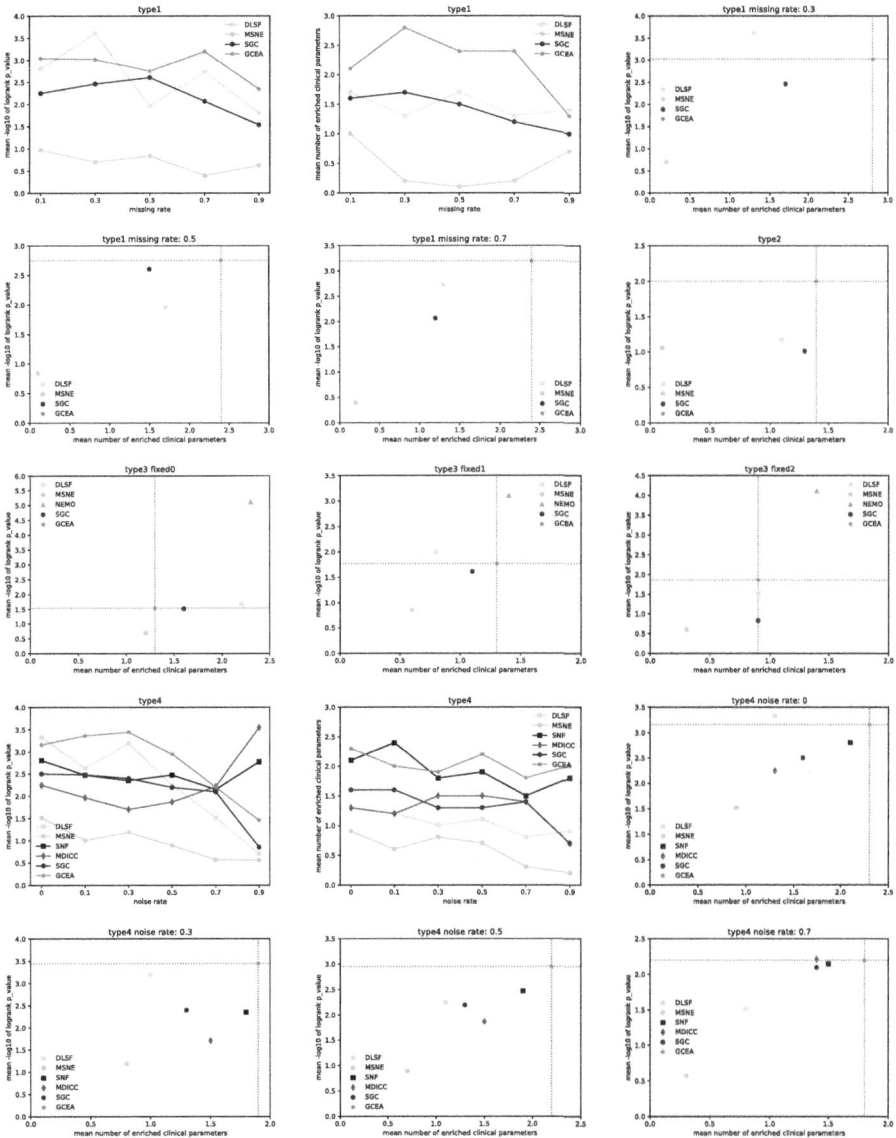

Fig. 4. Summary of evaluation results.

3.4 Ablation Study

In this section, we investigate the impact of instance-level contrastive loss (ICH) on the performance of GCEA. From Fig. 5, it can be seen that when the model excludes ICH, survival p-values and the number of enriched clinical labels decrease. This suggests that without ICH, the features extracted by the model lack sufficient representativeness, resulting in reduced reliability of the final clustering results. Therefore, the presence of ICH can significantly enhance the capacity of representation learnings in our model.

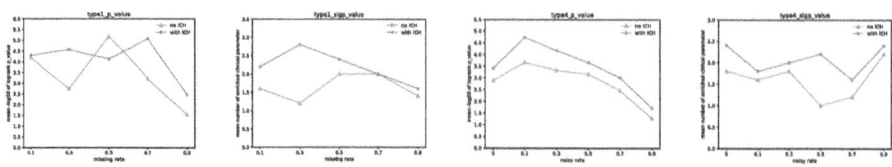

Fig. 5. Effect of ICH.

3.5 Hyperparameter Analysis

In this section, we discuss the effects of two primary hyperparameters, τ_I and τ_C, on the model performance. Figure 6 shows that, when changing the two hyperparameters, τ_I and τ_C, the evaluation results of the model remain relatively close, with noticeable differences only under certain missing rates. In noisy data, when both hyperparameters are set to 0.5, the model's clinical enrichment metrics outperform other parameter settings. These results indicate that while different values of hyperparameters might impact model's performance under certain conditions, GCEA is relatively stable with respect to various hyperparameter settings in general.

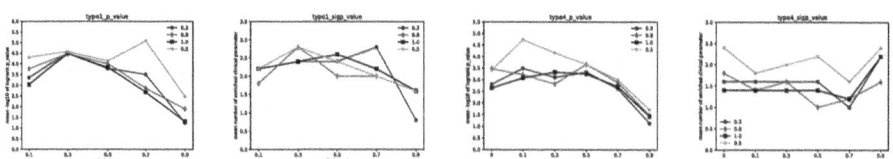

Fig. 6. Evaluation results under different hyperparameters.

3.6 Convergence Analysis

In this section, we analyze the convergence of the GCEA algorithm in practice. Figure 7 displays the variation of the model loss with the number of iterations, where the x-axis denotes the number of iterations, and the y-axis represents the total loss. As shown in Fig. 7, the model exhibits a similar decreasing trend in loss values across different datasets. Specifically, the descent process of the loss values can be roughly divided into two stages. Initially, the first half demonstrates a rapid decline, indicating that the model rapidly adapts to the data and achieves significant improvements during the initial phase. As training progresses, the rate of decrease in loss values gradually diminishes, leading to a plateau in the latter half, indicating that the model is gradually approaching local optimal solutions.

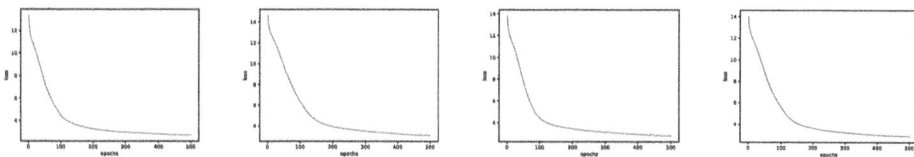

Fig. 7. The trend of loss with respect to epochs.

4   Conclusion

In this study, we propose a method called GCEA which utilizes a contrastive-enhanced autoencoders for cancer subtype identification based on partial multi-omics data. GCEA adaptively generates missing omics through adversarial generative learning, enabling it to handle various missing data situations. In addition, two contrastive components equipped with the autoencoders improve the learning capacity of latent representations and result in efficient clustering. Therefore, GCEA is a powerful multi-omics clustering method that can flexibly adapt to different incomplete multi-omics clustering scenarios. However, the GCEA algorithm still has shortcomings. The model may be sensitive to the choice of hyperparameters, leading to variations in performance based on their selection. Moreover, when there is only one omics available, the data generated by GCEA might be unreliable for clustering tasks under certain circumstances. In the future, we will explore more robust generative modules to further improve overall performance and applicability in real-world settings.

Acknowledgments. This work is partially supported by the National Natural Science Foundation of China (No. 62372279) and Natural Science Foundation of Shandong Province, China (No. ZR2023MF119).

References

1. Bailey, P., Chang, D., Nones, K. et al.: Genomic analyses identify molecular subtypes of pancreatic cancer. Nature **531**, 47–52 (2016). Author, F., Author, S.: Title of a proceedings paper. In: Editor, F., Editor, S. (eds.) CONFERENCE 2016, LNCS, vol. 9999, pp. 1–13. Springer, Heidelberg (2016)
2. Jahid, M. J., Huang, T. H., Ruan, J.: A personalized committee classification approach to improving prediction of breast cancer metastasis. Bioinformatics **30**(13), 1858–1866 (2014). Author, F.: Contribution title. In: 9th International Proceedings on Proceedings, pp. 1–2. Publisher, Location (2010)

3. Parker, J. S., Mullins, M., Cheang, M. C., Leung, S., et al.: Supervised risk predictor of breast cancer based on intrinsic subtypes. J. Clin. Oncol. **27**(8) (2009)
4. Prasad, V., Fojo, T., Brada, M.: Precision oncology: origins, optimism, and potential. Lancet Oncol. **17**(2), e81–e86 (2016)
5. The Cancer Genome Atlas Research Network: Comprehensive genomic characterization defines human glioblastoma genes and core pathways. Nature **455**, 1061–1068 (2008)
6. Wang, D., Gu, J.: Integrative clustering methods of multi-omics data for molecule-based cancer classifications. Quant.Biol. 58–67 (2016)
7. Huang, S., Chaudhary, K., Garmire, L.X.: More is better: recent progress in multi-omics data integration methods. Front. Gen. (2017)
8. Mitra, S., Saha, S., Hasanuzzaman, M.: Multi-view clustering for multi-omics data using unified embedding. Sci. Rep. **10**(1) (2020)
9. Wang, B., Mezlini, A., Demir, F., et al.: Similarity network fusion for aggregating data types on a genomic scale. Nat. Methods **11**, 333–337 (2014)
10. Wenjing, S., Weiwen, W., Daoqing, D.: Subtype-WESLR: identifying cancer subtype with weighted ensemble sparse latent representation of multi-view data. Briefings Bioinform. **23**(1)
11. Xu, H., Gao, L., Huang, M., Duan, R.: A network embedding based method for partial multi-omics integration in cancer subtyping. Methods **192**, 67–76 (2021)
12. Yang, H., Chen, R., Li, D., Wang, Z.: Subtype-GAN: a deep learning approach for integrative cancer subtyping of multi-omics data. Bioinformatics **37**(16), 2231–2237 (2021)
13. Yang, B., Xin, T.T., Pang, S.M., Wang, M., Wang, Y.J.: Deep subspace mutual learning for cancer subtypes prediction. Bioinformatics **37**(21), 3715–3722 (2021)
14. Yang,B.,Yang,Y.,Su,X.: Deep structure integrative representation of multi-omics data for cancer subtyping. Bioinformatics **38**(13), 3337–3342 (2022)
15. Rappoport, N., Shamir, R.: NEMO: cancer subtyping by integration of partial multi-omic data. Bioinformatics **35**(18), 3348–3356 (2019)
16. Michelucci, U.: An introduction to autoencoders. arXiv preprint https://arxiv.org/abs/2201.03898 (2022)
17. Li, Y., Hu, P., Liu, Z., Peng, D., Zhou, J.T., Peng, X.: Contrastive clustering. In: Proceedings of the AAAI Conference on Artificial Intelligence, pp. 8547–8555 (2021)
18. Liu, C., Wu, S., Li, R., Jiang, D., Wong, H.S.: Self-supervised graph completion for incomplete multi-view clustering. IEEE Trans. Knowl. Data Eng. (2023)
19. Rappoport, N., Shamir, R.: Multi-omic and multi-view clustering algorithms: review and cancer benchmark. Nucleic Acids Res. **46**(20), 10546–10562 (2018)
20. Kleinbaum, D.G., Klein, M.: Kaplan-Meier survival curves and the log-rank test. In: Kleinbaum, D.G., Klein, M. (eds.) Survival Analysis: A Self-learning Text, pp. 55–96. Springer, New York (2012). https://doi.org/10.1007/978-1-4419-6646-9_2
21. Wen, J., Xu, Y., Liu, H.: Incomplete multiview spectral clustering with adaptive graph learning. IEEE Trans. Cybernet. **50**(4), 1418–1429 (2018)
22. Zhang, C., Chen, Y., Zeng, T., Zhang, C., Chen, L.: Deep latent space fusion for adaptive representation of heterogeneous multi-omics data. Briefings Bioinform. **23**(2) (2022)
23. Yang, Y., Tian, S., Qiu, Y., Zhao, P., Zou, Q.: MDICC: novel method for multi-omics data integration and cancer subtype identification. Briefings Bioinform. **23**(3) (2022)

GGANet: A Model for the Prediction of MiRNA-Drug Resistance Based on Contrastive Learning and Global Attention

Zimai Zhang[1,2], Bo-Wei Zhao[1], Yu-An Huang[3], Zhu-Hong You[3], Lun Hu[1], Xi Zhou[1(✉)], and Pengwei Hu[1(✉)]

[1] Xinjiang Technical Institute of Physics and Chemistry, Chinese Academy of Sciences, Urumqi 810011, China
[2] The School of Software, Xinjiang University, Urumqi 810046, China
[3] Northwestern Polytechnical University, Xian 710072, China

Abstract. MicroRNAs (miRNAs) play crucial roles in organisms, and recent studies confirm their link to various diseases. The regulatory mechanisms and influence of miRNAs are current research hotspots. Biological experiments require significant time and resources, so we propose a novel model based on the global attention network graph (GGANet), considering multiple features of miRNAs and drugs. It uses clustering contrast learning to enhance information aggregation. (1) We fused multiple features for miRNAs and drugs during initialization to better represent node information. (2) Clustering comparison learning helps nodes learn differences and similarities in hidden features. (3) A global transformer module was used, which can pay attention to local node information while also utilizing the global graph attention mechanism. The model achieved an AUC of 0.9779, AUPR of 0.9771, and F1-score of 0.9615, demonstrating excellent link prediction performance and robustness.

Keywords: MiRNA-drug Resistance · Global-graph Attention · Feature Contrast Learning

1 Introduction

MicroRNAs (miRNAs) are non-coding RNAs of 19–25 nucleotides long and do not participate in protein transcription [2]. Recent years have shown that miRNAs significantly influence disease onset and development by regulating gene expression and controlling miRNA production or degradation, indirectly affecting various body functions [3]. For example, miR-124, abundant in the brain, is crucial for nervous system development and is linked to rectal, breast, and gastric cancers. miR-21 and miR-146a are key in immune regulation and inflammatory responses, making them therapeutic targets. Recent research shows promising results in targeting these miRNAs for cancer treatment [4, 5]. Studies increasingly show miRNAs' involvement in various diseases, making miRNA regulation a popular topic in biological research. Excitingly, drugs can directly control diseases by targeting miRNAs, offering a promising avenue for future research and treatment [6]. For example, miR-124 is significantly down-regulated in gastric and pancreatic cancers [4]. Over-expression of miR-210 in HUVECs increases Notch 1 protein, regulating angiogenesis and vascular maturation in post-ischemic brain tissue [7]. Additionally, inhibiting miR-211 expression can reduce airway inflammation in allergic asthma [8]. Exploring the relationship between drugs and miRNAs is a promising research area.

AI's importance is growing in fields like medicine, achieving remarkable results. A model called SANE is proposed it is powerful for drug repositioning of COVID-19 by Su X et al. [27], providing a new perspective on drug repositioning. However, traditional deep learning struggles with non-Euclidean problems, such as drug repositioning, drug association prediction, and some molecular relationship tasks, unlike its success in the image domain. The advent of graph neural networks has propelled advancements in related tasks [1]. There have been many studies on this, such as the PBMDA model proposed by You Z H et al. [20] and the MotifMDA model for predicting the association between miRNAs and diseases proposed by Zhao B W et al. [22]. Luo X et al. [23] proposed the PASNVGA model, while You Z. H. et al.[21] proposed the PCA-EELM for predicting protein-protein interactions. Zhao B W et al. [24] proposed the iGRLDTI model for predicting drug-target interactions. Hu P et al. [25] propose LMNLMI to predict lncRNA-miRNA interactions, and MFDR [30] to identify drug-target interactions. Li G et al. [26] proposed M6A-DCR to interpret RNA m6A modification sites by finding consensus regions and using deep learning for high-quality embeddings.

This paper proposes a novel model for predicting drug-miRNA resistance. Our model is the first to integrate multiple features from miRNAs and drugs using structural information to construct a comprehensive drug-miRNA graph. We extracted three features each from miRNAs and drugs. Two methods were used for feature aggregation: clustering comparison learning to enable nodes to learn from different nodes, and a new global graph attention mechanism called GOAT [12] for node feature learning. In essence, our key contributions are outlined below.

 (i) We leveraged the characteristic attributes of multiple miRNAs and drugs to integrate the miRNA information with the structural information of the drugs.
(ii) Node information aggregation is proposed using cluster-contrast learning and a new global-graph attention mechanism called GOAT. This new mechanism can effectively allow nodes to learn information about nodes that are more different and more distant from them.

2   Materials and Methods

2.1   Dataset

To evaluate the model's performance, we gathered pre-validated miRNA-drug relationships from ncDR (ncRNA in Drug Resistance, http://www.jianglab.cn/ncDR/) [13], obtaining 97 drugs to 706 miRNA resistance relationships. We used 3218 relationship pairs as positive cases and randomized miRNA-drug pairs without any relationship as negative cases. To proceed to the next step, we need to obtain the molecular structures of the drugs, We obtained the SMILES of 97 drugs from PubChem (https://pubchem. ncbi.nlm.nih.gov/) to characterize them. The nucleotide sequences of 706 miRNAs were obtained from miRbase (http://www.mirbase.org/) as attribute features.

Drug Representation Methaod. We employed three drug feature learning methods to initialize our features. Firstly, we encoded the drug substructure fingerprints based on base pairs and then converted them into binary vectors. This allowed us to obtain the D_F and efficiently learn the structural similarities between drugs. Secondly, we utilized the Graph Isomorphism Network (GIN) to obtain D_{GIN}, a graph neural network capable of representing graphs through unsupervised learning. This network is adept at extracting feature vectors from molecular structure graphs efficiently [14].

Weave modules: The system utilizes methods that represent small molecules as undirected graphs of atoms [15]. We have used this novel approach of graph convolution-based feature learning. It has positively and effectively contributed to the model training process. The resulting feature vector is denoted as D_{weave}. The method defines four passes: atom to atom, pair to pair, atom to pair and pair to atom.

MiRNA Representation Method. We have also obtained the characteristics of miR-NAs in three. First, we obtained miRNA gene expression profiles (M_{exp}) from the microRNA.org database. Secondly, we obtained a functional similarity matrix of miR-NAs computed by Gene Ontology (GO) terms [16]. By selecting the necessary miRNAs, We can get our desired functional similarity matrix M_{go}. Thirdly, we utilized the classical k-mer method to obtain the feature vector M_k of miRNAs [17].

As the initialization of the drug and miRNA representations resulted in feature representations of different dimensions, which is not conducive to model training, we used PCA (Principal Component Analysis) for dimensionality unification [18]. We unify all three representations of the drug and the three representations of the miRNA into 64 dimensions and then splice them together to get the final drug representation D_{result} and miRNA representation M_{result}.where $\oplus$ indicates splicing.

$$D_{result} = D_F \oplus D_{GIN} \oplus D_{weave} \tag{1}$$

$$M_{result} = M_{exp} \oplus M_{go} \oplus M_K \tag{2}$$

2.2 Methods

For the first time, we propose a method of aggregating multiple pieces of information about miRNA and drug structures to analyze and initialize node signatures from various perspectives. MiRNA and drug are represented as $M = \{m\}$ and $D = \{d\}$, respectively. The interaction matrix between them is denoted as $S \in \{0,1\}|D|\times|M|$, indicating the resistance relationship between the drug and the miRNA. If a resistance relationship exists between miRNA m and drug d, $S_{m,d} = 1$, and $S_{m,d} = 0$ otherwise. Using this information, we can initialize our heterogeneous graph $G = \{V, E\}$. The graph consists of nodes V, which is the union of sets M and D, $V = \{M \cup D\}$. The total number of nodes is represented by n. The edges of the graph are represented by E, $E = \{(m,d)| m \in M, d \in D, S_{m,d} = 1\}$ The feature matrix is represented by $X \in \mathbb{R}^{n\times f}$, with f being the dimensionality of the original features and d being the hidden layer dimensions. Mini-batch of nodes $X_B \in \mathbb{R}^{b\times f}$, b denotes batch size.

Clustering Comparative Learning. Cluster comparison learning by clustering samples followed by positive and negative sample selection Samples in the same cluster are positive, while those in different clusters are negative, using unsupervised, self-supervised learning. This approach reduces the number of comparisons and the model's computational complexity., preventing the selection of negative samples similar to positive ones.

During this stage of learning features by clustering comparisons, this model utilizes InfoNCE to minimize the loss between positive samples. The formula for this is as follows:

$$L_{Info} = -\frac{1}{N} \sum_{i=1}^{n} \log \left(\frac{exp\left(\frac{q_i \cdot k_{i+}}{\tau}\right)}{\sum_{j=1}^{N} exp\left(\frac{q_i \cdot k_{j-}}{\tau}\right)} \right) \tag{3}$$

where n represents the total count of samples, q_i is the eigenvector of the current sample, k_{i+} is the eigenvector of the positive sample corresponding to sample i, and k_{j-} is the eigenvector of the negative sample corresponding to sample i. τ is the temperature coefficient, a hyperparameter in the InfoNCE loss function.

After X is learned by clustering comparisons, we can get a new X_F (Fig. 1).

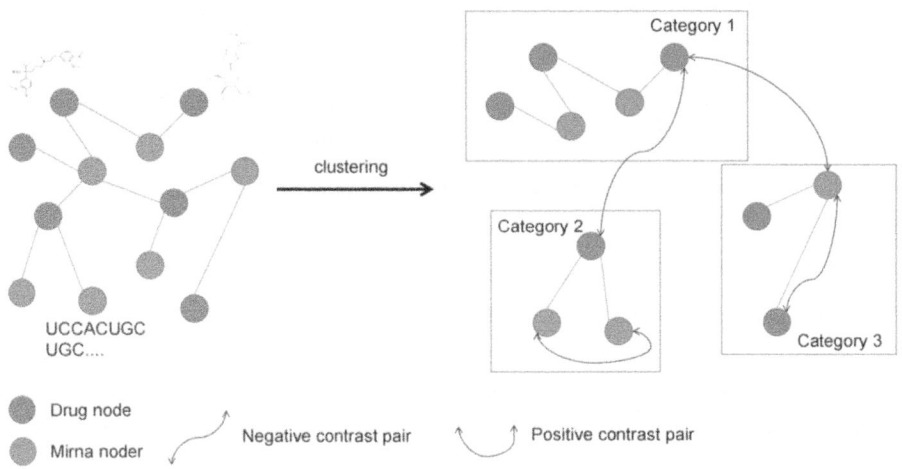

Fig. 1. Clustering Comparison Learning.

GOAT. GOAT (A scalable global graph transformer) is a global graph attention mechanism [12]. GOAT is an attention mechanism that approximates global attention while incorporating local attention. To address the high quadratic complexity of global attention, GOAT uses a projection matrix P for dimensionality reduction, lowering the feature matrix to a low-dimensional space and reducing computation costs. For any linear layer, there are weight matrices $W_K, W_Q, W_V \in \mathbb{R}^{f \times d}$. there exists projection matrices P_A, $P_V \in \mathbb{R}^{n \times k}$.

$$\Pr(\left\| \mathrm{Softmax}(SP_A)P_V^T X_F W_V - \mathrm{Softmax}(S)X_F W_V \right\|_F$$
$$\leq \varepsilon \|\mathrm{Softmax}(S)\|_F \|X_F W_V\|_F) > 1 - O(1/n) \tag{4}$$
$$\text{With } S = X_B W_Q (X_F W_K)^T / \sqrt{d} \, and \, k = O\left(\log(n)/\varepsilon^2\right)$$

We believe that the scheme provided by the above formula makes global attention on the big picture possible. For this model, we use node2vec node embedding as our position embedding. And thus get X_{pos}. We update the X_F definition:

$$X_F = X_F \oplus X_{pos} \tag{5}$$

GOAT initializes the $\tilde{K}$ and $\tilde{V}$ matrices using the method of Van Den Oord et al. (2017) et al. [19]. The sparse index matrix P is computed by the K-Means algorithm and is used as the projection matrix. P is represented by $p \in \mathbb{R}^{N \times k}$, where k is the dimension in the low-dimensional space. Each row of P is a one-hot vector that represents the clustering center to which the node is assigned. GOAT defines codebook:

$$C = diag^{-1}(1_n P)P^T X_F \tag{6}$$

Each row in the codebook represents the center of a cluster. It is possible to calculate $\tilde{K}$ and $\tilde{V}$ from C.

$$\tilde{K} = diag^{-1}(1_n P)P^T W_K = CW_K \tag{7}$$

$$\tilde{V} = diag^{-1}(1_n P)P^T W_V = CW_V \tag{8}$$

As an approximate global attention mechanism. GOAT is an approximate global attention mechanism with a bounded error in the real output X_G. To quantify this error, we define:

$$\tilde{X} = Pdiag^{-1}(1_n P)P^T X_F = PC \tag{9}$$

As the approximate X_F using the codebook. The attention matrix for a batch X_B is defined by the parameterized function f_W.

$$f_W(Y) = Softmax(\frac{X_B W_Q (Y W_K)^T}{\sqrt{d}}) \tag{10}$$

If the function f_W has a Lipschitz constant with an upper bound of lip(f_W) and a quantization error of ε, then the estimation error is bounded as,

$$\left\| \widetilde{X_B^{out}} - X_B^{out} \right\|_F \leq \varepsilon \left[1 + O(lip(f_w)) \right] \|A_B\|_F \cdot \|X\|_F \cdot \|W_V\|_F \tag{11}$$

With $\tilde{X}_B^{out} = \tilde{A}_B \tilde{X} W_V, X_B^{out} = A_B X W_V, A_B = f_W(X), \tilde{A}_B = f_W(\tilde{X})$.

Quantization error ε is defined as $\|X - \tilde{X}\|_F \leq \varepsilon \|X\|_F$.

In the above theorem, X_B^{out} represents the approximate full graph attention. This requires numerous matrix multiplication computations, making it expensive. However, GOAT proves that it can be computed using the lower dimensional $\tilde{K}$ and $\tilde{V}$.

$$\tilde{X}_B^{out} = Softmax(X_B W_Q \tilde{K}^T / \sqrt{d} + log(1_n P))\tilde{V} \qquad (12)$$

The algorithm does not perform explicit calculations for $\tilde{K}$ and $\tilde{V}$ in each iteration. Instead, it uses an approach called EMA (Exponential Moving Average) to cache and update the codebook in real time using batch statistics. In the global attention scheme, absolute position coding is required because it needs to be concatenated with the hidden features.

GOAT includes a local attention module that allows nodes to focus on their 1-hop neighbors for enhanced local learning. To support mini-batch training, GOAT uses neighbor sampling (NS). For the local module, relative position-coding distinguishes neighbors at varying distances from the source node, and local attention scores are calculated within the softmax function is

$$S_{ij} = \frac{X_{F,i} W_Q (X_{F,j} W_K)^T}{\sqrt{d}} + b_{D(i,j)} \qquad (13)$$

$b_{D(i,j)}$ is an offset calculated as the shortest distance between node i and node j over D(i,j).

We get X_{G1} from the global module and X_{G2} from the local module, which is then combined to give the final feature matrix X_{FIN}.

$$X_{FIN} = X_{G1} \oplus X_{G2} \qquad (14)$$

Figure 2 is an illustration of how the GOAT is sampled and how it is propagated forward.

Model Training. This model is designed to predict potential drug-miRNA resistance, given a dataset of drug-miRNA resistance, which includes the drug-miRNA relationship spectrum, as well as the feature vectors of the drug and the feature vectors of the miRNA. Next, we first use clustering comparison learning to update the representation of the nodes and subsequently continue to use GOAT to continue learning the features of the nodes. Lastly, a multilayer perceptron (MLP) is deployed to forecast potential relationships of resistance. During training, the model's predicted score undergoes comparison with the

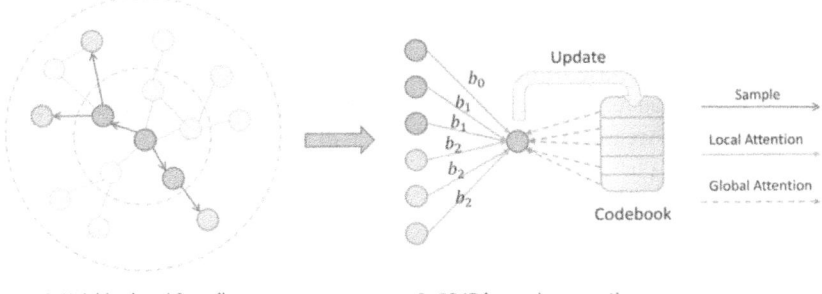

1. Neighborhood Sampling   2. GOAT forward propogation

Fig. 2. (1) represents the local neighborhood sampling process of GOAT. (2) represents the forward propagation and update codebook process of GOAT, where bl denotes the trainable positional bias for neighbors at distance l.

actual label, and the binary cross-entropy (BCE) loss function is employed to determine the loss value (Fig. 3).

$$Loss = -(y log(p) + (1 - y) log(1 - p)) \tag{15}$$

where y denotes the genuine label, taking a value of either 0 or 1, and p signifies the model's predicted output score. Finger 3 is the general flowchart of the GGANet.

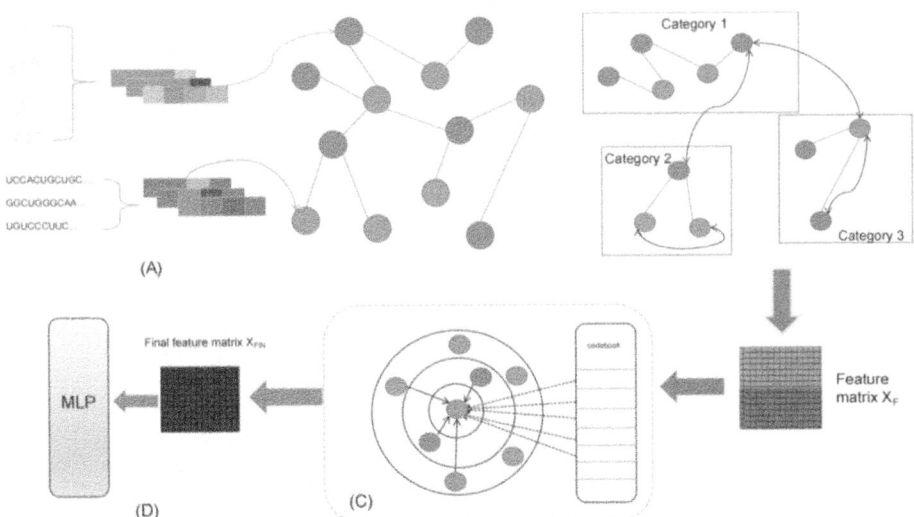

Fig. 3. The flowchart presents the overall framework, which consists of four stages. Stage (A) involves the fusion of multiple features of the drug and miRAN. Stage (B) is the clustering comparison learning stage. Stage (C) is the Global Attention stage, also known as GOAT. Finally, Stage (D) involves prediction with MLP after obtaining the final feature matrix.

3 Result

During the evaluation phase of our comparison test, we selected six metrics to assess the model's performance: accuracy, precision, recall, F1-score, AUC (Area Under the Curve), and AUPR (Area Under the Precision-Recall Curve).We plotted the ROC-AUC curve and the AUPR curve.

To demonstrate our model's superiority, we compared it with five advanced models using six metrics. All models were tested with a test set having a balanced ratio of 1 positive sample to 1 negative sample. Table 1 shows that our model outperforms the others across all indicators. The AUC and AUPR values of 0.9779 and 0.9771, respectively, highlight its effectiveness in discriminating between positive and negative cases. Accuracy, F1-score, and Recall improved by 3.61%, 3.88%, and 4.45%, respectively, showing better classification and reduced misclassification of positive examples. These results suggest our model excels in predicting drug-miRNA resistance relationships. Figure 4 displays the ROC-AUC and AUPR curves, with training sessions indicated at the bottom right.

Table 1. Comparison results with 5 state-of-the-art models on AUC, AUPR, Accuracy, Precision, F1-score, Recall.

Model	AUC	AUPR	Accuracy	Precision	F1-score	Recall
GCFMCL [28]	0.9638	0.9623	0.9027	0.8865	0.9047	0.9237
GAM-MDR [9]	0.9757	0.9717	0.9246	**0.9455**	0.9227	0.9011
AMMGC [11]	0.9435	0.9480	0.8839	0.8786	0.8890	0.8997
NASMDR [10]	0.9457	0.9423	0.9077	0.8867	0.9104	0.9353
GATECDA [29]	0.9041	0.8988	0.8337	0.8619	0.8271	0.7950
GGANet	**0.9779**	**0.9771**	**0.9607**	0.9441	**0.9615**	**0.9798**

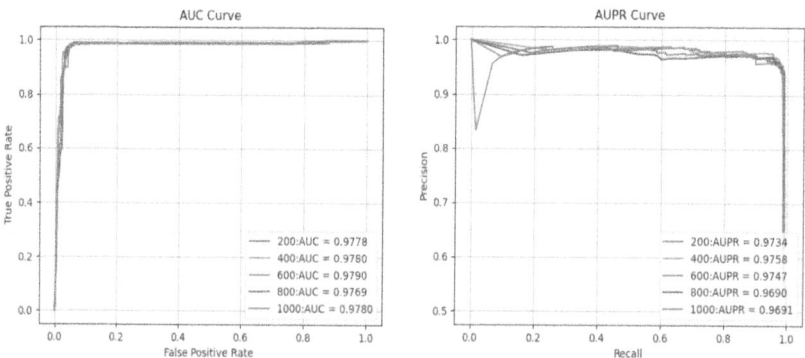

Fig. 4. Roc-AUC curves and AUPR curves.

The GOAT module has three options: full, global, and local. "Full" uses both local and global attention mechanisms, "global" uses only the global attention mechanism, and "local" uses only the local attention mechanism. The model was evaluated with six metrics, as shown in Table 2. AUC and ROC curves were also plotted for all three options. Table 2 indicates that the "full" option performed best overall, considering both distant and near node information.

Table 2. Comparison results for three options.

Option	AUC	AUPR	Accuracy	Precision	F1-score	Recall
Local	0.9769	0.9698	0.9243	0.8829	0.9297	**0.9839**
Global	0.9677	0.9757	0.9194	0.9217	0.9246	0.9362
Full	**0.9779**	**0.9771**	**0.9607**	**0.9441**	**0.9615**	0.9798

To validate whether GOAT is superior to the normal graph attention layer, we compared the two models while retaining the clustering contrast learning module. GGANet$_{GAT}$ model used the ordinary multi-head attention layer with 8 heads and a dropout of 0.6. Table 3 displays the results, which clearly indicate that the GOAT module outperforms the ordinary graph attention layer in this task (Fig. 5).

Hyperparametric Experiments. We investigated our model's hyperparameters, focusing on the "skip" value in the 1-hop of GOAT. Only the skip value was changed, with increments of 2, as shown in Table 4. The results indicate that changes in skip value have minimal impact on AUC and AUPR, demonstrating the model's robustness. Table 4 shows the best performance occurs when the skip value is 10.

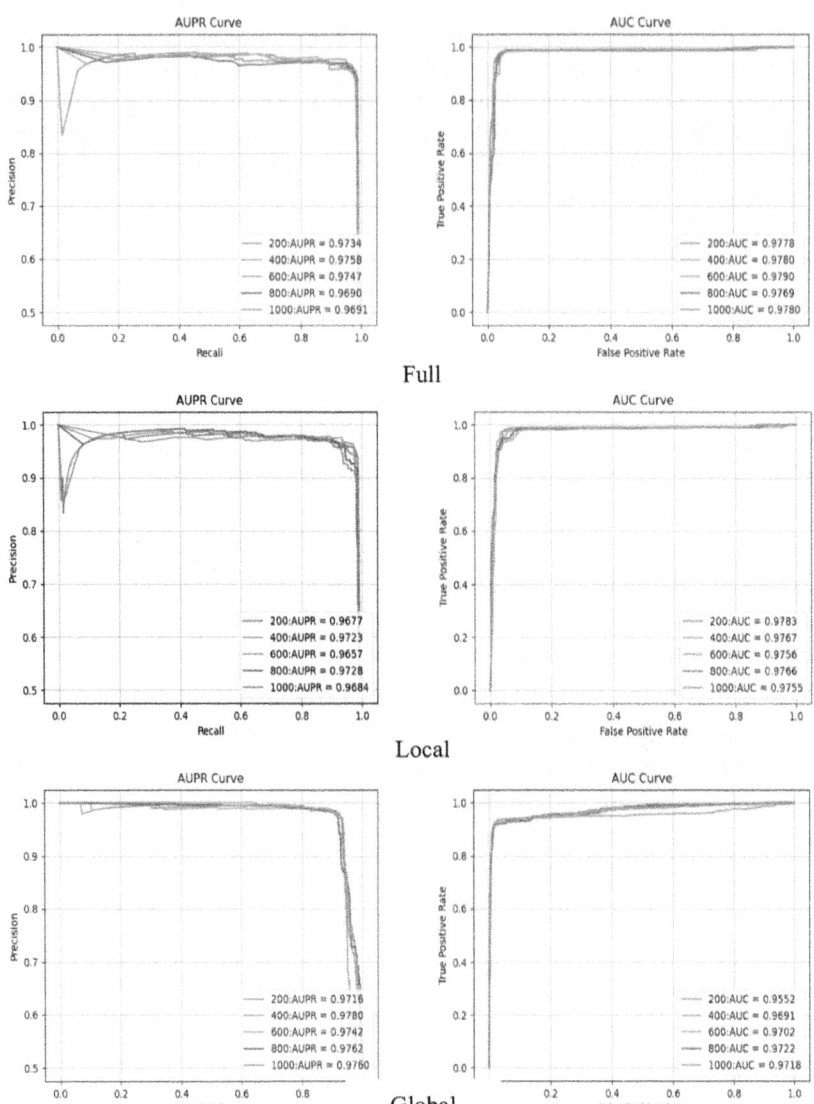

Fig. 5. Three options of ROC-AUC and AUPR curves.

Table 3. Comparison results with normal graph attention layer on AUC, AUPR, Accuracy, Precision, F1-score, and Recall.

Model	AUC	AUPR	Accuracy	Precision	F1-score	Recall
GGANet$_{GAT}$	0.9001	0.8872	0.7838	0.8942	0.7486	0.6439
GGANet	**0.9779**	**0.9771**	**0.9607**	**0.9441**	**0.9615**	**0.9798**

Table 4. Results for the different skip on AUC, AUPR, Accuracy, Precision, F1-score, and Recall.

Skip	AUC	AUPR	Accuracy	Precision	F1-score	Recall
2	0.9751	0.9602	0.9386	0.8983	0.9415	0.9891
4	**0.9803**	0.9692	0.9658	0.9490	0.9664	0.9844
6	0.9784	0.9714	0.9588	0.9377	0.9598	0.9829
8	0.9704	0.9541	0.9440	0.9073	0.9464	**0.9891**
10	0.9716	**0.9717**	**0.9673**	**0.9602**	**0.9676**	0.9751

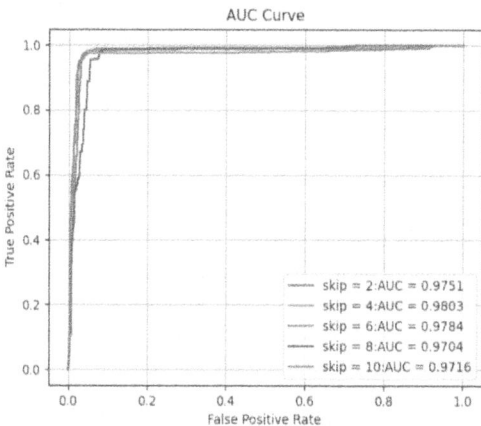

Fig. 6. ROC-AUC curves for different skip values.

4 Conclusion

MicroRNA-drug association prediction is increasingly important due to reduced resource costs. We introduce the GGANet model for forecasting potential interactions between miRNAs and drugs. GGANet fuses multiple miRNA and drug features into a final node feature matrix using cluster comparison learning and the Global Attention Module (GOAT). This approach enhances node differences and similarities, with GOAT further refining node representations. These refined features are then predicted using a Multi-layer Perceptron (MLP). We also analyzed the excellence of the GOAT module in detail and compared it with GOAT using a common graph attention layer. After the comparison experiments and a series of related experiments, we believe that GGANet has excellent predictive power in real-world tasks (Fig. 6).

Acknowledgments. This work was supported in part by the National Natural Science Foundation of China (62302495 and 62373348), in part by the Natural Science Foundation of Xinjiang Uygur Autonomous Region under grant (2023D01E15), in part by the Xinjiang Tianchi Talents Program (E33B9401) and in part by the Tianshan Talent Training Program (2023TSYCLJ0021).

References

1. Scarselli, F., Gori, M., Tsoi, A.C., et al.: The graph neural network model. IEEE Trans. Neural Networks **20**(1), 61–80 (2008)
2. Deng, L., Fan, Z., Xiao, X., et al.: Dual-channel heterogeneous graph neural network for predicting microRNA-mediated drug sensitivity. J. Chem. Inf. Model. **62**(23), 5929–5937 (2022)
3. Bartel, D.P.: MicroRNAs: genomics, biogenesis, mechanism, and function. Cell **116**(2), 281–297 (2004)
4. Ghafouri-Fard, S., Shoorei, H., Bahroudi, Z., et al.: An update on the role of miR-124 in the pathogenesis of human disorders. Biomed. Pharmacother. **135**, 111198 (2021)
5. Olivieri, F., Prattichizzo, F., Giuliani, A., et al.: MiR-21 and miR-146a: the microRNAs of inflammaging and age-related diseases. Ageing Res. Rev. **70**, 101374 (2021)
6. Abba, M.L., Patil, N., Leupold, J.H., et al.: MicroRNAs as novel targets and tools in cancer therapy. Cancer Lett. **387**, 84–94 (2017)
7. Bavelloni, A., Ramazzotti, G., Poli, A., et al.: MiRNA-210: a current overview. Anticancer Res **37**(12), 6511–6521 (2017)
8. Qin, H., Xu, B., Mei, J., et al.: Inhibition of miRNA-221 suppresses the airway inflammation in asthma. Inflammation **35**, 1595–1599 (2012)
9. Zhou, Z., Du, Z., Jiang, X., et al.: GAM-MDR: probing miRNA–drug resistance using a graph autoencoder based on random path masking. Briefings Funct. Genomics elae005 (2024)
10. Zheng, K., Zhao, H., Zhao, Q., et al.: NASMDR: a framework for miRNA-drug resistance prediction using efficient neural architecture search and graph isomorphism networks. Briefings in Bioinform. **23**(5), bbac338 (2022)
11. Niu, Y., Song, C., Gong, Y., et al.: MiRNA-drug resistance association prediction through the attentive multimodal graph convolutional network. Front. Pharmacol. **12**, 799108 (2022)
12. Kong, K., Chen, J., Kirchenbauer J, et al. GOAT: a global transformer on large-scale graphs. In: International Conference on Machine Learning, pp. 17375–17390. PMLR (2023)
13. Dai, E., Yang, F., Wang, J., et al.: NcDR: a comprehensive resource of non-coding RNAs involved in drug resistance. Bioinformatics **33**(24), 4010–4011 (2017)
14. Xu, K., Hu, W., Leskovec, J., et al.: How powerful are graph neural networks?. arXiv preprint arXiv:1810.00826 (2018)
15. Kearnes, S., McCloskey, K., Berndl, M., et al.: Molecular graph convolutions: moving beyond fingerprints. J. Comput. Aided Mol. Des. **30**, 595–608 (2016)
16. Yang, Y., Fu, X., Qu, W., et al.: MiRGOFS: a GO-based functional similarity measurement for miRNAs, with applications to the prediction of miRNA subcellular localization and miRNA–disease association. Bioinformatics **34**(20), 3547–3556 (2018)
17. Chor, B., Horn, D., Goldman, N., et al.: Genomic DNA k-mer spectra: models and modalities. Genome Biol. **10**, 1–10 (2009)
18. Smith, L.I.: A Tutorial on Principal Components Analysis (2002)

19. Van Den Oord, A., Vinyals, O.: Neural discrete representation learning. In: Advances in Neural Information Processing Systems, vol. 30 (2017)

20. You, Z.H., Huang, Z.A., Zhu, Z., et al.: PBMDA: a novel and effective path-based computational model for miRNA-disease association prediction. PLoS Comput. Biol. **13**(3), e1005455 (2017)

21. You, Z.H., Lei, Y.K., Zhu, L., et al.: Prediction of protein-protein interactions from amino acid sequences with ensemble extreme learning machines and principal component analysis. In: BMC Bioinformatics. BioMed Central, vol. 14, pp. 1–11 (2013). https://doi.org/10.1186/1471-2105-14-S8-S10

22. Zhao, B.W, He, Y.Z, Su, X.R, et al.: Motif-Aware miRNA-Disease association prediction via hierarchical attention network. IEEE J. Biomed. Health Inform. (2024)

23. Luo, X., Wang, L., Hu, P., et al.: Predicting protein-protein interactions using sequence and network information via variational graph autoencoder. IEEE/ACM Trans. Comput. Biol. Bioinform. **20**, 3182–3194 (2023)

24. Zhao, B.W, Su, X.R, Hu, P.W, et al.: iGRLDTI: an improved graph representation learning method for predicting drug–target interactions over heterogeneous biological information network. Bioinformatics **39**(8), btad451 (2023)

25. Hu, P., Huang, Y.A., Chan, K.C.C., et al.: Learning multimodal networks from heterogeneous data for prediction of lncRNA–miRNA interactions. IEEE/ACM Trans. Comput. Biol. Bioinf. **17**(5), 1516–1524 (2019)

26. Li, G., Zhao, B., Su, X., et al.: Discovering consensus regions for interpretable identification of RNA N6-methyladenosine modification sites via graph contrastive clustering. IEEE J. Biomed. Health Inform. **28**, 2362–2372 (2024)

27. Su, X., You, Z., Wang, L., et al.: SANE: a sequence combined attentive network embedding model for COVID-19 drug repositioning. Appl. Soft Comput. **111**, 107831 (2021)

28. Wei, J., Zhuo, L., Zhou, Z., et al.: GCFMCL: predicting miRNA-drug sensitivity using graph collaborative filtering and multi-view contrastive learning. Briefings Bioinform. **24**(4), bbad247 (2023)

29. Deng, L., Liu, Z., Qian, Y., et al.: Predicting circRNA-drug sensitivity associations via graph attention auto-encoder. BMC Bioinform. **23**(1), 160 (2022)

30. Hu, P.W, Chan, K.C.C., You, Z.H.: Large-scale prediction of drug-target interactions from deep representations. In: 2016 International Joint Conference on Neural Networks (IJCNN), pp. 1236–1243. IEEE (2016)

Multi-atlas Hypergraph Fusion Based on Brain Regions Overlap Amount for Diagnosis of ASD

Huajian Wang[1], Xiaochen Mu[2], Tengfei Zhang[2], Jianan Ning[3], and Yuefeng Ma[2(✉)]

[1] The College of Engineering, Qufu Normal University, Rizhao, China
[2] The School of Computer Science, Qufu Normal University, Rizhao, China
`rzmyf1976@163.com`
[3] Qingdao Branch of China Unicom, Qingdao, China

Abstract. Recently, the accurate classification of autism spectrum disorder (ASD) has attracted a lot of attention in human brain analysis. However, the existing diagnosis method only focus on single atlas or relationship among brain regions, neglecting the high-order and spatial relationship among the brain regions in different atlases. To tackle this weakness, in this paper, we propose a multi-atlas hypergraph fusion method based on brain regions (BRs) overlap amount to diagnosis ASD/TC. We calculate the correlations of BRs for each subject and formally introduce the measure of multi-atlas overlap (MAO), brain regions overlap amount (BRsOA). Then, we fuse multiple hypergraphs calculated from multi-atlas by MAO and generate a multi-atlas hypergraph (MAHG). In the final, the MAHG is used to perform the classification task. The experimental results on Autism Brain Imaging Data Exchange (ABIDE) verify that our proposed method outperforms classical methods in ASD/TC classification.

Keywords: Autism spectrum disorder · multi-atlas overlap · brain regions overlap amount · multi-atlas hypergraph

1 Introduction

Autism Spectrum Disorder (ASD) [1] is a common neurodevelopmental disorder. The individuals with ASD typically have the symptoms including repetitive behaviors and several impairment of intellect [11]. Early diagnosis and treatments can help to improve the quality of the life of the patients. However, ASD assessment remains a challenging and arduous task [9].

Recently, the significant progress has been made in the diagnosis of ASD based on functional resonance imaging fMRI (fMRI) [13]. As one of the neuroimaging techniques, fMRI is widely used to explore functional connectivity (FC) [15] changes. For example, [3] calculated FC in two frequency bands based on Dosenbach atlas and used SVM to diagnose ASD. [20] proposed a connectivity-based graph attention network for autism diagnosis using FC patterns obtained from fMRI. [8] developed a framework that combines an autoencoder with a single-layer perceptron (SLP) to enhance feature extraction

© The Author(s), under exclusive license to Springer Nature Singapore Pte Ltd. 2024
D.-S. Huang et al. (Eds.): ICIC 2024, LNBI 14881, pp. 276–286, 2024.
https://doi.org/10.1007/978-981-97-5689-6_24

quality and optimize model parameters to improve classification results on ASD classification. [12] designed a ROI-aware graph convolutional (Ra-GConv) layer that leverages the topological and functional information of fMRI data to diagnose ASD. [19] presented a prior brain structure learning-guided multi-view graph convolutional neural network (MVS-GCN), which collaborates the graph structure learning and multi-task graph embedding learning to identify the potential functional subnetworks. Although some performance for ASD diagnosis has been achieved, how to utilize the information from multi-atlas for exploring relationships between brain regions is still a challenge.

To address the limitations of existing approaches, we explore relationships among multiple atlases using hypergraph [21] structure, which provides a more adaptable representation of higher-order connections and complex structures. But how to use a hypergraph to represent all the atlases information of a subject becomes the next challenge. To this end, we first focus on the original principles of multi-atlas then we find that although the constructions of different atlases is obtained by different brain partitioning templates, these representations of brain are analysis of the same human brain in essence. As a result, there are necessarily spatial overlaps between brain regions in different atlases (automated anatomical labeling (AAL), craddock 200 (CC200), eickhoff-zilles (EZ), harvard-oxford (HO), and talaraich and tournoux (TT)). And the amount of overlap directly correlates with the correlation of corresponding brain regions in different atlases. Thus, we introduce the measure of multi-atlas overlap (MAO), brain regions overlap amount (BRsOA), to perform the multi-hypergraph fusion from multi-atlas.

In this article, to improve the performance of diagnosis and interpretability, we propose an effectively diagnostic method called Multi-atlas hypergraph fusion based on brain regions overlap amount for diagnosis of ASD. Specifically, the first step involves calculating the correlations of BRs for each subject. Subsequently, we generate the single-atlas hypergraphs of the brain connectivity networks (BCN) based on multiple atlases respectively. Next, leveraging the BRsOA, we construct a new hypergraph, called Multi-atlas Hypergraph (MAHG). The hypergraph represents the overlapping information of multi-atlas hypergraphs. To verify the effectiveness of our method, we apply SVM for the final classification in ABIDE and the experimental results on the ABIDE II dataset demonstrate that our proposed method outperforms the SOTA methods and improve the interpretability of classification.

2   Related Work

In this section, we give a brief review of three areas closely related to our research, which are multi-modal data application in ASD diagnosis and multi-hypergraph representation learning.

2.1   Multi-modal Data Application in ASD Diagnosis

In the present day, most research on ASD diagnosis methods tend to focus on single modality. However, existing datasets related to ASD, such as Autism Brain Imaging Data Exchange (ABIDE) [7], already include up to several different brain parcellation schemes to represent the time-series data of these subjects. Recently, researchers have also proposed using multi-modal data for ASD classification. [14] presented a generic framework that leverages both image and non-image information. [4] introduced a novel adversarial learning-based node-edge graph attention network (AL-NEGAT) for ASD identification based on multimodal MRI data. [18] proposed a multi-atlas (CC200, AAL and Dosenbach) deep feature representation method(AIMAFE). And they carry out the ASD classification by a multilayer perceptron (MLP). [17] proposed a new method (MAGCN) for automatic diagnosis of ASD based on six atlas. They extracted multiple feature representations based on FC of different brain atlases of each subject and used a stacking ensemble learning method to combine different feature representations. Although the studies mentioned above have all utilized multiple atlases on ABIDE for ASD diagnosis, They have not taken into account the relationships between atlases when using different atlases FC data, which resulted in insufficient information fusion.

2.2 Multi-hypergraph Representation Learning

In the analysis of multi-modal data, the multi-hypergraph learning will play a crucial role in understanding the structures and features of the data. Here are some related literature: [2] proposed a multi-hypergraph fusion method to leverage multiple existing features. They performed the fusion of multi-hypergraph by a regularization on a fully connected graph. [16] introduced a robust embedding based framework with multi-hypergraph fusion and they generated the Laplacian matrix of the fused hypergraph by transition probabilities fusion and stationary distribution. [22] presented a multi-modal representation learning and adversarial hypergraph fusion (MRL-AHF) framework. They constructed two hypergraphs from latent representations, and an adversarial network utilizing graph convolution is used to minimize the distribution of hyperedge features.

Commonly employed methods, such as regularization or matrix factorization, while successful in certain contexts, fail to be capable of diagnosing brain diseases based on fMRI. This limitation becomes particularly evident when attempting to fusion brain networks derived from multi-atlas.

3   Method

The overall flowchart of our proposed method for automatic diagnosis of ASD is shown in Fig. 1. Our proposed method mainly consists of three steps: 1) calculate the single-atlas hypergraphs for each subject, 2) multi-hypergraph fusion based on the brain regions overlap amount, and 3) classification for ASD. Next, we will describe our proposed method in detail.

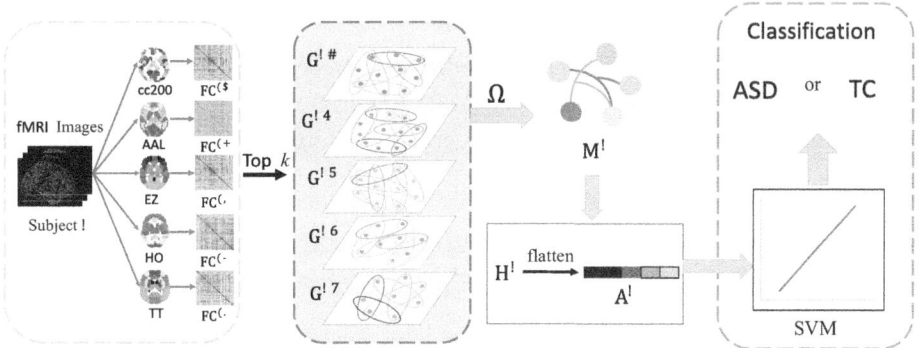

Fig. 1. Overview of the proposed method. In MAHG, for convenience, we use a larger node to represent brain region nodes from the same atlas, which one node color corresponds to one single-atlas hypergraph respectively.

3.1 Constructions of Single-Atlas Hypergraphs

For subject i, the single-atlas FC network can be generated based on single-atlas. In this step, we calculate the time-series correlation between each pair of brain regions in one atlas by Pearson correlation coefficient (PCC) as follows:

$$PCC\left(X_{k.}^i, X_{z.}^i\right) = \frac{E\left(X_{k.}^i X_{z.}^i\right) - E\left(X_{k.}^i\right)E\left(X_{z.}^i\right)}{\sqrt{E\left(\left(X_{k.}^i\right)^2\right) - E^2\left(X_{k.}^i\right)}\sqrt{E\left(\left(X_{z.}^i\right)^2\right) - E^2\left(X_{z.}^i\right)}} \tag{1}$$

where the $X_{k.}^i$ and $X_{z.}^i$ are the time-series of brain regions k and z respectively, and $E(\cdot)$ is the mathematical expectation. Then, we obtain the single-atlas FC^{i1} generated from the cc200 atlas, the FC^{i2} generated from the AAL atlas, the FC^{i3} generated from the EZ atlas, the FC^{i4} generated from the HO atlas, and the FC^{i5} generated from the TT atlas.

In next step, we construct the corresponding single-atlas hypergraph based on the single-atlas FC^{ij} to represent the high-order brain network. For the j th atlas with n_j brain regions, we select the top k most correlative nodes for the n th node and use a hyperedge to connect the top k nodes and node n. As a result, a single-atlas hypergraph G^{ij} with n_j hyperedges and n_j nodes is constructed. By performing the above procedure, we construct five single-atlas hypergraphs G^{i1}, G^{i2}, G^{i3}, G^{i4} and G^{i5}.

A hypergraph G with α nodes and ε hyperedges can be formally defined as a triple $G = (V, E, H)$, where V represents the set of nodes which includes α nodes, E represents the set of hyperedges which includes ε hyperedges, and the incidence matrix $H \in \{0, 1\}^{\alpha \times \varepsilon}$ is used to represent the relationships between nodes connected by hyperedges in the hypergraph. The definition of the incidence matrix H is as follows:

$$H(v, e) = \begin{cases} 1, & \text{if } v \in e \\ 0, & \text{otherwise} \end{cases} \tag{2}$$

Based on the general definition of hypergraph and Eq. (2), we generate five single-atlas hypergraphs incidences H^{i1}, H^{i2}, H^{i3}, H^{i4} and H^{i5} for subject i.

3.2 Construction of MAHG Based on BRsOA

Definition 1. Spatial Overlap Degree. Let A^l and A^j be two atlases, supposed that $V^l = \{V_1^l, \ldots, V_{n_l}^l\}$ is extracted based on A^l for a subject where V_ε^l denotes the voxels set of brain region ε in the lth atlas, we regard the intersection between V_ε^l and V_η^j, denotes as $\Gamma_{\varepsilon\rho}^{lj}$. Then, we give the measure of multi-atlas overlap, BRsOA, as follows:

$$\Gamma_{\varepsilon\rho}^{lj} = \frac{\left|V_\varepsilon^l \cap V_\eta^j\right|}{\left|V_\eta^j\right|} \tag{3}$$

where $|\cdot|$ is the number of the set.

Theoretically, we can compare any pair of brain regions between any two atlases to determine the degree of MAO. However, this would introduce deviation and significantly increase computational complexity. To this end, we choose the cc200 atlas as the reference atlas. Then, we compared other atlases to the reference atlas, and calculated the BRsOA between brain regions in reference atlas and other atlas. This approach conveniently achieved the multi-hypergraph fusion based on MAO.

For each node η in reference atlas cc200, we calculate the brain region nodes of other atlases that have spatial overlap with η. These nodes can form a hyperedge together with η. Then, we obtain η hyperedges because the cc200 atlas contains n_1 brain regions, and each of which overlaps with the brain regions in the other four atlases. Based on the definition of hypergraph and analysis above, we aim to describe the spatial overlap relationships between brain regions in these five atlases by introducing a novel matrix $\mathcal{C}$, which can be constructed as follows:

$$\mathcal{C} \in R^{t \times n_1} = \begin{bmatrix} M_{11} & \cdots & M_{1n_1} \\ M_{21} & & M_{2n_1} \\ \vdots & & \vdots \\ \vdots & & \vdots \\ M_{t1} & \cdots & M_{tn_1} \end{bmatrix} \tag{4}$$

where the t is the sum of the number of brain region nodes in the five atlases. The elements in matrix $\mathcal{C}$ are the values of spatial overlap. In order to clearly represent the connectivity of brain regions in different atlases, we need to set a threshold ε to binarize matrix $\mathcal{C}$. Through our experiment, we have verified that When $\mu = 0.1$, it could accurately reflect the spatial overlap among brain regions. Thus, we set the threshold $\mu = 0.1$ to binarize the matrix $\mathcal{C}$ into $\mathcal{C}^i$ which includes 0 or 1, as shown in Eq. (5).

$$C_{mn}^i = \begin{cases} 1, & \text{if } C_{mn} > \varepsilon \\ 0, & \text{if } C_{mn} < \varepsilon \end{cases} \tag{5}$$

Then, a MAHG can be defined as $M^i = (V^i, E^i, H^i)$, V^i is a node set that represents all brain regions of five atlases. E^i denotes a hyperedges set that contains $t \times (t + n_1)$

hyperedges. $H^i \in \{0, 1\}^{t \times (t+n_1)}$, and the incidence matrix H^i of the SIHG can be constructed as:

$$H^i = \left[H^{i \cdot} C^i \right] \in R^{t \times (t+n_1)} \tag{6}$$

$$H^{i \cdot} = \begin{bmatrix} H^{i1} & O & O & O & O \\ O & H^{i2} & O & O & O \\ O & O & H^{i3} & O & O \\ O & O & O & H^{i4} & O \\ O & O & O & O & H^{i5} \end{bmatrix} \tag{7}$$

At this point, we have obtained the SIHG and its incidence matrix for each subject which can effectively represents topological information for each subject. This is because the H^i is not only represents the time-series correlations between brain regions within individual atlases but also fuses information from different atlases. Next, we will further process the obtained incidence matrices of all subjects and perform classification by using SVM. The specific details will be presented in the following subsection.

3.3 Classification Based on SVM

To simplify the complex structure of the incidence matrices of the spatial integration hypergraphs for easier processing, we used a "flatten" operation to convert the M^i into one-dimensional arrays A_i, facilitating subsequent processing or classification.

To better utilize the structural representation of each subject's MAHG for ASD classification, our research employs an SVM model using different kernel functions: linear kernel, polynomial kernel, and gaussian kernel [5]. The mathematical expression of the linear kernel function is as follows,

$$K(x_i, x_r) = x_i^{\mathrm{T}} \cdot x_r + c \tag{8}$$

where c is a constant term, often set to 1, the vectors x_i, x_r represent the one-dimensional representations of the spatial integration hypergraphs for subjects i and j, respectively.

The mathematical expression of the polynomial kernel function is as follows,

$$K(x_i, x_r) = \left(\gamma x_i^{\mathrm{T}} \cdot x_r + c \right)^d \tag{9}$$

where the parameter γ is used to scale the inner product, and the parameter d represents the degree of the polynomial, which plays a crucial role in controlling the mapping to a higher-dimensional feature space.

The mathematical expression of the Gaussian kernel (RBF) is as follows,

$$K(x_i, x_r) = e^{\frac{\|x_i - x_r\|^2}{2\sigma^2}} \tag{10}$$

where the σ represents the width parameter of the Gaussian kernel, regulating the effective range of the Gaussian kernel function.

4 Experiments

4.1 Dataset and Image Preprocessing

The fMRI image data used in this study is provided by ABIDE, which contains 1112 subjects including 539 subjects with ASD and 573 subjects with TC. And we exclude the subjects with missing time-series of brain regions. In this study, we select 732 subjects from 17 international sites as experimental subjects which includes 399 subjects with ASD and 333 subjects with TC. In addition, we conducted research on ASD classification using five commonly used brain atlases include AAL, CC200 [6], EZ, HO, and TT. For more details about the dataset, please visit http://preprocessed-connectomes-project.org/abide/index.html.

4.2 Experimental Settings

In order to ensure fairness and consistency of experimental comparisons with baseline methods, we applied fMRI data from the same subjects and used a unified threshold k ($k = 45$, $\mu = 0.1$). For a fair experimental comparison with different classifiers, a standardized test configuration was employed.

4.3 Performance Evaluation Metrics

We use four metrics to evaluate the performance of the model: accuracy (ACC), sensitivity (SEN), specificity (SPE), and area under curve (AUC). AUC indicates the area under the curve in the Receiver Operating Characteristic (ROC). Let TP be true positives, FP be false positives, FN be false negatives, and TN be true negatives, the formulas of each evaluation metric can be calculated as follows:

$$ACC = \frac{TP + TN}{TP + TN + FP + FN} \tag{11}$$

$$SEN = \frac{TP}{TP + FN} \tag{12}$$

$$SPE = \frac{TN}{TN + FP} \tag{13}$$

4.4 Results and Analysis

Performance on Existing Methods. We conduct a comparative analysis against several SOTA methods including FC+MLP, FC+SVM and four deep learning methods [10, 12, 17, 19]. The Table 1 shows the outstanding classification performance of our proposed method. Notably, the best SPE achieved by our method was significantly higher than that of the existing methods. This further underscores the effectiveness of our novel composition method for ASD/TC classification.

Table 1. The performance of different methods for ASD/TC classification.

Method	ACC (%)	SEN (%)	SPE (%)	AUC
FC+MLP	66.6	68.1	63.2	0.646
FC+SVM	69.6	71.7	63.2	0.661
BrainGNN	59.8	61.7	60.2	0.611
DBN	73.2	**72.2**	72.2	0.730
MVS-GCN	71.6	66.8	65.1	0.731
MAGE	73.5	67.3	64.1	0.715
SOIH	**76.2**	65.8	**84.2**	**0.743**

Performance on Different Classifiers. To show the effectiveness of the SVM used in this paper for ASD/TC classification, we performed a comparative analysis against various standard classifiers. These include the polynomial SVM (P-SVM), linear SVM (L-SVM), gaussian kernel SVM (G-SVM), RF, ridge Classifier (Ridge), logistic regression (LR) and MLP. The results are shown in Table 2. It can be observed that the P-SVM classifier within our proposed framework achieved the best classification performance. The comparative experimental results in this subsection further underscore the effectiveness of our proposed method for ASD/TC classification.

Table 2. The performance of different classifiers.

Classifier	ACC (%)	SEN (%)	SPE (%)	AUC
RF	63.7	55.0	80.0	0.626
G-SVM	68.6	43.6	**88.1**	0.659
MLP	71.1	62.3	78.4	0.706
L-SVM	71.8	64.1	78.9	0.714
Ridge	71.9	64.5	78.9	0.715
LR	72.7	62.2	80.7	0.715
P-SVM	**76.2**	**65.8**	84.2	**0.743**

Comparison with Different Representations of Brain Networks. To prove the effectiveness of our multi-atlas hypergraph fusion method, we compared it with traditional BCN representation methods based on a single atlas. We used P-SVM to classify ASD on five atlases, and the results of the comparative experiment are shown in Table 3.

Interpretability Analysis. We conducted comparative experiments and analysis on variable μ and k with different values, and the experimental results are shown in Table 4 and Table 5, respectively. We chose $\mu = 0.1$ and $k = 45$ to achieve better classification performance. Additionally, we do not consider using a smaller μ ($\mu < 0.1$) suitable for our method because an excessively small μ implies low overlap between brain regions across different atlases, to the extent that their overlap can be neglected.

We also identified several brain regions are associated with ASD, including posterior cingulate cortex, amygdala, angular gyrus, anterior cingulate cortex and precuneus. Our work simultaneously identifying these paramount brain regions associated with ASD based on multi-atlas overlap. The analysis result once again confirms the critical role of hypergraph method in elucidating the complexity of ASD etiology.

Table 3. The classification metrics value of different representations.

	ACC (%)	SEN (%)	SPE (%)	AUC
G_{AAL}	70.0	64.7	75.4	0.705
G_{cc200}	71.1	55.7	82.1	0.704
G_{EZ}	69.1	64.1	73.7	0.689
G_{HO}	74.1	61.3	83.9	0.736
G_{TT}	68.6	**69.1**	73.7	0.737
SIHG	**76.2**	65.8	**84.2**	**0.743**

Table 4. The performance of different k.

k	ACC (%)	SEN (%)	SPE (%)	AUC
35	72.4	62.6	83.6	0.732
55	72.0	63.5	82.9	0.734
65	72.7	61.5	83.3	0.723
45	**76.2**	**65.8**	**84.2**	**0.743**

Table 5. The performance of different ε.

μ	ACC (%)	SEN (%)	SPE (%)	AUC
0.2	72.9	62.1	82.2	0.734
0.25	73.2	62.1	80.3	0.728
0.3	72.5	61.4	80.6	0.704
0.15	73.6	63.1	82.6	0.734
0.1	**76.2**	**65.8**	**84.2**	**0.743**

5 Conclusion

This paper proposes a novel multi-hypergraph fusion method based on the multi-atlas overlap of brain regions across any two pairwise atlases. We introduce the multi-atlas overlap and calculate the brain regions overlap amount in different atlases using voxel information from fMRI data. We use a multi-atlas hypergraph (MAHG) enriched with spatial overlap information to represent the brain network of each subject. The experimental results demonstrate that this method is effective for the classification task of ASD. Additionally, we identified several important brain regions associated with ASD. With the continuous advancement of science and technology, we believe that data types related to brain diseases will become more diverse. Therefore, our proposed method is not only effective for the automatic diagnosis of ASD/TC in clinical practice but also holds significant potential for the future.

Acknowledgments. This work was partly supported by the Shandong Natural Science Foundation under the ZR2021MF124 Project.

Disclosure of Interests. The authors have no competing interests to declare that are relevant to the content of this article.

References

1. American Psychiatric Association: Diagnostic and Statistical Manual of Mental Disorders: DSM-5, vol. 5. American Psychiatric Association Washington, DC (2013)
2. An, L., Chen, X., Yang, S., Li, X.: Person re-identification by multi-hypergraph fusion. IEEE Trans. Neural Netw. Learn. Syst. **28**(11), 2763–2774 (2016)
3. Chen, H., et al.: Multivariate classification of autism spectrum disorder using frequency-specific resting-state functional connectivity—a multi-center study. Prog. Neuro-Psychopharmacol. Biol. Psychiatry **64**, 1–9 (2016)
4. Chen, Y., et al.: Adversarial learning based node-edge graph attention networks for autism spectrum disorder identification. IEEE Trans. Neural Netw. Learn. Syst. (2022)
5. Cortes, C., Vapnik, V.: Support-vector networks. Mach. Learn. **20**, 273–297 (1995)
6. Craddock, R.C., James, G.A., Holtzheimer III, P.E., Hu, X.P., Mayberg, H.S.: A whole brain fMRI atlas generated via spatially constrained spectral clustering. Human Brain Mapp. **33**(8), 1914–1928 (2012)
7. Di Martino, A., et al.: The autism brain imaging data exchange: towards a large-scale evaluation of the intrinsic brain architecture in autism. Mol. Psychiatry **19**(6), 659–667 (2014)
8. Eslami, T., Mirjalili, V., Fong, A., Laird, A.R., Saeed, F.: ASD-DiagNet: a hybrid learning approach for detection of autism spectrum disorder using fMRI data. Front. Neuroinform. **13**, 70 (2019)
9. Farooq, M.S., Tehseen, R., Sabir, M., Atal, Z.: Detection of autism spectrum disorder (ASD) in children and adults using machine learning. Sci. Rep. **13**(1), 9605 (2023)
10. Huang, Z.A., Zhu, Z., Yau, C.H., Tan, K.C.: Identifying autism spectrum disorder from resting-state fMRI using deep belief network. IEEE Trans. Neural Netw. Learn. Syst. **32**(7), 2847–2861 (2020)
11. Kong, Y., Gao, J., Xu, Y., Pan, Y., Wang, J., Liu, J.: Classification of autism spectrum disorder by combining brain connectivity and deep neural network classifier. Neurocomputing **324**, 63–68 (2019)

12. Li, X., et al.: BrainGNN: interpretable brain graph neural network for fMRI analysis, vol. 74, p. 102233. Elsevier (2021)
13. Liu, J., et al.: Applications of deep learning to MRI images: a survey. Big Data Mining Analytics 1(1), 1–18 (2018)
14. Parisot, S., et al.: Disease prediction using graph convolutional networks: application to autism spectrum disorder and Alzheimer's disease. Med. Image Anal. **48**, 117–130 (2018)
15. Power, J.D., et al.: Functional network organization of the human brain. Neuron **72**(4), 665–678 (2011)
16. Wang, K.: Robust embedding framework with dynamic hypergraph fusion for multi-label classification. In: 2019 IEEE International Conference on Multimedia and Expo (ICME), pp. 982–987. IEEE (2019)
17. Wang, Y., Liu, J., Xiang, Y., Wang, J., Chen, Q., Chong, J.: MAGE: automatic diagnosis of autism spectrum disorders using multi-atlas graph convolutional networks and ensemble learning. Neurocomputing **469**, 346–353 (2022)
18. Wang, Y., Wang, J., Wu, F.X., Hayrat, R., Liu, J.: AIMAFE: autism spectrum disorder identification with multi-atlas deep feature representation and ensemble learning. J. Neurosci. Methods **343**, 108840 (2020)
19. Wen, G., Cao, P., Bao, H., Yang, W., Zheng, T., Zaiane, O.: MVS-GCN: a prior brain structure learning-guided multi-view graph convolution network for autism spectrum disorder diagnosis. Compu. Biol. Med. **142**, 105239 (2022)
20. Yin, W., Li, L., Wu, F.X.: A graph attention neural network for diagnosing ASD with fMRI data. In: 2021 IEEE International Conference on Bioinformatics and Biomedicine (BIBM), pp. 1131–1136. IEEE (2021)
21. Zhou, D., Huang, J., Schölkopf, B.: Learning with hypergraphs: clustering, classification, and embedding. In: Advances in Neural Information Processing Systems, vol. 19 (2006)
22. Zuo, Q., Lei, B., Shen, Y., Liu, Y., Feng, Z., Wang, S.: Multimodal representations learning and adversarial hypergraph fusion for early Alzheimer's disease prediction. In: Ma, H., et al. (eds.) Pattern Recognition and Computer Vision: 4th Chinese Conference, PRCV 2021, Beijing, China, 29 October–1 November 2021, Proceedings, Part III 4. pp. 479–490. Springer, Cham (2021). https://doi.org/10.1007/978-3-030-88010-1_40

Multi-input Deep Learning Model for RP Diagnosis Using FVEP and Prior Knowledge

Yuguang Chen[1,2,3,4,5,6], Mei Shen[2,3,4,5,6], Dongmei Lu[2,3,4,5,6], Jun Lin[9],
Jiaoyue Hu[2,3,4,5,6,7(✉)], Shiying Li[2,3,4,5,6(✉)], and Zuguo Liu[1,2,3,4,5,6,7,8(✉)]

[1] Institute of Artificial Intelligence, Xiamen University, Xiamen 361102, Fujian, China
zuguoliu@xmu.edu.cn
[2] Department of Ophthalmology, Xiang'an Hospital of Xiamen University, Xiamen 361005,
Fujian, China
mydear_22000@163.com, shiying_li@126.com
[3] Fujian Provincial Key Laboratory of Ophthalmology and Visual Science, Xiamen 361005,
Fujian, China
[4] Fujian Engineering and Research Center of Eye Regenerative Medicine, Xiamen 361005,
Fujian, China
[5] Eye Institute of Xiamen University, Xiamen 361005, Fujian, China
[6] School of Medicine, Xiamen University, Xiamen 361005, Fujian, China
[7] Xiamen University Affiliated Xiamen Eye Center, Xiamen 361005, Fujian, China
[8] Department of Ophthalmology, The First Affiliated Hospital of University of South China,
Hengyang 421001, Hunan, China
[9] Department of Ophthalmology, Yongchuan District People's Hospital of Chongqing,
Chongqing 402160, China

Abstract. Retinitis Pigmentosa (RP) is a hereditary disease characterized by progressive damage to the visual pathway, ultimately leading to vision loss. Flash Visual Evoked Potential (FVEP) serves as an effective tool for diagnosing RP, and automatic classification of FVEP using deep learning can alleviate the workload of doctors and improve work efficiency. This study proposed a multi input neural network for RP and other anomaly recognition: MGPResNet. One branch of the model conducts full connection on manually crafted features to integrate them, while the other branch adopts a 1D ResNet as its basic architecture, it incorporates global convolutional blocks and pyramid pooling blocks to extract features from FVEP waveforms at deeper levels and different scales. Subsequently, the features extracted by the two branches are concatenated, followed by full connection and activation layers to output the classification probabilities. The model was validated on the FVEP datasets of two hospitals. The proposed method demonstrated excellent accuracy on clinical datasets, with an accuracy of 96.80%, average precision of 96.52%, average recall of 96.47%, and average F1_score of 96.49%. It validated the significant potential of deep learning in the analysis of visual electrophysiological signals, provided an important foundation and new insights for the future use of deep learning techniques in clinical diagnosis and treatment.

Keywords: FVEP · Retinitis Pigmentosa · Deep learning · Out-of-distribution detection

D.-S. Huang et al. (Eds.): ICIC 2024, LNBI 14881, pp. 287–299, 2024.
https://doi.org/10.1007/978-981-97-5689-6_25

1 Introduction

Retinitis Pigmentosa(RP) is a hereditary retinal disease classified as a type of retinal dystrophy. The progression of RP leads to the gradual death of photoreceptor cells in the retina, resulting in progressive vision loss and potentially blindness. Currently, the diagnosis of RP typically involves electrophysiological tests, including VEP and ERG. The interpretation of these test results requires specialized knowledge and experience from ophthalmologists. However, traditional diagnostic methods require a significant amount of time and effort from doctors and demand a high level of expertise. Therefore, the development of artificial intelligence-based models to learn the features of visual electrophysiological data and establish automatic disease diagnosis models is particularly important, it will help alleviate the workload of doctors, improve diagnostic efficiency, and to some extent alleviate the problem of uneven distribution of medical resources, providing patients with more timely treatment and management.

Flash visual evoked potential (FVEP) is a cluster of electrical signals generated in the cerebral cortex in response to visual stimuli, reflects the transmission of visual information from the retina to the visual cortex of the brain [1]. As shown in Fig. 1, in the FVEP waveform, N1, P1, N2, P2, N3, and P3 are characteristic waveforms generated by flash stimulation. By measuring the peak time and amplitude of N2 and P2 waves, it is possible to determine whether the FVEP waveform is normal and thus evaluate the functional status of the visual system. In clinical practice, FVEP is used for diagnosing diseases such as ocular trauma, optic neuritis, optic nerve dysplasia, glaucoma, and pseudoblindness [1]. Mohammed et al. used FVEP to evaluate the final visual prognosis of patients with indirectly traumatic optic neuropathy [2]. LD et al. used B-ultrasound, VEP, and perioperative video endoscopy to evaluate the visual prognosis of patients with artificial corneas [3]. Zhang et al. found that patients with retinitis pigmentosa (RP) exhibited prolongation of the peak time and significant reduction in amplitude of the P2 wave [4].

Fig. 1 Waveform and characteristic waves of FVEP

The FVEP waveform in RP patients exhibits significant differences compared to those in healthy individuals, but lacks specificity. In fact, the FVEP waveforms of many other types of diseases also differ significantly from those of healthy individuals. Therefore, relying solely on the recognition of FVEP waveforms between RP patients and healthy individuals may lead to misdiagnosis, such as misclassifying abnormal FVEP waveforms of other types as RP disease. Additionally, current research on disease classification based on visual electrophysiological signals lacks the application of deep learning methods. Existing methods typically only utilize the electrical signals as input without incorporating other potential features, resulting in relatively low accuracy.

This study proposed an innovative approach by introducing two input branches, combining manually extracted statistical features of FVEP wave as prior knowledge, and integrating FVEP waveforms as inputs to achieve more comprehensive disease classification. This model not only identifies abnormal FVEP in RP but also recognizes abnormal FVEP waveforms of other types. Specifically, the method employs two input branches to process FVEP waveforms and manually extracted features separately. For FVEP waveforms, the proposed model utilizes global and local convolutions of different sizes to extract multi-scale features from the waveform. For manually extracted features, including age, visual acuity of the subject, and characteristics of six waveforms, we integrate their features through fully connected layers. Next, the features extracted from the two branches are fused together through feature concatenation, forming a more comprehensive feature representation. Finally, classification is performed through fully connected layers to achieve disease diagnosis of visual electrophysiological signals. This multi-branch design significantly improves the model's accuracy by simultaneously considering multiple aspects of waveform and subject information, fully exploiting multi-source information, and incorporating manually extracted statistical features as prior knowledge to make the model more discriminative and generalizable.

2 Related Works

Currently, deep learning has been widely applied in disease diagnosis and emotion recognition based on electrocardiograms (ECG) and electroencephalograms (EEG), providing more accurate and reliable tools for medical diagnosis. In 2022, Shuoxuan Zhang et al. proposed an ECG signal classification model based on bidirectional LSTM (Bi-LSTM) model, fully utilizing the characteristics of Bi-LSTM that can integrate contextual information. In the model, Bi-LSTM layer, fully connected layer, softmax layer, and classification layer were used to achieve binary classification of normal signals and atrial fibrillation signals [5]. M. Mohamed Suhail et al. developed an accurate ECG classification and monitoring system for heart disease classification using long short-term memory (LSTM) and gated recurrent neural network (GRNN) techniques [6]. Surbhi Bhatia et al. developed a new deep convolutional neural network (CNN) and bidirectional LSTM (BLSTM) model to automatically divide ECG heartbeats into five different groups according to the ANSI-AAMI standard [7]. In 2023, Giovanni Bortolan et al. used three-dimensional display of the entire ECG signal, observed the area constraints of the leads, obtained spatiotemporal images, and combined three-dimensional ECG display with deep learning methods to identify cardiac abnormalities from a variable number of leads [8]. Adel A. Ahmed et al. used a 1D-CNN for arrhythmia classification and trained and validated the model on the MIT-BIH dataset [9]. Ahmed S. Sakr et al. proposed a novel end-to-end deep learning model (ECG-COVID) based on ECG for detecting COVID-19 [10]. Mona Algarni et al. proposed a deep learning-based EEG signal emotion recognition method, classifying emotions into three categories: arousal, valence, and liking, using stacked Bi-LSTM) to identify human emotions [11]. Majid Maf et al. diagnosed attention deficit hyperactivity disorder (ADHD) based on EEG using high-dimensional neural networks with four and six dimensions [12].

However, compared to ECG and EEG, the amount of data available for visual electrophysiology is relatively limited, resulting in far fewer studies on deep learning of visual electrophysiological signals than those on ECG and EEG. Nicholas Waytowich et al. used a compact CNN with automatic feature extraction to decode steady-state visual evoked potential (SSVEP) signals [13]. Liang N et al. proposed a multi-input neural network based on convolution and confidence branches (MCAC-Net), which accurately identifies retinitis pigmentosa (RP) based on flash visual evoked potential (FVEP) and performs excellently in identifying out-of-distribution data as well [14], MCAC-Net is the latest deep learning model for classifying RP disease. Despite the relatively limited amount of visual electrophysiological data and less attention compared to ECG and EEG, research on visual electrophysiological signals is crucial for the diagnosis of ophthalmic diseases. Therefore, in-depth research on visual electrophysiological signals, especially in the field of artificial intelligence, is imperative.

3 Method

3.1 Feature Engineering

Although deep learning neural network models have excellent performance in feature extraction, automatically capturing key features in time series data, the introduction of prior knowledge still plays an important role in model prediction. By integrating prior knowledge as manually extracted features with features automatically extracted by deep learning models, the performance of the model can be significantly improved.

In the medical field, previous studies showed that RP disease leads to delays in the peak latency of the N and P waves of FVEP [4]. Additionally, age, and visual acuity are also associated with RP disease and abnormalities outside RP to a certain extent. This study conducted a statistical analysis to assess the correlation between age, visual acuity, and RP disease. Furthermore, this study analyzed the statistical features of FVEP waveforms, including wavelet entropy, spectral distance, mean peak value, standard deviation of amplitude, standard deviation of amplitude difference, standard deviation of amplitude after P2, standard deviation of amplitude difference after P2. The point-biserial correlation coefficient (PBC) was used to evaluate the correlation between these features and RP disease, and the results are shown in the Table 1.

The absolute values of the PBCs between these features and RP are all higher than 0.3, espacial the indicating a strong correlation between these features and RP disease, particularly the visual acuity, standard deviation of amplitude and standard deviation of amplitude after P2. These features were selected manually as prior knowledge, combined with features learned by deep learning models, to provide more comprehensive and targeted information to the model, thereby improving the predictive performance for RP disease.

Table 1. The PBCs between features and RP disease

Features	PBC
Age	0.42
Visual acuity	-0.57
Wavelet entropy	0.32
Mean absolute difference	0.40
Spectral distance	-0.31
Mean peak value	-0.37
Standard deviation of amplitude	-0.50
Standard deviation of amplitude difference	-0.34
Standard deviation of amplitude after P2	-0.63
Standard deviation of amplitude difference after P2	-0.39

3.2 Data Preprocess

In order to further enhance the performance and generalization ability of the model, data preprocessing was conducted, including resampling, data augmentation, and normalization. FVEP data from different hospitals and different instruments often have different sampling rates. In order to achieve the universality and robustness of the model, it is necessary to resample these data to have the same sampling rate and sequence length. This study adopts the method of downsampling after linear interpolation to achieve data augmentation which helps preserve the characteristics of the original data, making the newly generated data similar to the original data.

Different instruments record visual electrophysiological signals using different units to measure the electrical potentials. As a result, the recorded signals are in different ranges, leading to significant differences in scales between features. This study employed the Min-Max normalization method to normalize the dataset, scaling the data to the range of [0, 1].

3.3 Model Structure

The model used in this study is a multi-input residual neural network called Multiple inputs ResNet with Global convolution and Pyramid pooling (MGPResNet).

As shown in Fig. 2, this is a multi-input residual neural network (MGPResNet) with global convolution, local convolution, and pyramid pooling. The model consists of an input layer, a global convolutional layer, residual convolutional layers, a pyramid pooling module, feature concatenation, and an output layer. The input layer incorporates two branches for FVEP waveform data and manually extracted features, including age, vision, feature waves, and statistical FVEP features. Fully connected layers with neuron counts of 32, 64, 128, and 256 handle feature integration. The global convolutional layer convolves FVEP waveforms with 320-sized filters and downsamples via max-pooling. The pyramid pooling module performs pooling operations on feature maps to capture multi-scale features. The residual convolutional layers include five layers with residual blocks, each with two convolutional layers, batch normalization, ReLU activation, and downsampling. Feature concatenation combines global and local features, and links FVEP waveform features with other features, followed by dropout and regularization. The output layer employs softmax activation for final classification probabilities.

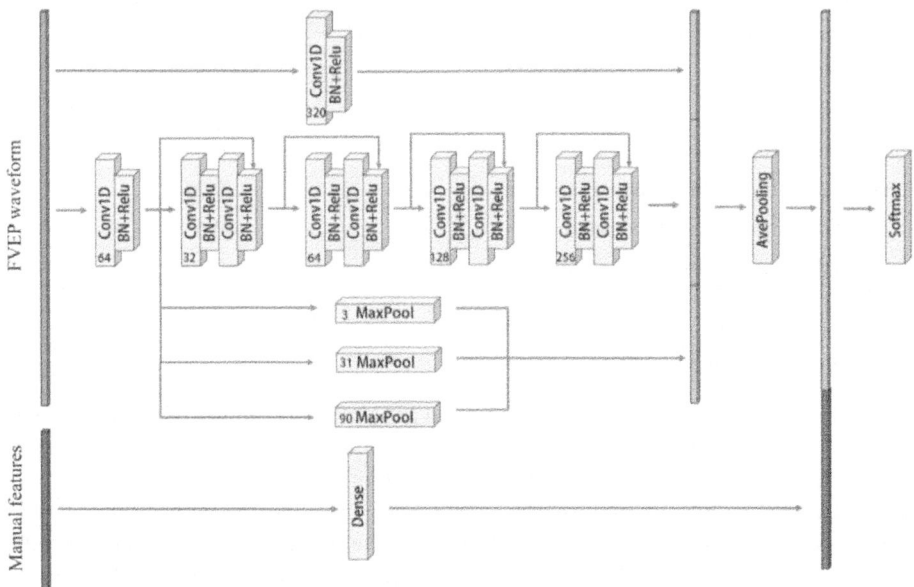

Fig. 2 Model structure of MGPResNet

Global Convolutional. Global convolution operates on the spatial dimension of the entire input feature map to capture overall feature information. In 1D time series data, global convolution operates on the entire time sequence, enabling the model to better understand the features of the entire sequence. Additionally, it integrates features from all positions into a global feature representation through a global average pooling layer.

Let the input feature map be denoted as X, with a size of H × W × C, where H represents height, W represents width, and C represents the number of channels. Let the convolution kernel be denoted as K, with a size of H × W × C × D, where D represents the number of convolution kernels. The output feature map Y of global convolution has a size of 1 × 1 × D, meaning that each convolution kernel produces a scalar value after operating on the entire input feature map. Suppose for the d-th channel in the output feature map, the corresponding convolution kernel is K_d then the pixel value $Y_{1,1,d}$ in the output feature map can be calculated using the Eq. 1, where $X_{i,j}$ represents the value in the d channel of the input feature map at the i row and j column, and $K_{i,j,d}$ represents the weight value of the d channel in the convolution kernel K_d at the i row and j column.

$$Y_{1,1,d} = \sum_{i=1}^{H} \sum_{j=1}^{w} X_{i,j} * K_{i,j,d} \tag{1}$$

Pyramid Pooling Module. The pyramid pooling module [15] typically consists of multiple pooling operations with different sizes, each of which downsamples the input feature map to a different extent. By using the pyramid pooling module, the model can capture feature information at different scales, enhance the model's receptive field, and improve its generalization ability.

Let X be the input feature map with a size of L × C, where L represents length, and C represents the number of channels. Assuming that the pyramid pooling module divides the input feature map into N regions of different scales and performs pooling operations on each region separately to obtain pooled feature representations. For the i-th region, let its size be Li. Assuming the pooling operation uses the pooling function fi, the result of the pyramid pooling $Y_{i,c}$ for the c-th channel of the input feature map X can be calculated using Eq. 2, here, $X_{i,c}$ represents the values of the c-th channel of the input feature map X within the i-th region.

$$Y_{i,c} = f_i(X_{i,c}) \tag{2}$$

The function f_i represents the pooling function selected for the i-th region, which can be either max pooling or average pooling. The final output feature map Y of the pyramid pooling module is obtained by concatenating the pooling results of all regions. The calculation formula is Eq. 3, Here, $Y_1 \ldots Y_n$ represent the pooling results of different regions, and $[Y_1, Y_2, Y_3, \ldots Y_n]$ denotes the concatenation of all pooling results along the channel dimension to obtain the feature map.

$$Y = [Y_1, Y_2, Y_3, \ldots Y_n] \tag{3}$$

4   Experimental Design

4.1   Evaluation Indicator

The experimental results are evaluated by four metrics: Accuracy, Precision, Recall, and F1_score. For multi-class tasks, the averages of Precision, Recall, and F1_Score across all classes were compute to assess the model's classification performance. And the ROC curve and PR curve are also used to evaluate the performance of the model.

4.2 Experimental Design

The first part of the experiments compares the proposed MGPResNet with four other models: MFCN, MDenseNet, MECGNet, MResNet,and MCAC-Net [14]. These four models are added multi-input branches based on the FCN [16], ECGNet [17], DenseNet [18], and ResNet [19] models, respectively. Among them, ResNet serves as the foundational model for MGPResNet.

The ablation experiments consist of two parts. Firstly, compared the classification performance of the FCN, DenseNet, ECGNet, and ResNet models before and after adding the manual feature branch to assess the impact of the manual feature branch. Then the pyramid pooling module and global convolution were added to ResNet and MResNet, and the resulting changes in classification performance were evaluated.

4.3 Dataset

The FVEP datasets used in this study was retrospectively collected from two hospitals and comprised four subsets: the Normal FVEP dataset, the RP dataset, and two datasets containing abnormalities outside RP. The Normal FVEP dataset and RP dataset were collected using Brand 1 electrophysiological examination instruments at Hospital 1, spanning from July 1, 2012, to March 1, 2020. Abnormality outside RP datasets were collected from Hospital 2 using electrophysiological examination instruments of Brand 1 and Brand 2, respectively, from July 1, 2018, to March 1, 2023. Each subject underwent two examinations for both left and right eyes. Each record includes the subject's age, visual acuity and characteristic waves annotated independently by three experienced ophthalmologists.

Table 2. Dataset

	Subjects	Total number	Trainset	Testset
Normal data	2403	9328	8800	528
RP data	329	1233	1000	233
Abnormalities data outside RP	396	2470	2200	270
Total	3128	13031	12000	1031

The details and division of the training and testing sets is shown in Table 2. After mixing the Normal FVEP, RP disease FVEP and abnormality FVEP outside RP, they were randomly divided into training and testing sets. 70% of the data in the training set is used for training, while the remaining 30% is used for validation During training.

5 Results

5.1 Results of Classification Experiments

Table 3. Results of detecting RP and abnormalities outside RP

Model	detecting RP			detecting abnormalities outside RP		
	Precision	Recall	F1_score	Precision	Recall	F1_score
MFCN	87.34	88.46	87.90	99.26	100	99.63
MDenseNet	87.14	89.74	88.42	99.26	99.63	99.45
MECGNet	70.86	84.19	76.95	**100**	**100**	**100**
MResNet	88.03	88.03	88.03	**100**	**100**	**100**
MGPResNet	**92.77**	**93.16**	**92.96**	99.63	98.89	99.26

The Table 3 presents the performance of MFCN, MDenseNet, MECGNet, MResNet, and the proposed MGPResNet model in identifying RP and abnormalities outside RP. These four models all incorporate manual feature input branches, and four metrics are utilized to comprehensively evaluate their performance. In terms of RP recognition, the proposed MGPResNet model outperforms the other models across all three metrics, with a precision of 92.77%, recall of 93.16%, and F1_score of 92.96%. Regarding the identification of abnormalities outside RP, the performance of the proposed MGPResNet model is slightly lower than that of the other models, but the difference is minimal.

Table 4 shows the average metrics of the five models in the classification experiments. The proposed MGPResNet model outperforms the other models across all four average metrics, with accuracy of 96.80%, average precision of 96.52%, average recall of 96.47%, and average F1_score of 96.49%.

Table 4. Results of classification

Model	Accuracy	Precision	Recall	F1_score
MFCN	94.48	93.94	94.26	94.10
MDenseNet	94.38	93.86	94.37	94.11
MECGNet	88.57	87.52	88.30	87.88
MResNet	94.57	94.24	94.24	94.24
MCAC-Net[14]	90.7	94.4	94.0	94.2
MGPResNet	**96.80**	**96.52**	**96.47**	**96.49**

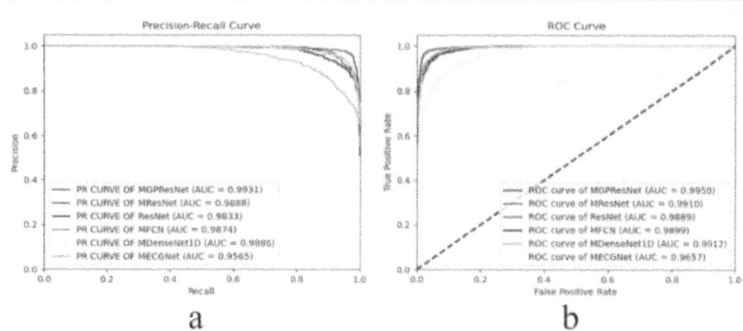

Fig. 3 PR curve and ROC curve

The Fig. 3 respectively display the PR curves and ROC curves of the five models. In the PR and ROC curve plotting, the micro-average method is utilized, which summarizes all classes' true positives, false positives, and thresholds, then computes the overall false positive rate (FPR) and true positive rate (TPR), finally generating the aggregated PR curve and ROC curve, and calculating the overall AUC value. As shown in Fig. 3, the PR curve of the MGPResNet is closest to the upper right corner, with an AUC-PR of 0.9931, the ROC curve is nearest to the upper left corner, with an AUC-ROC of 0.9950, indicating that the MGPResNet performs the best among all models.

5.2   Results of Ablation Experiments

The Table 5 summarizes ablation experiment results, comparing FCN, DenseNet, ResNet, and ECGNet with and without multi-branch inputs. It is easily to see that models with multi-branch inputs outperform the single-input models across all metrics.

As shown in the Table 6, ResNet serves as a submodel of the proposed MGPResNet. GPResNet and MGPResNet have added pyramid pooling modules and global convolution based on ResNet and MResNet, respectively. A comparison reveals the superiority of the pyramid pooling module and global convolution. It indicates the superiority of the proposed multi-branch input structure, pyramid pooling module, and global convolution module.

Table 5. Results of classification

Model	Accuracy	Precision	Recall	F1_score
FCN	88.57	90.15	84.62	86.11
MFCN	94.48	93.94	94.26	94.10
ECGNet	87.11	86.14	87.48	86.70
MECGNet	88.57	87.52	88.30	87.88
DenseNet	88.37	87.52	88.30	87.88
MDenseNet	94.38	93.86	94.37	94.11
ResNet	92.64	91.73	94.19	92.60
MResNet	94.57	94.24	94.24	94.24

Table 6. Results of ablation experiments of pyramid pooling module and global convolution

Model	Accuracy	Precision	Recall	F1_score
ResNet	92.64	91.73	94.19	92.60
GPResNet	94.89	94.22	95.07	94.64
MResNet	94.57	94.24	94.24	94.24
MGPResNet	**96.80**	**96.52**	**96.47**	**96.49**

6 Discussion

This study proposed a multi-input residual convolutional neural network, MGPResNet, which utilized both the FVEP signal waveforms and their statistical features to identify RP diseases and abnormalities outside RP by FVEP. The model employed global convolution to capture global features of FVEP and utilized residual blocks to deepen the network for extracting deeper features. Additionally, a pyramid pooling module are introduced to capture features at different scales. These features were then concatenated to form a multi-level and multi-scale feature representation of FVEP data. Subsequently, manual features of FVEP, including the wavelet entropy, spectral distance, mean peak value, standard deviation of amplitude, standard deviation of amplitude difference, standard deviation of amplitude after P2, standard deviation of amplitude difference after P2, along with age and visual acuity, are computed and passed through fully connected layers. Finally, the features extracted from manual features are concatenated with the multi-level and multi-scale features extracted based on FVEP waveforms, followed by fully connected and SoftMax layers to obtain classification probabilities. The model was tested on clinical datasets, demonstrated an average accuracy of 96.80%, average precision of 96.52%, average recall of 96.47%, and average F1_score of 96.49%. Comparison with four classification models and the latest deep learning model for classifying RP further validated the superiority of this model.

7 Conclusion

The multi-input neural network proposed in this study simultaneously took manual features and FVEP waveforms as inputs. By incorporating global convolution and pyramid pooling modules to capture features at different scales, it achieved excellent results in the detection of RP diseases and abnormalities outside RP. This approach significantly reduces the workload for medical personnel and enhances work efficiency, demonstrating the role and potential of deep learning in the detection of FVEP diseases.

While the RP disease identification algorithm designed based on visual electrophysiological signals and deep learning in this study demonstrated excellent performance on FVEP dataset collected from clinical settings, there are still some limitations. The dataset used in this study is relatively homogeneous and lacks diversity, with insufficient data volume and imbalanced samples in FVEP data. More experiments and cross-validation with larger and more diverse datasets are needed to ensure the robustness and reliability of the model.

Acknowledgments. This study was supported from the National Natural Science Foundation of China(81974138, SL; 82271054, ZL; U20A20363, JH), Scientific and technological projects with combination of medicine and engineering in Xiamen of China(3502Z20224030, SL), Nature Science Foundation of Fujian Province of China(2022J01110650, SL), and National Key Research &Development Program of China (2018YFA0107301, SL).

Disclosure of Interests. It is now necessary to declare any competing interests or to specifically state that the authors have no competing interests.

References

1. Odom, J.V., Bach, M., Brigell, M., et al.: ISCEV standard for clinical visual evoked potentials: (2016 update). Doc Ophthalmol. **133**(1), 1–9 (2016)
2. Mohammed, M.A., Mossallam, E., Allam, I.Y.: The Role of the Flash Visual Evoked Potential in Evaluating Visual Function in Patients with Indirect Traumatic Optic Neuropathy. Clin Ophthalmol. **15**, 1349–55 (2021)
3. Silva, L.D., Santos, A., Hirai, F., et al.: B-scan ultrasound, visual electrophysiology and perioperative videoendoscopy for predicting functional results in keratoprosthesis candidates. Br. J. Ophthalmol. **106**(1), 32–36 (2022)
4. Zhang, M., Ouyang, W., Wang, H., Meng, X., Li, S., Yin, Z.Q.: Quantitative assessment of visual pathway function in blind retinitis pigmentosa patients. Clin. Neurophysiol. **132**(2), 392–403 (2021)
5. Shuoxuan, Z.: A Classification Scheme for ECG Signals Based on Bidirectional LSTM Model. Adv. Compute. Signals Syst. **6**(5) (2022)
6. Suhail, M.M., Razak, T.A.: Cardiac disease classification from ecg signals using hybrid recurrent neural network method. Adv. Eng. Softw., **174** (2022)
7. Bhatia, S., Pandey, S.K., Kumar, A., Alshuhail, A.: Classification of Electrocardiogram Signals Based on Hybrid Deep Learning Models. Sustainability **14**(24) (2022)
8. Bortolan, G.: 3D ECG display with deep learning approach for identification of cardiac abnormalities from a variable number of leads. Physiol Meas, **44**(2) (2023)
9. Ahmed, A.A., Ali, W., Abdullah, T.A.A., Malebary, S.J.: Classifying cardiac arrhythmia from ECG signal using 1D CNN deep learning model. Mathematics **11**(3) (2023)

10. Sakr, A.S., Plawiak, P., Tadeusiewicz, R., Plawiak, J., Sakr, M.: Hammad M: ECG-COVID: An end-to-end deep model based on electrocardiogram for COVID-19 detection. Inf. Sci. (N Y), 619:324–39 (2023)

11. Algarni, M., Saeed, F., Al-Hadhrami, T., Ghabban, F., Al-Sarem, M.: Deep Learning-Based Approach for Emotion Recognition Using Electroencephalography (EEG) Signals Using Bi-Directional Long Short-Term Memory (Bi-LSTM). Sensors (Basel) **22**(8) (2022)

12. Mafi, M., Radfar, S.: High Dimensional Convolutional Neural Network for EEG Connectivity-Based Diagnosis of ADHD. J. Biomed. Phys. Eng. **12**(6), 645–54(2022)

13. Waytowich, N., Lawhern, V.J., Garcia, J.O., et al.: Compact convolutional neural networks for classification of asynchronous steady-state visual evoked potentials. J Neural Eng. **15**(6), 066031(2018)

14. Liang, N., Wang, C., Li, S., Xie, X., Lin, J., Zhong, W.: The classification of flash visual evoked potential based on deep learning. BMC Med Inform Decis. Mak. **23**(1), 13(2023)

15. Zhao, H., Shi, J., Qi, X., Wang, X., Jia, J.: Pyramid Scene Parsing Network. arXiv preprint arXiv: 1612.01105 (2016)

16. Long, J., Shelhamer, E., Darrell, T.: Fully convolutional networks for semantic segmentation. IEEE Trans. Pattern Anal. Mach. Intell. **39**(4), 640–651 (2015)

17. Zeyang, Z., Ziheng, Z., Cao, Z., et al.: ECGNet: an efficient network for detecting premature ventricular complexes based on ECG images. IEEE Trans. Bio-medical Eng. (2022)

18. Huang, G., Liu, Z„ Laurens, V.D.M., Weinberger, K.Q.: Densely Connected Convolutional Networks. arXiv preprint arXiv: 1608.06993 (2016)

19. He, K., Zhang, X., Ren, S., et al.: Deep residual learning for image recognition. In: Proceedings of the IEEE Conference on Computer Vision and Pattern Recognition, pp. 770–778 (2016)

Quantitative Analysis of Lower Limb Alignments for Joint Disease Patient Based on Template Registration

Yibo Liu and Xiankun Zhang[✉]

College of Artificial Intelligence, Tianjin University of Science and Technology, Tianjin 300222, China
zhxkun@tust.edu.cn

Abstract. Numerous studies show that accurate reconstruction of lower limb alignments is one of the main determinants of the long-term success of Total knee arthroplasty (TKA). The purpose of this paper is to explore a new 3D CT measurement method base on template registration to extract the lower limb alignments quickly and accurately. By constructing statistical shape model of leg and using non-rigid template matching method, we can accurately extract key anatomical structure feature points of lower limb. A bidirectional-section method is proposed to locate the femoral head center. Then, the feature points are connected to obtain the femoral mechanical axis and the tibial mechanical axis. The proposed method is verified in 20 cases of CT data. The experimental results demonstrate that the average of the feature point extraction error is less than 3 mm, the average angle error between the mechanical axes and the ground-truth is less than 0.6°. Compared with the benchmark method, the measurement accuracy of the feature points is improved by 39.75%, the processing speed increased by 37.50%. The proposed method can extract lower limb alignments accurately, it will help the long-term success of TKA surgery.

Keywords: Total knee arthroplasty · lower limb alignment · statistical shape model · non-rigid template matching · bidirectional-section method

1 Introduction

TKA is a successful and effective method for the treatment of end-stage osteoarthritis, the prosthesis survival rate can reach 5–10 years [1]. However, the postoperative complications of patients are still as high as 15%–30% [2]. Studies show that after TKA surgery, the stability and durability of the knee joint are closely related to the alignment of the lower limb alignments during the operation [3, 4]. The lower limb alignments are also the positioning reference for doctors during the entire surgical resection. The recovery and accuracy of the lower limb alignments are recognized as an important factor in reducing the failure of the prosthesis and improving the postoperative efficacy [5]. The lower limb alignments include the femoral mechanical axis (FMA) and the tibial mechanical axis (TMA), the former is the line connecting the femoral head center and the intercondylar apex [6], the latter connecting the tibial spine center and the ankle center.

In general, the lower limb weight-bearing full-length positive X-ray film is often used to extract the lower limb alignments in the TKA surgery [7]. However, it is difficult to obtain a standard positive X-ray film due to many factors such as projection angle and body position. Coupled with the doctors finding the center point of the hip, knee, and ankle on the X-ray image by experience, the subjective error is too large to guarantee the accuracy of the osteotomy position and angle. Patrick Salviaet al. [8] added landmark points to determine the lower limb alignments based on knee CT image data. This method is also commonly used in clinical practice, it is easy to operate but has a large human influence. The inaccurate judgment of inexperienced doctors on bone landmarks will cause greater error. Su et al. [9] reported the method of calculating the femoral head center using the spherical center difference of the hip joint based on the CT image. Piotr Zarychtaet et al. [10] detected the femoral head center based on fuzzy connectivity (FC). However, it is not reliable to simulate the circumference of the entire femoral head by extracting part of the boundary of the femoral head. The reason is that the femoral head is not a regular sphere, and it cannot be simply regarded as a circle on the X-ray film.

In this paper, a method for measuring lower limb alignments based on 3D leg template registration is proposed. To improve the non-rigid registration accuracy, we performed statistical shape model training on all the legs. A bidirectional-section method is proposed to obtain the femoral head center based on the reference point on the surface of the femoral head. It will solve the problem of doctor's reliance on subjective experience in extracting lower limb alignments. The proposed method demonstrates its superiority in terms of accuracy and reliability compared with the manual measurement method. If the proposed method is used in locating the lower limb alignments during the operation, it will help the TKA surgery achieve long-term success.

2   Proposed Method

The flow chart of the lower limb alignment extraction method is shown in Fig. 1. The proposed method includes the following four steps: 1) bone segmentation in the CT images is performed using graph cut segmentation algorithm, and the leg mesh surface is generated; 2) by constructing statistical shape models of all leg bones, we can obtain a standard leg template, and then register the leg template to the leg to be measured by non-rigid registration algorithm GMMReg; 3) the feature points of the legs are obtained based on the index value of the ground-truth feature points of the leg template; 4) we proposed a bidirectional-section method to extract the coordinates of the femoral head center, and the lower limb alignments are obtained.

2.1   Standard Leg Template Generation and Template Registration

The Staismo algorithm, a framework for PCA based statistical models [11], is used to train the statistical shape models of all the legs. Hence, the mean shape of all the preoperative legs can be obtained, and it can be utilized as the standard leg template. The standard leg template and other preoperative legs have different image spaces. The

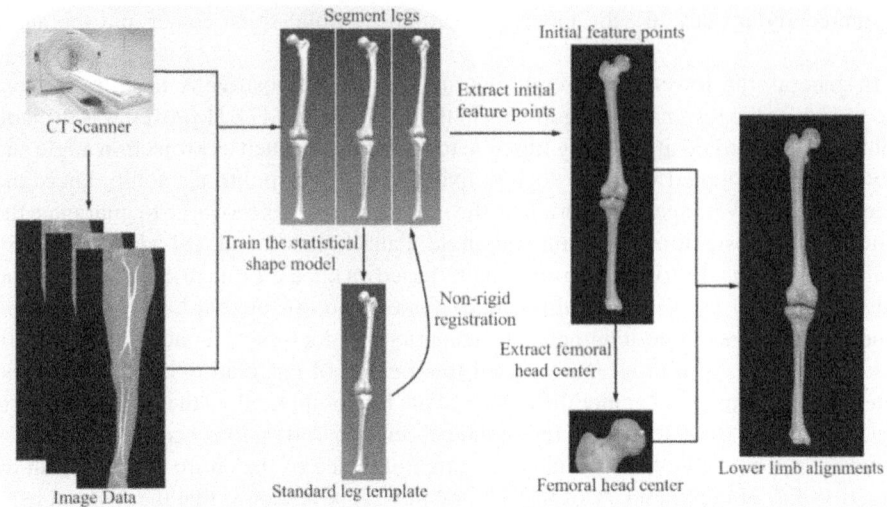

Segment legs

Initial feature points

Extract initial
feature points

CT Scanner

Train the statistical
shape model

Non-rigid
registration

Extract femoral
head center

Lower limb alignments

Image Data   Standard leg template   Femoral head center

Data Preparation and Template Registration   **Lower Limb Alignments Extraction**

Fig. 1. Workflow of the lower limb alignments extraction

leg template is sequentially registered to all the preoperative legs by using the GMMReg algorithm [12]. The GMMReg method is a non-rigid registration method and mainly uses the Gaussian mixture model to represent the input point set. The problem of point set registration is reformulated as the problem of aligning two Gaussian mixtures, and the statistical discrepancy measure between the two corresponding mixtures is minimized. To align mixture densities, the closed-form expression was used to calculate the L2 distance between two Gaussian mixtures. Given two point sets, the scene set S and the model set M, GMMReg method finds the parameter θ of a parametrized spatial transformation family T which minimizes the following cost function:

$$d_{L_2}(S, M, \theta) = \int (gmm(S) - gmm(T(M, \theta)))^2 dx \tag{1}$$

where $gmm(S)$ refers to the Gaussian mixture density constructed from a point set S. The above function can be treated as a special case of the density power divergence:

$$d_\omega(a, b) = \int \left\{ \frac{1}{\omega} a^{1+\omega} - \frac{1+\omega}{\omega} ab^\omega + b^{1+\omega} \right\} dx \tag{2}$$

where a and b are two density functions. It is a Bregman divergence function controlled by parameter ω. . The Kullback-Leibler (KL) divergence can be obtained when ω approach 0: $d_0(a, b) = lim_{\omega \to 0} d_\omega(a, b) = \int a(x) log\{a(x)/b(x)dx\}$. When $\omega = 1$, $d_1(a, b) = \int \{b(x) - a(x)\}^2 dx$ becomes the L2 distance between the densities. For $0 < \omega < 1$, the class of density power divergences provides a smooth bridge between the KL divergence and the L2 distance. In this way, all the preoperative legs to be measured are replaced by the leg template after registration. Figure 2 shows the non-rigid registration process of the leg template with the preoperative legs. The leg template (blue) and target leg (red) before non-rigid registration is shown in Fig. 2(a). Figure 2(b) shows the result after registration. It can be seen that the leg template is fused with the target leg.

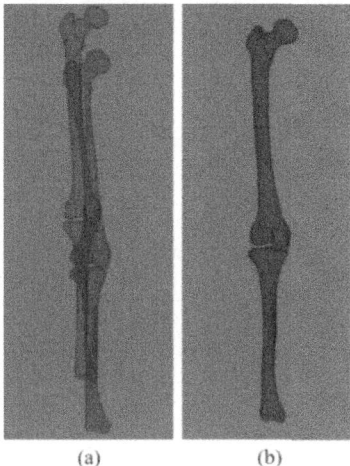

(a)   (b)

Fig. 2. The 3D-3D non-rigid registration based on the GMMReg method. (a) The leg template (blue color) and the target leg (red color) before registration; (b) The leg template is registered to the target leg (Colour figure online)

2.2   Feature Points Extraction

The feature points of the leg are the following four points: the femoral head center, the intercondylar apex, the tibial spine center, and the ankle center. The ground-truth feature points do not belong to the point set of the leg template, we need to calculate the index values corresponding to the four feature points. The method of obtaining the index values of the ground-truth feature points on the leg template is as follows: extract n points ($100 \leq n \leq 1000$) near the ground-truth feature point on the leg template and find the point closest to the feature point among n points, the index value of the closest point on the leg template is obtained by using the Visualization Toolkit (VTK) [13]. This index value can be considered as the index value of the ground-truth feature point. The range of n takes into account the efficiency and accuracy of the algorithm. It is difficult to ensure the accuracy if the number of points is too small, the processing speed will slow down if there are too many points.

After the leg template is registered to the leg to be measured, the shapes of the two are almost the same. The coordinates of the new feature points can be obtained according to the index values of the four feature points on the leg template after spatial transformation. The new feature points can be regarded as the initial feature points of the leg to be measured. In Fig. 3(a), the green points are the ground-truth feature points of the leg template. Figure 3(b) shows the initial feature points (the yellow points) of the leg to be measured obtained by using index values of the leg template feature points.

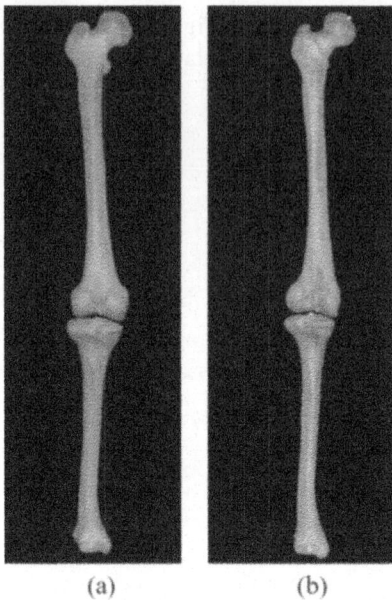

Fig. 3. The ground-truth for the feature points and the obtained initial feature points. (a) The ground-truth for the feature points of the leg template; (b) The obtained initial feature points of the leg to be measured

2.3   Extraction of the Femoral Head Center

The four initial feature points of the leg to be measured have been obtained, but the femoral head center is not in the correct position, as shown in Fig. 3(b). The reason is that the reconstructed leg 3D model only has a surface layer consisting of point clouds, and there are no points inside. The index value corresponds to a point in the model surface, thus, the femoral head center obtained according to the index value is not at the sphere center, but on the sphere surface. To solve this problem, the bidirectional-section method is used to obtain the true femoral head center. Specifically, the feature point on the surface of the femoral head sphere is used as a reference point (the yellow point in Fig. 4). Transversely cut the femoral head to get the section, and then the center point of the section is obtained by using VTK. Then use the center point of the transverse section as a reference point, longitudinally cut the femoral head to obtain the center point of the longitudinal section, this point is the femoral head center (the red point in Fig. 4). Finally, the femoral head center is connected with the intercondylar apex to obtain the FMA, the TMA is obtained by connecting the tibial spine center and the ankle center. The method of obtaining the feature points and mechanical axes of the postoperative legs is the same as that of the preoperative legs.

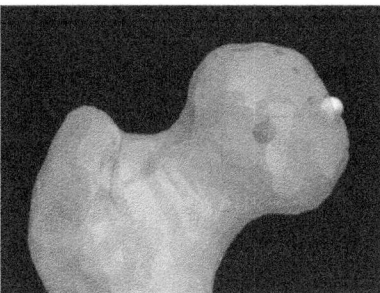

Fig. 4. Obtain the center point of the femoral head by bidirectional dissection

3   Experiments and Results

3.1   Experimental Setup

Twenty patients were included in this study. Each patient had a leg CT scan with a width of 0.75 mm (Siemens Medical) before and after unilateral TKA surgery. The scan ranged from middle pelvis to toes. The dimensions of the CT images used in the experiment are $220 \times 200 \times 880$. Each case contains preoperative and postoperative CT images and the time interval between CT scans before and after TKA surgery is about 14 months. The proposed method was implemented by the C + + programming language under the windows platform. To evaluate the accuracy of the proposed method in the extraction of lower limb alignments, four orthopedists were invited to operate the proposed method, the Elastix CT registration method [14], and the benchmark method. The benchmark method means the doctor manually selects points on the CT or X-ray image to determine the bony landmarks, and then obtain the lower limb alignments. We referred to these four doctors as doctors A-D in the paper. Doc. A, Doc. B and Doc. C have 3 years of experience, Doc. D has 10 years of experience.

The proposed method was applied to the 20 patients (40 legs) operated by Doc. A, Doc. B performed the Elastix CT registration method on the data. To distinguish the manual measurement results of orthopedists with different years of experience, Doc. C and Doc. D performed the benchmark method on the data, respectively. The same observers measured the feature points three times for each leg on different days, and the mean values were calculated. The measurement results include the distance error of four feature points, the angle error of the FMA and the TMA. In addition, we also counted the accuracy of the non-rigid registration.

3.2   Lower Limb Alignment Detection Accuracy

Figure 5 shows the error distribution of the non-rigid registration of the leg template with the pre and postoperative legs. It can be seen from the figure that the absolute value of the maximum distance error of the non-rigid registration is less than 4 mm, the distance error of most corresponding point pairs is less than 2 mm. The average TRE of the preoperative and postoperative legs is 1.73 mm and 1.93 mm, respectively. And the

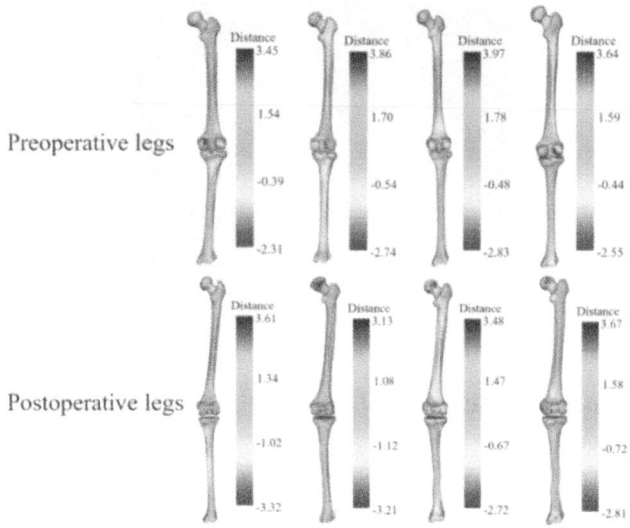

Fig. 5. The error distribution of the non-rigid registration of the leg template with the pre and postoperative legs

average consumption time of the preoperative and postoperative legs is 2.27 min and 2.26 min, respectively.

The distance error (RMS) of the four feature points of the proposed method, the Elastix CT registration method, and the benchmark method is shown in Fig. 6 and Fig. 7. The angle error of the mechanical axes (FMA and TMA) of these methods is shown in Fig. 8. The experimental results demonstrate that in terms of positioning the feature points of the legs, the proposed method provides a better and more stable result among these methods. Compared with the Elastix CT registration method operated by Doc. B and the benchmark method operated by Doc. C and Doc. D, the positioning accuracy of the feature points is improved by 25.47%, 39.75% and 31.28%, respectively. While the accuracy of the mechanical axes is improved by 39.07%, 54.61% and 46.13% for the combined pre- and post-operative results, respectively. It can be seen that although the measurement results of a more experienced specialist doctor (Doc. D) are more accurate than the doctor with less clinical experience (Doc. C) based on the benchmark method. However, even experienced orthopedists, their manual measurement results are not as good as the proposed method. It is clinically considered that the angle error of the lower limb alignments is acceptable within 3° [15]. The angle error of the FMA and TMA obtained by the proposed method is less than 0.6°, so the proposed method meets the clinical requirements of lower limb alignments detection.

Both the proposed method and the Elastix CT registration method are based on template registration, and the total consumption time of both is about 5 min. The processing time on the baseline method is about 8 min.

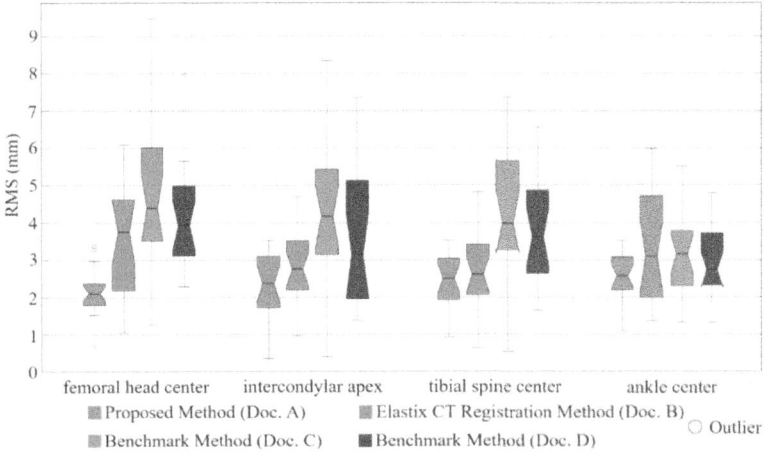

Fig. 6. The RMS of the feature points of the preoperative legs

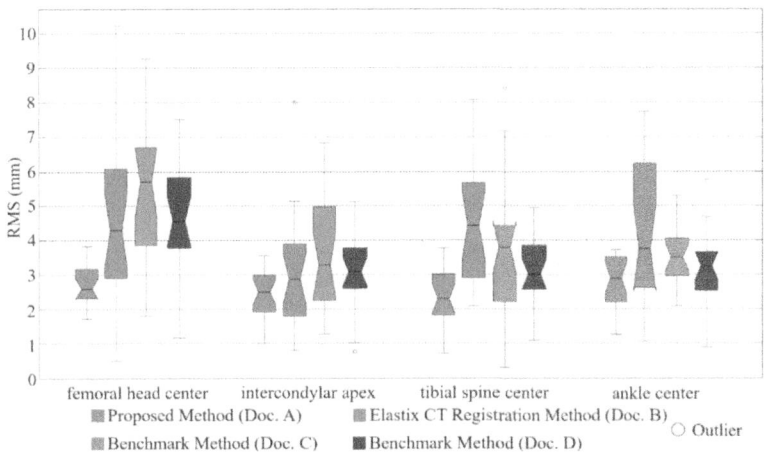

Fig. 7. The RMS of the feature points of the postoperative legs

4   Conclusions

In this work, we propose a method for measuring lower limb alignments based on 3D leg template registration. The proposed method can be used to guide the osteotomy position and angle of TKA surgery, and to evaluate the effect after TKA surgery. These processes were completed by introducing the standardized leg template and the template registration. The proposed method has high accuracy and with high processing speed, compared with the Elastix CT registration method and the benchmark method. Extracting the lower limb alignments by the proposed method can get rid of the problem that depends on the experience of the doctor, which helps TKA surgery achieve long-term success.

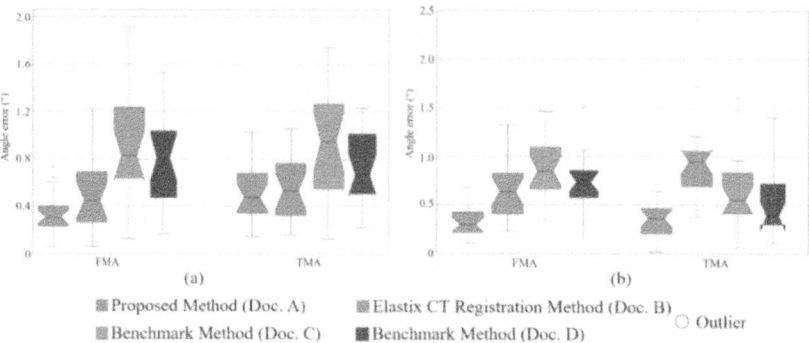

Fig. 8. Results of the angle error of the mechanical axes. (a) Preoperative legs; (b) Postoperative legs

Acknowledgments. The authors gratefully acknowledge the support provided by National Training Program of Innovation and Entrepreneurship for Undergraduates (grant number:202310057360).

Disclosure of Interests. The authors declare that there are no conflicts of interest related to this article.

References

1. Meding, J.B., Meding, L.K., Ritter, M.A., Keating, E.M.: Pain relief and functional improvement remain 20 years after knee arthroplasty. Clin Orthop Relat R. **470**, 144–149 (2012)
2. Wainwright, C., Theis, J.C., Garneti, N., Melloh, M.: Age at hip or knee joint replacement surgery predicts likelihood of revision surgery. J. Bone Joint Surg. **93**, 1411–1415 (2011)
3. Siston, R.A., et al.: Averaging different alignment axes improves femoral rotational alignment in computer-navigated total knee arthroplasty. JBJS **90**, 2098–2104 (2008)
4. Catani, F., et al.: Alignment deviation between bone resection and final implant positioning in computer-navigated total knee arthroplasty. JBJS **90**, 765–771 (2008)
5. Bargren, J.H., Blaha, J.D., Freeman, M.A.: Alignment in total knee arthroplasty: correlated biomechanical and clinical observations. Clin Orthop Relat R. **173**, 178–183 (1983)
6. Moreland, J.R., Bassett, L.W., Hanker, G.J.: Radiographic analysis of the axial alignment of the lower extremity. J. Bone Joint Surg. **69**, 745–749 (1987)
7. Wang, D., Nung, L.N., Jing, Y.S.: Radiological lower limb measurement of total knee arthroplasty. Chin J Orthop. **25**, 565–567 (2005)
8. Salvia, P., et al.: Knee kinematics validation of a re-orientation technique of knee axis. In: 9th Symposium on the 3D Analysis of Human Movement, France (2006)
9. Su, Y., Xu, Z., Li, H., Qin, K.: Measurement method in virtual surgical planning for congenital dislocation of the hip. J Comp Aid Des Comp Grap. **16**, 1159–1163 (2004)
10. Zarychta, P., Konik, H., Zarychta-Bargieła, A.: Computer assisted location of the lower limb mechanical axis. In Information Technologies in Biomedicine: Third International Conference, ITIB, Gliwice, Poland, 629565_1_En_17_JobSheet_200 June 2012. Proceedings, Springer Berlin Heidelberg, pp. 93–100 (2012)

11. Lüthi, M., et al.: Statismo-A framework for PCA based statistical models. Insight J. **1**, 1–18 (2012)
12. Jian, B., Vemuri, B.C.: Robust point set registration using gaussian mixture models. IEEE T Pattern Anal. **33**, 1633–1645 (2011)
13. Schroeder, W. J., Lorensen, B., Martin, K.: The visualization toolkit: an object-oriented approach to 3D graphics. Kitware (2004)
14. Klein, S., Staring, M., Murphy, K., Viergever, M., Pluim, J.: Elastix: a toolbox for intensity-based medical image registration. IEEE T Med Imaging. **29**, 196–205 (2009)
15. Rivière, C., et al.: Alignment options for total knee arthroplasty: a systematic review. Orthop Traumatol-Surg. **103**, 1047–1056 (2017)

Spatial Domain Identifying: Graph Attention Network with Two Different Decoders

Yi Liu🆔 and Quan Zou$^{(\boxtimes)}$ 🆔

Institute of Fundamental
and Frontier Sciences, University of Electronic Science and Technology of China,
Chengdu 611731, China
zouquan@nclab.net

Abstract. Exploring spatial domains to investigate tissue structures is a chance provided by spatial transcriptome technology, while also a significant challenge in spatial transcriptomics research. Current approaches only focus on spatial gene expression and cannot simultaneously incorporate spatial location information. Graph deep learning models can simultaneously encode node features and positional information. However, during decoding, most of the models still only focus on reconstructing feature information, ignoring positional information. Here, we propose a new method, DeepDomain, which aims to improve the latent representation of nodes by jointly reconstructing gene expression profiles and spatial neighborhood networks using a deep graph attention network with two distinct decoders. Utilizing enhanced spatial latent representations to identify spatial domains in three datasets, DeepDomain achieved higher accuracy in evaluation metrics and a better description of organizational structure when compared to existing methods.

Keywords: Spatial domain identification · Deep learning · Graph attention network

1 Introduction

The availability of spatial transcriptome data with slice images and slice site information, the combination of which makes spatial domain identification gradually become a research hotspot for spatial transcriptome data. The identification of spatial domains enables us to have the opportunity to slice spatial regions, which can help us better understand the biological functions of different regions of the tissue [1, 2], and study the underlying organization of cancer and normal tissue [1, 3, 4]. Developing spatial domain identification methods applicable to spatial transcriptome data has become an essential task.

Rational classification of spatial spots in the spatial transcriptome data is an effective way of spatially identifying spatial domains. In scRNA-seq analysis, it is a common task to classify the different cells and identify functionally similar cells [5, 6]. However, since spatial transcriptome data can provide the spatial location of each spot, which makes the data information we can utilize richer, it is inaccurate to classify the spots in the spatial

D.-S. Huang et al. (Eds.): ICIC 2024, LNBI 14881, pp. 310–320, 2024.
https://doi.org/10.1007/978-981-97-5689-6_27

transcriptome dataset solely based on gene expression data through scRNA-seq clustering methods only. Several studies have recently been dedicated to spatial transcriptome data for spatial domain identification. SpaGCN [7], STAGATE [8], and conST [9] apply the graph neural networks to reconstruct spot gene expression to re-embedding spots, SpaceFlow [10] extracts low-dimensional embedding of spatial spots by capturing spatial structural loss through graph neural networks, and BayesSpace [11] considers spot neighborhood information using Markov random fields to identify the spatial domain. However, methods that only consider the reconstruction of spot gene expression ignore the similarity between neighboring spots, and methods that only consider spot neighborhood information do not make full use of gene expression information. While traditional graph neural networks are claimed to be able to capture both location information and feature information, through input of the node feature information and location information simultaneously, they only consider reconstructing the feature information when defining the loss function and decoder, which may lead to missing information.

To this end, we proposed the DeepDomain based on the graph neural networks that simultaneously exploit the similarity between neighboring spots and the gene expression of the spot itself to identify spatial domains more accurately. Unlike the existing methods and traditional graph neural networks, the DeepDomain uses the graph attention network [12] with two different decoders. One for reconstructing spatial spot gene expression profile, and another for reconstructing spatial neighborhood network. By inputting both feature and location information simultaneously and reconstructing the two parts of the information, the DeepDomain enables to reduce the omission of information during the training process. After integrate both the spatial gene expression information and the topology of the spatial neighborhood network, the DeepDomain can obtain a more comprehensive low-dimensional representation of the spots.

We applied the DeepDomain on three datasets containing manual annotations and compared the results with existing methods, and the different accuracy calculation metrics demonstrated the superiorities of the DeepDomain over existing methods in spatial domain identification.

2   Materials and Methods

2.1   Overview of DeepDomain

Spatially resolved transcriptomics data contains the spatial expression profile and the spatial locations of spots. DeepDomain combines the graph attention network and inner product decoder to learn a more comprehensive latent representation of spots, considering information about the spot itself and neighboring spots from the spatially resolved transcriptomics data (see Fig. 1).

Specifically, the neighboring spot pairs can be summarized by a spatial neighbor network constructed by the K-nearest neighbor (KNN) algorithm. The nodes in the network represent the spots and the edges indicate the neighboring spot pairs. The spatial neighbor network can be represented by an adjacency matrix A. Then, the spot gene expression matrix X and spatial adjacency matrix A are inputs to a two-layer graph attention encoder to get latent representations Z of spots. Next, DeepDomain feeds Z into two decoders simultaneously, including the graph attention decoder and

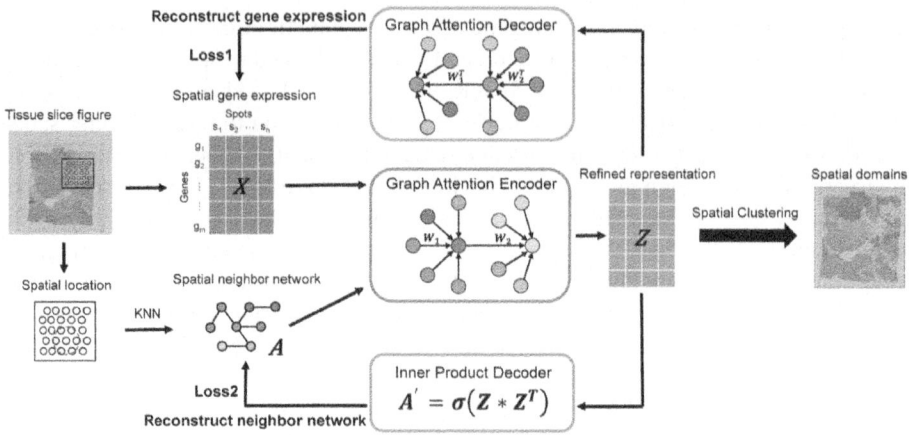

Fig. 1. Workflow of DeepDomain.

the inner product decoder. The graph attention decoder is used to reconstruct the spot gene expression matrix X and the inner product decoder is used to reconstruct the spatial neighbor network A. Finally, the latent representation Z could be extracted for spatial domain identification using the existing spatial clustering algorithms, such as mclust [13].

2.2 Datasets

We applied DeepDomain to three spatial resolved transcriptomics datasets.

First, the human dorsolateral prefrontal cortex (DLPFC) dataset (http://research.libd. org/spatialLIBD/), which includes 12 human DLPFC tissue slices, and each slice was manually annotated by the original author, including the DLPFC layers and white matter (WM) [14]. The number of spots in the 12 slices ranged from 3460 to 4789.

Second, the human breast cancer dataset, which includes 3789 spots with 36601 genes and can be obtained from 10x Genomics Data Repository (https://www.10xgenomics.com/resources/datasets/human-breast-cancer-block-a-section-1-1-standard-1-1-0). Each spot was manually annotated with 20 regions.

Third, the mouse brain's anterior section dataset, which includes 2695 spots with 32285 genes and can be downloaded from 10x Genomics Data Repository (https://www.10xgenomics.com/resources/datasets). Each spot was manually annotated by the Allen Brain Atlas reference (https://mouse.brain-map.org/static/atlas), including 52 regions.

2.3 Data Pre-processing

We used the Python SCANPY [15] package to log-transform raw gene expression counts and normalize them. For all datasets, we selected the top 3000 highly variable genes to be input into the DeepDomain model.

2.4   Construction of the Spatial Neighbor Network

The spot location information provided by spatially resolved transcriptomics gives us more chances to explore different spatial domains. For a given spot, there is usually a similarity between the spot and its spatially neighbor spots. To make better use of this similarity, we constructed the spatial neighbor network by calculating the Euclidean distance using spatial location information. We choose k spots with the closest distance to the given spot as its nearest neighbors. Then, the spatial neighbor network can be further represented by its adjacency matrix A, if spot j is one of the neighbors of spot i, $A_{ij} = 1$, otherwise 0.

2.5   Graph Neural Networks for Latent Representation Learning

The DeepDomain model learns the latent representations of spots based on graph neural networks. The graph neural networks of DeepDomain contain three parts: a graph attention encoder, a graph attention decoder, and an inner product decoder.

Graph Attention Encoder. To take into account information from both the spot itself and its spatial location, we use the graph attention network as a graph attention encoder [12]. The graph attention network not only focuses on the information of individual nodes, but also focuses on the graph structure and aggregates the information from neighbor nodes, which enables the DeepDomain model to better learn the latent representation of each node.

The input to the first layer of the graph attention encoder is the spot gene expression matrix X and the adjacency matrix A.

Assuming that the number of encoder layers is L, the output of the encoder in layer $k(k \in \{1,2,\ldots,L-1\})$ of spot i is defined as:

$$Z_i^{(k)} = \sigma\left(\sum_{j \in N_i} \alpha_{ij}^{(k)}\left(W^{(k)} X_i^{(k-1)}\right)\right),\tag{1}$$

where $Z_i^{(k)}$ is the latent representation of spot i in the k-th layer, σ is the activation function ELU, N_i is the neighbors of spot i that can be provided by the adjacency matrix A, $W^{(k)}$ is the k-th layer's trainable weight matrix, $X_i^{(k-1)}$ is the representation of spot i in the $(k$-1)-th layer and the initial representation of spot i $X_i^{(0)}$ is the gene expressions of spot i. $\alpha_{ij}^{(k)}$ is the coefficients calculated according to the attention mechanism and we follow Velickovic et al. [12]. $\alpha_{ij}^{(k)}$ is expressed as:

$$\alpha_{ij}^{(k)} = \text{Softmax}_j\left(e_{ij}^{(k)}\right) = \frac{\exp\left(e_{ij}^{(k)}\right)}{\sum_{l \in N_i} \exp\left(e_{il}^{(k)}\right)},\tag{2}$$

$$e_{ij}^{(k)} = \text{Sigmoid}\left(a_s^{(k)^T}\left(W^{(k)} X_i^{(k-1)}\right) + a_v^{(k)^T}\left(W^{(k)} X_j^{(k-1)}\right)\right),\tag{3}$$

where $a_s^{(k)^T}$ and $a_v^{(k)^T}$ are the trainable weight vectors, Sigmoid and Softmax are the sigmoid and softmax activation functions respectively.

In the L-th layer of the encoder, we do not use the attention layer and the activation function, and the output is defined as:

$$Z_i^{(L)} = W^{(L)} X_i^{(L-1)}, \tag{4}$$

Graph Attention Decoder. The structure of the decoder is symmetrical to the encoder. The input to the first layer of the graph attention decoder is the latent representation of each node $Z^{(L)}$, the output of layer L of the encoder. The graph attention decoder is used for spot gene expression reconstruction. The output of the decoder in layer $k(k \in \{1,2,\ldots,L-1\})$ of spot i is defined as:

$$H_i^{(k)} = \sigma\left(\sum_{j \in N_i} \hat{\alpha}_{ij}^{(k)} \left(\widehat{W}^{(k)} H_i^{(k-1)}\right)\right), \tag{5}$$

where $H_i^{(k)}$ is the latent representation of spot i in the k-th layer, σ is the activation function ELU, N_i is the neighbors of spot i that can be provided by the adjacency matrix A, $\widehat{W}^{(k)}$ is the k-th layer's trainable weight matrix, $H_i^{(k-1)}$ is the representation of spot i in the $(k$-1)-th layer and the initial representation of spot i $H_i^{(0)}$ is the output of layer L of the encoder $Z^{(L)}$. $\hat{\alpha}_{ij}^{(k)}$ is the coefficients calculated according to the attention mechanism and can be calculated by Eq. (2).

In the last layer of the decoder, we do not use the attention layer and the activation function, and the output is defined as:

$$H_i^{(L)} = \widehat{W}^{(L)} H_i^{(L-1)}, \tag{6}$$

Inner Product Decoder. The Inner product decoder is used to ensure that the original graph topology is preserved while reconstructing gene expression, reducing the loss of spatial neighbor network. The input to the first layer of the graph attention decoder is the latent representation of each node $Z^{(L)}$, the output of layer L of the encoder. The output of the decoder is defined as:

$$\widehat{A}_i^{(L)} = \text{Sigmoid}\left(Z^{(L)} Z^{(L)^T}\right), \tag{7}$$

where $Z^{(L)^T}$ is the transpose of $Z^{(L)}$, and Sigmoid is the sigmoid activation function.

Loss Function. The DeepDomain model consists of two loss components, including the loss of the spot gene expression reconstruction and the loss of spatial neighbor network reconstruction. The final loss is defined as:

$$\text{Loss} = \alpha * \text{Loss}_1 + \beta * \text{Loss}_2, \tag{8}$$

The final training objective is to minimize the final loss Loss, where α and β are the tradeoff parameters for the two loss parts respectively, Loss_1 and Loss_2 are the loss of the spot gene expression reconstruction and the loss of spatial neighbor network reconstruction respectively. Loss$_1$ is the loss of the spot gene expression reconstruction. Assuming that the number of spots is M, Loss$_1$ is defined as:

$$\text{Loss}_1 = \sum_{i=1}^{M} \|X_i - H_i^{(L)}\|_2, \tag{9}$$

Loss$_2$ is the binary cross entropy between $\widehat{A}_i^{(L)}$ and the adjacency matrix A.

2.6 Evaluation Metrics

We assessed the ability of the DeepDomain model to recognize spatial domains using the following metrics: Adjusted Rand Index (ARI) and Normalized Mutual Information (NMI). ARI and NMI can all be used to measure the similarity between predicted spatial domain labels and real annotated region labels, with higher values of the metrics being preferred. Both they can be calculated by using the Python package scikit-learn [16].

Assuming that there are number of n spots involved in classification, K_P number of classification labels in the prediction cluster P, and K_T number of classification labels in the real annotated cluster T. Using n_{ij} to represent the number of spots included in the i-th prediction cluster and the j-th real annotated cluster together, a_i to represent the number of spots in the i-th prediction cluster, and b_j to represent the number of spots in the j-th real annotated cluster.

ARI can be calculated as:

$$\text{ARI} = \frac{\sum_{ij}\binom{n_{ij}}{2} - \left[\sum_i\binom{a_i}{2}\sum_j\binom{b_j}{2}\right]/\binom{n}{2}}{\frac{1}{2}\left[\sum_i\binom{a_i}{2}\sum_j\binom{b_j}{2}\right] - \left[\sum_i\binom{a_i}{2}\sum_j\binom{b_j}{2}\right]/\binom{n}{2}}, \tag{10}$$

The value of ARI is in the range of $[-1,1]$, with larger values indicating that the prediction results are closer to the true results.

NMI can be calculated as:

$$\text{NMI} = \frac{\sum_{i=1}^{K_P}\sum_{j=1}^{K_T}|P_i - T_j|\log\frac{P_i\bigcap T_j}{|P_i|\times|T_j|}}{\max\left(-\sum_{i=1}^{K_P}|P_i|\log\frac{P_i}{n}, -\sum_{j=1}^{K_T}|T_j|\log\frac{T_j}{n}\right)}, \tag{11}$$

The value of NMI is in the range of $[0,1]$, with larger values indicating that the prediction results are closer to the true results.

3 Results and Discussion

3.1 Identification of Spatial Domains in Human DLPFC Dataset

We applied the DeepDomain model on each of the 12 slices of the human DLPFC dataset to test whether the DeepDomain model can identify different spatial domains. Each slice has corresponding annotations for the DLPFC layer. For each slice, we applied it to the DeepDomain model with 500 iterations. After gaining the latent representations, we used the mclust [13] algorithms to further obtain the predicted spatial domain labels.

Based on the existing annotation labels of 12 slices, we calculated the similarity between the predicted labels of the DeepDomain model using the ARI and NMI metrics. The average ARI value for the 12 slices is 0.46, and the average NMI value is 0.61. We then visualized the average ARI and NMI values for the 12 slices (see Fig. 2A).

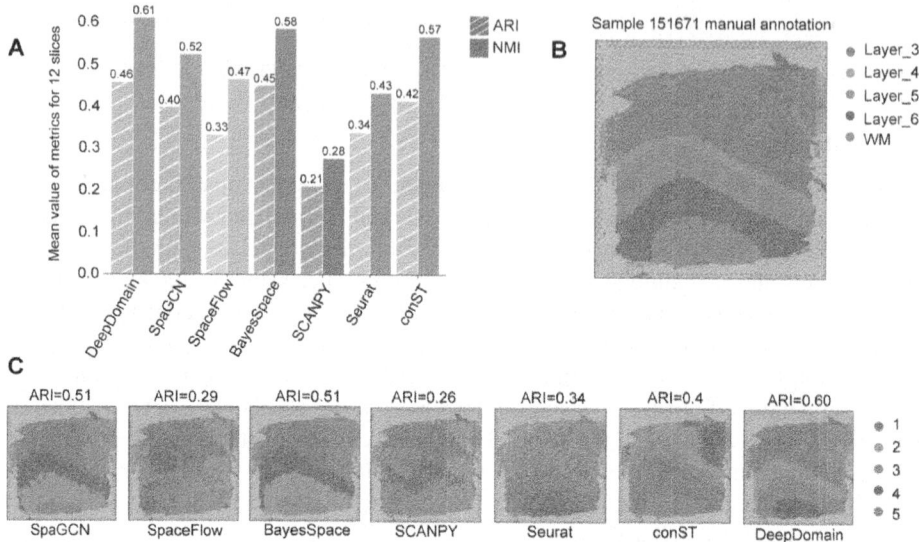

Fig. 2. DeepDomain improves the identification of spatial domains in the human DLPFC dataset. (A) Histogram of the mean accuracy of the 7 methods for spatial domain identification on 12 slices of the DPLFC dataset. (B) Visualization of the manual annotation from the original study of slice 151671. (C) Visualization of the results by the 7 spatial domain identification methods.

To further demonstrate the ability of the DeepDomain to identify spatial domains, we compared the results of the DeepDomain with the results of existing algorithms (see Fig. 2A), including algorithms BayesSpace [11], SCANPY [15], SpaceFlow [10], Seurat [17], conST [9], and SpaGCN [7], each of which uses its default parameters for spatial domain prediction. Compared with the above six methods, the DeepDomain obtained the highest scores of ARI and NMI. Although the ARI score of BayesSpace is close to the DeepDomain, the NMI score is lower than the DeepDomain.

We made a more specific visual comparison of slice 151671. We visualized the manual annotations for slice 151671, where each layer of the slice has a distinct boundary and the slice has 5 layers (see Fig. 2B). We further visualized the predicted results of the seven methods (see Fig. 2C). There are no obvious boundaries in the visualization of the results of SCANPY, Seurat, and SpaceFlow (see Fig. 2C). The three algorithms also have significantly lower ARI scores than the DeepDomain, while the ARI scores of SCANPY, Seurat, and SpaceFlow are 0.26, 0.34, and 0.29, respectively, and the ARI score of DeepDomain is 0.6. Although the SpaGCN, conST, and BayesSpace algorithms are bounded by distinct boundaries, their boundary layering results are not very similar to the manual annotations (see Fig. 2B and Fig. 2C), resulting in their ARI scores being lower than the DeepDomain (0.51 for SpaGCN, 0.4 for conST, 0.51 for BayesSpace).

We recorded the loss values (see Fig. 3) of the DeepDomain in slice 151671 of each iteration and visualized the loss values. Both the gene expression reconstruction loss and the adjacency reconstruction loss decrease with the number of iterations, and eventually, the final loss reaches convergence.

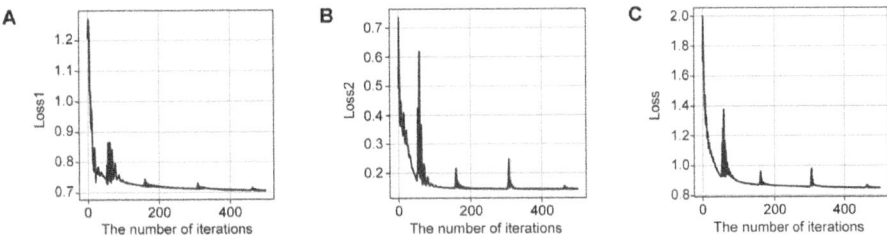

Fig. 3. Visualization of loss values during iterations for the DeepDomain applied to slice 151671. (A) Visualization of the loss values for the spot gene expression reconstruction. (B) Visualization of the loss values for the spatial neighbor network reconstruction. (C) Visualization of the total loss values for two parts of the loss.

Not only the steadily decreasing loss values but also the results of comparison with other existing methods show that the DeepDomain can identify the spatial domain well and has a better performance of higher prediction accuracy compared to existing methods.

3.2 Identification of Spatial Domains in Human Breast Cancer Dataset

In the second case, we used the human breast cancer dataset to further demonstrate the spatial domain identification capability of the DeepDomain. After 500 iterations of training, the reconstruction loss for both gene expression and the adjacency network both decreased steadily to convergence. According to the existing manual annotation, the human breast cancer dataset contains 20 spatial domains [9]. Based on these manually annotated labels, we calculated the accuracy of the DeepDomain prediction results (see Table 1), while the ARI score is 0.53 and the NMI score is 0.69, illustrating the ability of the DeepDomain algorithm to identify different spatial domains.

We also compared the results of the DeepDomain with six existing methods (BayesSpace [11], SCANPY [15], SpaceFlow [10], Seurat [17], conST [9], and STAGATE [8]) and calculated the ARI and NMI scores from the different methods (see Table 1). The ARI score of the DeepDomain was significantly higher than other methods, while the conST gained the poorest ARI score. The NMI score of the DeepDomain was 0.69, which was also higher than other methods, and SCANPY gained the poorest NMI score.

3.3 Identification of Spatial Domains in Mouse Brain Dataset

To illustrate the generality of the application of the DeepDomain to different species, we chose the mouse brain's anterior section dataset to compare the spatial domain identification of different methods on mouse species.

Table 1. The accuracy of the different method results applied to human breast cancer dataset

Methods	ARI	NMI
DeepDomain	0.53	0.69
STAGATE	0.46	0.68
BayesSpace	0.47	0.62
SCANPY	0.43	0.6
SpaceFlow	0.49	0.66

By referring to the annotated labels, we quantified the results of the DeepDomain algorithm versus six other methods (BayesSpace [11], SCANPY [15], SpaceFlow [10], Seurat [17], conST [9], and STAGATE [8]) on the mouse brain's anterior section data, recording their metric scores (see Table 2). We found that DeepDomain gained the highest ARI score among all the compared methods. Although the STAGATE algorithm gained the highest NMI score, the DeepDomain is only 0.01 lower than the NMI score of the STAGATE algorithm. Furthermore, the DeepDomain algorithm has the highest ARI score on this mouse brain dataset, and both accuracy metrics are obviously higher than the STAGATE algorithm on the human breast cancer dataset.

Overall, the results from all three datasets above support that the DeepDomain algorithm outperforms the existing algorithms in the superiority of spatial domain identification of different species.

Table 2. The accuracy of the different method results applied to mouse brain dataset

Methods	ARI	NMI
DeepDomain	0.37	0.72
STAGATE	0.35	0.73
BayesSpace	0.34	0.68
SCANPY	0.29	0.66
SpaceFlow	0.3	0.7

4 Conclusions

There are two important features contained in spatial transcriptome data, including the gene expression of the spot and the spatial location information of the spot, where the spatial location of the spot can be further translated into a spot neighborhood network. In this work, we proposed DeepDomain, an innovative computational approach designed for the analysis of spatial domains, tailored specifically for spatial transcriptome data. By utilizing graph neural networks, DeepDomain retains the characteristics of traditional graph neural networks, being able to encode both gene expression information and spatial location information. By using two different decoders, DeepDomain takes gene expression profile and spatial location network reconstruction into account, being able to more effectively incorporate the importance of neighboring nodes into the spatial neighborhood network, boosting the capability of nodes' latent representation learning, addressing the shortcomings of traditional graph neural network learning tasks. Given the unique characteristics of spatial transcriptome data, the DeepDomain simultaneously considers two important features associated with each spot, including gene expression levels and the topological characteristics within the spatial neighbor network. By minimizing the reconstruction loss of the two aspects of information, DeepDomain achieves low-dimensional feature extraction for each spot. By testing on three distinct spatial transcriptome datasets with manual annotations, DeepDomain accurately identifies spatial domains within tissue slices and outperforms existing methods.

Acknowledgments. The work was supported by the National Natural Science Foundation of China (No. 62131004), the National Key R&D Program of China (2022ZD0117700).

References

1. Rao, A., Barkley, D., Franca, G.S., Yanai, I.: Exploring tissue architecture using spatial transcriptomics. Nature **596**(7871), 211–220 (2021). https://doi.org/10.1038/s41586-021-03634-9

2. Wang, X., et al.: Three-dimensional intact-tissue sequencing of single-cell transcriptional states. Science (New York, NY) **361**(6400) (2018). https://doi.org/10.1126/science.aat5691

3. Baron M., et al.: The stress-like cancer cell state is a consistent component of tumorigenesis. Cell Syst. **11**(5), 536–46.e7 (2020). https://doi.org/10.1016/j.cels.2020.08.018

4. Moncada, R., et al.: Integrating microarray-based spatial transcriptomics and single-cell RNA-seq reveals tissue architecture in pancreatic ductal adenocarcinomas. Nature Biotechnol. **38**(3), 333–42 (2020). https://doi.org/10.1038/s41587-019-0392-8

5. Chen, J., Xu, H., Tao, W., Chen, Z., Zhao, Y., Han, J.J.: Transformer for one stop interpretable cell type annotation. Nat. Commun. **14**(1), 223 (2023). https://doi.org/10.1038/s41467-023-35923-4

6. Li, X., et al.: Deep learning enables accurate clustering with batch effect removal in single-cell RNA-seq analysis. Nat. Commun. **11**(1), 2338 (2020). https://doi.org/10.1038/s41467-020-15851-3

7. Hu, J., et al.: SpaGCN: integrating gene expression, spatial location and histology to identify spatial domains and spatially variable genes by graph convolutional network. Nat. Methods **18**(11), 1342–1351 (2021). https://doi.org/10.1038/s41592-021-01255-8

8. Dong, K., Zhang, S.: Deciphering spatial domains from spatially resolved transcriptomics with an adaptive graph attention auto-encoder. Nat. Commun. **13**(1), 1739 (2022). https://doi.org/10.1038/s41467-022-29439-6

9. Yongshuo, Z., Tingyang, Y., Xuesong, W., Yixuan, W., Zhihang, H, Yu L.: conST: an interpretable multi-modal contrastive learning framework for spatial transcriptomics. bioRxiv. 2022.01.14.476408 (2022). https://doi.org/10.1101/2022.01.14.476408

10. Ren, H., Walker, B.L., Cang, Z., Nie, Q.: Identifying multicellular spatiotemporal organization of cells with SpaceFlow. Nat. Commun. **13**(1), 4076 (2022). https://doi.org/10.1038/s41467-022-31739-w

11. Zhao, E., Stone, M.R., Ren, X., Guenthoer, J., Smythe, K.S., Pulliam, T., et al.: Spatial transcriptomics at subspot resolution with BayesSpace. Nat. Biotechnol. **39**(11), 1375–1384 (2021). https://doi.org/10.1038/s41587-021-00935-2
12. Velickovic, P., Cucurull, G., Casanova, A., Romero, A., Lio, P., Bengio, Y.J.S.: Graph Attention Netw. **1050**(20), 10–48550 (2017)
13. Scrucca, L., Fop, M., Murphy, T.B., Raftery, A.E.: Mclust 5: clustering, classification and density estimation using gaussian finite mixture models. R J. **8**(1), 289–317 (2016)
14. Maynard, K.R., et al.: Transcriptome-scale spatial gene expression in the human dorsolateral prefrontal cortex. Nat. Neurosci. **24**(3), 425–436 (2021). https://doi.org/10.1038/s41593-020-00787-0
15. Wolf, F.A., Angerer, P., Theis, F.J.: SCANPY: large-scale single-cell gene expression data analysis. Genome Biol. **19**(1), 15 (2018). https://doi.org/10.1186/s13059-017-1382-0
16. Pedregosa, F., et al.: Scikit-learn: machine learning in Python. J. Mach. Learn. Res. **12**, 2825–2830 (2011). https://doi.org/10.48550/arXiv.1201.0490
17. Satija, R., Farrell, J.A., Gennert, D., Schier, A.F., Regev, A.: Spatial reconstruction of single-cell gene expression data. Nat. Biotechnol. **33**(5), 495–502 (2015). https://doi.org/10.1038/nbt.3192

stMCFN: A Multi-view Contrastive Fusion Method for Spatial Domain Identification in Spatial Transcriptomics

Jing Jing[1] (ID), Ying-Lian Gao[2], Yue Gao[1], Dao-Hui Ge[1], Chun-Hou Zheng[1], and Jin-Xing Liu[1,3](✉) (ID)

[1] School of Computer Science, Qufu Normal University, Rizhao 276826, Shandong, China
sdcavell@126.com
[2] Qufu Normal University Library, Qufu Normal University, Rizhao 276826, Shandong, China
[3] School of Health and Life Sciences, University of Health and Rehabilitation Sciences, Qingdao 266113, Shandong, China

Abstract. Rapid advances in spatial transcriptomics allow for sequencing gene expression profiles while preserving spatial information. Many spatial clustering methods based on graph neural networks have been proposed, but they often cannot adaptively learn the complex relationships between gene expression and spatial information, which makes it challenging to identify spatial domains effectively. In this paper, we propose stMCFN, a multi-view contrastive fusion network based on graph autoencoders for spatial domain identification. To fully mine the potential structural information in spatial transcribed data, multiple views are constructed using gene expression and spatial information for self-supervised contrastive learning of graph autoencoders without data enhancement. This method generates shared embedding from different views and obtains clustering-friendly feature representation through an attention-based feature fusion module to adaptively integrate embedding from different views. Finally, the experimental results on the human dorsolateral prefrontal cortex dataset and the mouse brain anterior dataset show that stMCFN's ability to identify spatial domains is significantly superior to other state-of-the-art methods. The code is available at https://github.com/JING-ING/stMCFN.

Keywords: Spatial Transcriptomics · Contrastive Learning · Multi-view Graph · Spatial Domain Identification · Clustering

1 Introduction

Single-cell RNA sequencing (scRNA-seq) has greatly promoted the study of cell function in tissues [1, 2], but lacks spatial context. Recently, spatial transcriptomics (ST) technology [3] fills this gap by combining gene expression with spatial position information, enhancing our understanding of cell interactions and tissue architecture [4]. Various ST techniques, such as seqFISH + [5], MERFISH [6], Slide-seqV2 [7], 10x Visium [4], and Stereo-seq [8], have been developed to provide the spatial location for detailed analysis of tissue microstructure.

© The Author(s), under exclusive license to Springer Nature Singapore Pte Ltd. 2024
D.-S. Huang et al. (Eds.): ICIC 2024, LNBI 14881, pp. 321–331, 2024.
https://doi.org/10.1007/978-981-97-5689-6_28

Identifying spatial domains is crucial in ST data analysis, which is essentially a clustering task based on gene expression and spatial location information. Early methods like k-means [9] and Louvain [10] algorithm used only gene expression data, neglecting spatial information. Recent methods improve spatial domain identification by integrating both types of information. For example, stLearn [11] combines morphological features from histology images with gene expression to cluster similar spots, while DeepST [12] uses denoising and variational graph autoencoders to enhance latent representations. STAGATE [13] employs a graph attention autoencoder to aggregate information from neighboring spots for more comprehensive identification. Spatial-MGCN [14] uses graph convolutional network (GCN) encoders to extract and merge embeddings from feature and spatial graphs for identifying spatial domains. GraphST [15] employs graph neural networks with self-supervised contrastive learning to cluster spatial spots. However, many contrastive learning-based methods generate the corrupted graph by perturbing features and removing edges, relying heavily on data augmentation. ConSpaS [16] integrates local and global similarities to decipher the spatial domain by integrating graph autoencoders and augmentation-free contrastive learning. Although these methods achieve good performance, there are still some limitations. They often use a single similarity measure for constructing adjacency matrices based on spatial location, which may not fully capture the data's structural information. Additionally, many techniques rely on basic concatenation or stacking operations to integrate diverse information, which can introduce noise and reduce the effectiveness of subsequent analyses.

To solve the mentioned challenges, we propose a multi-view contrastive fusion method called stMCFN, which is used to identify spatial domains. Distinct graph structures are constructed according to the similarity of gene expression and spatial position respectively, and combined with gene expression to form multiple views. The self-supervised contrastive learning method can make spatially adjacent spots have similar representations [17]. Therefore, the consistent features are then learned from the complementary information of different views through a multi-view contrastive learning model without data augmentation, which makes the representation more comprehensive and discriminative. In addition, we construct a feature fusion module to extract and adaptively fuse the embedding representation from different views, which makes the final representation more robust for clustering tasks. Experiments on datasets from the 10x Visium platform show that stMCFN outperforms other state-of-the-art methods in spatial clustering.

2 Method

2.1 Overview of stMCFN

The overall framework of stMCFN is shown in Fig. 1. stMCFN first constructs two graph structures based on gene expression and spatial location respectively through different similarity measures, which form two views with gene expression. Next, the graph autoencoder aims to learn each view-specific spot representation by aggregating neighbor representations. Meanwhile, the model captures consistent embedding representations from two different views through a contrastive learning module without data augmentation. Furthermore, considering the importance of different embedding sub-spaces, we employ a feature fusion module based on multi-head attention to adaptively aggregate the embeddings learned from different views. The latent embedding is optimized by jointly minimizing reconstruction loss and contrastive learning loss, and finally clustered by performing k-means.

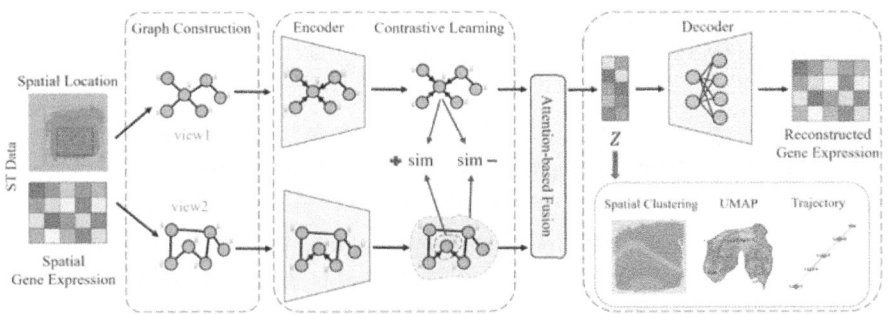

Fig. 1. The overall framework of the stMCFN method. (Note that the green and pink circles indicate positive and negative samples paired with a given spot, respectively.)

2.2 Graph Construction

In general, a graph $G(X, A)$ in the graph network consists of a feature matrix representing the attributes of nodes and an adjacency matrix representing the structure of the graph. For most spatial clustering methods, the feature matrix is the normalized gene expression matrix X, while the adjacency matrix is usually constructed by Euclidean distance. In this study, to make full use of the spatial location information and gene expression information of the ST data, we construct graph structures (adjacency matrix) from two different perspectives.

Firstly, based on the spatial position of the organization, the spatial adjacency matrix $A^{(1)}$ is constructed by measuring the similarity between spots through Euclidean distance. If the distance d between spot i and spot j is less than the pre-defined radius r, then spot i is the neighbor of spot j, that is, $A_{ij}^{(1)} = A_{ji}^{(1)} = 1$; otherwise 0.

From another perspective, to fully explore the potential structural information of gene expression data itself, the k-nearest-neighbor (KNN) was used to construct the feature adjacency matrix $A^{(2)}$ according to the cosine similarity of gene expression. Specifically, for a given spot, the first k spots are selected as neighbors by sorting the computed cosine similarity. $A_{ij}^{(2)} = A_{ji}^{(2)} = 1$ if the spot i is the neighbor of spot j, otherwise it is 0.

Finally, two adjacency matrices are given to capture different similarity structures between spots. They and the gene expression matrix are used as inputs of the model.

2.3   Multi-view Contrastive Learning Network

Graph Autoencoder

The proposed model is built using an autoencoder (encoder and decoder). We use a GCN as the encoder, which can map the graph to the low-dimensional latent space H while preserving the feature information and topological structure of the spots. For the encoder, a separate graph convolution operation is performed on each view, which consists of two convolutional layers that iteratively aggregate and update spot representations through propagation rules. We use adjacency matrices $\tilde{A}^{(m)}$ constructed by two similarity measures and a gene expression matrix as input. The l th convolutional layer can be expressed as follows:

$$H_{(l+1)}^{(m)} = GCN\left(H_{(l)}^{(m)}, A^{(m)}\right) = \sigma\left(\tilde{D}^{(m)-\frac{1}{2}} \tilde{A}^{(m)} \tilde{D}^{(m)-\frac{1}{2}} H_{(l)}^{(m)} W_{(l)}^{(m)}\right), \qquad (1)$$

where $\tilde{D}^{(m)}$ represents the corresponding degree matrix. $W^{(m)}$ is the trainable weight parameter. $\sigma(\cdot)$ is the nonlinear activation function ReLU. $H_{(0)}^{(m)}$ is the preprocessed gene expression matrix X. $H_{(l)}^{(m)}$ and $H_{(l+1)}^{(m)}$ denote the input and output of the l th layer of the GCN. $H^{(m)}$ is the potential representation embedding vector learned from different views, in which $m = 1, 2$.

Multi-view Contrastive Learning

To learn valuable information from different views to achieve unified feature representation, we use contrastive learning to optimize spot embedding without data augmentation. The contrastive learning learns discriminative representations by comparing positive samples with negative samples [17]. Specifically, for the embedding of a given spot in the current view, the embedding of the same spot in another view is considered a positive sample, while the rest of the embeddings of spots are considered negative samples. Through this way, we pull the same embedding spot in different views close and push the different embedding spots away. The multi-view contrastive learning loss can be defined as follows:

$$L_{con} = -\log \frac{\exp(sim(h_i^{(1)}, h_i^{(2)}) \big/ \tau)}{\sum_{k=1}^{N} \exp(sim(h_i^{(1)}, h_k^{(2)}) \big/ \tau)}, \qquad (2)$$

where $sim(\cdot)$ is the cosine similarity metric and N is the number of spots. τ is a temperature parameter. The final contrastive loss is expressed as the average of the contrastive loss for all spots:

$$L_{con} = \frac{1}{N} \sum_{i=1}^{N} -\log \frac{\exp(sim(h_i^{(1)}, h_i^{(2)}) \big/ \tau)}{\sum_{k=1}^{N} \exp(sim(h_i^{(1)}, h_k^{(2)}) \big/ \tau)}. \qquad (3)$$

2.4 Attention-Based Feature Fusion Module

Although more discriminative embedding representations can be learned through the above process, it does not ensure that features from different views are equally important for clustering tasks. Inspired by natural language processing [18], we construct a feature fusion module based on the multi-head attention mechanism. This module learns the weights of embeddings from different views, allowing for their adaptive fusion into a unified representation. This can be formalized as follows:

$$Q_h = W_h^q H^{(2)}, K_h = W_h^k H^{(1)}, V_h = W_h^v H^{(1)}, \tag{4}$$

$$\alpha_h = softmax\left(Q_h \cdot K_h^T\right), \tag{5}$$

$$Z = W \cdot \|_{h=1}^{H}(\alpha_h \cdot V_h) + H^{(1)}, \tag{6}$$

where the query Q_h is calculated by embedding $H^{(1)}$ and h represents the number of heads. The key K_h and value V_h are calculated by embedding $H^{(2)}$. α_h represents the attention score, which is calculated by the dot product operation. W_h^q, W_h^k, W_h^v and W all represent learnable weight parameters. $\|$ represent the concatenation operation, which is used to concatenate output representations from different heads. The final embedding representation Z is used as input to the decoder and clustering tasks.

2.5 Decoder

The decoder of the model adopts a two-layer linear decoder whose dimension is symmetric with the GCN encoder. The fused potential embedding Z is decoded into a reconstructed feature matrix $\hat{X} = D(Z)$. The reconstruction loss aims to minimize the error between the reconstructed gene expression matrix and the input gene expression matrix by the mean squared error loss function, which is defined as follows:

$$L_{re} = \|X - \hat{X}\|_F^2. \tag{7}$$

2.6 Overall Loss Function

To sum up, the reconstruction loss and contrastive learning loss are combined to form the final loss function, which is used to optimize spot embedding and can be expressed as follows:

$$L_{total} = L_{re} + \alpha L_{con}, \tag{8}$$

where α is a hyperparameter used to balance the reconstruction loss and contrastive learning loss.

3   Experiments and Results

3.1 Datasets and Evaluation Metrics

To effectively evaluate the clustering performance of stMCFN, we collect the LIBD human dorsolateral prefrontal cortex (DLPFC) dataset [19] from the 10x Visium platform as the benchmark dataset. This dataset contains 12 tissue slices, each with up to six cortical layers and a white matter layer, all of which have been manually annotated by Maynard et al. [20]. Each slice contains 3460 to 4789 spots, along with 33538 genes. In addition, the MBA dataset containing 2695 spots and 32285 genes was collected and divided into 52 distinct regions by manual annotation.

For all data samples, we first filtered genes expressed in fewer than 100 cells. Next, the top 3000 highly variable genes were screened out using the Seurat algorithm [21] and retained for subsequent analysis. Following this, SCANPY [22] was employed to normalize and scale the filtered gene expression count matrix to reduce the influence of outlier values on the data. Finally, the preprocessed data served as the input of the stMCFN model.

We use the two most representative evaluation metrics to evaluate the clustering performance of stMCFN and other spatial clustering methods. These two indicators are the adjusted rand index (ARI) [23] and the normalized mutual information (NMI) [24].

The ARI can be used to measure the agreement between the results of two clusters, ranging from -1 to 1, and it can be calculated as:

$$ARI = \frac{\mathbf{max}(RI) - E(RI)}{RI - E(RI)}, \tag{9}$$

where RI is the unadjusted Rand index and $E(RI)$ represents the average expected value of RI when clustering is completely randomly assigned.

The NMI is used to measure the similarity between the true labels and clustering results, which can be represented as:

$$NMI = \frac{MI(P,T)}{\sqrt{H(P)H(T)}}, \tag{10}$$

where T represents the real label of the spot, P represents the result of clustering, $MI(P,T)$ represents the mutual information between P and T, and $H(\cdot)$ denotes the entropy function. The value range of NMI is [0,1], with higher values representing better clustering performance.

3.2 Experimental Implementation

We implemented the stMCFN method and other comparison methods on an NVIDIA RTX 2080 GPU by PyTorch. The Adam optimizer is used to optimize the algorithm with a learning rate of 0.001. The input dimension of the autoencoder model is 3000, while the dimensions of its hidden layer are 128 and 64, respectively. The important parameters of the stMCFN model are the weight coefficient α and the temperature parameter τ of the contrastive learning loss, and the number of heads of multi-head attention h. We performed parametric analysis experiments to find optimal parameters within a specific range, where $\alpha \in \{0.1, 0.2,..., 0.9\}$, $\tau \in \{0.1, 0.2,..., 0.9, 1\}$ and $h \in \{1, 2, 4, 6, 8, 10\}$. Finally, $\alpha = 0.5$, $\tau = 1$, and $h = 8$ are determined to be the optimal hyperparameters. According to our experience, the nearest neighbor number k and radius r used in constructing different neighbor matrices are set to 14 and 450, respectively. In the experiments, clustering was performed using the k-means algorithm. Finally, the model is trained until clustering performance stops improving within a maximum of 200 epochs. In this study, for mouse brain fine-grained tissue, τ is set to 0.5, and gene filtering operation is not performed in the preprocessing stage.

3.3 Experimental Results and Analysis

We compared the clustering performance of stMCFN with six state-of-the-art spatial clustering methods, including stLearn, STAGATE, DeepST, GraphST, Spatial-MGCN, and ConSpaS, applied to the DLPFC dataset. Figure 2 shows that stMCFN achieved the best overall clustering performance across 12 slices of the DLPFC dataset, with a median ARI of 0.614 and a median NMI of 0.712. Notably, stMCFN reached an ARI of 0.851 on slice 151672.

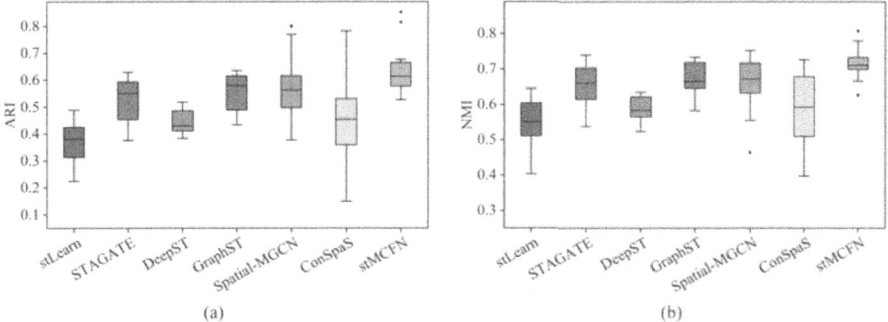

Fig. 2. Comparison of clustering performance between baseline methods and stMCFN on 12 slices of the DLPFC dataset.

Furthermore, taking the spatial domain visualization results of slice 151672 in Fig. 3 as an example, STAGATE, DeepST, and GraphST can only accurately identify the regions of WM, Layer 6, and Layer 5, while ConSpaS and stMCFN can clearly distinguish Layers 3 and 4. However, the spatial domain identification by stMCFN embeddings is closer to the manually annotated spatial domain. These findings indicate that stMCFN outperforms other methods in extracting and integrating feature representations from multi-view graph data, effectively promoting spatial clustering tasks.

Further, to verify the effectiveness of stMCFN in identifying complex tissue structures, it was applied to the MBA dataset. As shown in Table 1, the ARI value of stMCFN is significantly better than other methods, at 0.431. Although the NMI is slightly lower than that of STAGATAE, at 0.713. However, stMCFN can clearly identify continuous cerebral cortex (CTX) regions in Fig. 4, closely matching the actual annotations. These results demonstrate the effectiveness of stMCFN in identifying fine-grained spatial domains.

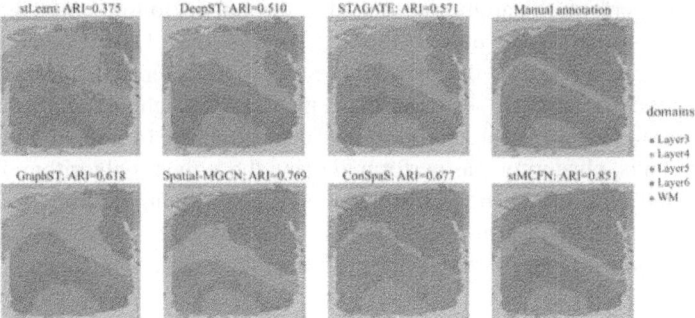

Fig. 3. Results of manual annotation and spatial domain identification for all methods and on slice 151672.

Table 1. Comparison of clustering performance on MBA dataset.

Method	ARI	NMI
STAGATE	0.372	0.721
GraphST	0.357	0.688
stMCFN	**0.431**	**0.713**

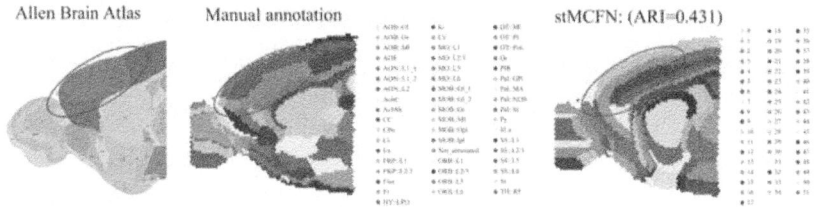

Fig. 4. The corresponding Allen Reference Atlas of the MBA dataset (left), manual annotation results and spatial domain identified by stMCFN (right).

4   Ablation Experiment

To evaluate the effectiveness of each component of the stMCFN model, we conducted ablation experiments, as shown in Fig. 5 (a). We evaluated three variants with individual component removals—reconstruction loss, contrastive learning loss, and attention fusion module. The fourth variant lacked both contrastive learning loss and the attention fusion module, the latter replaced by simple stacking operation. The clustering performance is the worst when the contrastive learning and attention fusion modules are removed simultaneously, which highlights its importance in feature learning. Obviously, all components work together to enable stMCFN to achieve optimal clustering results.

Further, to verify the benefits of multi-view fusion, we compared single-type and multi-type graph structures on 12 slices of DLPFC. Specifically, view1 + view1 represents both using the spatial adjacency graph, view2 + view2 represents both using the feature adjacency graph, and view1 + view2 combines the two graph structures. The results in Fig. 5 (b) show that although the method using only the spatial adjacency graph performs well, clustering performance can be enhanced by using multiple graph structures, which indicates that a more comprehensive feature representation can be learned through multiple graph structures.

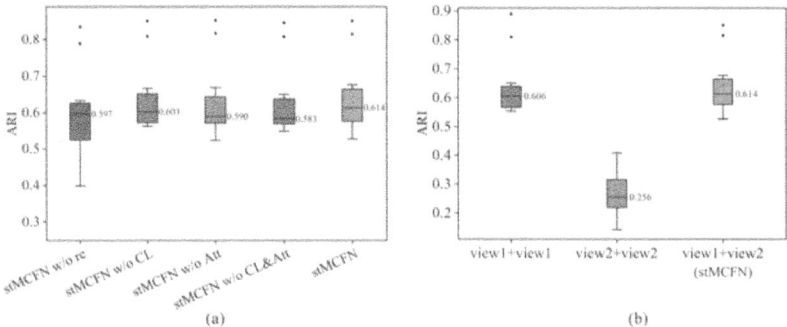

(a)   (b)

Fig. 5. Performance comparison of various variants on 12 slides of the DLPFC dataset.

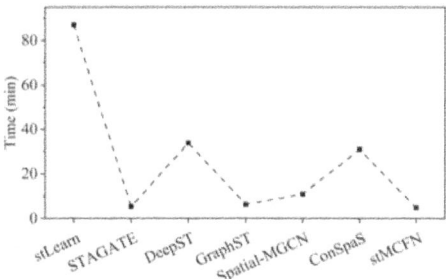

Fig. 6. The total running time (in minutes) of different methods on 12 slides of the DLPFC dataset.

Finally, we also compared the total running time of stLearn, STAGATE, DeepST, GraphST, Spatial-MGCN, ConSpaS, and stMCFN on 12 slices. These methods are all executed on the GPU. It can be seen from Fig. 6 that our method and STAGATE take the shortest time and have higher operating efficiency.

5 Conclusion

The spatial clustering task is the most basic and important process for analyzing ST data. In this paper, a multi-view contrastive learning fusion network called stMCFN is proposed, which can combine multiple views to extract key features for spatial domain identification of ST data. The strength of stMCFN lies in its ability to perform contrastive learning without data augmentation operations like shuffling. In addition, we also introduce an attention mechanism to fuse the features of the encoder output from the multi-view contrastive network. This approach adjusts the significance of the potential representations from different views based on the attention coefficients, thereby enhancing the effectiveness of potential representations.

To assess the performance of stMCFN, we compare it with several state-of-the-art spatial clustering methods on the widely used DLPFC and MBA datasets. Our experimental results show that stMCFN achieves the highest ARI scores and best performance in most slices. In summary, stMCFN shows excellent ability in spatial clustering tasks on different tissues. However, this paper only uses the 10x Visuim dataset, and we plan to explore its application on high-resolution ST datasets in the future.

Acknowledgments. This work was supported in part by the National Natural Science Foundation of China under Grant No. 62172254.

References

1. Kolodziejczyk, A.A., Kim, J.K., Svensson, V., et al.: The technology and biology of single-cell RNA sequencing. Mol. Cell **58**(4), 610–620 (2015)
2. Shapiro, E., Biezuner, T., Linnarsson, S.: Single-cell sequencing-based technologies will revolutionize whole-organism science. Nat. Rev. Genet. **14**(9), 618–630 (2013)
3. Marx, V.: Method of the Year: spatially resolved transcriptomics. Nat. Methods **18**(1), 9–14 (2021)
4. Ståhl, P.L., Salmén, F., Vickovic, S., et al.: Visualization and analysis of gene expression in tissue sections by spatial transcriptomics. Science **353**(6294), 78–82 (2016)
5. Lubeck, E., Coskun, A.F., Zhiyentayev, T., et al.: Single-cell in situ RNA profiling by sequential hybridization. Nat. Methods **11**(4), 360–361 (2014)
6. Chen, K. H., Boettiger, A. N., Moffitt, J. R., et al.: Spatially resolved, highly multiplexed RNA profiling in single cells. Science **348**(6233), aaa6090 (2015)
7. Stickels, R.R., Murray, E., Kumar, P., et al.: Highly sensitive spatial transcriptomics at near-cellular resolution with Slide-seqV2. Nat. Biotechnol. **39**(3), 313–319 (2021)
8. Chen, A., Liao, S., Cheng, M., et al.: Spatiotemporal transcriptomic atlas of mouse organogenesis using DNA nanoball-patterned arrays. Cell **185**(10), 1777–1792. e1721 (2022)
9. Arthur, D., Vassilvitskii, S.: K-means++ the advantages of careful seeding. In: Proceedings of the Eighteenth Annual ACM-SIAM Symposium on Discrete Algorithms, pp. 1027–1035 (2007)
10. Blondel, V.D., Guillaume, J.-L., Lambiotte, R., et al.: Fast unfolding of communities in large networks. J. Stat. Mech: Theory Exp. **2008**(10), P10008 (2008)
11. Pham, D., Tan, X., Xu, J., et al.: stLearn: integrating spatial location, tissue morphology and gene expression to find cell types, cell-cell interactions and spatial trajectories within undissociated tissues. bioRxiv, 2020.2005. 2031.125658 (2020)
12. Xu, C., Jin, X., Wei, S., et al.: DeepST: identifying spatial domains in spatial transcriptomics by deep learning. Nucleic Acids Res. **50**(22), e131–e131 (2022)
13. Dong, K., Zhang, S.: Deciphering spatial domains from spatially resolved transcriptomics with an adaptive graph attention auto-encoder. Nat. Commun. **13**(1), 1739 (2022)

14. Wang, B., Luo, J., Liu, Y., et al.: Spatial-MGCN: a novel multi-view graph convolutional network for identifying spatial domains with attention mechanism. Briefings Bioinform. **24**(5), bbad262 (2023)
15. Long, Y., Ang, K.S., Li, M., et al.: Spatially informed clustering, integration, and deconvolution of spatial transcriptomics with GraphST. Nat. Commun. **14**(1), 1155 (2023)
16. Wu, S., Qiu, Y., Cheng, X.: ConSpaS: a contrastive learning framework for identifying spatial domains by integrating local and global similarities. Briefings Bioinform. **24**(6), bbad395 (2023)
17. You, Y., Chen, T., Sui, Y., et al.: Graph contrastive learning with augmentations. In: Advances in neural information processing systems, pp. 5812–5823. (2020)
18. Vaswani, A., Shazeer, N., Parmar, N., et al.: Attention is all you need. Adv. Neural Inform. Process. Syst. **30**, (2017)
19. Pardo, B., Spangler, A., Weber, L.M., et al.: SpatialLIBD: an R/Bioconductor package to visualize spatially-resolved transcriptomics data. BMC Genomics **23**(1), 434 (2022)
20. Maynard, K.R., Collado-Torres, L., Weber, L.M., et al.: Transcriptome-scale spatial gene expression in the human dorsolateral prefrontal cortex. Nat. Neurosci. **24**(3), 425–436 (2021)
21. Satija, R., Farrell, J.A., Gennert, D., et al.: Spatial reconstruction of single-cell gene expression data. Nat. Biotechnol. **33**(5), 495–502 (2015)
22. Wolf, F.A., Angerer, P., Theis, F.J.: SCANPY: large-scale single-cell gene expression data analysis. Genome Biol. **19**, 1–5 (2018)
23. Hubert, L., Arabie, P.: Comparing partitions. Jour. Classifi. **2**, 193–218 (1985)
24. Strehl, A., Ghosh, J.: Cluster ensembles---a knowledge reuse framework for combining multiple partitions. J. Mach. Learn. Res. **3**, 583–617 (2002)

Pattern Recognition

3D Partial U-Net: A Lightweight ConvNet for Head and Neck Lymph Node Segmentation

Fei Wu[1], Hao Chen[1], Quan Li[2(✉)], and Tao Peng[1(✉)]

[1] School of Future Science and Engineering, Soochow University, Suzhou, China
sdpengtao401@gmail.com
[2] Center of Stomatology, The Second Affiliated Hospital of Soochow University,
Suzhou 215004, China
lq2009@suda.edu.cn

Abstract. Accurate lymph node (LN) segmentation plays a crucial role in tumor diagnosis and treatment for head and neck cancer patients. Automatic LN segmentation remains challenging due to large size variation, extracapsular extension, and similar appearance to surrounding vessels on computed tomography (CT). The morphology and volume of LNs are crucial for diagnosis, while the current 3D convolution-based methods are well used for boundary extraction. However, existing volumetric convolutional neural networks (ConvNets) are parameter-heavy, computationally complex and slow to use when dealing with 3D medical images. To this end, we introduce a lightweight volumetric ConvNet termed 3D Partial U-Net to facilitate LN segmentation applications on head and neck cancer patients. Our approach introduces a combination of volumetric partial convolution and pointwise convolution, which significantly reduces the parameter count and computational complexity. A light boundary refinement output module with a large kernel depth-wise separable convolution placed at the end of our model is proposed to enhance the precision of LN segmentation. Compared to the symmetric U-Net, our model features a larger encoder and a smaller decoder, empowering it with enhanced capability to learn complex LN features. We evaluate the model performances on 678 LNs from 123 head and neck cancer patients. The results of experiments showcase that 3D Partial U-Net surpasses existing approaches with the fewest floating-point operations (FLOPs) and parameters, achieving state-of-the-art. The source code with our proposed 3D Partial U-Net is available at https://github.com/cii030/3D-Partial-U-Net.

Keywords: Head and Neck Lymph Node · 3D Medical Image Segmentation · Lightweight

1 Introduction

Head and neck (H&N) cancer ranks as the sixth most ordinary form of malignancy globally [1], with cervical lymph node (LN) involvement in these patients increasing the risk of distant metastases and death [2]. Therefore, the accurate segmentation of LNs is crucial in clinical radiotherapy treatment, aiming to enhance the probability of

D.-S. Huang et al. (Eds.): ICIC 2024, LNBI 14881, pp. 335–346, 2024.
https://doi.org/10.1007/978-981-97-5689-6_29

successful cure while mitigating toxicity for H&N cancer patients [3]. However, manual annotation of LNs highly demands for labor and consumes a great deal of time, with experts spending approximately 50 min per patient case [4]. This increases the waiting time for patients and restricts the frequency of adjustments to radiotherapy treatment plans. Therefore, there is a significant demand for automated segmentation of LNs. Computed tomography (CT) scanning has been a vital imaging modality for diagnosing head and neck cancer due to its rich informational content and minimal patient risk. [5] However, this task is challenging as LNs can appear similar to surrounding tissues and have variable morphology in CT (as shown in Fig. 1).

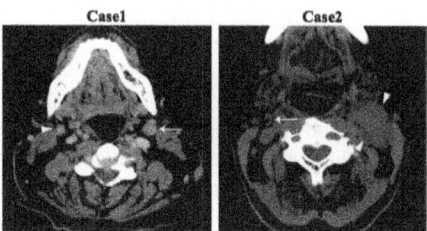

Fig. 1. Example of two head and neck CT images [6], containing LNs of various shapes and sizes. In case1, the arrow points to an enlarged rounded LN and the arrowhead points to a small node that exhibits a rounded shape with irregular margins. In case2, the arrowheads points to a large conglomerate nodal mass and the arrow points to a sub-centimeter-sized necrotic LN.

Recently, the advancement of deep learning methods has achieved significant success in automating the segmentation of organs at risk (OAR) [7]. In particular, U-Net [8], with its distinctive encoder-decoder architecture and skip connection, can effectively utilize a limited number of supervised datasets to extract detailed features, thereby achieving excellent segmentation performance, therefore achieves great success in medical image segmentation. Li et al. [9] introduced MIUNet to develop a domain-adaptive vascular segmentation algorithm. It demonstrated robustness to etiological noise and domain shifts between different datasets. Furthermore, they proposed the FreeSDG [10] to address the problem of annotation scarcity in medical image segmentation.

Meanwhile, 3D variants of U-Net are widely applied in LN segmentation tasks. The morphology and volume of LNs are crucial for diagnosis, requiring the use of 3D convolution for feature extraction. Oda et al. [11] utilized 3D U-Net [12] for segmenting lymph nodes and other anatomical structures in enhanced chest CT scans. Strijbis et al. [13] combined multi-view voxel classification networks with 3D U-Net to achieve precise segmentation of individual LNs, and the planning target volumes were extrapolated. Xu et al. [14] introduced a deep dilated convolutional encoder-decoder architecture using a cosine-sine loss function for LN segmentation on positron emission tomography (PET)/CT images to address the challenge of class imbalance in LN segmentation. Chen et al. [2] proposed an attention-guided classification scheme that simultaneously performs LN segmentation and classification, where region of interest (ROI) containing the LNs and the surrounding tissues served as the input, yielding promising results. Yet the models in studies as mentioned earlier encountered the following issue: they result in significant computational costs and memory consumption, particularly in terms

of graphics processing unit (GPU) memory usage, due to the 3D convolution operation. Motivated by this issue, we aim to explore a lightweight model with excellent
segmentation performance, making it more feasible for practical applications in mobile
devices.

Recently, Chen et al. [15] reexamined popular operators to increase floating-point
operations per second (FLOPS) and designed FasterNet, achieving significantly faster
processing speeds than other methods across various devices, while maintaining accuracy
in diverse 2D vision tasks. Liu et al. [16] reflected the design spaces and tested the
limits of what a pure ConvNet can achieve. Consequently, they designed ConvNeXt and
demonstrated that pure ConvNets can achieve comparable or even superior performance
to Swin Transformers [17] while retaining the simplicity and efficiency of standard
Convolutional Neural Networks (ConvNets). These works provide a solid foundation
for us to design a lightweight yet high-precision model.

In this work, we introduce a lightweight volumetric ConvNet referred to as 3D Partial
U-Net for accurate cervical LN segmentation. The first module is the volumetric partial
convolution (VPConv), which performs 3D convolution operations only on a subset of
channels, thereby reducing computation and parameter count. Then, we custom-design
a modern convolutional block, wherein VPConv is followed by two pointwise convolutions to expand the receptive field. This design employs fewer activation functions and
normalization layers. We design a light boundary refinement output module (LBROM),
which first employs a transpose convolution to upsample to the desired number of channels for the output and utilizes a combination of a large-kernel depth-wise convolution
and two pointwise convolutions, along with a residual connection. This module refines
the edges further before the output, effectively enhancing the segmentation results.

2 Methods

2.1 Network Design

We propose 3D Partial U-Net, a lightweight network architecture with an asymmetric
encoder-decoder structure as illustrated in Fig. 2. The encoder consists of three layers,
composed of 1, 2, and 2 convolution blocks proposed by us, called 3D Partial U-Net
Block, while the bottleneck comprises four blocks, and the decoder has three layers,
each consisting of a single block. At the beginning of the model, we put a stem cell,
which is convolutional layer of $1 \times 1 \times 1$ used to downsample the input images into
an appropriate feature map size. Unlike the concatenation in skip connection of U-Net
[8], we opt for the use of addition, significantly reducing computational complexity. The
downsampling stage employs batch normalization (BN) followed by a convolution with
$2 \times 2 \times 2$ kernel size and 2 stride size, while the upsampling stage utilizes BN followed
by a transposed convolution with $2 \times 2 \times 2$ kernel size and 2 stride size. This approach
significantly reduces information loss. At the end of the model, we place the LBROM
to reduce the channel count in order to refine the edges further. The channel counts for
each layer are represented as hyperparameters, labeled from C_1 to C_4. In experiments
involving the 3D Partial U-Net, we set $C_1=32$, $C_2=64$, $C_3=128$, and $C_4=256$, unless
otherwise stated.

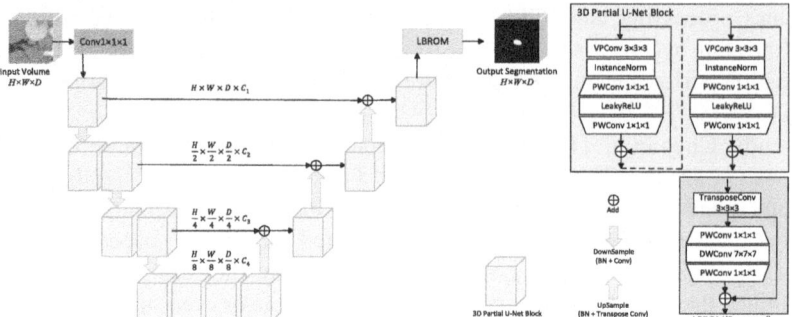

Fig. 2. Overview architecture of our proposed 3D Partial U-Net.

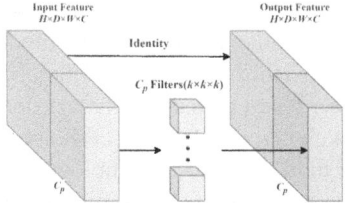

Fig. 3. Illustration of VPConv: Only (c_p/c) of the channels undergo convolution, while the rest remain unchanged. In our work, c_p equals to one quarter of c.

2.2 Volumetric Partial Convolution

Inspired by Reference [15], the volumetric partial convolution (VPConv) used in our convolutional block is designed to reduce both memory access and computational redundancy (Fig. 3). The VPConv utilizes 3D convolution to extract spatial features from selected input channels, capturing crucial patterns and structures in the data. For the remaining channels, they remain unchanged to preserve the integrity of the original input data's information. This channel selection mechanism enables VPConv to effectively reduce computational load and memory requirements while preserving critical information. For a 3D input image, the FLOPs of a single VPConv are only:

$$h \times w \times d \times k^2 \times c_p^2 \tag{1}$$

where h, w, d represent the length, width, and depth of the volumetric image, k represents the kernel size, and c_p represents the channel number affected by the 3D Conv. In practical implementations, there is typically a ratio $r = c_p/c = 1/4$, resulting in VPConv having FLOPs only $1/16$ of those in a conventional Conv. Meanwhile, the memory access of VPConv is:

$$h \times w \times d \times 2c_p + k^2 \times c_p^2 \tag{2}$$

The memory access quantity for VPConv is only $1/4$ of that for conventional convolution, as the remaining $(c - c_p)$ channels do not participate in the computation, eliminating the need for memory access.

The VPConv convolves only c_p channels while leaving the rest $(c - c_p)$ channels unchanged, but this does not mean that the remaining channels are useless. Subsequent pointwise convolution (PWConv) layers fully utilize these channels, allowing feature information to flow through all channels.

2.3 3D Partial U-Net Block

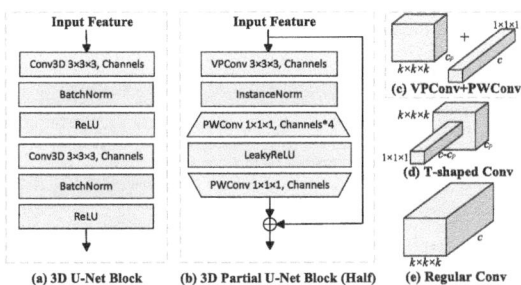

Fig. 4. Comparison of 3D U-Net Block (a) and half of 3D Partial U-Net Block (b). VPConv followed by PWConv (c) resembles T-shaped Conv (d), which allocates more computational resources to the central position compared to regular Conv of 3D U-Net (e).

Based on VPConv, we propose a modern Conv block called 3D Partial U-Net Block. In each our Conv block, there is a VPConv followed by a PWConv layer. Together, they constitute an inverted residual block. The middle layer experiences a fourfold increase in channel count, and a shortcut is incorporated to fuse input features. Such structure allows for independent expansion and compression of each channel to enrich the representation of feature. The combination of PWConv and VPConv on the input feature map resembles a T-shaped Conv in terms of its effective receptive field as shown in Fig. 4. In contrast to a standard convolution that uniformly processes a patch, it focuses more on the central position, which has been proven effective in our LN segmentation task. Activation and normalization layers are crucial in ConvNets, but excessive usage can limit feature diversity, thereby slowing down computation speed. Therefore, inspired by ConvNeXt [16], our block employs fewer activation functions and normalization layers. We use activation functions only between PWConv layers and normalization layers only between VPConv and the first PWConv layer. In comparison to the common U-Net [8] architecture, we replace ReLU with LeakyReLU [18] to address the issue of neuron deactivation caused by the hard truncation of negative gradients, resulting in improved performance on our small dataset, as shown in Fig. 4. Instance normalization (IN) [19] calculates normalization statistics considering individual samples and all elements within a single channel. It is a normalization algorithm more suitable for scenarios requiring higher demands on individual pixels, such as our LN segmentation task. Therefore, we replace BN with IN in our convolutional blocks.

2.4 Light Boundary Refinement Output Module

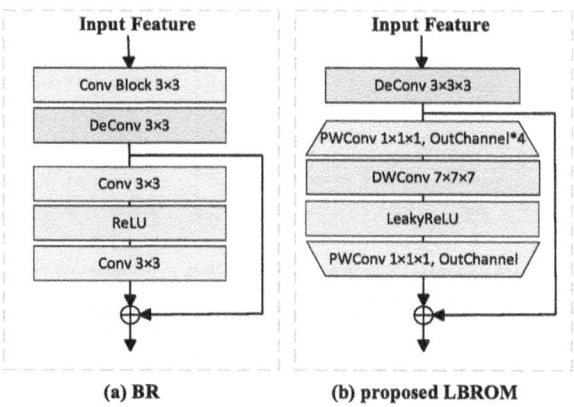

(a) BR   **(b) proposed LBROM**

Fig. 5. Boundary refinement (a) and our proposed light boundary refinement output module (b). We have simplified and improved it by significantly reducing parameters and FLOPs.

The boundary refinement (BR) [20] module is constructed based on the residual structure (Fig. 5 (a)) and is employed for both predicting and refining the segmentation score maps. To find a simple boundary refinement module for easy training, we proposed a light boundary refinement output module (LBROM) (Fig. 5 (b)) to replace the BR module, which is simpler than the BR module. Due to the significant increase in FLOPs when adding convolutional blocks at the upper layers of the model, we remove the convolutional block before the transpose convolution. Additionally, we replace the convolution after the transpose convolution with a combination of pointwise convolutions and depth-wise convolution. Especially, we employ DWConv with $7 \times 7 \times 7$ kernel size and 3 padding size to serve as a shifted window, evenly dividing the volumetric image. After testing in our model, with an input size of $32 \times 64 \times 64 \times 48$ and an output size of $1 \times 64 \times 64 \times 48$, the parameters and FLOPs of the LBROM are both approximately 1/13 of the BR block.

3 Experiments and Results

3.1 Dataset

We use a clinical dataset consisting of head and neck cancer patients operated on between 2009 and 2018 at our institute. This dataset includes 123 patients (total of 678 LNs) with oropharyngeal squamous cell carcinoma (OPSCC). All the patients underwent contrast-enhanced CT imaging and subsequent surgical neck dissection. The contrast-enhanced CT images had a slice thickness of 3 mm and a pixel spacing of 1.1719 mm. A radiation oncologist delineated a total of 678 lymph nodes. The size range of the total node is in the range of [20 mm^3, 50876 mm^3].

3.2 Implementation Details

We randomly select 476 and 67 nodes for training and validation. Another 135 nodes are remained for testing. We use a patch size of $64 \times 64 \times 48$ voxels to cover the largest LNs as the input of all the methods, which contains the LN and its surrounding tissues. Due to the limited data for training (476 nodes), we augment raw training data to 2100 volumes of LNs, where each volume is generated by rotating along a randomly selected axis (i.e., x, y, z-axis) with a random angle selected from the range of -30° to 30°.

We develop 3D Partial U-Net using Pytorch framework. We utilize a loss combined of Generalised Dice Loss (GDL) [21] and Focal Loss [22] to train 3D Partial U-Net.

The formulation of loss L between the prediction $\hat{y}$ and the target y is as follow:

$$L = GDL(\hat{y}, y) + Focal(\hat{y}, y) \tag{3}$$

Furthermore, we utilize the AdamW optimizer with the initial learning rate set to 0.0003. Considering that small nodes are harder to distinguish than large nodes, to test the performance of the models on smaller nodes, we split all the testing nodes into three groups, including small nodes (volume ≤ 404 mm^3), large nodes (volume > 404 mm^3), and all nodes, where 404 mm^3 represents the median volume of all nodes. We focus on mean Dice Similarity Coefficient (DSC) $\pm$ standard deviation (STD) and mean Intersection over Union (IoU) $\pm$ STD of segmenting the nodes as well as the parameter count and computational complexity (in FLOPs). All experiments are conducted on a computer equipped with a 13th Gen Intel Core i7-13700K processor and a GeForce RTX 4090 GPU with 24 GB of memory..

3.3 Ablation Study

In this section, we perform an ablation study to illustrate our contribution. As shown in Table 1 and Table 2, we first start with the original 3D Partial U-Net. Then, we remove LBROM, resulting in a slight decrease in segmentation accuracy. Next, we replace LBROM in the original network with BR. It can be observed that the parameters and FLOPs of model3 are higher than those of model1, but the segmentation accuracy of both models is similar. It is worth noting that using LBROM is more effective for segmenting small lymph nodes. Next, we replace the 3D Partial U-Net block with the 3D U-Net block (as shown in Fig. 4), resulting in a significant increase in model parameter count and FLOPs. However, there is a slight decrease in accuracy, demonstrating the effectiveness of our own convolutional blocks.

3.4 Analysis on Number of Channels

The number of channels serves as the primary hyperparameter in the 3D Partial U-Net, influencing parameters count, complexity, and model performance. In Table 2, we present additional experiments demonstrating two different configurations of the 3D Partial U-Net. Model5 (3D Partial U-Net-L) follows $C_1=64$, $C_2=128$, $C_3=256$, and $C_4=512$ while Model6 (3D Partial U-Net-M) follows $C_1=16$, $C_2=32$, $C_3=128$, and $C_4=160$. It is noticeable that increasing the number of channels further enhances accuracy but also adds to computational complexity. Conversely, decreasing the number of channels reduce performance slightly, but it leads to a lightweight model with a relatively minor impact on performance.

Table 1. Summary of different models in the ablation study (Model1 ~ 4) and analysis on number of channels (Model5 ~ 6).

Model	Architecture
Model1	3D Partial U-Net
Model2	3D Partial U-Net – LBROM
Model3	3D Partial U-Net – LBROM + BR
Model4	Replace 3D Partial U-Net blocks with 3D U-Net blocks
Model5	$C_1=64$, $C_2=128$, $C_3=256$, $C_4=512$
Model6	$C_1=16$, $C_2=32$, $C_3=128$, $C_4=160$

Table 2. Comparison of ablation experiment segmentation results.

Methods	#Params	FLOPs	All Nodes		Large Nodes		Small Nodes	
			DSC ↑	IoU ↑	DSC ↑	IoU ↑	DSC ↑	IoU ↑
Model1	1.83M	7.82G	0.743 ± 0.113	0.604 ± 0.132	0.716 ± 0.117	0.569 ± 0.128	0.766 ± 0.107	0.632 ± 0.130
Model2	1.83M	7.51G	0.738 ± 0.102	0.596 ± 0.124	0.728 ± 0.092	0.580 ± 0.107	0.747 ± 0.112	0.609 ± 0.137
Model3	1.86M	11.73G	0.742 ± 0.107	0.601 ± 0.126	0.725 ± 0.112	0.579 ± 0.125	0.755 ± 0.101	0.618 ± 0.125
Model4	18.28M	43.97G	0.737 ± 0.109	0.594 ± 0.127	0.734 ± 0.092	0.587 ± 0.109	0.739 ± 0.122	0.600 ± 0.141
Model5	7.24M	29.81G	0.749 ± 0.098	0.608 ± 0.119	0.739 ± 0.101	0.595 ± 0.115	0.756 ± 0.097	0.619 ± 0.122
Model6	0.96M	3.15G	0.722 ± 0.108	0.576 ± 0.128	0.706 ± 0.092	0.554 ± 0.107	0.735 ± 0.119	0.595 ± 0.141

3.5   Performance Comparison with SOTA

We evaluate the performance of 3D Partial U-Net by conducting comparisons with recent widely used volumetric medical image segmentation models, such as 3D U-Net [12], SegResNet [23], Attention U-Net [24], DenseVNet [25], UNETR [26], MedNeXt [27], 3D UX-Net [28] and PMSFNet [29].

The results are shown in Table 3. It can be observed that 3D Partial U-Net achieves state-of-the-art performance on all nodes and small nodes, with only a slight lag in performance on large nodes. Compared to 3D UX-Net [28], which achieves the second-best segmentation performance on all nodes, our model reduces FLOPs by fifteen times and parameters by thirty times. Note that the compared lightweight model DenseVNet [25] has the fewest parameters and FLOPs, but it performs the worst. Another lightweight model PMSFNet [29] achieves the best accuracy on large nodes, but it performs poorly on small nodes. Figure 6 shows the slice-based and case-based visualization of each model.

Table 3. Performance comparison with SOTA approaches for 3D medical image segmentation.

Methods	#Params	FLOPs	All Nodes		Large Nodes		Small Nodes	
			DSC ↑	IoU ↑	DSC ↑	IoU ↑	DSC ↑	IoU ↑
3D U-Net [12]	12.70M	28.19G	0.683 ± 0.160	0.537 ± 0.158	0.680 ± 0.129	0.529 ± 0.135	0.683 ± 0.185	0.544 ± 0.177
SegResNet [23]	4.70M	13.61G	0.697 ± 0.118	0.547 ± 0.132	0.720 ± 0.105	0.572 ± 0.117	0.678 ± 0.126	0.528 ± 0.121
Att U-Net [24]	16.03M	48.10G	0.646 ± 0.166	0.497 ± 0.162	0.689 ± 0.133	0.540 ± 0.139	0.610 ± 0.183	0.462 ± 0.174
DenseVNet [25]	0.87M	1.39G	0.640 ± 0.146	0.488 ± 0.154	0.663 ± 0.155	0.514 ± 0.160	0.621 ± 0.138	0.466 ± 0.147
UNETR [26]	101.6M	58.71G	0.683 ± 0.154	0.537 ± 0.153	0.691 ± 0.116	0.538 ± 0.121	0.677 ± 0.181	0.535 ± 0.177
MedNeXt [27]	11.64M	38.14G	0.708 ± 0.129	0.562 ± 0.137	0.721 ± 0.096	0.572 ± 0.113	0.697 ± 0.153	0.554 ± 0.155
3D UX-Net [28]	53.00M	105.2G	0.732 ± 0.120	0.591 ± 0.137	0.706 ± 0.101	0.554 ± 0.114	0.754 ± 0.131	0.621 ± 0.148
PMSFNet [29]	2.27M	5.61G	0.712 ± 0.112	0.564 ± 0.130	**0.726 ± 0.081**	**0.576 ± 0.096**	0.699 ± 0.132	0.554 ± 0.153
Our Method	**1.83M**	**7.82G**	**0.743 ± 0.113**	**0.604 ± 0.132**	0.716 ± 0.117	0.569 ± 0.128	**0.766 ± 0.107**	**0.632 ± 0.130**

344   F. Wu et al.

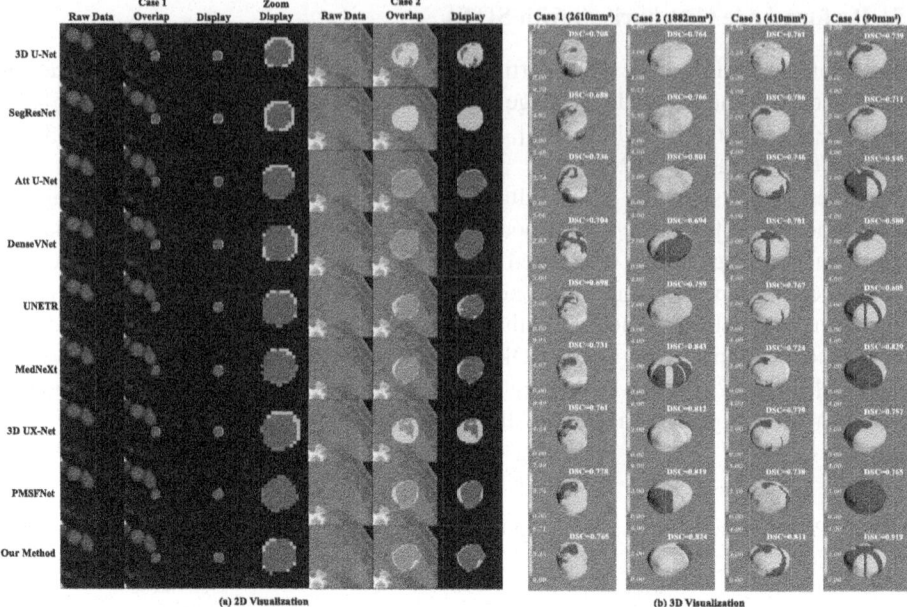

(a) 2D Visualization   (b) 3D Visualization

Fig. 6. (a) Slice-based comparison in experiments of different models. Blue areas show the overlap between the segmented result and ground truth. Red areas show segmented result not containing the ground truth. The green areas are opposite red areas. (b) Visualizing the 3D surface distance (in voxels) between the segmented surface and the ground truth in experiments of different models for four selected cases. Different colors represent difference surface distances. Color bar shows the range of Euclidean distance.

4 Conclusion

In this work, we have proposed a new lightweight model 3D Partial U-Net for head and neck LN segmentation. Specifically, we propose VPConv and design a 3D Partial U-Net Block to replace conventional convolution block. The unique asymmetric encoder-decoder structure we proposed, along with LBROM, significantly enhances the ability to segment small LNs, while there is a slight deficiency in the ability to segment large LNs, which is an area for future improvement. We validate 3D Partial U-Net on our own LN dataset. Our method achieves reduced computational complexity and the fewest parameters with state-of-the-art accuracy.

However, we only segment LNs from CT scans because it is the most commonly employed imaging modality for initial diagnosis. Nevertheless, in radiation therapy, positron emission tomography (PET) is also routinely used for staging and identifying treatment targets. In the future, we intend to validate our method across various imaging modalities, including PET. Since malignant LNs typically exhibit higher uptake values, the addition of PET can enhance the performance of all models. Furthermore, our model can be potentially utilized for general 3D medical image segmentation. We will explore its application in more medical scenarios to address complex challenges in healthcare in future work [30].

Acknowledgments. This work was supported by the China Postdoctoral Science Foundation (No. 2023M742568).

References

1. Chi, C., et al.: The clinical characteristics and prognostic nomogram for head and neck cancer patients with bone metastasis. J. Oncol.. **2021**, 1–12 (2021)
2. Chen, L. et al.: Attention guided lymph node malignancy prediction in head and neck cancer. Inter. J. Radiation Oncolo. *Biology* Phys. **110**(4), 1171–1179 (2021)
3. Navran, A., et al.: The impact of margin reduction on outcome and toxicity in head and neck cancer patients treated with image-guided volumetric modulated arc therapy (VMAT). Radiother. Oncol. **130**, 25–31 (2019)
4. van der Veen, J., et al.: Deep learning for elective neck delineation: more consistent and time efficient. Radiother. Oncol. **153**, 180–188 (2020)
5. Peng, T., et al.: A multi-center study of ultrasound images using a fully automated segmentation architecture. Pattern Recogn. **145**, 109925 (2024)
6. Hoang, J.K., et al.: Evaluation of cervical lymph nodes in head and neck cancer with CT and MRI: tips, traps, and a systematic approach. Am. J. Roentgenol. **200**(1), W17–W25 (2013)
7. Peng, T., et al.: H-SegMed: a hybrid method for prostate segmentation in TRUS images via improved closed principal curve and improved enhanced machine learning. Int. J. Comput. Vision **130**(8), 1896–1919 (2022)
8. Ronneberger, O. et al.: U-Net: convolutional networks for biomedical image segmentation. In: Navab, N. et al. (eds.) MICCAI 2015. LNCS, vol. 9351, pp. 234–241. Springer, Cham (2015). https://doi.org/10.1007/978-3-319-24574-4_28
9. Li, H. et al.: Domain adaptive retinal vessel segmentation guided by high-frequency component. In: Antony, B. et al. (eds.) OMIA 2022. LNCS, vol. 13576, pp. 115–124. Springer, Cham (2022). https://doi.org/10.1007/978-3-031-16525-2_12
10. Li, H. et al.: Frequency-Mixed single-source domain generalization for medical image segmentation. In: Greenspan, H. et al. (eds,) MICCAI 2023. LNCS, vol. 14225. pp. 127–136. Springer, Cham (2023). https://doi.org/10.1007/978-3-031-43987-2_13
11. Oda, H. et al.: Dense volumetric detection and segmentation of mediastinal lymph nodes in chest CT images. In: Medical Imaging 2018: Computer-Aided Diagnosis (2018)
12. Çiçek, Ö. et al.: 3D U-Net: learning dense volumetric segmentation from sparse annotation. In: Ourselin, S. et al. (eds.) MICCAI 2016. LNCS, vol. 9901, pp. 424–432. Springer, Cham (2016). https://doi.org/10.1007/978-3-319-46723-8_49
13. Strijbis, V.I.J., et al.: Deep learning for automated elective lymph node level segmentation for head and neck cancer radiotherapy. Cancers **14**(22), 5501 (2022)
14. Xu, G., et al.: DiSegNet: a deep dilated convolutional encoder-decoder architecture for lymph node segmentation on PET/CT images. Comput. Med. Imaging Graph. **88**, 101851 (2021)
15. Chen, J. et al.: Run, don't walk: chasing higher FLOPS for faster neural networks. In: CVPR2023, pp. 12021–12031 (2023)
16. Liu, Z., Mao, H., Wu, C.-Y., Feichtenhofer, C., Darrell, T., Xie, S.: A ConvNet for the 2020s. In: CVPR 2022, pp. 11976–11986 (2022)
17. Liu, Z. et al.: Swin Transformer: hierarchical vision transformer using shifted windows. In: ICCV 2021, pp. 10012–10022 (2021)
18. Maas, A. et al.: Rectifier nonlinearities improve neural network acoustic models. In: ICML 2013 (2013)
19. Ulyanov, D. et al.: Instance Normalization: The Missing Ingredient for Fast Stylization. arXiv preprint arXiv:1607.08022 (2016)

20. Islam, M. et al.: Learning where to look while tracking instruments in robot-assisted surgery. In: Shen, D., et al. (eds.) MICCAI 2019. LNCS, vol. 11768, pp. 412–420. Springer, Cham (2019). https://doi.org/10.1007/978-3-030-32254-0_46

21. Sudre, C.H. et al.: generalised dice overlap as a deep learning loss function for highly unbalanced segmentations. In: Cardoso, M., et al. (eds.) Deep Learning in Medical Image Analysis and Multimodal Learning for Clinical Decision Support. LNCS, vol. 10553, pp. 240–248. Springer, Cham (2017). https://doi.org/10.1007/978-3-319-67558-9_28

22. Lin, T.-Y. et al.: Focal loss for dense object detection. In: ICCV 2017, pp. 2980–2988 (2017)

23. Myronenko, A.: 3D MRI brain tumor segmentation using autoencoder regularization. In: Crimi, A., et al. (eds.) Brainlesion: Glioma, Multiple Sclerosis, Stroke and Traumatic Brain Injuries. BrainLes 2018. LNCS, vol 11384, pp. 311–320. Springer, Cham (2019). https://doi.org/10.1007/978-3-030-11726-9_28

24. Oktay, O. et al.: Attention U-Net: learning where to look for the pancreas. In: Medical Imaging with Deep Learning (2018)

25. E. Gibson. et al.: Automatic multi-organ segmentation on abdominal CT With dense V-networks. IEEE Trans. Medical Imaging $37(8)$, 1822–1834 (2018)

26. Hatamizadeh, A. et al.: UNETR: transformers for 3d medical image segmentation. In: WACV 2022, pp. 574–584 (2022)

27. Roy, S. et al.: MedNeXt: Transformer-driven Scaling of ConvNets for Medical Image Segmentation. In: Greenspan, H., et al. (eds.) MICCAI 2023. LNCS, vol. 14223, pp.405–415. Springer, Cham. (2023). https://doi.org/10.1007/978-3-031-43901-8_39

28. Lee, H. et al.: 3D UX-Net: a large kernel volumetric ConvNet modernizing hierarchical transformer for medical image segmentation. In: ICLR 2023 (2023)

29. Zhong, J. et al.: PMFSNet: Polarized Multi-scale Feature Self-attention Network For Lightweight Medical Image Segmentation arXiv preprint arXiv:2401.07579 (2024)

30. Peng, T. et al.: Organ boundary delineation for automated diagnosis from multi-center using ultrasound images. Expert Syst. Appl. **238**, 122128 (2024)

A Classification Method for Diabetic Retinopathy Based on Self-supervised Learning

Fei Long, Haoren Xiong, and Jun Sang$^{(\boxtimes)}$

School of Big Data and Software Engineering, Chongqing University, Chongqing 401331, China
jsang@cqu.edu.cn

Abstract. Diabetic retinopathy, a prevalent eye disease in diabetic patients, poses a high risk of blindness. Current computer-aided diagnostic methods require extensive labeled datasets, which are labor-intensive and time-consuming. Given the scarcity of large labeled datasets in medical imaging, self-supervised learning presents a promising alternative, capable of leveraging unlabeled data to improve diagnostic accuracy. Addressing this challenge, our paper introduces a classification method for diabetic retinopathy using self-supervised learning. The method encompasses three stages: data preprocessing, self-supervised learning, and classification. In preprocessing stage, fundus images are resized to 255×255 pixels, divided into nine sub-images each, and labeled with serial numbers. Additionally, masks are applied to sub-image edges to minimize model reliance on edge features. The self-supervised learning stage employs a VGG16-based network with nine branches to learn intrinsic features of fundus images, thus decreasing the requirement for labeled samples. The network inputs are these sub-images, shuffled in order, with output being their sequence numbers. This stage produces a pre-trained network. For the classification stage, this pre-trained network is further fine-tuned using a small labeled dataset (300 images), modifying the final fully connected layer for either binary or five-category classification. Our approach has demonstrated impressive results: 92.5% accuracy for binary and 66.7% for five-category classification on APTOS dataset, and 92.7% for binary and 62.4% for five-category classification on the Kaggle EyePACS dataset. These outcomes underscore the substantial potential of self-supervised learning in diabetic retinopathy diagnosis, offering a significant reduction in the dependency on extensive labeled datasets and thereby enhancing the diagnostic process.

Keywords: Deep Learning · Self-supervised Learning · Diabetic Retinopathy · Fundus Image

1 Introduction

Diabetic retinopathy is a widespread disease among people with diabetes. About one-third of people with diabetes have some kind of eye disease, and diabetic retinopathy accounts for the majority of these eye diseases. [1] There are two categories of diabetic retinopathy: non-proliferative and proliferative. The category of non-proliferative conditions encompasses varying degrees of severity, namely mild, moderate, and severe

D.-S. Huang et al. (Eds.): ICIC 2024, LNBI 14881, pp. 347–357, 2024.
https://doi.org/10.1007/978-981-97-5689-6_30

[2]. Non-proliferative diabetic retinopathy, characterized by its relatively lower disease severity, presents a treatment protocol that is generally simple. However, if the disease is left untreated, it may develop into proliferative, with new vessels on the disc, vitreous hemorrhage, and retinal detachment, which can seriously affect the patient's vision and lead to complete blindness. The diagnosis of diabetic retinopathy currently relies on a combination of fundus image and fundus fluorescein angiography analysis by experienced physicians. Computer-aided diagnosis can allow the doctor to lighten the burden.

Supervised learning has made significant progress in the diagnosis of diabetic retinopathy. However, an extant challenge remains; supervised learning is contingent upon extensive labeled data. In practical scenarios, access to such voluminous manually annotated datasets is often limited. Therefore, an organization or an individual intending to train a model using the datasets at their disposal would inevitably face considerable time and labor costs in data annotation. The escalation of these costs becomes pronounced with the augmentation of data size, thereby underscoring the urgency of employing self-supervised learning methodologies in model training.

Here is the contribution of our method:

- Introduced a self-supervised learning-based classification method for diabetic retinopathy, significantly reducing reliance on labeled samples and streamlining the diagnostic process.
- Introduced a multi-branch network architecture based on VGG16 during the self-supervised learning phase, effectively extracting intrinsic features of fundus images, and providing an accurate pre-trained model for the classification phase.
- Implemented a methodology comprising data preprocessing, self-supervised learning, and classification stages, achieving binary classification accuracies of 92.5% on the APTOS dataset and 92.7% on the Kaggle EyePACS dataset.

2   Related Works

Currently, self-supervised learning has started to be applied in the field of medical images. Since there is not much literature on self-supervised learning-based diagnostic methods for diabetic retinopathy, some self-supervised learning-based medical imaging diagnostic methods will be introduced here. Jiuwen Zhu et al. proposed Rubik's Cube + [3] method. After segmenting 3D images, they performed a series of operations on the image, such as disordering, masking, and rotating, then feeding them into a network, and applied the network to classify cerebral hemorrhage after the network completed the sorting task. Liang Chen et al. [4] used image context restoration for self-supervised learning, which first extracts medical image features and then tries to reconstruct the images. They trained the model by comparing the original image with the generated image and finally achieved 89.39% precision in the classification of standard scan planes of fetal 2D ultrasound images.

3   Proposed Method

In this section, we will introduce the proposed method for diabetic retinopathy diagnosis based on self-supervised in detail.

3.1 Self-supervised Learning Stage

In the self-supervised learning stage. We refer to the work of V. Vives-Boix et al. [5] to determine the best model choice. The study compares in detail the classification results of several classical models. After comparing the results with other work, VGG16 was finally chosen as the backbone network.

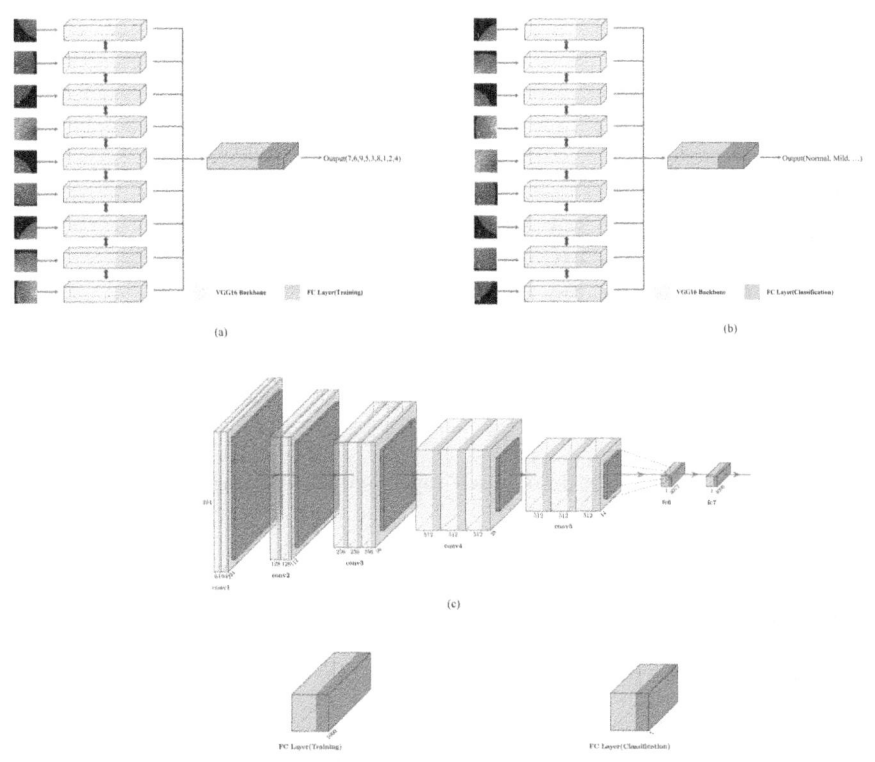

Fig. 1. Model's architecture detail.

In the self-supervised learning stage, this one-dimensional vector represents the prediction of the image sequence, and in the classification stage, it represents the prediction of the disease type, the disease is categorized into two or five. Figure 1 shows the detail of model's architecture,(a)represents the architecture of self-supervised learning stage. (b)shows the architecture of classification stage. (c) shows the detail of VGG16 backbone. (d)represents the the architecture of final fully-connected layer in self- supervised learning stage. (e)represents the the architecture of final fully-connected layer in classification stage.

This stage generates a random sequence of a thousand different sub-image numbers, and then selects one of these sequences at random for the sub-images to be input into each network branch in that order. Figure 2 shows the sub-image and the shuffled sub-image. Each sub-image is fed into a VGG16 backbone network. After feature extraction, each backbone network will output a 1 × 512 dimensional vector through the fully connected layer, which will eventually be concatenated into a 1-dimensional vector and fed into the fully connected layer. The final fully connected layer will output a 1 × 9 vector, which is a sequence that represents the network's prediction of where the sub-image should be in the original image. By keeping the model's sequence predictions close to the true labels, the model is able to learn the image features. Figure 2 shows the detail of self-supervised learning stage.

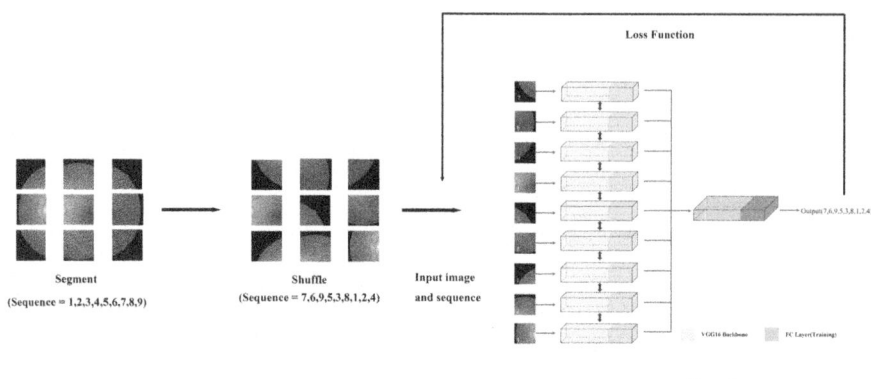

Fig. 2. The detail of self-supervised learning stage.

3.2 Classification Stage

After completing self-supervised learning stage, we used the pre-trained network obtained from the previous stage, the parameters of the network are frozen except for the last fully connected layer. The model itself ultimately yields results for the prediction order of sub-images, whereas disease is categorized into only two (i.e., normal and diabetic retinopathy) or five (i.e., normal, mild, moderate, severe, and proliferative). Therefore, the output vector dimension of the last fully connected layer was changed from 1 × 9 (due to the numbers of the sub-images) to 1 × 2 (for binary classification) and 1 × 5 (for quintuple classification). The structure of the last fully connected layer was modified to ensure the accuracy of the classification task. The parameters of the model were frozen except for the last fully connected layer. In order to pursue the best results, 300 images are used for model fine-tuning in this paper, following the methodology of Hongyu Zhou et al. [6].

In the classification stage, the sub-image order is no longer disrupted. Instead, the original images are preprocessed and then input into the model in order. The results are fully presented and analyzed in the subsequent sections. Figure 3 shows the details of classification stage. Each sub-image is fed into a VGG16 backbone network, each VGG16 backbone will output a 1 × 512-dimensional vector. These nine vectors are then concatenated into a single one-dimensional vector, which is input into a fully connected layer and outputs a vector representing the prediction of the disease class.

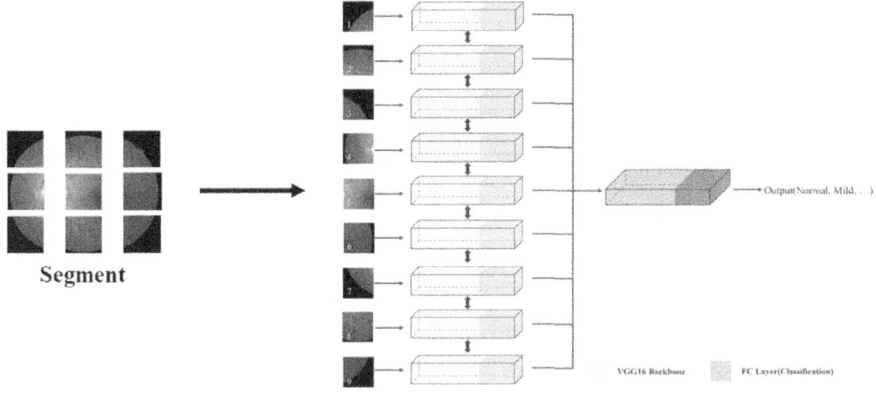

Fig. 3. The detail of classification stage.

4 Experiments

This section will introduce the the details of the experiment, the results of the two datasets, and the comparison with other experiments. The experimentation was carried out on a platform comprising an Intel i5-10400F, Nvidia GTX 1660Ti, 16GB RAM platform with Python version 3.7 and Pytorch version 1.7.0. The optimizer for our model is Adam, the learning rate is 0.001, and the batch size is 128.

4.1 Result of Kaggle ATPOS 2019 Dataset

This is the result tested on the Kaggle ATPOS 2019 dataset.

To ensure the robustness of our experiments, K-Fold validation was employed for the model assessment. A random subset of images, comprising 10% of the total dataset, was selected for fine-tuning the model, while the remaining 90% of images were employed for model validation.

From 2-class validation, we can see that the proposed method achieves the highest result of 92.5% accuracy. The comparison result is shown in Table 1. The confusion matrix results for the binary classification task exhibit high accuracy. However, from the perspective of the five classification tasks in Table 2, it has achieved 66.7% of the best results. It is still a certain distance from ideal accuracy. Inspecting the confusion matrix depicted in Fig. 4 provides insight into this phenomenon: the proficiency in categorizing the three classes of moderate, proliferate, and severe lacks the requisite strength. This deficiency emerges as a consequence of the intrinsic similarity among these three categories within the images, thereby giving rise to a propensity for confusion.

4.2 Experiment on Kaggle EyePACS Dataset

To augment the breadth of our experimental findings, an additional evaluation was conducted on the Kaggle EyePACs dataset. Corresponding to our APTOS experiments, analogous binary and five-category classification were undertaken on the Kaggle dataset.

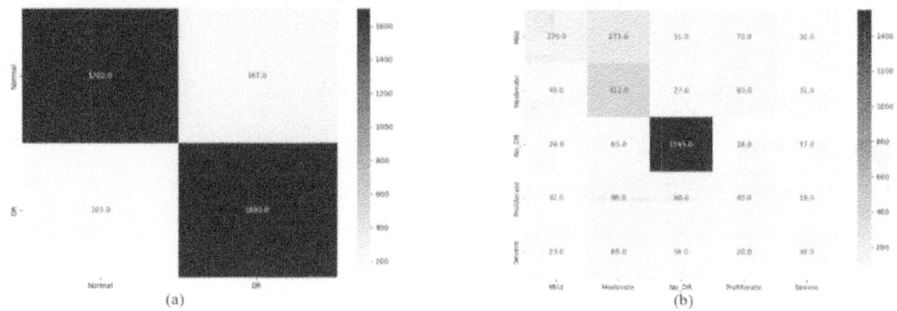

Fig. 4. Confusion matrix of Kaggle ATPOS 2019 dataset

Table 1. 10-fold binary-classification results of Kaggle ATPOS 2019 dataset

Fold	Accuracy	Error Rate
1	0.91	0.090
2	0.879	0.121
3	0.915	0.085
4	0.901	0.099
5	0.868	0.132
6	0.882	0.118
7	0.901	0.099
8	0.891	0.109
9	0.922	0.078
10	0.925	0.075

The following is the result of our test. From the results, the results on this dataset are not as good as those on APTOS. The model achieves a maximum accuracy of 92.7% on the binary-classification task and 62.4% on the five-category task. The accuracy rate has a particular gap compared with the previous experimental results. There are two reasons for this: Firstly, several disease stages (Mild, Moderate, Severe) are quite close in fundus images, and there may be some difficulties in classification. Secondly, the dataset comes from different devices, and there are a lot of noisy, out-of-focus pictures, which also cause certain difficulties for image classification. Figure 5 is the confusion matrix of Kaggle EyePACS dataset. Table 3 is the results of Kaggle EyePACS dataset.

Table 2. 10-fold five-category results of Kaggle ATPOS 2019 dataset

Fold	Accuracy	Error Rate
1	0.646	0.354
2	0.647	0.353
3	0.636	0.364
4	0.667	0.333
5	0.661	0.339
6	0.624	0.376
7	0.650	0.350
8	0.644	0.356
9	0.662	0.338
10	0.642	0.358

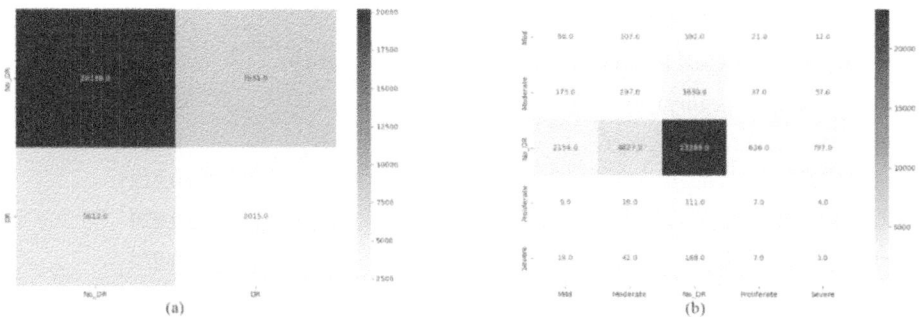

Fig. 5. Confusion matrix of Kaggle EyePACS dataset

Table 3. Results of Kaggle EyePACS dataset

Results	Accuracy
Binary-classification	0.927
Five-category classification	0.624

4.3 Comparison with Other Methods

On the ATPOS dataset, this paper has compared our result with the supervised learning method results. The methods are quite close in the two classification results, but there is a certain gap between the classic traditional model and our methods in the five-category classification task. This is also determined by the difficulty of the classification task itself. The two classification task only need to find "abnormal" pictures, while the five-category classification task requires us to classify accurately. Table 4 is the comparison with the supervised learning methods on Kaggle APTOS 2019 dataset.

On the Kaggle dataset, this also made a comparison. Our method is still very close to supervised learning on the binary-classification task, but on the five-category classification task, our results still have a particular gap compared with the complex and novel network. We attribute this result to the fact that the EyePACs dataset contains a large amount of data in different situations, and the features of these data are not common during fine tuning. In addition, there is a significant gap between the amount of data trained and the amount of data contained in EyePACs during model training. Table 5 is the comparison with the supervised learning methods on Kaggle EyePACS dataset.

This paper not only completed the experiment on the two datasets but also used other self-supervised methods to complete the experiment on the two datasets and compare them. Table 6 and Table 7 are the comparison results with other methods.

Table 4. Comparison with the five-category classification supervised learning methods on Kaggle APTOS 2019 dataset

Five-category Supervised Learning Method(Kaggle APTOS 2019)	Accuracy
VGG16 [5] (Five-category)	0.7271
InceptionV3 [5] (Five- category)	0.7217
Resnet50 [5] (Five- category)	0.7135
VGG16 [5] (Binary)	0.929
InceptionV3[5](Binary)	0.935
Resnet50 [5] (Binary)	0.845
Custom CNN [5](Binary)	0.864
Custom CNN [8] (Binary)	0.890
Custom CNN [9] (Binary)	0.907
Proposed Method (Five- category)	0.667
Proposed Method (Binary)	0.927

Table 5. Comparison with the five-category classification supervised learning method on Kaggle EyePACS dataset

Five-category Supervised Learning Method(Kaggle EyePACS)	Accuracy
VGG16noFC [10] (Five-category)	0.83
Custom CNN [11] (Five-category)	0.75
ShallowNet + MI [10] (Five-category)	0.87
Ensemble Models [12] (Five-category)	0.808
Inception V3 [13] (Five-category)	0.632
Custom CNN [14] (Five-category)	0.631
Custom CNN [15] (Five-category)	0.699
Custom CNN [16] (Binary)	0.95
Custom CNN [17] (Binary)	0.945
VGG16 [18] (Binary)	0.836
Custom CNN [19] (Binary)	0.910
Custom CNN [8] (Binary)	0.840
Proposed Method (Five-category)	0.624
Proposed Method (Binary)	0.927

Table 6. Comparison with the self-supervised learning method on Kaggle APTOS 2019 dataset

Self-Supervised Learning Method (Kaggle APTOS 2019)	Accuracy
SimCLR (Binary-classification)	0.743
SimCLR (Five-category classification)	0.582
SimSiam (Binary-classification)	0.766
SimSiam(Five-category classification)	0.602
Proposed Method (Binary-classification)	0.925
Proposed Method (Five-category classification)	0.642

Table 7. Comparison with the self-supervised learning method on Kaggle EyePACS dataset

Self-Supervised Learning Method (Kaggle EyePACS)	Accuracy
SimCLR (Binary-classification)	0.731
SimCLR (Five-category classification)	0.614
SimSiam (Binary-classification)	0.734
SimSiam (Five-category classification)	0.735
Proposed Method (Binary-classification)	0.927
Proposed Method (Five-category classification)	0.624

5   Conclusion

In this experimental endeavor, a self-supervised diagnosis of diabetic retinopathy was accomplished through the training of a deep neural network. After that, we transferred the model to a classification task, modified the last three fully connected layers, and fine-tuned it with a small amount of data. Across the Kaggle ATPOS 2019 dataset, attaining a 92% accuracy in binary classification and 66% accuracy in the five-category classification was realized. A parallel achievement was recorded in the Kaggle EyePACS dataset, with binary classification yielding a 92% accuracy rate and five-category classification yielding 62% accuracy.

Significantly, we observed that an overly simplistic proxy task hinders the model from sufficiently capturing intricate details, while excessively complex tasks introduce ambiguity, thereby impeding the model's comprehension of the task. This underscores the imperative of judiciously selecting proxy tasks and model structures aligned with real-world scenarios when applying self-supervised learning to medical imaging. Striking the right balance between proxy task complexity and network architecture stands as two critical facets warranting attention and exploration in the future.

References

1. Tsiknakis, N., et al.: Deep learning for diabetic retinopathy detection and classification based on fundus images: a review. Comput. Biol. Med. **135**, 104599 (2021). https://doi.org/10.1016/J.COMPBIOMED.2021.104599
2. Ting, D.S.W., et al.: Development and validation of a deep learning system for diabetic retinopathy and related eye diseases using retinal images from multiethnic populations with diabetes. JAMA **318**, 2211–2223 (2017). https://doi.org/10.1001/JAMA.2017.18152
3. Zhu, J., Li, Y., Hu, Y., Ma, K., Zhou, S.K., Zheng, Y.: Rubik's Cube+: a self-supervised feature learning framework for 3D medical image analysis. Med. Image Anal. **64**, 101746 (2020). https://doi.org/10.1016/J.MEDIA.2020.101746
4. Chen, L., Bentley, P., Mori, K., Misawa, K., Fujiwara, M., Rueckert, D.: Self-supervised learning for medical image analysis using image context restoration. Med. Image Anal. **58**, 101539 (2019). https://doi.org/10.1016/J.MEDIA.2019.101539
5. Vives-Boix, V., Ruiz-Fernández, D.: Diabetic retinopathy detection through convolutional neural networks with synaptic metaplasticity. Comput. Methods Programs Biomed. **206**, 106094 (2021). https://doi.org/10.1016/j.cmpb.2021.106094

6. Zhou, H.-Y., Lu, C., Yang, S., Han, X., Yu, Y.: Preservational Learning Improves Self-supervised Medical Image Models by Reconstructing Diverse Contexts, pp. 3499–3509 (2021)

7. Kumar, N.S., Ramaswamy Karthikeyan, B.: Diabetic retinopathy detection using CNN, transformer and MLP based architectures. In: ISPACS 2021 - International Symposium on Intelligent Signal Processing and Communication Systems: 5G Dream to Reality, Proceeding (2021). https://doi.org/10.1109/ISPACS51563.2021.9651024

8. Lahmar, C., Idri, A.: Deep hybrid architectures for diabetic retinopathy classification. Comput. Methods Biomech. Biomed. Eng. Imaging Vis. **11**, 166–184 (2023). https://doi.org/10.1080/21681163.2022.2060864

9. El-Ateif, S., Idri, A.: Single-modality and joint fusion deep learning for diabetic retinopathy diagnosis. Sci. Afr. **17**, e01280 (2022). https://doi.org/10.1016/J.SCIAF.2022.E01280

10. Chen, W., Yang, B., Li, J., Wang, J.: An approach to detecting diabetic retinopathy based on integrated shallow convolutional neural networks. IEEE Access. **8**, 178552–178562 (2020). https://doi.org/10.1109/ACCESS.2020.3027794

11. Pratt, H., Coenen, F., Broadbent, D.M., Harding, S.P., Zheng, Y.: Convolutional neural networks for diabetic retinopathy. Proc. Comput. Sci. **90**, 200–205 (2016). https://doi.org/10.1016/J.PROCS.2016.07.014

12. Qummar, S., et al.: A deep learning ensemble approach for diabetic retinopathy detection. IEEE Access. **7**, 150530–150539 (2019). https://doi.org/10.1109/ACCESS.2019.2947484

13. Wang, X., Lu, Y., Wang, Y., Chen, W.B.: Diabetic retinopathy stage classification using convolutional neural networks. In: Proceedings - 2018 IEEE 19th International Conference on Information Reuse and Integration for Data Science, IRI 2018, pp. 465–471 (2018). https://doi.org/10.1109/IRI.2018.00074

14. Khaled, O., El-Sahhar, M., El-Dine, M.A., Talaat, Y., Hassan, Y.M.I., Hamdy, A.: Cascaded architecture for classifying the preliminary stages of diabetic retinopathy. In: ACM International Conference Proceeding Series, pp. 108–112 (2020). https://doi.org/10.1145/3436829.3436854

15. Kumar, S.: Diabetic retinopathy diagnosis with ensemble deep-learning. ACM Inter. Conf. Proc. Ser. (2019). https://doi.org/10.1145/3387168.3387206

16. Ghosh, R., Ghosh, K., Maitra, S.: Automatic detection and classification of diabetic retinopathy stages using CNN. In: 2017 4th International Conference on Signal Processing and Integrated Networks, SPIN 2017, pp. 550–554 (2017). https://doi.org/10.1109/SPIN.2017.8050011

17. Xu, K., Feng, D., Mi, H.: Deep Convolutional Neural Network-Based Early Automated Detection of Diabetic Retinopathy Using Fundus Image. Molecules 2017, vol. 22, pp. 2054. 22, 2054 (2017). https://doi.org/10.3390/MOLECULES22122054

18. García, G., Gallardo, J., Mauricio, A., López, J., del Carpio, C.: Detection of Diabetic Retinopathy Based on a Convolutional Neural Network Using Retinal Fundus Images. LNCS(LNAI and LNBI), vol. 10614, pp. 635–642 (2017). https://doi.org/10.1007/978-3-319-68612-7_72

19. Kolla, M., Venugopal, T.: Efficient classification of diabetic retinopathy using binary CNN. In: Proceedings of 2nd IEEE International Conference on Computational Intelligence and Knowledge Economy, ICCIKE 2021, pp. 244–247 (2021). https://doi.org/10.1109/ICCIKE51210.2021.9410719

Attention Fusion Network for Fine-Grained Sleep Apnea Detection Using Respiratory Signals

Di Wu[1], Yong Fan[2], Zhenchao Ouyang[3], Ke Lan[4], Xiaoli Liu[2], Hong Liang[2], and Zhengbo Zhang[2(✉)]

[1] School of Biological Science and Medical Engineering, Beihang University, Beijing 100083, China
[2] Center for Artificial Intelligence in Medicine, The General Hospital of PLA, Beijing 100853, China
zhengbozhang@126.com
[3] Zhongfa Aviation Institute, Beihang University, Hangzhou 311115, China
[4] Beijing SensEcho Science and Technology Co., Ltd., Beijing 100041, China

Abstract. Increasing life pressures have led to the manifestation of various sleep-related symptoms, among which sleep apnea is one of the most prevalent. Recent researchers have developed numerous methods to assist in the diagnosis of the Apnea-Hypopnea Index in clinical settings, such as morphology, machine learning and deep learning. However, these methods do not possess the capability for precise event localization and do not offer comprehensive performance. Therefore, this paper proposes a fine-grained sleep apnea detection neural network (FG-AFSAN) that is based on respiratory signals, which can localize each event accurately, and incorporate an attention fusion mechanism to assess the severity of sleep apnea syndrome. We evaluated our model using both public and clinic datasets, and the results demonstrate that our model achieves a comparable mean Average Precision of 79.69% in event localization, an accuracy of 71% in AHI predictions, which highlights its potential in future clinical applications for sleep apnea screening.

Keywords: Sleep apnea detection · Respiratory signals · Attention fusion network · Deep learning

1 Introduction

Sleep apnea (SA) is a sleep disorder characterized by temporary pauses or stops in breathing during sleep. These episodes, often accompanied by loud snoring or choking sensations, can have significant long-term effects on both mental and physiological health [1]. According to statistical data, SA affects more than 50% of adults [2], highlighting its prevalence in the population. As a result, widespread screening and timely intervention for individuals with SA are crucial. Identifying and addressing SA cases promptly can significantly improve the overall well-being and quality of life for affected individuals.

D. Wu and Y. Fan—Contributing equally to this work.

D.-S. Huang et al. (Eds.): ICIC 2024, LNBI 14881, pp. 358–369, 2024.
https://doi.org/10.1007/978-981-97-5689-6_31

Currently, overnight polysomnography (PSG) is the gold standard for diagnosing SA. Trained clinicians assess the severity of sleep apnea in patients by calculating the Apnea-Hypopnea Index (AHI) according to established guidelines. However, the complexity associated with wearing PSG devices and the potential for annotation errors have limited the widespread use of PSG in large-scale screening and rapid diagnosis. To improve the efficiency and reduce uncertainty in diagnosing sleep apnea, researchers are increasingly directing their efforts toward creating automatic detection algorithms that rely on smart wearable devices.

The rapid development of smart devices enables the diagnosis of SA through the utilization of diverse, high-quality physiological data. The cardiorespiratory coupling hypothesis suggests that analyzing electrocardiogram (ECG) data during sleep can reveal patterns associated with sleep apnea [3–6]. Furthermore, quantitative assessment of sleep apnea has been demonstrated to be possible by analyzing changes in peripheral oxygen saturation (SpO2) [7–12]. A similar study [13] monitored nasal and oral airflow during sleep to detect respiratory disorders linked to sleep apnea. Additionally, characteristic patterns related to sleep apnea can be detected by analyzing the spectrum and amplitude of ribcage and abdominal movements during sleep [14]. This approach offers the advantage of convenience and is not constrained by individual differences.

A rule-based algorithm [15] offers simplicity and interpretability for detecting sleep apnea from respiratory signals. However, it may require threshold adjustments when applied to different datasets. In machine learning approaches, manually designed features such as respiratory cycle and respiratory rate are extracted [16]. These features, combined with other frequency and time domain features, are utilized in machine learning models, including SVM, Random Forest, AdaBoost and XGBoost, for the classification of sleep apnea [17, 18].

Advances in deep learning techniques enable automatic feature extraction from signals, eliminating the requirement for manual feature selection. Convolutional neural networks (CNNs) are employed to fuse local features from multiple channels of respiratory signals for classification purposes [19]. In addition, architectures like long short-term memory (LSTM) and bidirectional LSTM (Bi-LSTM) have been proposed to capture long temporal dependencies [13, 20, 21]. Shen et al. [22] combined CNN with LSTM to capture both local and long-term dependencies, yielding a more robust model. These methods primarily focus on SA classification rather than accurately detecting the start time and duration of the SA events, which can be regarded as coarse-grained approaches. Tang [8] proposed a two-stage detection algorithm. It initially locates SA using an event detection network and then classifies the selected signal segments using an event classification network. Besides, the development of deep learning has shown that attention mechanisms, as effective feature selection strategies [23]. Considering that SA causes significant morphological changes in signals, attention mechanisms can aid networks in focusing on crucial signal components.

To improve the localization and classification capabilities of sleep apnea events, this paper presents a single-stage SA detection network specifically designed for analyzing one-dimensional respiratory signals, the contributions of this paper are summarized as follows:

- We present the fine-grained sleep apnea event detection network (FG-AFSAN), which consists of two key components: the adaptive preprocessing module (APM) and the fused attention mechanism (FAM). The APM utilizes end-to-end parameter learning to effectively filter and normalize signals, thereby improving model generalization. Meanwhile, the FAM leverages position and channel attention mechanisms to enhance the model's capability in capturing critical signal features at the onset and conclusion of sleep apnea. This integration optimizes both low-level morphological features and high-level semantic features.
- Experimental results on both publicly available datasets and local collected datasets demonstrate that the proposed method can provide finer-grained detection results while maintaining a comparable level of accuracy.

2 Methods

Figure 1 presents the comprehensive architecture of FG-AFSAN proposed in this paper, comprising the adaptive preprocessing module, feature extraction module, and detection module. The adaptive preprocessing module is responsible for filtering and normalizing raw signals, optimizing their quality. The feature extraction module integrates two attention mechanisms: position attention and channel attention. Channel attention determines the presence of SA in the signal segment, while the position attention mechanism focuses on identifying abrupt pattern features before and after the occurrence of SA. The detection module utilizes connecting layers to predict the positions of events and provide additional confidence information through regression.

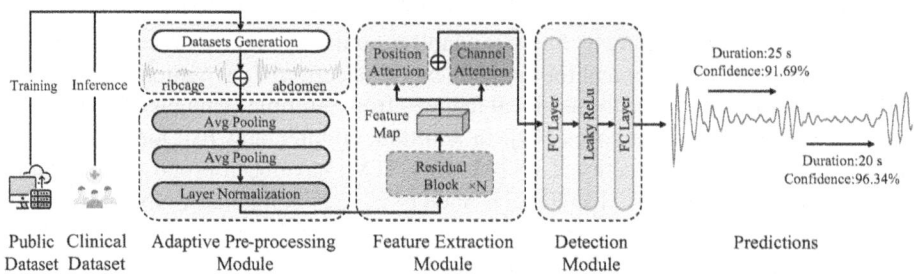

Fig. 1. Workflow of the proposed sleep apnea detection and localization model.

As shown in Fig. 1, our network undergoes distinct training and inference stages. Before the execution of the two stages, dataset construction and augmentation techniques are applied. Publicly available datasets are utilized for training the model during the training stage. In the inference stage, the complete recording of each individual is divided into fixed segments. Finally, the model generates predictions for each segment.

2.1 Adaptive Preprocessing Module

In order to achieve end-to-end processing and improve generalization in the detection pipeline, we introduced an adaptive preprocessing module (APM), shown in Fig. 1.

The APM effectively handles motion artifacts, baseline drift, and random noise using a learning-based approach, eliminating the need for explicit filter construction or linear shifts. It consists of two average pooling layers for smoothing and global signal filtering, respectively. Additionally, layer normalization is employed to mitigate baseline drift in the signals.

Typically, removing baseline drift from signals involves the use of Eq. 1.

$$x = x' - \mu_0 \tag{1}$$

where x' and x are the raw signal and the corrected signal, respectively, and μ_0 denotes the mean value of signal.

The computation of layer normalization [24, 25] is formed as:

$$x = f\left(\frac{\mathbf{g}}{\sqrt{\sigma^2 + \varepsilon}} \odot (x' - \mu) - \mathbf{b}\right) \tag{2}$$

where $f(\cdot)$ is the layer normalization operation, $\mathbf{g}$ and $\mathbf{b}$ are learnable parameters, σ and μ represent the standard deviation and mean, respectively. Therefore, layer normalization can be considered to be able to perform baseline correction.

2.2   Fused Attention Module

While the APM extracts certain morphological features from the signal, directly inputting it into the detection model poses significant challenges. To address this challenge, this paper introduces a two-part feature extraction module. The feature extraction module consists of a ResNet-like network as the backbone, comprising multiple repeated residual blocks. The backbone's role is to extract high-dimensional features from the signal. The output feature map of the backbone is denoted as X_b, with dimensions $X_b \in \mathbb{R}^{C \times L}$. Here, C represents the number of channels, and L represents the length.

Subsequently, the feature map X_b is fed into both the position attention block and the channel attention block independently. In the position attention module, X_b is processed through three mapping layers, yielding three feature maps: X_b^q, X_b^k and X_b^v, each with dimensions of $C \times L$. These feature maps capture positional information. The position attention weights, represented as A_p, are computed using Eq. 3, with the softmax function normalizing the weights for each position. Consequently, A_p has dimensions $A_p \in \mathbb{R}^{L \times L}$.

$$A_p = softmax\left((X_b^q)^T \cdot X_b^k\right) \tag{3}$$

The output of the position attention block is $E_p = \alpha \cdot (X_b^v \cdot A_p) + X_b$, where $E_p \in \mathbb{R}^{C \times L}$.

For the channel attention module, the channel attention weights A_c can be calculated using Eq. 4, where the softmax function normalizes the attention weights for each channel, and $A_c \in \mathbb{R}^{C \times C}$.

$$A_c = softmax\left(X_b \cdot X_b^T\right) \tag{4}$$

The output of the channel attention module is $E_c = \beta \cdot (A_c \cdot X_b) + X_b$, where $E_c \in \mathbb{R}^{C \times L}$. The overall fused attention module's output is $E = E_p + E_c$, where $E \in \mathbb{R}^{C \times L}$.

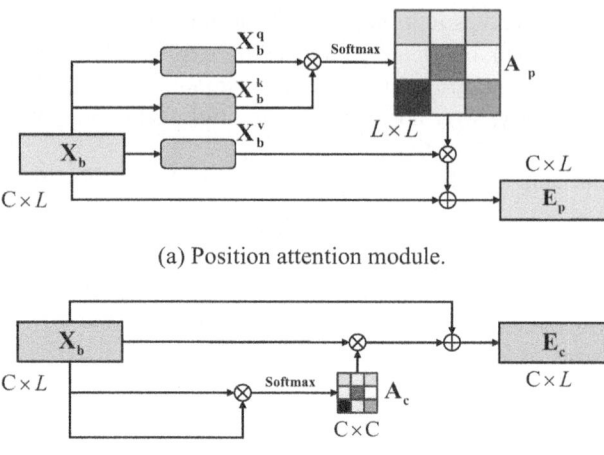

(a) Position attention module.

(b) Channel attention module.

Fig. 2. Illustration of attention mechanisms.

2.3 Detection Modul

The detection module comprises connected layers and activation function layers that enable the determination of the start and end times of SA, along with the probability of abnormal events, through forward propagation (Fig. 2).

As shown in Fig. 3, a signal segment is divided into S non-overlapping cells. Each cell predicts the relative center point coordinates (x') with respect to its left boundary, the duration length (w'), and the probability of a center point falling within the cell. If a target falls into the i-th cell, the predicted relative center point coordinates within the entire time window are calculated as $\hat{e}^x = (x' + i)/S$. The duration length can be determined using $\hat{e}^w = w'/S$. The output of the detection module, represented as Y, has dimensions $Y \in \mathbb{R}^{S \times (P + (B \times 3))}$. Here, Y contains P probabilities of an event predicted as different classes, in this study we set $P = 1$ for the purpose of just detecting the abnormal event. And B represents the number of bounding boxes which predicted by each cell to avoid missed detections when multiple events exist during a segment. Each bounding box contains center point coordinates, duration, and confidence score.

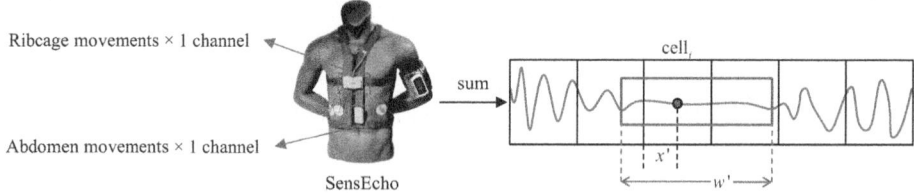

Fig. 3. Sleep apnea location method used in detection module.

Furthermore, the employed loss function in this paper adopts a multi-task learning strategy to separate fine-grained information related to events from the output [26]. The loss function independently calculates the boundary loss, loss for predicting the presence of an event, and loss for predicting the absence of an event, as depicted in Eq. 5. Each part of the loss is calculated using mean square error and summed up through a linear weighting scheme. Where $\lambda_1, \lambda_2, \lambda_3$ denotes the weights of each loss, in this paper we use 5, 0.5, 0.5.

$$\mathcal{L} = \lambda_1 \cdot \mathcal{L}_{boundary} + \lambda_2 \cdot \mathcal{L}_{event} + \lambda_3 \cdot \mathcal{L}_{no-event} \tag{5}$$

The boundary loss function omits the boundary information and only uses the predicted center point position $\hat{e}^x$ and duration $\hat{e}^w$ with the true values for calculation. Equation 6 outlines the calculation process for the boundary loss function component, where $\mathbb{I}$ represents the indicator function, indicating that the boundary loss is only calculated for cells where a target exists.

$$\mathcal{L}_{boundary} = \sum_{i=0}^{S} \sum_{j=0}^{B} \mathbb{I}_{ij}^{obj} \left(\sqrt{e_i^x} - \sqrt{\hat{e}_i^x} \right)^2 + \sum_{i=0}^{S} \sum_{j=0}^{B} \mathbb{I}_{ij}^{obj} \left(\sqrt{e_i^w} - \sqrt{\hat{e}_i^w} \right)^2 \tag{6}$$

Events related loss contains the event loss and the no-event loss, which are represented by Eq. 7 and Eq. 8, respectively. Both losses should be calculated for each cell, regardless of whether a target is predicted. Here, c_i and $\hat{c}_i$ denotes the predicted and ground truth confidence score of the i-th cell.

$$\mathcal{L}_{event} = \sum_{i=0}^{S} \sum_{j=0}^{B} \mathbb{I}_{ij}^{event} (c_i - \hat{c}_i)^2 \tag{7}$$

$$\mathcal{L}_{no-event} = \sum_{i=0}^{S} \sum_{j=0}^{B} \mathbb{I}_{ij}^{no-event} (c_i - \hat{c}_i)^2 \tag{8}$$

3 Experiment and Analysis

3.1 Generation of Datasets

The data utilized in this paper comprises a public dataset (UCDDB) and a clinic dataset. The UCDDB dataset consists of PSG recordings from 25 patients randomly selected from the sleep disorders clinic at St. Vincent's University Hospital [27]. Clinical experts performed detailed annotations of SA events during the sleep process of each patient in the dataset. The clinical dataset, recruited by the Chinese PLA General Hospital, consists of 115 participants, including both individuals without sleep disorders and those with sleep disorders. PSG recordings were obtained, along with the start and end times of SA events labeled by clinical experts in respiratory medicine. This paper focuses solely on utilizing the chest respiratory movement and abdominal respiratory movement signals. The sampling frequencies of these signals in the two datasets are 8 Hz and 128 Hz, respectively. To align with the input size requirements of the network, the signals are uniformly resampled to 100 Hz. Additionally, the chest and abdominal respiratory signals are summed to mitigate interference from body movements prior to input.

Let $X(t)$ be the respiratory signal recorded during an entire night's sleep, containing N sleep apnea events denoted as $(e_1, e_2, \ldots, e_N)$. Each sleep apnea event e_k, where $k = 1, \ldots, N$, is characterized by its center position relative to $X(0)$, denoted as o_k, and the duration of the event represented as e_k^w. To ensure a fixed input size for the network, the respiratory signal is divided into time windows of size D, ensuring that each window encompasses at least one sleep apnea event. This process involves selecting o_k as the center and expanding the window by $D/2$ on both sides. Consequently, each sample, denoted as s_k, can be expressed as $s_k = X(o_k - D/2 : o_k + D/2)$. The label for sample s_k follows the format $[label, D/2, e_k^w]$.

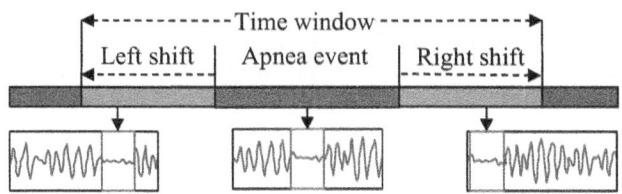

Fig. 4. Illustration of dataset and augmentation.

As the dataset generation method described above, the center positions of events in each sample are fixed to be at the center of the time window. However, the center positions of events can occur at any position within the time window. To enhance the generalization of our model, a time window random shift method is proposed here to generate complex datasets. This method relies on randomly changing the relative position of sleep apnea events with respect to the time window. Figure 4 provides an illustrative diagram of the dataset construction and the dataset augmentation method.

3.2 Experimental Setting

The performance evaluation of the FG-AFSAN in detecting sleep apnea events involved a comparison with standard network architectures (Simple CNN, LSTM, and Complex CNN). Ablation experiments were also conducted to assess the effectiveness of the fused attention mechanism. The evaluation comprised per-segment and per-recording experiments. The per-segment experiment measured the model's accuracy in identifying sleep apnea events within each signal segment, using the mean average precision (mAP) as the evaluation metric. The per-recording experiment focused on the model's diagnostic capability for classifying the severity of sleep apnea in patients. This assessment relied on the calculation of the Apnea-Hypopnea Index (AHI) using Eq. 9. Patients with an AHI less than 5 were classified as normal, while those with an AHI greater than or equal to 5 were diagnosed as abnormal.

$$AHI = \frac{N_{events}}{H} \tag{9}$$

where N_{events} represents the total number of sleep apnea events that occur during sleep, and H represents the duration of sleep unit in hours.

The AASM [28] defines a sleep apnea event as lasting at least 10 s. To capture multiple SA events, a 60-s time window is chosen and divided into 6 non-overlapping cells. Given the resampled signal at a frequency of 100 Hz, the input size of the network is $(1, 1800)$, and the output dimensions are $(1, S, C + B * (2 + 1)) = (1, 6, 7)$.

The networks in this paper were implemented in PyTorch, using the Adam optimizer with an initial learning rate of 10^{-3}. A multistepLR technique was employed for dynamic learning rate adjustment. During training, a 5-fold cross-validation was used, while during inference, all available clinical data was used to evaluate the model performance.

3.3 Experimental Results

Adaptive Preprocessing Module. We evaluated the adaptive preprocessing module (APM) by selecting signal segments with random noise and baseline drift. As illustrated in Fig. 5, the blue line represents the original signal, while the red line represents the output of the APM. The APM effectively smoothed out random noise (around 20 s and 30 s), mitigated motion artifacts (from 40 s to 45 s), corrected baseline drift (from 50 s to 60 s) and preserved signal details relevant to apnea events (from 10 s to 40 s), which demonstrating its effectiveness in enhancing the quality of the input signal.

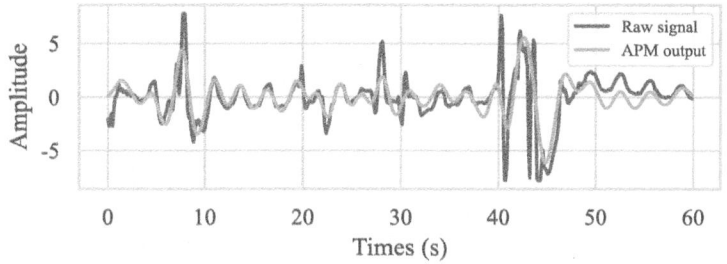

Fig. 5. Output of adaptive preprocessing module.

Per-segment. Figure 6 displays the mean average precision (mAP) results of various models on the UCDDB dataset. Notably, FG-AFSAN outperforms all the compared algorithms, demonstrating the effectiveness of the Fused Attention-based attention mechanism in sleep apnea detection. FG-AFSAN achieves an impressive mAP of 79.69% at the IOU threshold of 0.5, surpassing the other comparative algorithms by at least 4.97%. And with the increasing of IOU threshold, our model still has a higher mAP than other models.

Per-recording. For the AHI prediction experiment, Table 1 presents the classification results of different models on the clinical dataset. The Simple CNN exhibits the lowest performance, achieving accuracy and F1 score of only 0.44 and 0.36, respectively. It appears that simply stacking convolutional modules does not effectively extract features relevant to sleep apnea events or provide clinically useful screening capabilities. Therefore, considering that some sleep apnea events have a longer duration, we further adopted LSTM to improve the model's ability to capture the correlation between long sequences,

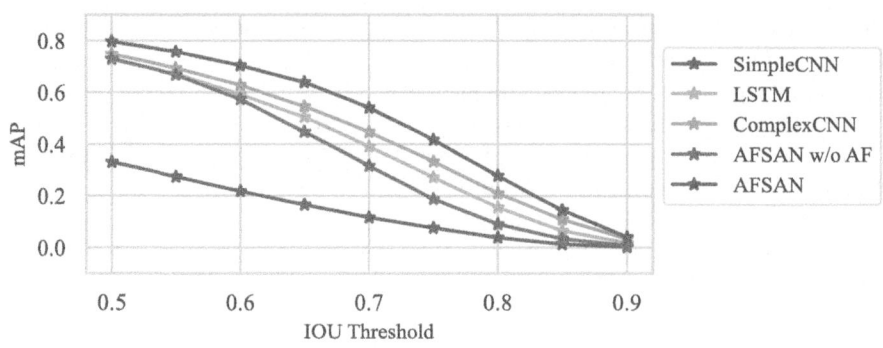

Fig. 6. mAP of different models on UCDDB dataset.

Table 1. Comparison of different network backbone on clinical dataset.

Methods	Accuracy	F1 Score
Simple CNN	0.44	0.36
LSTM	0.51	0.36
Complex CNN	0.54	0.40
Ours w/o AF	0.57	0.49
Ours	0.71	0.71

resulting in an accuracy improvement of 0.07 compared to Simple CNN. Given the complex morphology of sleep apnea events, we also utilized convolutional kernels of varying sizes to enhance the model's perception and increased the network's depth to improve its classification and detection capabilities, leading to further enhancements in accuracy and F1 score. Our model without the Attention Fusion mechanism, slightly outperforms the Complex CNN, owing to the integration of residual structures in the feature extraction layer, which allows for increased network depth while mitigating the vanishing gradient problem [29]. Lastly, our proposed model significantly improves accuracy and F1 score by introducing position and channel attention mechanisms, achieving scores of 0.71 for both metrics. This suggests that these attention mechanisms enhance the model's ability to pinpoint event start and end positions and extract event features across different dimensions. And Fig. 7 presents the relationship between true AHI and prediction results on the clinical dataset, which suggests AFSAN has an accurate detection for sleep apnea.

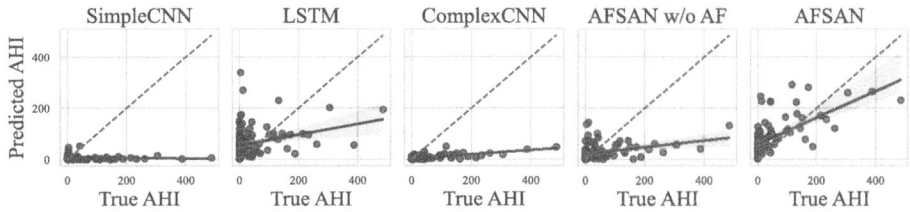

Fig. 7. Predictions of AHI on clinical dataset.

4 Conclusion

This article introduces FG-AFSAN, a novel network for detecting sleep apnea events using a fusion attention mechanism. It efficiently identifies and analyzes fine-grained abnormal respiratory events through thoracic and abdominal motion signals. We apply an adaptive preprocessing module to filter and normalize the signals, enhancing the accuracy of the detection process. By incorporating position and channel attention mechanisms, we further improve the localization of sleep apnea events. Our experimental results validate the effectiveness of this approach. Moving forward, our future work involves optimizing the network structure to enhance the classification of different sleep respiratory events, while also collecting additional clinical data for validation purposes.

Acknowledgments. This work was supported by the National Natural Science Foundation of China (Grant No. 62171471).

References

1. Gutiérrez-Tobal, G.C., Álvarez, D., Crespo, A., Del Campo, F., Hornero, R.: Evaluation of machine-learning approaches to estimate sleep apnea severity from at-home oximetry recordings. IEEE J. Biomed. Health Inform. **23**(2), 882–892 (2018)
2. Van Steenkiste, T., et al.: Portable detection of apnea and hypopnea events using bio-impedance of the chest and deep learning. IEEE J. Biomed. Health Inform. **24**(9), 2589–2598 (2020)
3. Hu, S., et al.: Semi-supervised learning for low-cost personalized obstructive sleep apnea detection using unsupervised deep learning and single-lead electrocardiogram. IEEE J. Biomed. Health Inform. **27**(11), 5281–5292 (2023). https://doi.org/10.1109/JBHI.2023.3304299
4. Yue, H., et al.: Validity study of a multiscaled fusion network using single-lead electrocardiogram signals for obstructive sleep apnea diagnosis. J. Clin. Sleep Med. **19**(6), 1017–1025 (2023)
5. Sharaf, A.I.: Sleep apnea detection using wavelet scattering transformation and random forest classifier. Entropy **25**(3), 399 (2023)
6. Setiawan, F., Lin, C.W.: A deep learning framework for automatic sleep apnea classification based on empirical mode decomposition derived from single-lead electrocardiogram. Life **12**(10), 1509 (2022)
7. Sharma, P., Jalali, A., Majmudar, M., Rajput, K.S., Selvaraj, N.: Deep-learning based sleep apnea detection using spo2 and pulse rate. In: 2022 44th Annual International Conference of the IEEE Engineering in Medicine & Biology Society (EMBC), pp. 2611–2614. IEEE (2022)

8. Tang, X., Zhao, L., Shuai, Y., Li, Z., Wang, X.: An one-dimensional signal based object detection network for apnea and hypopnea locating. In: Fourteenth Inter-national Conference on Digital Image Processing (ICDIP 2022). vol. 12342, pp. 1132–1140. SPIE (2022)
9. Massie, F., Vits, S., Khachatryan, A., Van Pee, B., Verbraecken, J., Bergmann, J.: Central sleep apnea detection by means of finger photoplethysmography. IEEE J. Trans. Eng. Health Med. **11**, 126–136 (2023)
10. Albuhayri, M.A.: Cnn model for sleep apnea detection based on spo2 signal. Comput. Inform. Science **16**(1), 1–39 (2023)
11. Paul, T., et al.: Ecg and spo2 signal-based real-time sleep apnea detection using feed-forward artificial neural network. In: AMIA Annual Symposium Proceedings, vol. 2022, pp. 379. American Medical Informatics Association (2022)
12. Sharma, M., Kumbhani, D., Yadav, A., Acharya, U.R.: Automated sleep apnea detection using optimal duration-frequency concentrated wavelet-based features of pulse oximetry signals. Appli. Intell., 1–13 (2022)
13. Yang, W., Fan, J., Wang, X., Liao, Q.: Sleep apnea and hypopnea events detection based on airflow signals using lstm network. In: 2019 41st Annual International Conference of the IEEE Engineering in Medicine and Biology Society (EMBC), pp. 2576–2579. IEEE (2019)
14. Ng, A.S., Chung, J.W., Gohel, M.D., Yu, W.W., Fan, K.L., Wong, T.K.: Evaluation of the performance of using mean absolute amplitude analysis of thoracic and abdominal signals for immediate indication of sleep apnoea events. J. Clin. Nurs. **17**(17), 2360–2366 (2008)
15. Sannino, G., De Falco, I., De Pietro, G.: An automatic rules extraction approach to support osa events detection in an mhealth system. IEEE J. Biomed. Health Inform. **18**(5), 1518–1524 (2014)
16. Yan, X., Wang, L., Zhu, J., Wang, S., Zhang, Q., Xin, Y.: Automatic obstructive sleep apnea detection based on respiratory parameters in physiological signals. In: 2022 IEEE International Conference on Mechatronics and Automation (ICMA), pp. 461–466. IEEE (2022)
17. Lin, Y.Y., Wu, H.T., Hsu, C.A., Huang, P.C., Huang, Y.H., Lo, Y.L.: Sleep apnea detection based on thoracic and abdominal movement signals of wearable piezo-electric bands. IEEE J. Biomed. Health Inform. **21**(6), 1533–1545 (2016)
18. Lv, X., Li, J., Yan, Q.: Automated detection of sleep apnea from abdominal respiratory signal using hilbert-huang transform. In: Bioinformatics Research and Applications: 16th International Symposium, ISBRA 2020, Moscow, Russia, December 1–4, 2020, Proceedings 16. pp. 364–371. Springer (2020). https://doi.org/10.1007/978-3-030-57821-3_35
19. Haidar, R., McCloskey, S., Koprinska, I., Jeffries, B.: Convolutional neural net-works on multiple respiratory channels to detect hypopnea and obstructive apnea events. In: 2018 International Joint Conference on Neural Networks (IJCNN), pp. 1–7 (2018). https://doi.org/10.1109/IJCNN.2018.8489248
20. Van Steenkiste, T., Groenendaal, W., Deschrijver, D., Dhaene, T.: Automated sleep apnea detection in raw respiratory signals using long short-term memory neural networks. IEEE J. Biomed. Health Inform. **23**(6), 2354–2364 (2018)
21. ElMoaqet, H., Eid, M., Glos, M., Ryalat, M., Penzel, T.: Deep recurrent neural networks for automatic detection of sleep apnea from single channel respiration signals. Sensors **20**(18), 5037 (2020)
22. Shen, F., Cheng, S., Li, Z., Yue, K., Li, W., Dai, L., et al.: Detection of snore from OSAHS patients based on deep learning. J. Healthcare Eng. **2020** (2020)
23. Fu, J., Liu, J., Tian, H., Li, Y., Bao, Y., Fang, Z., Lu, H.: Dual attention network for scene segmentation. In: Proceedings of the IEEE/CVF Conference on Computer Vision and Pattern Recognition, pp. 3146–3154 (2019)
24. Ba, J.L., Kiros, J.R., Hinton, G.E.: Layer normalization. arXiv preprint-arXiv:1607.06450 (2016)

25. Wang, Z., Zhang, J., Zhang, X., Chen, P., Wang, B.: Transformer model for functional near-infrared spectroscopy classification. IEEE J. Biomed. Health Inform. **26**(6), 2559–2569 (2022)
26. Segal, Y., Fuchs, T.S., Keshet, J.: Speechyolo: Detection and localization of speech objects. arXiv preprint arXiv: 1904.07704 (2019)
27. Heneghan, C.: St. vincent's university hospital/university college dublin sleep apnea database. Vincent's university hospital/University College Dublin sleep apnea database (2011)
28. Adult Obstructive Sleep Apnea Task Force of the American Academy of Sleep Medicine, A.O.S.A.T.F.: Clinical guideline for the evaluation, management and long-term care of obstructive sleep apnea in adults. J. Clin. Sleep Med. **5**(3), 263–276 (2009)
29. He, K., Zhang, X., Ren, S., Sun, J.: Deep residual learning for image recognition. In: Proceedings of the IEEE Conference on Computer Vision and Pattern Recognition, pp. 770–778 (2016)

Automatic Seizure Recognition Based on Data Enhancement and 1DCNN-BiLSTM Network Using EEG Signal

Wenrong Hu[1], Junliang Shang[1], Juan Wang[1], Jin-Xing Liu[2], Yuxia Wang[1], and Shasha Yuan[1(✉)]

[1] School of Computer Science, Qufu Normal University, Rizhao 276826, China
`jiayouyss@126.com`
[2] School of Health and Life Sciences, University of Health and Rehabilitation Sciences, Qingdao 266114, China

Abstract. Epilepsy is one of the most common life-threatening neurological disorders in the world. The uncertain nature of seizures poses a challenge in the detection and recognition of epilepsy. Aiming at the problem of imbalance between seizure and non-seizure data, an automatic seizure recognition based on data enhancement and 1DCNN-BiLSTM network using scalp EEG is proposed in this paper. The raw EEG signals are denoised by discrete wavelet transform(DWT), and three frequency band sub-signals are selected for restructuring. Then, three efficient data enhancement techniques, MixUp, SuperMix, and Co-MixUp are used to address data imbalance. The model uses a combination of Convolutional Neural Network (CNN) with attention mechanism and Bidirectional Long Short-Term Memory (BiLSTM) to capture key EEG feature capabilities. Finally, the classification results of the three EEG sub-signals are fused to recognize epileptic seizures. Evaluations and experiments are performed on two public EEG datasets, Bonn dataset and CHB-MIT dataset, using five-fold cross-validation to assess the generalization ability of the model. Experimental results show that data enhancement based on Co-MixUp is more suitable for 1DCNN-BiLSTM model.

Keywords: EEG · Seizure recognition · Data enhancement · Convolutional Neural Network · Bidirectional Long Short-Term Memory

1 Introduction

Epilepsy is a chronic neurological brain disease characterized by sudden onset, which is caused by abnormal brain electrical signals due to disturbed nerve cell activity in the brain [1]. To diagnose epileptic disorders caused by abnormal neuronal discharges, EEG signals were introduced [2]. The current diagnosis of epilepsy is usually by manually analyzing EEG signals. However, this method is not only time-consuming but also relies on the experience and knowledge of experts. Therefore, designing an accurate recognition method for epilepsy has a vital role in protecting patients' lives.

Over the past decade, the methods of using EEG signal to diagnose epileptic seizures have developed rapidly, mainly including traditional machine learning and deep learning. Sharma et al. used Gaussian model and combined with multiple support vector machine (MSVM) to complete epilepsy classification detection[3]. Among the latest developments, Sharan et al. extended the traditional method of seizure detection by using the original power spectrum of EEG signals and CNN [4]. In order to make up for the inability of CNN to capture global features [5], Wu et al. used the Long Short Term Memory (LSTM) to predict epileptic seizures end-to-end [6]. Due to the limitations of LSTM, Tuncer et al. used BiLSTM to mine EEG data information [7]. Later, Ahmad et al. integrated CNN with BiLSTM to effectively extract spatial and time series information [8]. Most studies are based on sufficient sample data, and the problem of data imbalance has brought certain challenges to the research. In view of the problem of data imbalance, Cubuk et al. have proposed a study using task-dependent transformations to enhance training data [9].

To further solve the problem of EEG signals data imbalance, this paper proposed an automatic seizure recognition model based on data enhancement and 1DCNN-BiLSTM network. The model extracts local features from the 1DCNN layer as initial input state embeddings for the BiLSTM layer, fuses the local and contextual features, and generates hybrid features. Then, the model uses the Attention layer to optimize the hybrid features. Finally, the model deploys a Full Connection Layer to map the output of the Attention mechanism to lower dimensions and then classify the final feature vector. This approach takes into account temporal characteristics and often achieves encouraging performance in the diagnosis of seizures.

The rest of this article is structured as follows. In Sect. 2, we describe the data set used in the experiment. Section 3 details the model in the study. Experimental results are presented in Sect. 4. Section 5 and Sect. 6 are the discussion and summary.

2 EEG Database

Two datasets, Bonn and CHB-MIT, were used to verify the effect of the model. The Bonn dataset contains EEG signals of healthy subjects and epileptic patients during seizure and non-seizure time [10]. The A and B contain EEG signal recordings from healthy subjects in different states. The C and D are intracranial EEG recordings of interictal periods obtained from the hippocampus in the epileptic focus and in the hemisphere contralateral to the focus. The CHB-MIT dataset contains long-term EEG data collected from 23 patients. The experiment selected 22 channels shared by most patients. For the lack of F4-C4 channels in chb21, only 21 channels are used.

2.1 Model Architecture

As shown in Fig. 1, the proposed network consists of the following five modules: (1) Preprocessing with DWT and data enhancement, (2) 1DCNN block, (3) BiLSTM block, (4) Self-Attention block, (5) Full Connection Layer block and (6) Discriminative Decision block.

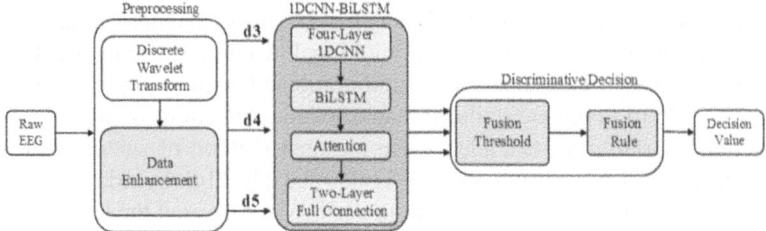

Fig. 1. Framework of the proposed automatic seizure recognition.

2.2 Preprocessing

Discrete Wavelet Transform.   We use DWT to decompose the signal into 4-8Hz (d5), 8-16Hz (d4), 16-32Hz (d3), 32-64Hz (d2), and 64-128Hz (d1). In order to ensure the integrity of information acquisition, we selected EEG data with a frequency range of 4-32HZ(d3-d5). Figure 2 shows signal reconstruction of EEG data.

Data Enhancement.   Data enhancement is an effective technique for improving classification accuracy to increase data by using transformations that preserve the input context in learning problems[9]. Mixup, Co-Mixup, and SuperMix are several typical techniques used for data enhancement in deep learning.

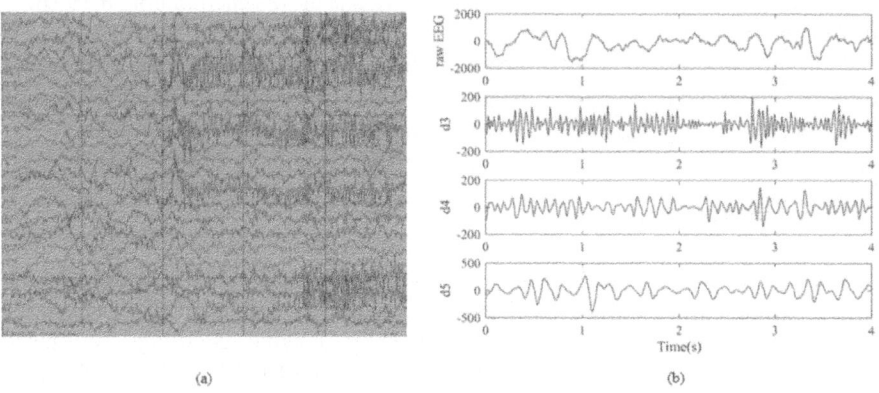

Fig. 2. Signal reconstruction of EEG signals after DWT processing. (a) the initial 22-channel 4s EEG signal on the CHB-MIT dataset. (b) the raw signal and three sub-signals after DWT

MixUp. While most data enhancement algorithms are only applied to image data, MixUp can be applied to sequential data. By randomly selecting vectors and corresponding labels of two training samples, MixUp uses linear interpolation to generate a new vector and corresponding labels as augmented data. To expand the training data and enhance the deep learning model generalization. The kernel formula of MixUp is defined as:

$$\tilde{x} = \lambda x_i + (1 - \lambda)x_j \tag{1}$$

Here, x_i, x_j are two randomly selected samples, and $\lambda \in [0, 1]$ [0, 1].

SuperMix. SuperMix is another advanced data enhancement technique that further develops the idea of Mixup [11]. SuperMix adds class information on top of MixUp, an approach designed to create more challenging training samples that improve the model's generalization and robustness. The mixing strategy is defined as:

$$\hat{x} = \sum_{i=1}^{n} x_i \odot w_i \tag{2}$$

where x_i is the clipped sample, w_i is the weight of each clipped sample, the operator $\odot$ denotes the element-wise product.

Co-MixUp. Considering that MixUp does not make full use of the rich informative supervision signals in the data, an extended Co-MixUp algorithm was proposed [12]. Co-MixUp not only considers the characteristics of the samples, but also the category they belong to, ensuring that the mixed samples are from the same class.

2.3 1DCNN-BiLSTM Network

We propose a progressive model to fuse the features of different layers. Figure 3 details 1DCNN-BiLSTM network.

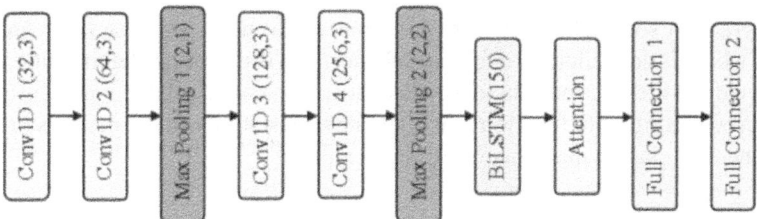

Fig. 3. Structure of 1DCNN-BiLSTM network.

1DCNN Block. CNNs have achieved good results in various tasks. The 4-layer 1DCNN used in the study can better learn the higher-order features in the data and enhance the data feature capture capability. Maximum pooling is attached to CNN combination to directly extract largest features to downsample output data, thus reducing the resolution of the output of the CNN convolutional layer and preventing overfitting while allowing better learning of multi-level local abstract features in continuous data for higher-level and more complex learning.

BiLSTM Block. The proposal of LSTM can better capture the long memory problem in temporal information processing, but it is unable to address encoding back-to-front information. In this experiment, BiLSTM is used for temporal weight sharing, which is combined with spatial weight sharing of CNN. BiLSTM is a combination of reverse LSTM and forward LSTM to relate contextual information, which can better perform bidirectional semantic dependency. Figure 4 are the structures of LSTM and BiLSTM.

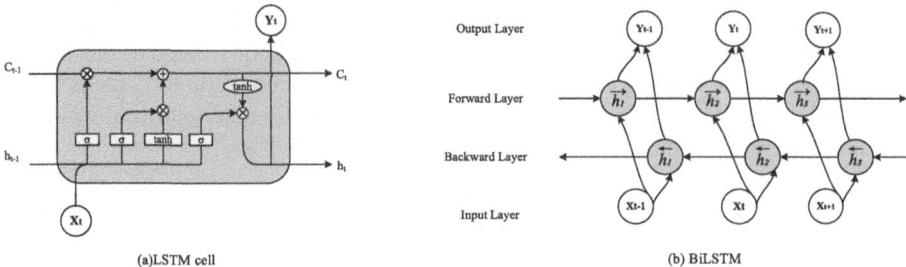

(a)LSTM cell   (b) BiLSTM

Fig. 4. BiLSTM structure. (a) the structure of LSTM cell. (b) the structure of BiLSTM.

Self-attention Block. The attention mechanism allows neural networks to process important information in the data automatically processing sequence data [13]. Combining the attention mechanism with BiLSTM can establish associations not only with elements at neighboring bit positions in the sequence but also with other relevant elements in the sequence, which can adaptively capture long-term dependencies.

Full Connection Layer Block. Two dense layers are used in this layer. The reason why two dense layers are used is because the sigmoid is too obvious for the function value to change, which affects the accuracy value of the neural network. The tanh function can keep the nonlinear fluctuation on input and output, which is in line with the neural network gradient solving and good fault tolerance. So the first dense layer has a tanh function, which is equivalent to a feature selector, and is useful for determining whether the extracted features are relevant to the classification category; the second dense layer is a classifier with a sigmoid function that maintains the stability of the model's prediction performance for classifying seizures or non-seizures.

2.4 Discriminative Decision

The signal probability value output by the model is calculated by difference. If the probability value is greater than the threshold, the label value is set to "1", otherwise the label value is set to "0". The threshold was selected as 0.75. When labeling the third-level sub-signals, at least one label value is "1", it can be identified as a seizure.

3 Experiment and Results

3.1 Evaluation Indicators

This paper mainly uses three evaluation metrics, accuracy, sensitivity, and specificity for the analysis of the experimental results.

$$Accuracy = \frac{TP + TN}{TP + FP + TN + FN} \tag{3}$$

$$Sensitivity = \frac{TP}{TP + FN} \tag{4}$$

$$Specificity = \frac{TN}{TN + FP} \tag{5}$$

where TP is the amount of seizures correctly detected by the model, TN is the amount of non-seizures correctly detected by the model, FP is the amount of the model identified non-seizures as seizures and FN is the number of the model identified seizures as non-seizures.

3.2 Experimental Results

Results on the Bonn Dataset. Considering the small number of dataset, the classification experiment with seven combination forms were designed. It can be seen from Table 1 that the sensitivity in case1, case2, and case5 reached 100%.

Results on the CHB-MIT Dataset. It can be clearly found in Table 2 that Co-MixUp brings batter performance that cannot be ignored. And, in patient chb06, the 100% of accuracy, sensitivity and specificity were achieved. In all patients, there were 20 patients with accuracy and sensitivity above 92.5%.

Table 1. Experimental results on the Bonn dataset

Name	Data	Accuracy(%)				Sensitivity(%)				Specificity(%)			
		None	MixUp	SuperMix	Co-MixUp	None	MixUp	SuperMix	Co-MixUp	None	MixUp	SuperMix	Co-MixUp
Case1	E-A	100.00	100.00	100.00	100.00	100.00	100.00	100.00	100.00	100.00	100.00	100.00	100.00
Case2	E-B	97.50	100.00	92.50	100.00	100.00	100.00	100.00	100.00	95.65	100.00	86.96	100.00
Case3	E-C	97.50	97.50	97.50	97.50	100.00	97.06	97.12	100.00	95.65	97.83	100.00	95.65
Case4	E-D	95.00	97.50	100.00	100.00	94.12	97.06	100.00	100.00	95.65	97.83	100.00	100.00
Case5	E-AB	98.33	100.00	100.00	100.00	100.00	100.00	100.00	100.00	97.44	100.00	100.00	100.00
Case6	E-CD	98.33	98.33	99.15	99.17	100.00	95.24	100.00	100.00	97.44	100.00	98.57	98.72
Case7	All	98.50	98.00	98.50	99.50	93.48	91.30	95.65	97.75	98.50	100.00	99.35	100.00

3.3 Ablation Experiments

To verify the CNN combination's contribution, comparative experiments are conducted using CNN-BiLSTM model, the CNN-Attention model and 1DCNN-BiLSTM. The parameters were set the same for all experimental models. 1DCNN-BiLSTM can significantly improve the ability of training the feature data, and show good performance in classification. Figure 5 shows ablation experiment of our proposed method.

Table 2. Experimental results on 24 patients in the CHB-MIT dataset

Case	Accuracy(%)				Sensitivity(%)				Specificity(%)			
	None	Mix Up	SuperMix	Co-MixUp	None	Mix Up	SuperMix	Co-MixUp	None	Mix Up	SuperMix	Co-MixUp
Chb01	99.66	99.18	95.98	99.65	95.45	97.27	95.45	98.18	99.91	99.27	96.00	99.73
Chb02	99.28	99.70	99.05	99.17	95.00	95.00	90.00	97.50	99.50	100.00	99.50	99.25
Chb03	99.16	99.29	98.52	97.67	97.00	97.00	97.00	98.00	99.20	99.40	98.60	97.65
Chb04	96.49	97.24	98.40	92.68	63.16	73.68	86.31	93.69	98.16	98.42	99.00	92.63
Chb05	99.51	99.57	99.89	99.56	97.78	98.52	97.78	100.00	99.59	99.63	100.00	99.54
Chb06	100.00	100.00	100.00	100.00	100.00	100.00	100.00	100.00	100.00	100.00	100.00	100.00
Chb07	99.70	99.70	99.82	99.88	98.75	98.75	97.32	100.00	99.75	99.75	99.94	93.96
Chb08	97.64	99.21	96.71	94.22	88.70	88.70	79.57	99.57	98.09	99.74	97.57	91.31
Chb09	99.86	99.86	99.8	99.86	97.14	97.14	95.72	97.14	100.00	100.00	100.00	100.00
Chb10	99.87	99.87	99.58	99.90	97.78	97.78	99.78	97.78	100.00	100.00	99.67	100.00
Chb11	99.88	99.98	99.88	99.67	98.00	100.00	97.50	97.50	99.98	99.99	100.00	99.78
Chb12	92.86	92.86	98.57	98.22	90.48	90.48	98.10	97.62	95.24	95.24	99.05	98.81
Chb13	81.96	81.96	81.96	96.86	75.76	75.76	75.76	82.96	82.27	82.27	82.27	97.56
Chb14	70.65	92.73	97.14	84.42	72.73	85.46	68.18	87.88	70.45	100.00	98.87	84.09
Chb15	93.75	93.75	95.63	97.89	93.69	93.69	95.00	96.91	95.00	95.00	96.25	97.94
Chb16	90.00	87.50	95.98	98.22	85.00	75.00	87.50	87.50	95.00	100.00	99.38	98.75
Chb17	98.98	98.98	98.98	99.13	94.67	94.67	94.67	100.00	99.20	99.20	99.20	99.08
Chb18	98.13	98.13	99.70	98.99	97.50	97.50	96.88	95.00	98.75	98.75	99.69	99.19
Chb19	99.76	99.76	99.76	97.94	95.00	95.00	95.00	95.00	100.00	100.00	100.00	98.08
Chb20	98.67	98.67	98.67	98.67	98.67	98.67	98.67	98.67	98.67	98.67	98.67	98.67
Chb21	99.52	97.00	93.33	95.52	90.00	96.00	90.00	98.00	100.00	98.00	94.00	98.80
Chb22	99.43	99.43	99.43	98.76	98.00	98.00	98.00	98.00	99.59	99.59	99.59	98.80
Chb23	87.50	90.94	88.81	93.97	66.67	73.30	86.65	93.33	93.33	93.30	83.33	94.00
Chb24	85.36	96.30	92.98	86.47	74.07	59.26	66.67	84.44	85.93	98.15	94.33	86.44
Mean	95.31	96.74	97.02	96.97	90.04	90.69	91.06	95.61	96.15	98.10	97.29	97.05

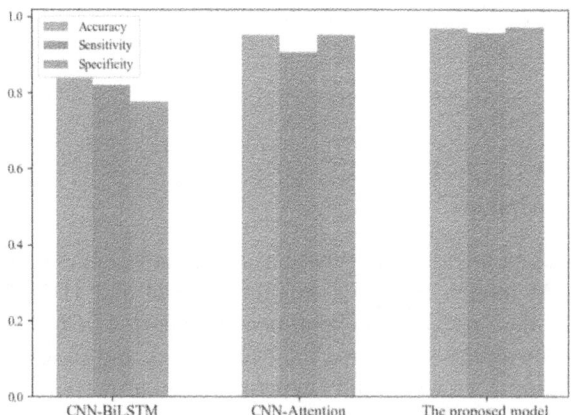

Fig. 5. The proposed model (CNN-BiLSTM-Attention) is compared with CNN-BiLSTM and CNN-Attention.

4 Discussion

The imbalance of epilepsy samples is also a problem often encountered in practical applications. To address this challenge, we propose a seizure recognition based on data enhancement and 1DCNN-BiLSTM network using scalp EEG. The evaluation results on the datasets show that 1DCNN-BiLSTM can learn effective feature representations from the data, and the model has stable performance of epilepsy recognition and classification detection. Due to the large amount of noise in some EEG data collected, the experimental effect of patients chb04, chb08, chb13, chb14, chb16, chb23 and chb24 was poor. The use of Co-MixUp can greatly improve the smoothing of adjacent samples and reduce the influence of noisy data on classification performance.

T-SNE (t-distributed stochastic neighbor embedding) is an effective way to reduce dimension visualization of 1DCNN-BiLSTM output features [14]. To validate the performance of 1DCNN-BiLSTM, t-SNE technique is mapping data to a two-dimensional space. The raw EEG signals and feature mappings of each layer are shown in Fig. 6. The model can distinguish the two types of data well.

Table 3. compares 1DCNN-BiLSTM with several available seizure recognition methods. All of these methods were evaluated in studies in the CHB-MIT. Our proposed method is more promising for recent studies in terms of average sensitivity and specificity. Although compared to method proposed by Shen et al. [15], their model has a higher sensitivity. However, our model does not use eigenvalue algorithm to extract features of different sub-bands. Low sensitivity is unacceptable in practical applications. The combined model method (CNN-LSTM, 1DCNN-BiLSTM) has obtained better results than single model method (1DCNN-Spectrogram, BiLSTM), and verified the feature fusion intensity of different models [16, 17]. Varli et al. [18] used the original EEG performed signal conversion for spectral analysis, but our method has advantages in accuracy and specificity. To sum up, our model has higher performance.

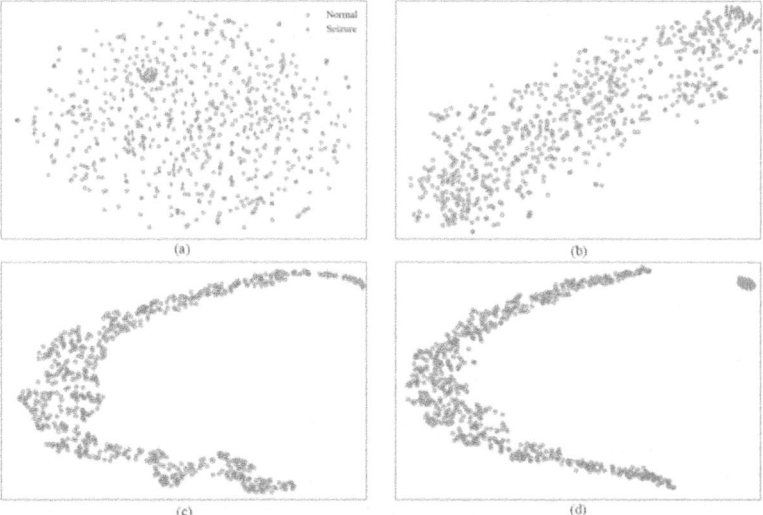

Fig. 6. T-SNE visualization of proposed model in chb06. (a) is the raw EEG data; (b), (c) and (d) are signal embeddings in the 1DCNN layer, BiLSTM layer and Attention layer respectively.

Table 3. Performance comparison of the other models on the CHB-MIT dataset.

Authors	Method	Accuracy(%)	Sensitivity(%)	Specificity(%)
Shen et al. [15]	SVM-RUSBoosted tree	96.38	96.15	96.76
Jana et al. [16]	1DCNN-Spectrogram	88.00	86.52	91.11
Yao et al. [17]	BiLSTM	83.89	83.72	84.06
Varli et al. [18]	CNN-LSTM	95.08	96.70	93.28
The proposed model	Co-mixup-1DCNN-BiLSTM	96.97	95.61	97.05

5 Conclusion

In this study, an automatic seizure recognition technique based on EEG data enhancement and 1DCNN-BiLSTM network was proposed. According to our results, the processing of unbalanced data with Co-MixUp data enhancement has good performance. The network has significant advantages in processing classification performance of more informative series data. It can learn effective feature representations which is important for doctors' clinical diagnosis. In future research, the database will be added to verify the experimental performance, and the algorithm will be improved to improve the recognition accuracy.

Acknowledgment. This work was supported by the Program for Youth Innovative Research Team in the University of Shandong Province in China (No.2022KJ179), and jointly supported by the National Natural Science Foundation of China (No. 62172253).

References

1. Gabeff, V., et al.: Interpreting deep learning models for epileptic seizure detection on EEG signals. Artif. Intell. Med. **117**, 102084 (2021)
2. Chu, H., Chung, C.K., Jeong, W., Cho, K.-H.: Predicting epileptic seizures from scalp EEG based on attractor state analysis. Comput. Methods Programs Biomed. **143**, 75–87 (2017)
3. Sharma, R., Chopra, K.: EEG-based epileptic seizure detection using GPLV model and multi support vector machine. J. Inf. Optim. Sci. **41**, 143–161 (2020)
4. Sharan, R.V., Berkovsky, S.: Epileptic seizure detection using multi-channel EEG Wavelet power spectra and 1-D convolutional neural networks. In: Annual International Conference of the IEEE Engineering in Medicine and Biology Society IEEE Engineering in Medicine and Biology Society Annual International Conference 2020, pp. 545–548 (2020)
5. Lu, G., Liu, Y., Wang, J., Wu, H.: CNN-BiLSTM-attention: a multi-label neural classifier for short texts with a small set of labels. Inf. Process. Manag. **60** (2023)

6. Wu, X., Yang, Z., Zhang, T., Zhang, L., Qiao, L.: An end-to-end seizure prediction approach using long short-term memory network. Front. Hum. Neurosci. **17** (2023)
7. Tuncer, E., Bolat, E. D.: Classification of epileptic seizures from electroencephalogram (EEG) data using bidirectional short-term memory (Bi-LSTM) network architecture. Biomed Signal Process Control **73** (2022)
8. Ahmad, I., Wang, X., Javeed, D., Kumar, P., Samuel, O.W., Chen, S.: A hybrid deep learning approach for epileptic seizure detection in EEG signals. IEEE J. Biomed. Health. Inf. (2023)
9. Cubuk, E.D., Zoph, B., Mané, D., Vasudevan, V., Le, Q. V.: AutoAugment: learning Augmentation Strategies From Data. In: 2019 IEEE/CVF Conference on Computer Vision and Pattern Recognition (CVPR), pp. 113–123 (2019)
10. Andrzejak, R.G., Lehnertz, K., Mormann, F., Rieke, C., David, P., Elger, C.E.: Indications of nonlinear deterministic and finite-dimensional structures in time series of brain electrical activity: dependence on recording region and brain state. Phys. Rev. E.**64**, 061907 (2001)
11. Dabouei, A., Soleymani, S., Taherkhani, F., Nasrabadi, N. M.:SuperMix: supervising the Mixing Data Augmentation. In: 2021 IEEE/CVF Conference on Computer Vision and Pattern Recognition (CVPR), pp. 13789–13798 (2021)
12. Kim, J.-H., Choo, W., Jeong, H., Song, H. O.: Co-Mixup: Saliency Guided Joint Mixup with Supermodular Diversity. ArXiv.abs/2102.03065 (2021)
13. Niu, D., Yu, M., Sun, L., Gao, T., Wang, K.: Short-term multi-energy load forecasting for integrated energy systems based on CNN-BiGRU optimized by attention mechanism. Appl Energy **313** (2022)
14. van der Maaten, L., Hinton, G.: Visualizing Data using t-SNE. J. Mach. Learn. Res. **9**, 2579–2605 (2008)
15. Shen, M. K., Wen, P., Song, B., Li, Y.: An EEG based real-time epilepsy seizure detection approach using discrete wavelet transform and machine learning methods. Biomed. Signal Process Control **77** (2022)
16. Jana, G.C., Sharma, R., Agrawal, A.: A 1D-CNN-spectrogram based approach for seizure detection from EEG Signal. Proc. Comput. Sci. **167**, 403–412 (2020)
17. Yao, X., Li, X., Ye, Q., Huang, Y., Cheng, Q., Zhang, G.-Q.: A robust deep learning approach for automatic classification of seizures against non-seizures. Biomed. Signal Process. Control **64**, 102215 (2021)
18. Varlı, M., Yılmaz, H.: Multiple classification of EEG signals and epileptic seizure diagnosis with combined deep learning. J. Comput. Sci. **67**, 101943 (2023)

DP-DCAN: Differentially Private Deep Contrastive Autoencoder Network for Single-Cell Clustering

Huifa Li[1], Jie Fu[1], Zhili Chen[1(✉)], Xiaomin Yang[2(✉)], Haitao Liu[1], and Xinpeng Ling[1]

[1] Shanghai Key Laboratory of Trustworthy Computing, East China Normal University, Shanghai 200241, China
zhlchen@sei.ecnu.edu.cn

[2] School of Medicine, Xinhua Hospital, Shanghai Jiao Tong University, Shanghai 200241, China
yangxiaomin@xinhuamed.com.cn

Abstract. Single-cell RNA sequencing (scRNA-seq) is important to transcriptomic analysis of gene expression. Recently, deep learning has facilitated the analysis of high-dimensional single-cell data. Unfortunately, deep learning models may leak sensitive information about users. As a result, Differential Privacy (DP) is increasingly being used to protect privacy. However, existing DP methods usually perturb whole neural networks to achieve differential privacy, and hence result in great performance overheads. To address this challenge, in this paper, we take advantage of the uniqueness of the autoencoder that it outputs only the dimension-reduced vector in the middle of the network, and design a Differentially Private Deep Contrastive Autoencoder Network (DP-DCAN) by partial network perturbation for single-cell clustering. Firstly, we use contrastive learning to enhance the feature extraction of the autoencoder. And then, since only partial network is added with noise, the performance improvement is obvious and twofold: one part of network is trained with less noise due to a bigger privacy budget, and the other part is trained without any noise. Experimental results of 8 datasets have verified that DP-DCAN is superior to the traditional DP scheme with whole network perturbation. The code is available at https://github.com/LFD-byte/DP-DCAN.

Keywords: scRNA-seq data · Autoencoder · Differential privacy · Contrastive learning

---

H. Li and J. Fu—Contributing equally to this work.

D.-S. Huang et al. (Eds.): ICIC 2024, LNBI 14881, pp. 380–392, 2024.
https://doi.org/10.1007/978-981-97-5689-6_33

1   Introduction

Single-cell RNA sequencing (scRNA-seq) has enabled unbiased, high-throughput, and high-resolution transcriptome analysis at the level of individual cells [5], playing a crucial role in identifying cell types through transcriptome analysis, investigating developmental biology, uncovering intricate diseases and predicting cell developmental pathways. Therefore, the accurate identification of cell types is a crucial step in scRNA-seq analysis. Clustering is a very powerful method for cell type annotation, as it can identify cell types in an unbiased manner. In the past years, deep embedding clustering methods have been used and advanced to analyze scRNA-seq data [4, 14, 17].

However, drawing meaningful insights in many of these approaches fundamentally relies upon the utilization of privacy-sensitive, training data associated with individuals. scRNA-seq data yields some important privacy concerns, including genetic conditions and predispositions to specific diseases [11]. In contrast to the modifiability of credit card numbers, scRNA-seq data remains unchangeable once revealed. The revelation of these holds the potential to detrimentally affect patients by influencing their prospects in employment and education. For example, a hospital wants to share AI models based on scRNA-seq data for learning and research. They must get patient consent and ensure models don't expose private health info. But deep learning models can memorize data, risking privacy when used. Privacy risks are also seen in membership inference attacks, which could reveal if a person's data was used to train a model.

It is thus apparent that the implementation of privacy-enhancing techniques is required to facilitate the training of models of sensitive scRNA-seq data. One of the state-of-the-art paradigms to prevent privacy disclosure in machine learning is differential privacy (DP) [1, 6, 7]. It operates on the principle of introducing controlled noise into the data or its analysis results, which is carefully calibrated to protect individual privacy without significantly compromising the data's utility for group-level inferences. Several studies have demonstrated that the implementation of differential privacy can effectively mitigate the inadvertent disclosure of private training data, including vulnerabilities to membership inference attacks [9] during the training process. There are a few works that take into account medical and biological data in differentially private learning [2, 10, 15, 18]. However, there is no differentially private learning methods have taken scRNA-seq data into account.

On the other hand, the random noise introduced by differential privacy leads to poor performance of current differential privacy deep learning models. Additionally, research indicates that the noise introduced by differential privacy increases as the model size grows [12]. Achieving good performance of trained models while ensuring differential privacy protection is a challenge, especially in autoencoders, as they are more sensitive to random noise [8].

In this paper, we first introduce a <u>D</u>ifferentially <u>P</u>rivate <u>D</u>eep <u>C</u>ontrastive <u>A</u>utoencoder <u>N</u>etwork for single-cell dimension reduction and clustering (DP-DCAN). The algorithm leverages autoencoders to learn discriminative features of instances and utilizes contrastive learning to enhance the extraction of features by contrasting the similarity and dissimilarity of samples under different clusters. In addition, we incorporate differential privacy into the process to protect the privacy of clustering outputs. However, adding differential privacy protection at the autoencoder stage is non-trivial because the random noise introduced by differential privacy affects the training effectiveness of the autoencoder. Therefore, based on the post-processing property of differential privacy, we designed a differential privacy method with partial network perturbation that can effectively reduce the impact of noise on our autoencoder model. The experimental results demonstrated that our algorithm DP-DCAN can achieve excellent performance under differential privacy protection on real scRNA-seq datasets.

2 Privacy Threat Model

In the context of deep learning models for scRNA-seq data sharing as shown in Fig. 1, it is presumed that the patients and medical center are entities considered trustworthy. Conversely, the querying party, which utilizes the shared model, is assumed to be honest but curious. Therefore, the querying party is curious about sensitive information in scRNA-seq data analysis. Also, considering the querying party has full background knowledge of genome data. Patients send biological samples to the medical center, where scRNA-seq data can be obtained by sequencing. Therefore, the scRNA-seq data of the medical center is reliable. In the scRNA-seq data sharing model, the specific privacy threat model is as follows: The querying party is honest but curious; the querying party can obtain all background knowledge about patients' scRNA-seq data; Patients, medical center, and querying party are rational.

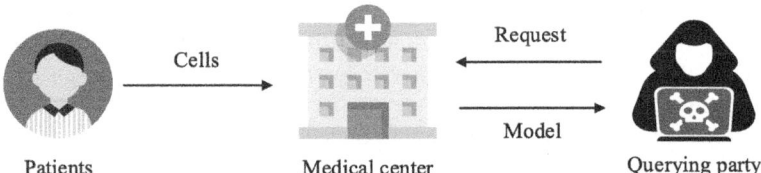

Patients   Medical center   Querying party

Fig. 1. Privacy threat model of scRNA-seq data deep learning.

3 Methodology

In this section, we first introduce the proposed method termed DP-DCAN. We then present our idea of a differentially private single-cell autoencoder network. And we introduce a contrastive learning module. Finally, we elaborate on the proposed loss function in DP-DCAN.

3.1 Overview

As shown in Fig. 2, DP-DCAN framework mainly consists of two key components: the private single-cell autoencoder network and the two-stage contrastive learning module. In the private single-cell autoencoder network, the autoencoder is trained to learn the essential feature representation of the scRNA-seq data while preventing leakage of sensitive information elegantly. In the two-stage contrastive learning module, we further enhance the autoencoder's understanding of feature representation by comparing the similarities and disparities between different samples and across various clusters.

The detailed procedure of DP-DCAN is outlined in Algorithm 1. The algorithm employs a two-stage training. In the first stage, the autoencoder maximizes the similarity between instances from the same cluster in compressed feature space as shown in Lines 4–8. The second stage enables the autoencoder to maximize the distance between different clusters as shown in Lines 11–15. The design of loss functions $\mathcal{L}_{instance}$ and $\mathcal{L}_{cluster}$ for two-stage training is shown in Sect. 3.4. In the two stages, Gaussian noise is added to the encoder network θ_e of autoencoder while optimizing the model in Lines 6 and 13. The decoder network θ_d doesn't be disturbed because it's useless during inference. Lines 7 and 14 keep track of the privacy loss of RDP and convert it to (ϵ, δ) in Line 16.

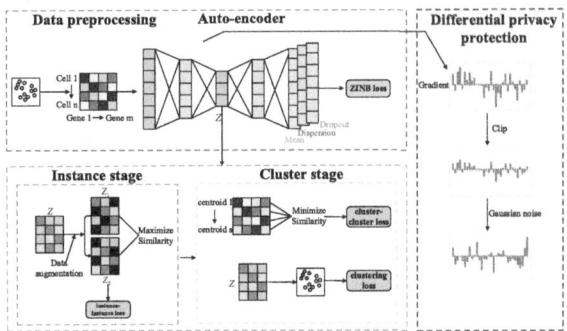

Fig. 2. The overview of DP-DCAN.

Algorithm 1 DP-DCAN framework

Inputs: Examples $X = \{x_1, x_2, \ldots, x_N\}$, learning rate η, lot size L, noise scale σ, clipping bound C.

Output: θ_e and privacy budget (ϵ, δ)

1:    Initialize autoencoder parameter $\theta = \{\theta_e, \theta_d\}$ randomly, $R_1 = 0, R_2 = 0$
2:    # Instance stage
3:    Design $\mathcal{L}_{instance}$ as shown in Equ. (6)
4:    **for** $t_1 \in [1, T_1]$ **do**
5:       $DPAN_{input} \leftarrow (\theta_e^{t_1}, \theta_d^{t_1}, \mathcal{L}_{instance}, X, L, C, \sigma, \eta)$
6:       $(\theta_e^{t_1+1}, \theta_d^{t_1+1}), R_1 \leftarrow DPAN(DPAN_{input})$
7:       $R_1 += R_1$
8:    **end for**
9:    # Cluster stage
10:   Design $\mathcal{L}_{cluster}$ as shown in Equ. (10)
11:   **for** $t_2 \in [T_1 + 1, T_2]$ **do**
12:      $DPAN_{input} \leftarrow (\theta_e^{t_2}, \theta_d^{t_2}, \mathcal{L}_{cluster}, X, L, C, \sigma, \eta)$
13:      $(\theta_e^{t_2+1}, \theta_d^{t_2+1}), R_2 \leftarrow DPAN(DPAN_{input})$
14:      $R_2 += R_2$
15:   **end for**
16:   Compute privacy budget ϵ by Theorem **1**
17:   **return** $\theta_e^{T_2}$ and (ϵ, δ).

3.2 Differentially Private Single-Cell Autoencoder Network

The trained encoder can map the high-dimensional to low-dimensional and the trained decoder doesn't work during inference. Moreover, the existing differential privacy methods usually perturb whole neural networks to achieve differential privacy and hence result in great performance overheads. Inspired by these, We proposed a differentially private deep autoencoder network that only adds noise to the gradient of the encoder while minimizing the empirical loss function for single-cell clustering. The network is based on a Zero-Inflated Negative Binomial (ZINB) loss function that can learn the hidden feature representation of scRNA-seq data. Within this framework, fully connected layers stacked in both the encoder and decoder facilitate the capture of intricate representation embeddings from scRNA-seq datasets.

Algorithm 2 outlines our training approach with parameters θ by minimizing the empirical loss function $\mathcal{L}(\theta)$. The encoder parameters θ_e are perturbed when optimizing. And the decoder parameters θ_d don't be perturbed. At each optimization step, we compute the gradient $g(x_i)$ including protected gradient $g_e(x_i)$ and exposed parameters $g_d(x_i)$ for a random subset of samples, clip the ℓ_2 norm of each gradient and compute the average in Lines 2–6. Then, we add noise to the gradient of θ_e in order to protect privacy with the smallest possible performance loss and take a step in the opposite direction of this average noisy gradient in Lines 7–10.

Algorithm 2 Differentially private autoencoder network (DPAN)

Inputs: Examples $X = \{x_1, x_2, ..., x_N\}$, learning rate η, lot size L, loss function $\mathcal{L}(\theta)$, noise scale σ, clipping bound C.

Output: $(\hat{\theta}_e, \hat{\theta}_d)$ and privacy budget RDP

  1:   Take a random sample $\mathcal{B}$ with probability L/N

  2:   **for** $x_i \in \mathcal{B}$ **do**

  3:      Compute $g_e(x_i), g_d(x_i) \leftarrow \nabla_\theta \mathcal{L}(\theta, x_i)$

  4:      $\bar{g}_e(x_i) \leftarrow g_e(x_i)/\max(1, \frac{\|g_e(x_i)\|_2}{C})$

  5:      $\bar{g}_d(x_i) \leftarrow g_d(x_i)/\max(1, \frac{\|g_d(x_i)\|_2}{C})$

  6:   **end for**

  7:   $\tilde{g}_e \leftarrow \frac{1}{L}\left(\sum_i^L \bar{g}_e(x_i) + \mathcal{N}(0, \sigma^2 C^2 I)\right)$

  8:   $\tilde{g}_d \leftarrow \frac{1}{L}\sum_i^L \bar{g}_d(x_i)$

  9:   $\hat{\theta}_e \leftarrow \theta_e - \eta\tilde{g}_e$

10:   $\hat{\theta}_d \leftarrow \theta_d - \eta\tilde{g}_d$

11:   Compute privacy budget RDP by Theorem **2**

12:   **return** $(\hat{\theta}_e, \hat{\theta}_d)$ and RDP.

3.3 Contrastive Learning Module

To capture the relationships between cells, We integrated a two-stage contrastive learning module into our model to enhance the autoencoder's ability to discern more discriminative features within instances and clusters. In the first stage, specific samples x are augmented into $\{x_1, x_2\}$ as positive examples. Then, the augmented data is mapped to a lower-dimensional space using the encoder. Inspired by SimSiam [3], which does not rely on negative pairs. We maximize the similarity between two augmentations of x in order to converge towards improved performance. This approach learns more robust and useful representations by enhancing the similarity of samples from the same class. Instance-level contrastive learning extracts distinctive features by minimizing differences between these instances. We minimize their negative cosine similarity:

$$\mathcal{L}_{ii} = -\frac{z_1}{\|z_1\|_2} \cdot \frac{z_2}{\|z_2\|_2}. \tag{1}$$

where z_1, z_2 are x_1, x_2 which are mapped to a low dimensional vector through an encoder, and $\|\cdot\|$ is ℓ_2-norm.

In the second stage, based on the centroids as representative positions for each cluster, cluster-level contrastive learning is applied to capture discriminative cluster features. Utilizing K-means, we acquire centroid positions. Cluster-level contrastive learning is committed to maximizing the distance between clusters, ensuring clusters are positioned as distantly from each other as feasible:

$$\mathcal{L}_{cc} = \frac{1}{s^2} \sum_{i=0}^{s} \sum_{j=0}^{s} \frac{c_i}{\|c_i\|_2} \cdot \frac{c_j}{\|c_j\|_2}. \tag{2}$$

where c_i is the centroid of cluster i, c_j is the centroid of cluster j and s is the number of clusters.

3.4 Joint Embedding and Clustering Optimization

We use a hybrid loss to optimize our proposed DP-DCAN during the two-stage training phase. Different from the two-stage training proposed by Wang [13] that can't compute the gradient for each individual sample in a batch which is not suitable for differential privacy, our method is designed to calculate the gradient by an individual sample. Specifically, in the first stage, we jointly combined the reconstruction loss and the instance-instance loss to minimize the distance of the instances of the same cluster. Reconstruction loss $\mathcal{L}_{zinb}$, instance-instance loss $\mathcal{L}_{ii}$ and hybrid loss $\mathcal{L}_{instance}$ are defined as follows:

$$\text{ZINB}(x|\pi, \mu, r) = \pi\delta_0(x) + (1-\pi)\text{NB}(x|\mu, r), \tag{3}$$

$$\text{NB}(x|\mu, r) = \frac{\Gamma(x+r)}{x!\Gamma(r)}(\frac{r}{r+\mu})^r(\frac{\mu}{r+\mu})^x, \tag{4}$$

$$\mathcal{L}_{zinb}(X, \hat{X}) = -\log(\text{ZINB}(x|\pi, \mu, r)), \tag{5}$$

$$\mathcal{L}_{instance} = \rho\mathcal{L}_{zinb} + (1-\rho)\mathcal{L}_{ii}. \tag{6}$$

where x is the expression value in cell, π is the probability of a dropout event, μ is the mean, and r represents the divergence of the negative binomial distribution. ρ is a weighted factor that balance the impact of $\mathcal{L}_{zinb}$ and $\mathcal{L}_{ii}$.

In the second stage, we employ reconstruction loss, clustering loss, and cluster-cluster loss to maximize the distance between clusters. Clustering loss represents the disparity between the clustering results and the true data distribution which is measured by Kullback–Leibler (KL) divergence. Clustering loss $\mathcal{L}_{cls}$, cluster-cluster loss $\mathcal{L}_{cc}$ and hybrid loss $\mathcal{L}_{cluster}$ are defined as follows:

$$q_{ij} = \frac{\left(1+\|z_i - \lambda_j\|^2\right)^{-1}}{\sum_{j'}\left(\left\|1+z_i - \lambda_{j'}\right\|^2\right)^{-1}}, \tag{7}$$

$$p_{ij} = \frac{q_{ij}^2/\sum_j q_{ij}}{\sum_{j'}\left(q_{ij''}^2/\sum_{j'} q_{ij'}\right)}, \tag{8}$$

$$L_{cls} = KL(P||Q) = \sum_i \sum_j p_{ij} \log \frac{p_{ij}}{q_{ij}}, \tag{9}$$

$$\mathcal{L}_{cluster} = \beta_1\mathcal{L}_{zinb} + \beta_2\mathcal{L}_{cls} + \beta_3\mathcal{L}_{cc}. \tag{10}$$

where q_{ij} is the soft label of the embedding point z_i. This label measures the similarity between z_i and cluster center λ_j by Student's t-distribution. The target distribution p_{ij} is obtained by squaring q_{ij} and then normalizing by the sum of the squared values across all clusters. β_1, β_2 and β_3 are three weighted factors that balance the impact of $\mathcal{L}_{zinb}$, $\mathcal{L}_{cls}$ and $\mathcal{L}_{cc}$. Their combined sum equals 1.

4 Privacy Analysis

Due to the latent representation being generated in the middle of the autoencoder during inference, we just need to implement differential privacy on the encoder to achieve privacy protection. The privacy guarantee of DP-DCAN is proved as follows:

Theorem 1 (Privacy loss of DP-DCAN). *The privacy loss of DP-DCAN satisfies:*

$$(\varepsilon, \delta) = (R1 + R2 + \log((\alpha - 1)/\alpha) - \big(\log \delta + \log \alpha\big)/(\alpha - 1), \delta). \tag{11}$$

where $0 < \delta < 1$, *R1 and R2 is the RDP of algorithm DPAN which is computed by* Theorem 2.

Theorem 2. *After T times of iterations, the RDP of training in DPAN satisfies:*

$$R_{train}(\alpha) = \frac{T}{\alpha-1}\ln[\sum_{i=0}^{\alpha}\binom{\alpha}{i}(1-q)^{\alpha-i}q^i\exp\big(\frac{i^2-i}{2\sigma^2}\big). \tag{12}$$

where $q = \frac{L}{N}$, σ is noise scale of training, and $\alpha > 1$ is the order.

The proof of Theorem 1 and Theorem 2 can see Appendix B.1 and Appendix B.2 separately.

5 Experiments

In this section, we will explore the impact of the scope and intensity of model perturbation on model performance through a clustering task on multiple real-world scRNA-seq datasets.

5.1 Datasets

We conduct extensive experiments on 8 real-world scRNA-seq datasets from various sequencing platforms. The detailed information is described in Appendix Table S1, showing the source organ, the platform, the number of cell types, the number of cells and the reference. All 8 datasets are from different species, including mouse and human, as well as from different organs, such as the brain and embryo. Specifically, the numbers of cells range from 56 to 4271, and genes range from 3840 to 57241. After filtering the genes that have no expression value and normalization, 2000 highly variable genes were selected as representative genes for each cell by scanpy package.

5.2 Implementation and Parameters Setting

In the proposed DP-DCAN framework, the sizes of the hidden fully connected layer in the encoder are set to (256, 64), the decoder is the reverse of the encoder, and the bottleneck layer (the latent space) has a size of 32. In our two-stage training process, the optimizer for the instance stage is Adam optimizer with setting $lr = 0.001$ and for the cluster stage is Adadelta with setting $lr = 1.0$. We set the batch size to around 0.1 of the dataset size. To achieve an appropriate trade-off between utility and privacy, the clipping bound $C = 0.1$ and $\delta = 10^{-5}$. Clustering performance is measured using metrics such as Normalized Mutual Information (NMI) and Adjusted Rand Index (ARI).

Table 1. Performance of our method and the other clustering baseline methods .

Metric	DP-ALG	CL	Biase	Yan	Li	Muraro	Klein	Romanov	Zeisel	Zhang
NMI	Non-private	scDC	96.52	39.91	**92.00**	88.71	90.25	**66.62**	**76.29**	**77.23**
		bmVAE	61.74	00.00	82.07	72.81	60.84	51.50	61.66	60.17
		DCAN	**100.00**	**84.35**	91.22	**90.22**	**95.76**	61.54	69.13	76.34
	DPAN	scDC	63.60	30.14	88.14	82.73	**93.28**	62.67	71.89	**75.48**
		bmVAE	00.00	30.28	68.13	64.06	61.82	52.34	45.44	62.97
		DCAN	**89.53**	**79.56**	**91.19**	**83.54**	85.66	**62.98**	**74.35**	72.53
ARI	Non-private	scDC	98.21	19.66	80.16	92.55	86.65	**63.75**	72.90	73.72
		bmVAE	34.91	00.00	69.55	49.06	37.24	25.22	32.19	30.43
		DCAN	**100.00**	**80.92**	**89.91**	**93.88**	**97.58**	61.57	**74.66**	**74.48**
	DPAN	scDC	49.46	09.76	77.08	86.67	**95.85**	55.77	56.12	70.25
		bmVAE	00.00	15.61	53.16	57.22	51.12	61.32	44.47	50.42
		DCAN	**93.40**	**68.19**	**90.27**	**88.58**	81.18	**68.72**	**74.17**	**70.87**

5.3 Experimental Evaluation

We evaluate the clustering performance on 8 scRNA-seq datasets in terms of both clustering methods (CL) and differential privacy algorithms (DP-ALG) shown in Table 1 and Table 2. DCAN represents our proposed deep contrastive learning autoencoder network for clustering, and the combination of our proposed privacy-preserving algorithm DPAN is DP-DCAN. To explore the ability of our proposed clustering method to extract features from cells, we compare DP-DCAN with two baseline methods for single-cell clustering scDC and bmVAE. For our proposed differential privacy method DPAN, we demonstrate its superiority by comparing it with DP-PSAC [16], which is a differentially private per-sample adaptive clipping algorithm based on a non-monotonic adaptive weight function. Each clustering method was run ten times to take the average. The values of the best performance metrics are bolded.

Clustering Methods Evaluation. Table 1 summarizes the clustering performance of the proposed DCAN and the baseline clustering methods on 8 scRNA-seq datasets when $\varepsilon = 8.0$. In the non-private clustering methods, DCAN achieves the best NMI and ARI on 4 and 6 of 8 scRNA-seq datasets, respectively. In the private clustering methods, DCAN achieves the best NMI and ARI on 6 and 7 of 8 scRNA-seq datasets, respectively. The non-private clustering methods tend to achieve better NMI and ARI on most scRNA-seq datasets than private clustering methods. This is because differential privacy algorithms introduce noise into the model preventing the model from optimizing in the true direction. However, appropriate noise may result in performance gains. For example, in the Li and Romanov dataset, the NMI and ARI of DCAN with DPAN are better than those without DPAN. This phenomenon could be attributed to the introduction of moderate noise, which potentially enhances the generalization performance of the model, and additional investigation is warranted to fully comprehend this effect. Furthermore, the clustering performance of deep embedded clustering with the contrastive learning module is better and more stable, which again proves the superiority of DCAN.

Table 2. Performance of our method and the other differential privacy baseline method.

Metric	CL	DP-ALG	Biase	Yan	Li	Muraro	Klein	Romanov	Zeisel	Zhang
NMI	scDC	DP-PSAC	32.13	29.33	36.60	51.53	61.87	29.69	45.21	33.46
		DPAN	**63.60**	**30.14**	**88.14**	**82.73**	**93.28**	**62.67**	**71.89**	**75.48**
	bmVAE	DP-PSAC	00.00	00.00	00.00	56.99	41.19	43.45	**47.24**	41.87
		DPAN	00.00	**30.28**	**68.13**	**64.06**	**61.82**	**52.34**	45.44	**62.97**
	DCAN	DP-PSAC	84.86	76.74	87.63	66.96	57.43	51.47	53.39	58.93
		DPAN	**89.53**	**79.56**	**91.19**	**83.54**	**85.66**	**62.98**	**74.35**	**72.53**
ARI	scDC	DP-PSAC	23.44	**11.98**	16.86	40.08	59.11	37.95	33.17	23.90
		DPAN	**49.46**	09.76	**77.08**	**86.67**	**95.85**	**55.77**	**56.12**	**70.25**
	bmVAE	DP-PSAC	00.00	00.00	00.00	54.69	45.66	58.54	41.23	34.60
		DPAN	00.00	**15.61**	**53.16**	**57.22**	**51.12**	**61.32**	**44.47**	**50.42**
	DCAN	DP-PSAC	78.70	62.59	78.28	70.33	45.40	61.40	62.21	47.62
		DPAN	**93.40**	**68.19**	**90.27**	**88.58**	**81.18**	**68.72**	**74.17**	**70.87**

Differential Privacy Methods Evaluation. Table 2 shows the impact of various differential privacy algorithms on the performance of different clustering methods. We can observe that our DPAN outperforms the other baseline differential privacy algorithm for clustering performance. For the 8 scRNA-seq datasets, the clustering methods with DPAN achieve the best NMI and ARI on more than 6 of these datasets, respectively. This is because the DPAN is able to mitigate the loss of performance by centrally allocating the privacy budget to the encoder part of the model. In clustering methods scDC and bmVAE, DPAN significantly maintains the ability of the model to learn the feature representation of the data, while DP-PSAC exhibits a tendency to excessively disrupt the model. In the proposed clustering method DCAN with DPAN has achieved the best performance across all datasets compared with DP-PSAC. In summary, we can conclude that DP-DCAN performs better than the other methods under two clustering metrics. To demonstrate the intuitive discrimination ability of DP-DCAN, we employ the Uniform Manifold Approximation and Projection (UMAP) technique with default parameters to project the scRNA-seq data into a 2D space on the Muraro datasets in Fig. 3 (A).

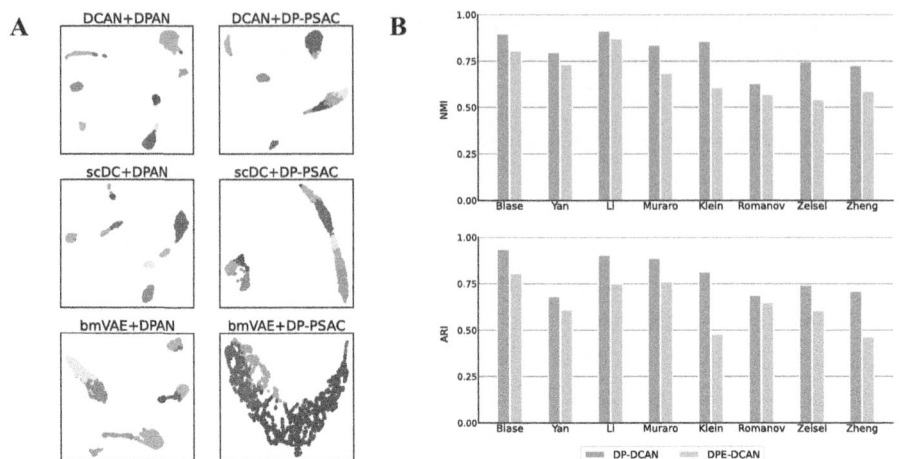

Fig. 3. (A) Comparison of clustering results with 2D visualization on the Muraro dataset. (B) Clustering performance of DP-DCAN and DPE-DCAN when $\varepsilon = 8.0$.

5.4 Partial Network Perturbation vs. Entire Network Perturbation

We conducted a comparative analysis of the proposed DP-DCAN and DPE-DCAN, the latter involving perturbation of the entire model during the training process. As depicted in Fig. 3 (B), we compare the proposed DP-DCAN and DPE-DCAN when the $\varepsilon = 8.0$. Obviously, DP-DCAN outperforms the DPE-DCAN for clustering performance for all scRNA-seq datasets. Even in the Klein dataset, the NMI and ARI scores of DP-DCAN are 25.25% and 33.35% higher than DPE-DCAN, respectively, indicating a significant performance gap. This finding underscores the importance of partial network perturbation. In the context of a designated privacy allocation, enhancing the safeguarding of the model's architecture necessitates the incorporation of increased perturbation within the gradients. Notably, the noise scale for DP-DCAN stands at 2.00, while for DPE-DCAN, it is set to 4.36. Considering the collective impact of noise intensity and the scope of noise impact, our method has exhibited exceptional efficacy across various scRNA-seq datasets.

5.5 Differential Privacy Budget Analysis

Table 3 shows the clustering performance of our method at different noise levels. In this experiment, we vary the privacy budget $\varepsilon \in \{4.0, 6.0, 8.0\}$ on 8 scRNA-seq datasets. As depicted in Table 3, we observe that as the privacy budget (privacy protection level) decreases, there is a general trend of decreasing NMI and ARI across the datasets. Therefore, there is a trade-off between privacy protection and performance. However, in the Biase dataset, the NMI and ARI violate this trend. Maybe this can be attributed to the introduction of moderate noise. In summary, DP-DCAN demonstrates strong robustness to different noise levels.

Table 3. Clustering performance of DP-DCAN on different privacy budgets.

Metric	ϵ	σ	Biase	Yan	Li	Muraro	Klein	Romanov	Zeisel	Zhang
NMI	8.0	2.00	89.53	79.56	91.19	83.54	85.66	62.98	74.35	72.53
	6.0	2.49	96.52	77.49	90.62	83.47	85.61	62.13	74.48	72.53
	4.0	3.46	92.41	72.10	90.58	82.80	85.30	61.73	74.91	67.93
ARI	8.0	2.00	93.40	68.19	90.27	88.58	81.18	68.72	74.17	70.87
	6.0	2.49	98.21	69.42	89.28	88.41	81.10	68.23	75.09	68.12
	4.0	3.46	94.39	57.58	88.45	87.71	80.84	68.12	75.67	59.78

6   Conclusion

In this work, We propose DP-DCAN, a differentially private deep contrastive autoencoder network for single-cell clustering. Our scheme utilizes contrastive learning to enhance the feature extraction of the autoencoder. To protect single-cell privacy, we incorporate differential privacy for autoencoder during training. Furthermore, we design the partial network perturbation method for antoencoder. This perturbation method reduces the model dimension of the perturbation and ensures the utility. We conduct a comprehensive privacy analysis of our approach using RDP and validate our scheme through extensive experiments. The results indicate that DP-DCAN can achieve excellent performance under differential privacy protection. In future work, we will consider the possibility of partial network perturbation for neural networks other than the autoencoder, and achieve more efficient corresponding DP schemes.

Acknowledgments. This work is supported by the Natural Science Foundation of Shanghai (Grant No. 22ZR1419100), the National Natural Science Foundation of China Key Program (Grant No. 62132005), and CAAI-Huawei MindSpore Open Fund (Grant No. CAAIXSJLJJ-2022-005A).

References

1. Abadi, M., et al.: Deep learning with differential privacy. In: Proceedings of the 2016 ACM SIGSAC Conference on Computer and Communications Security, pp. 308–318 (2016)
2. Chen, J., Wang, W.H., Shi, X.: Differential privacy protection against membership inference attack on machine learning for genomic data. In: BIOCOMPUTING 2021: Proceedings of the Pacific Symposium, pp. 26–37. World Scientific (2020)
3. Chen, X., He, K.: Exploring simple Siamese representation learning. In: Proceedings of the IEEE/CVF Conference on Computer Vision and Pattern Recognition, pp. 15750–15758 (2021)
4. Eraslan, G., Simon, L.M., Mircea, M., Mueller, N.S., Theis, F.J.: Single-cell RNA-seq denoising using a deep count autoencoder. Nat. Commun. **10**(1), 390 (2019)
5. Flores, M., et al.: Deep learning tackles single-cell analysis—a survey of deep learning for scRNA-seq analysis. Brief. Bioinform. **23**(1), bbab531 (2022)
6. Fu, J., et al.: Differentially private federated learning: a systematic review. arXiv preprint arXiv:2405.08299 (2024)

7. Fu, J., et al.: DPSUR: accelerating differentially private stochastic gradient descent using selective update and release. arXiv preprint arXiv:2311.14056 (2023)

8. Ha, T., Dang, T.K., Dang, T.T., Truong, T.A., Nguyen, M.T.: Differential privacy in deep learning: an overview. In: 2019 International Conference on Advanced Computing and Applications (ACOMP), pp. 97–102. IEEE (2019)

9. Leemann, T., Pawelczyk, M., Kasneci, G.: Gaussian membership inference privacy. In: Advances in Neural Information Processing Systems, vol. 36 (2024)

10. Liu, H., Wu, Z., Peng, C., Lei, X., Tian, F., Lu, L.: Adaptive differential privacy of character and its application for genome data sharing. In: 2019 International Conference on Networking and Network Applications (NaNA), pp. 429–436. IEEE (2019)

11. Oestreich, M., Chen, D., Schultze, J.L., Fritz, M., Becker, M.: Privacy considerations for sharing genomics data. EXCLI J. **20**, 1243 (2021)

12. Shen, Y., Wang, Z., Sun, R., Shen, X.: Towards understanding the impact of model size on differential private classification. arXiv preprint arXiv:2111.13895 (2021)

13. Wang, J., Xia, J., Wang, H., Su, Y., Zheng, C.H.: scDCCA: deep contrastive clustering for single-cell RNA-Seq data based on auto-encoder network. Brief. Bioinform. **24**(1), bbac625 (2023)

14. Wang, J., et al.: ScGNN is a novel graph neural network framework for single-cell RNA-Seq analyses. Nat. Commun. **12**(1), 1882 (2021)

15. Wu, X., Wei, Y., Mao, Y., Wang, L.: A differential privacy DNA motif finding method based on closed frequent patterns. Clust. Comput. **22**(Suppl. 2), 2907–2919 (2019)

16. Xia, T., et al.: Differentially private learning with per-sample adaptive clipping. In: Proceedings of the AAAI Conference on Artificial Intelligence, vol. 37, pp. 10444–10452 (2023)

17. Yan, J., Ma, M., Yu, Z.: bmVAE: a variational autoencoder method for clustering single-cell mutation data. Bioinformatics **39**(1), btac790 (2023)

18. Yilmaz, E., Ji, T., Ayday, E., Li, P.: Genomic data sharing under dependent local differential privacy. In: Proceedings of the Twelfth ACM Conference on Data and Application Security and Privacy, pp. 77–88 (2022)

FD-SDG: Frequency Dropout Based Single Source Domain Generalization Framework for Retinal Vessel Segmentation

Boyang Li[1,2], Haojin Li[1,2], Yule Zhang[1,2], Heng Li[1,2(✉)], Jiangyu Chen[1,2], Fuhai Pan[3], Jianwen Chen[3(✉)], and Jiang Liu[1,2]

[1] Research Institute of Trustworthy Autonomous Systems, Southern University of Science and Technology, Shenzhen 518055, China
lih3@sustech.edu.cn
[2] Department of Computer Science and Engineering, Southern University of Science and Technology, Shenzhen 518055, China
[3] Department of Orthopedics Medicine, Southern University of Science and Technology Hospital, Shenzhen 518055, China
469353224@qq.com

Abstract. Single-source domain generalization aims to enhance model performance on unseen target domain test sets using only a single source domain dataset, typically by mitigating domain shifts between domains. In retinal vessel segmentation tasks, differences in dataset composition, such as variations in the proportions of different diseases and imaging noise levels, are considered significant sources of domain shift. However, few previous studies have delved into the mechanisms through which this type of domain shift influences model performance. In this study, we hypothesize that disparities in dataset composition could manifest as differences in distribution patterns of frequency domain features, rendering the model susceptible to overfitting specific patterns. Building on this hypothesis, we propose a novel Frequency Dropout based Single Source Domain Generalization (FD-SDG) framework that employs a Frequency Dropout Randomization mechanism to disentangle complex co-adaptive relationships among features from different frequency bands, thereby enhancing the model's robustness to variable frequency domain noise patterns in the sample space. Additionally, we introduce a Salient Structure Representation Normalization mechanism to align post-perturbation data features in the feature space using invariant anatomical structures. Through comparison experiments and ablation studies conducted on multiple sets of fundus images across-domain experiments, our method achieves state-of-the-art performance, underscoring its high generalizability and robustness.

Keywords: fundus image · vessel segmentation · single-source domain generalization · frequency dropout · domain randomization

---

B. Li and H. Li—Contributing equally to this work

1 Introduction

In retinal vessel segmentation tasks, generalizability plays a crucial role in determining the usability of segmentation models. Given that vascular annotation of fundus images is costly and raises privacy concerns, available well-labeled segmentation datasets tend to be small. Segmentation methods trained on a single dataset often struggle to perform satisfactorily on other test sets. Researchers have attributed this performance degradation to the presence of domain shifts between datasets. To address this problem, single-source domain generalization (SDG) aims to mitigate the performance deterioration caused by domain shifts, thus enhancing model performance on potential test sets (target domains) when only a single training set (source domain) is available [1–3].

Domain randomization, as an implementation of domain generalization, extends the distribution boundaries of the source domain(s) by augmenting data, thereby adapting to the diverse data in the unseen target domain [4, 5]. Some studies suggest broadening the training set across the frequency domain enhances model generalizability and robustness [6, 7]. Typically, these methods rely on spectral transformations and involve randomized perturbations in the frequency domain space. These perturbations encompass various techniques such as high-frequency filtering [1, 8], band-specific amplitude substitution [9], phase perturbations [10] and others. While these studies are based on the variations in information across different frequency bands, providing a foundation for subsequent research, they do not extensively explore the impact of image features from different frequency bands on the cross-domain performance of the model.

Based on the aforementioned investigations, we propose a new hypothesis: variations in the distribution patterns of frequency domain features, or "task-relevant information", within frequency bands constitute a significant aspect of the observed domain shift between datasets. Enhancing the segmentation model's robustness to such differences in distribution patterns can consequently enhance cross-domain performance. In fundus datasets, differences in the occurrence frequencies of low-frequency interferences like cataracts and high-frequency noises stemming from irrelevant disease foci may vary across datasets [11]. This variability renders the model training susceptible to overfitting, as illustrated in Fig. 1, wherein this discrepancy manifests as differences between the distributions of frequency-domain features. As discussed in [4], such domain-dependent shifts in features serve as one of the key mechanisms for domain shift resulting from disparities in frequency domain feature distributions.

In this paper, we present FD-SDG, a Fourier-based retinal vessel segmentation framework. We suggest that deep learning models may experience overfitting due to the distribution of features in the frequency domain, potentially stemming from complex co-adaptive relationships among features across different frequency bands. Motivated by the dropout solution addressing co-adaptive relationships among neurons in deep learning models [12, 13], we introduce a Frequency Dropout Randomization (FDR) mechanism to facilitate image broadening. This technique deactivates frequency bands of image frequency domain information, with uniform and random band deactivation across frequencies due to the challenge of effectively modeling the frequency domain feature distribution across various medical image types. Concurrently, we propose a Salient Structure Representation Normalization (SSRN) mechanism to align the augmented feature space, enhancing the model's capability to extract critical anatomical structure information. Through this approach, we enhance the robustness of FD-SDG to frequency domain feature differences between datasets, thereby steadily improving the model's generalization performance and increasing the framework's applicability.

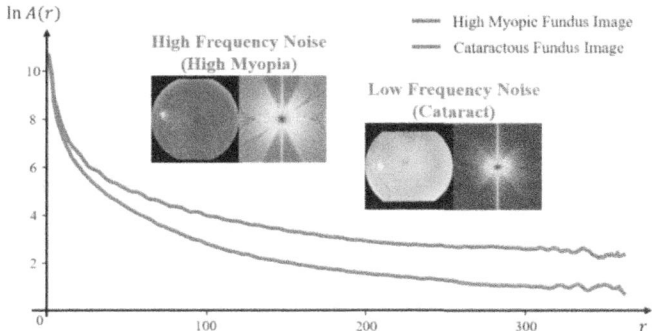

Fig. 1. Plot of the average logarithm of amplitude spectrum vs. frequency band. Variations in the composition of retinal image datasets can lead to differences in the distribution patterns of frequency domain features, thereby inducing domain shifts.

Our contributions can be summarized as follows:

1. We introduce a novel single-source generalizable retinal vessel segmentation framework, named FD-SDG, aimed at mitigating domain shifts arising from the diverse distribution patterns of frequency across datasets.
2. We propose two key mechanisms to enhance the framework's robustness: 1) FDR mechanism, which improves the model's resilience to dataset-specific frequency-domain information by applying random dropout across different frequency bands of the input images; 2) SSRN mechanism, which constrains feature distribution post-generalization through a structural reconstruction task, enhancing the model's resistance to frequency domain perturbations.
3. Through cross-domain experiments performed between multiple retinal vessel segmentation datasets, we constructed comparison experiments and ablation studies to validate the state-of-the-art level of generalization and noise immunity that our model possesses.

2 Proposed Method

In the context of SDG, a source domain dataset D_s with N labelled samples $\left\{\left(x_i^s, y_i^s\right)\right\}_{i=1}^N$ is given, where $y_i^s \in \mathbb{R}^{H \times W}$ is corresponding label to input data $x_i^s \in \mathbb{R}^{H \times W \times 3}$. A target dataset can be denoted as $D_t \in \left\{x_i^t\right\}_{i=1}^N$, which lacks segmentation labels. Generally, the goal of SDG is to enable the model trained on D_s to perform as well as on unseen target data domains D_t^k, where k denotes the number of target domains.

As shown in Fig. 2, to combat the performance decline of the model trained on single-source clinical data, an SDG algorithm has been proposed for retinal vessel segmentation. The Frequency Dropout Randomization (FDR) mechanism augments single-source domain data through randomization filtering on frequency view. Simultaneously,

Salient Structure Representation Normalization (SSRN) and attention mechanism are coupled to a multi-flow segmentation network, learning the features of fundus anatomical structures. In the following, we will sequentially introduce the mechanisms involved in our proposed framework, presenting each one separately.

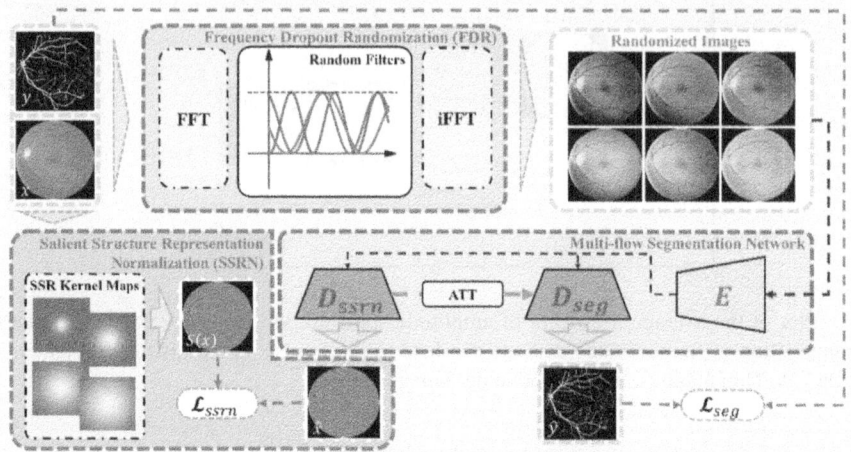

Fig. 2. The overview of our proposed FD-SDG. The FDR enhances the source domain data through a random dropout mechanism based on frequency domain filtering. Besides, SSRN aligns the feature representation of the augmented image through self-supervised salient structure reconstruction. The multi-flow segmentation architecture ingests the randomized data and utilizes intermediate features for segmentation tasks via the attention mechanism, achieving an efficient single-source domain generalization network.

2.1 Frequency Dropout Randomization

As previously discussed, enhancing the segmentation model's resilience to particular frequency-domain feature distribution patterns contributes to enhancing the model's generalization capability. Hence, we utilize Frequency Dropout Randomization as a form of domain randomization implementation. FDR randomly deactivates specific frequency characteristic information, thereby diminishing sensitivity to correlated noise. This process reduces the model's dependence on particular frequency modes and enhances its generalization capacity.

The Fast Fourier transform $\mathcal{F}(\cdot)$ is employed to extract the frequency representation of a image x, which is:

$$\mathcal{F}(x)(m, n) = \sum_{h=0}^{H-1} \sum_{w=0}^{W-1} x(h, w) e^{-i2\pi \left(m\frac{h}{H} + n\frac{w}{W} \right)} \tag{1}$$

Accordingly, frequency-view can be transferred to the original feature space through an inverse Fast Fourier Transform, denoted as $\mathcal{F}^{-1}(\cdot)$.

Aiming to suppress different frequency band information randomly, we establish a band-pass filter with randomized parameters to manipulate the frequency domain representation, thereby achieving a frequency-based random dropout.

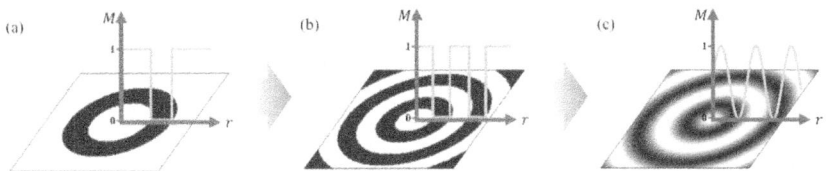

Fig. 3. Graphical representations of (a) Vanilla Frequency Filter, $f_{vanilla}$, (b) Parallel Frequency Filter, $f_{parallel}$ and (c) Smooth Trigonometric Frequency Filter, f_{smooth}, alongside their corresponding filter maps m are presented together.

Specifically, a frequency filter is applied to modulate the spectral content of the image by attenuating information within specific frequency bands. Filtering an image in the frequency domain is typically achieved by applying a map m, which is multiplied by the frequency-domain representation. But for simplicity, we represent this filtering process using a one-dimensional function. The frequency filter $f(\cdot)$ is defined as a function with the frequency band as the independent variable (i.e., the Euclidean distance from the center of the spectrum) and its value domain ranging between [0, 1]. A value closer to 0 indicates stronger suppression of the corresponding frequency band by the filter. In Fig. 3, we depict three variants of the filter function, denoted as $f_{vanilla}$, $f_{parallel}$, and f_{smooth}, to illustrate our design concepts.

Analogous to the implementation of traditional dropout mechanism [14], an intuitive design strategy entails configuring the filter m to function as an ideal band-pass filter. We first introduce a **Vanilla Frequency Filter** function that suppresses information within a single part of the image frequency spectrum stochastically, as shown in Fig. 3(a), which is:

$$f_{vanilla}(r) = \begin{cases} 0, for \alpha \leq r \leq \beta, \\ 1, otherwise, \end{cases} \tag{2}$$

where α and β define the boundary of dropout band.

Within dropout methodologies, it's typical to drop out multiple nodes concurrently within the model architecture [15]. Drawing inspiration from this principle and extending upon the foundation of $f_{vanilla}$, we introduce the **Parallel Frequency Filter** function. This design aims to achieve robust and efficient randomization through parallel dropout across multiple frequency bands. As shown in Fig. 3(b), it can be formulated as follows:

$$f_{parallel}(r) = \sum_{I=1}^{T} (f_{vanilla}(r - i*d)) - (T - 1)*d, where T \in \mathbb{N}^+. \tag{3}$$

where the shape of the filter function is jointly defined by the parameters T and d.

However, given that both $f_{vanilla}$ and $f_{parallel}$ are ideal band-pass filters, they are susceptible to introducing artifacts in the fundus image due to the ringing phenomenon [16], which can adversely impact the segmentation task. To mitigate this issue, we ultimately selected the **Smooth Trigonometric Frequency Filter** function for use in FDR. As depicted in Fig. 3(c), this filter function can be viewed as a smooth version of $f_{parallel}$, which can be formulated as:

$$f_{smooth}(r) = \frac{1}{2}(\sin(\omega * r + \phi) + 1), \tag{4}$$

where ω and ϕ can be randomly assigned to enhance the data space. The comprehensive procedure of FDR can be formalized as:

$$FDR(x) = \mathcal{F}^{-1}(\mathcal{F}(x) \odot m_{smooth}), \tag{5}$$

where m_{smooth} denotes a randomly generated trigonometric frequency filter map corresponding to f_{smooth} and $\odot$ denotes the Hadamard product.

In conclusion, our proposed FDR mechanism reduces the segmentation model's dependency on band-specific information by employing random and smooth filtering in the frequency domain. This enhances data diversity, making our segmentation framework more adaptable to various unseen retinal vessel datasets.

2.2 Salient Structure Representation Normalization

The FDR mechanism intentionally diversifies the training data by employing randomized band suppression, with the objective of achieving domain generalization through data augmentation. However, this approach also presents a challenge to the feature extraction capability of segmentation models. In previous studies of DG, feature normalization serves as a prevalent method for aligning feature distributions [17]. In parallel, we introduce Salient Structure Representation Normalization to bolster model capabilities through supplementary self-supervised tasks.

In fundus images, anatomical structures hold significant diagnostic value. Despite the potential corruption of associated frequency bands by FDR, we anticipate the model's ability to restore them. Leveraging the image gradient properties of these structures, Gaussian filter $S(\cdot)$ is employed to extract salient structures in the original image, which is formulated as:

$$S(x) = x - x * g(r, \sigma), \tag{6}$$

where $g(\cdot)$ denotes a Gaussian kernel with radius r and spatial constant σ. The self-supervised task we have established necessitates the integration of a reconstruction submodel within the segmentation model, ensuring alignment between the submodel output $\hat{x}$ and $S(x)$. The corresponding loss function is defined as:

$$\mathcal{L}_{ssrn} = \hat{x} - S(x)_1. \tag{7}$$

Under the salient structure extractor and self-supervision, SSRN is naturally introduced, which not only learns salient clinical anatomical structure but also improves the generalizability of the model.

2.3 Multi-flow Segmentation Network

A multitasking model is required to incorporate the aforementioned mechanisms. Utilizing the common Unet-based segmentation model architecture as a foundation, we propose a multi-stream network to effectively merge the self-supervised reconstruction task with the segmentation task.

Specifically, our segmentation network can be divided into three parts: an encoder E and two decoders, D_{ssrn} and D_{seg}, bridged through skip connection and attention mechanisms. Skip connection links the E and D_{ssrn} and attention mechanism connected D_{ssrn} and D_{seg}. The Unet-like architecture composed of E and D_{ssrn} is used for the SSRN task, achieving feature reconstruction from $FDR(x)$ to $\hat{x}$ with the objective function formulated in Eq. 7. To ensure D_{seg} effectively utilizes the reconstructed image features from D_{ssrn}, we use a channel and spatial attention module [18] on the skip connection between D_{ssrn} and D_{seg}. This enhances the model's feature extraction capability.

Accordingly, given $\hat{y}$ as the segmentation output and y as the ground-truth segmentation mask of input data x, the cross-entropy loss is used to define the segmentation objective function:

$$\mathcal{L}_{seg} = \mathbb{E}\big[-ylog\hat{y} - (1 - y)\log(1 - \hat{y})\big], \tag{8}$$

and the overall objective function of our network is formalized as:

$$\mathcal{L}_{total} = \mathcal{L}_{seg} + \lambda\mathcal{L}_{ssrn}, \tag{9}$$

where λ denotes a hyper-parameter to balance.

3 Experiments

3.1 Experimental Settings

In this section, we conduct cross-dataset experiments for retinal vessel segmentation. We first validate the superiority of our proposed model over state-of-the-art DG baselines through comparative experiments, emphasizing generalization performance. Then, we conduct ablation studies to illustrate the rationale behind the modules and architecture.

We utilize two datasets for data training: 1) DRIVE [19] and 2) EyePACS [20]. DRIVE comprises 40 labeled fundus images, serving as the source domain in the cross-domain experiment. EyePACS is a large dataset of fundus photographs lacking segmentation labels, utilized in various multi-source domain generalization (DG) baselines. 3) IOSTAR [21], 4) LES-AV [22], and 5) CHASE_DB1 [23] datasets were utilized as multiple target domain datasets for cross-domain experiments, with data volumes of 30, 22 and 28, respectively. While LES-AV and CHASE_DB1 consist of fundus photography, IOSTAR comprises Scanning Laser Ophthalmoscopy (SLO) images. The type and number of disease occurrences vary across datasets.

The image data are resized to 512×512 pixels. For model training, we employ the Adam optimizer and the learning rate is initially set to 10^{-3} for the initial 20 epochs and then linearly decayed to 0 over the subsequent 30 epochs. An early stopping mechanism is introduced to select the model's test parameters. In the FDR mechanism, ω and ϕ

Table 1. Comparison of domain generalization results on vessels segmentation from fundus images.

Model	LES-AV			IOSTAR			CHASE_DB1		
	DICE	IoU	Mcc	DICE	IoU	Mcc	DICE	IoU	Mcc
FACT	0.584	0.412	0.565	0.672	0.506	0.654	0.380	0.235	0.390
SAN-SAW	0.629	0.459	0.599	0.617	0.446	0.585	0.470	0.307	0.450
FedDG	0.745	0.594	0.725	0.720	0.563	0.697	0.660	0.492	0.632
GIN-IPA	0.609	0.461	0.596	0.651	0.489	0.629	0.480	0.359	0.457
SLAug	0.678	0.513	0.662	0.650	0.482	0.634	0.580	0.408	0.567
FreeSDG	0.795	0.659	0.778	0.736	0.582	0.716	0.696	0.533	0.671
FD-SDG	**0.754**	**0.605**	**0.735**	**0.743**	**0.591**	**0.722**	**0.707**	**0.547**	**0.683**

are randomly sampled from $[0, 10]$ and $[0, 2\pi]$, respectively. The radius r and spatial constant σ in SSRN are configured at 27 and 9, while the weight λ in the total loss function $\mathcal{L}_{total}$ is defined as 1. In order to evaluate the performance of the model, three segmentation metrics are used, including the DICE coefficient, Intersection over Union (IoU), and Matthews Correlation Coefficient (Mcc).

3.2 Comparison Study

The efficacy of the proposed method is underscored through comparative analysis with various state-of-the-art DG benchmarks. Our comparison encompasses two Fourier-based DG approaches, namely FACT [9] and FedDG [25], as well as the semantic-aware DG method SAN-SAW [2424]. In addition, three SDG methods, SLAug [5], GIN-IPA [4], and Free-SDG [8] are included for a comprehensive evaluation. By conducting both quantitative assessments and visual examinations, we confirm the state-of-the-art cross-domain performance of our fundus vascular segmentation method.

Table 1 encapsulates the results of a quantitative comparative analysis between our proposed algorithm and existing methods, delineating performance across various evaluative metrics applied to three distinct fundus image datasets. FACT [9] augments the source domain data by replacing the full-band amplitude spectrum, but excessive information loss can degrade its performance. The special data normalization and whitening in SAN-SAW [24] aim to reduce differences in feature distribution between domains but significantly limit feature expressiveness, affecting segmentation performance. FedDG [25] employs a contrastive learning framework to share frequency representation information across multiple source domains and gives reasonable performance. GIN-IPA [4] introduced a causality-inspired data augmentation, but its training process instability suggests overly aggressive augmentation. SLAug [5] employs position scale enhancement combined with saliency balance fusion strategies, and FreeSDG [8] utilizes hybrid frequency domain representation with context-aware learning. By quantitative comparison, our FD-SDG bolsters the model generalization by performing dropout augmentation on the feature of different frequency bands, outperforming the best methods without additional data dependencies.

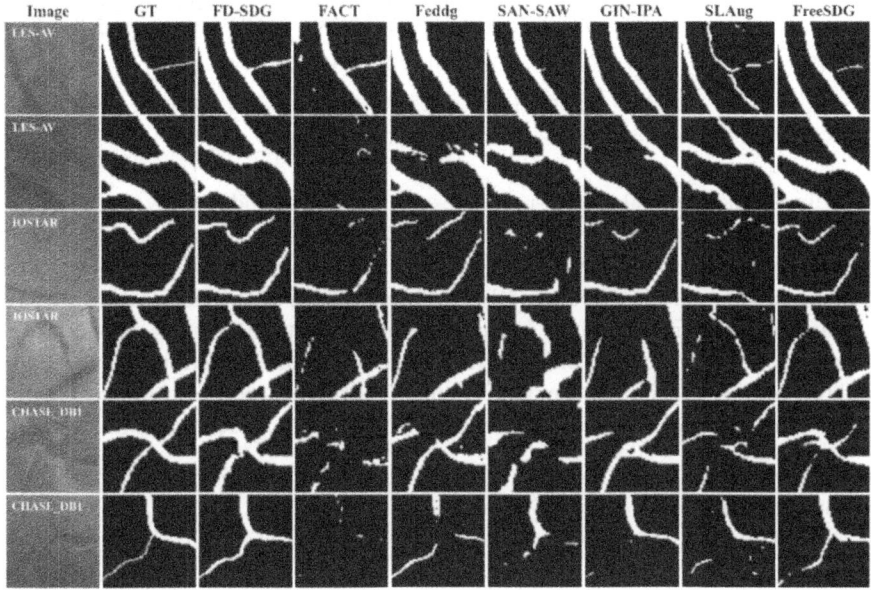

Fig. 4. Comparative evaluation of detailed vessel segmentation in three cross-domain datasets.

As shown in Fig. 4, visual comparison also corroborates the effectiveness of our proposed method. Noise from abnormal illumination, blurring, and lesions in fundus images affects segmentation algorithm performance, while dataset variations in quality degradation challenge model generalizability. FD-SDG counters these problems by employing the FDR mechanism to augment source domain data in frequency view and diminish domain discrepancies introduced by noise interference. In addition, the segmentation of small blood vessel structures, such as those depicted in datasets like CHASE_DB1 [23] and IOSTAR [21], tests the feature extraction prowess of the model.

The SSRN module, which imposes a consistency constraint on prominent structures, enables the model to learn more about anatomical information and accurately delineate small blood vessels. In contrast with other competing methods, FD-SDG clearly and comprehensively preserves the vascular structure of the fundus, exhibiting state-of-the-art robustness and effectiveness without reliance on external data.

3.3   Ablation Study

The effectiveness of each component within the FD-SDG framework is better highlighted by conducting ablation studies on the three main modules, with the results detailed in Table 2. Through the implementation of the FDR mechanism, source domain data is

Table 2. Ablation study results for three principal components.

Model			LES-AV			IOSTAR			CHASE_DB1		
FDG	SSRN	ATT	DICE	IoU	Mcc	DICE	IoU	Mcc	DICE	IoU	Mcc
			0.678	0.512	0.656	0.551	0.381	0.542	0.491	0.325	0.476
✓			0.693	0.530	0.678	0.697	0.535	0.673	0.686	0.522	0.662
✓	✓		0.717	0.558	0.702	0.720	0.563	0.699	0.701	0.540	0.679
✓	✓	✓	0.754	0.605	0.735	0.743	0.591	0.722	0.707	0.547	0.683

effectively augmented, enhancing robustness against noise induced by the absence of certain frequency band information. The SSRN captures the pivotal structural information within fundus blood vessel images, thereby facilitating improvements in downstream segmentation tasks. Additionally, the attention (ATT) mechanism optimizes the processing of reconstructed image features, leading to better utilization of salient structural information for the segmentation task. The synergistic integration of these three modules contributes to the stable enhancement of the model's generalization capabilities in single-domain generalization, as demonstrated by our proposed method.

4   Conclusion

In this paper, we introduce a novel frequency-dropout-based framework for retinal vessel segmentation, designed to address domain shifts arising from differences in frequency-domain distribution patterns across datasets. Our approach combines Frequency Dropout Randomization and Salient Structure Representation Normalization mechanisms within a multi-flow framework, delivering superior performance compared to other state-of-the-art methods. Additionally, we furnish comprehensive ablation experiments alongside visualization results to better elucidate our method. Overall, our framework enhances a single source domain without additional data dependencies, improves the cross-domain generalization of the model, and offers a promising solution for accurate and generalized vessel segmentation in fundus images.

Future studies should comprehensively explore the impact of frequency dropout as a domain augmentation technique on the generalization capabilities of diverse deep learning models. Considering the variance in information content across different frequency bands, it would be worthwhile to investigate the implementation of adaptive dropout, which can potentially enhance model robustness and performance.

Acknowledgments. This work was supported in part by General Program of National Natural Science Foundation of China (Grant No. 82272086), Shenzhen Science and Technology Program (JCYJ20210324103800001, JCYJ20220530112609022), Guangdong Basic and Applied Basic Research Fund (2022A1515010487), Nanshan District Healthcare Program (NSZD2023058) and SUSTech Undergraduate Innovation and Entrepreneurship (S202314325015). We appreciate Shenzhen DE Sci&Tech Company for their data support in this study.

References

1. Hu, S., Liao, Z., Zhang, J., Xia, Y.: Domain and content adaptive convolution based multi-source domain generalization for medical image segmentation. IEEE Trans. Med. Imaging **42**, 233–244 (2021)
2. Li, H., Li, H., Shu, H., Chen, J., Hu, Y., Liu, J.: Self-supervision boosted retinal vessel segmentation for cross-domain data. In: 2023 IEEE 20th International Symposium on Biomedical Imaging (ISBI), pp.1–5 (2023)
3. Li, H., et al.: RaffeSDG: Random Frequency Filtering enabled Single-source Domain Generalization for Medical Image Segmentation (2024)
4. Ouyang, C., et al.: Causality-inspired single-source domain generalization for medical image segmentation. IEEE Trans. Med. Imaging **42**, 1095–1106 (2021)
5. Su, Z., Yao, K., Yang, X., Wang, Q., Sun, J., Huang, K.: Rethinking data augmentation for single-source domain generalization in medical image segmentation. In: AAAI Conference on Artificial Intelligence (2022)
6. Li, H., et al.: A Generic fundus image enhancement network boosted by frequency self-supervised representation learning. Med. Image Anal. **90**, 102945 (2023)
7. Li, H., et al.: Enhancing and adapting in the clinic: source-free unsupervised domain adaptation for medical image enhancement. IEEE Trans. Med. Imaging **43**, 1323–1336 (2023)
8. Li, H., et al.: Frequency-mixed Single-source Domain Generalization for Medical Image Segmentation. *ArXiv*, abs/2307.09005 (2023)
9. Xu, Q., Zhang, R., Zhang, Y., Wang, Y., Tian, Q.: A fourier-based framework for domain generalization. IEEE/CVF Conf. Comput. Vis. Pattern Recogn. (CVPR) **2021**, 14378–14387 (2021)
10. Çugu, I., Mancini, M., Chen, Y., Akata, Z.: Attention consistency on visual corruptions for single-source domain generalization. IEEE/CVF Conf. Comput. Vis. Pattern Recogn. Workshops (CVPRW) **2022**, 4164–4173 (2022)
11. Li, H., Li, H., Qiu, Z., Hu, Y., Liu, J.: Domain Adaptive Retinal Vessel Segmentation Guided by High-frequency Component. OMIA@MICCAI (2022)
12. Hinton, G.E., Srivastava, N., Krizhevsky, A., Sutskever, I., Salakhutdinov, R.. Improving neural networks by preventing co-adaptation of feature detectors. ArXiv, abs/1207.0580 (2012)
13. Ou, M., et al.: MVD-Net: Semantic segmentation of cataract surgery using multi-view learning. In: 2022 44th Annual International Conference of the IEEE Engineering in Medicine & Biology Society (EMBC), pp. 5035–5038 (2022)
14. Srivastava, N., Hinton, G.E., Krizhevsky, A., Sutskever, I., Salakhutdinov, R.: Dropout: a simple way to prevent neural networks from overfitting. J. Mach. Learn. Res. **15**, 1929–1958 (2014)
15. Inoue, H.: Multi-Sample Dropout for Accelerated Training and Better Generalization. *ArXiv*, abs/1905.09788 (2019)
16. Vanmali, A.V., Kataria, T., Kelkar, S.G., Gadre, V.M.: Ringing artifacts in wavelet based image fusion: analysis, measurement and remedies. Inf. Fusion **56**, 39–69 (2020)
17. Zhao, L., Wang, L.: Task-specific inconsistency alignment for domain adaptive object detection. IEEE/CVF Conf. Comput. Vis. Pattern Recogn. (CVPR) **2022**, 14197–14206 (2022)
18. Woo, S., Park, J., Lee, J., Kweon, I.: CBAM: Convolutional Block Attention Module. *ArXiv*, abs/1807.06521 (2018)
19. Staal, J., Abràmoff, M.D., Niemeijer, M., Viergever, M.A., Ginneken, B.V.: Ridge-based vessel segmentation in color images of the retina. IEEE Trans. Med. Imaging **23**, 501–509 (2004)

20. Gulshan, V., et al.: Development and validation of a deep learning algorithm for detection of diabetic retinopathy in retinal fundus photographs. JAMA **316**(22), 2402–2410 (2016)
21. Zhang, J., Dashtbozorg, B., Bekkers, E.J., Pluim, J.P., Duits, R., Romeny, B.M.: Robust retinal vessel segmentation via locally adaptive derivative frames in orientation scores. IEEE Trans. Med. Imaging **35**, 2631–2644 (2016)
22. Orlando, J.I., Breda, J.B., Keer, K.V., Blaschko, M.B., Blanco, P.J., Bulant, C.A. Towards a glaucoma risk index based on simulated hemodynamics from fundus images. In: International Conference on Medical Image Computing and Computer-Assisted Intervention (2018)
23. Owen, C.G., et al.: Retinal arteriolar tortuosity and cardiovascular risk factors in a multi-ethnic population study of 10-year-old children; the child heart and health study in England (CHASE). Arterioscler. Thromb. Vasc. Biol. **31**, 1933–1938 (2011)
24. Peng, D., Lei, Y., Hayat, M., Guo, Y., Li, W.: Semantic-aware domain generalized segmentation. IEEE/CVF Conf. Comput. Vis. Pattern Recogn. (CVPR) **2022**, 2584–2595 (2022)
25. Liu, Q., Chen, C., Qin, J., Dou, Q., Heng, P.: FedDG: federated domain generalization on medical image segmentation via episodic learning in continuous frequency space. IEEE/CVF Conf. Comput. Vis. Pattern Recogn. (CVPR) **2021**, 1013–1023 (2021)

LightSnore-Net: A Lightweight Neural Network for Snoring Detection and Mitigation in Smart Pillows

Xin Luo[1], Zijun Mao[1(✉)], Suqing Duan[3], Xiankun Zhang[1], Chuanlei Zhang[1], and Haifeng Fan[1,2]

[1] College of Artificial Intelligence, Tianjin University of Science and Technology, Tianjin 300457, China
zijunmao@tust.edu.cn
[2] Yunsheng Intelligent Technology Co., Ltd., Tianjin 300457, China
[3] College of Electronic Information and Automation, Tianjin University of Science and Technology, Tianjin 300457, China

Abstract. Snoring, a common sleep disorder, is often delayed in diagnosis due to the inconvenience of polysomnography (PSG). Current anti-snoring devices on the market often employ invasive designs that not only compromise wearer comfort, but can also disrupt normal sleep patterns. To address this issue, our study leverages recent advances in deep learning and Internet of Things (IoT) sensor technologies to propose a novel snoring detection network, LightSnore-Net. This network, based on a lightweight attention mechanism, can be deployed in the STM32F103C8T6 microcontroller of a smart pillow to facilitate non-invasive snore monitoring in a home environment. By analysing the Mel Frequency Cepstral Coefficients (MFCC) features of snore sounds extracted from audio modules and integrating a lightweight channel attention mechanism with a convolutional neural network, LightSnore-Net is able to accurately identify key features of snore sounds. On public datasets, this model achieves 99% detection accuracy, significantly outperforming traditional methods and recent deep learning models. This research provides an efficient and accurate solution, paving a new way for early diagnosis and monitoring of snoring.

Keywords: Snore monitoring · Lightweight attention mechanisms · Deep learning · Smart pillows · Audio signal processing

1 Introduction

Snoring is widely recognised as a potential risk factor for cardiovascular disease, independent of other comorbidities [1]. As a core manifestation of sleep-disordered breathing, snoring indicates low ventilation events and partial upper airway obstruction during sleep. Patients with obstructive sleep apnoea (OSA) often have significant snoring. The severity of OSA is typically classified using clinical thresholds to assess health status and treatment needs [2]. The physiological consequences of snoring and upper airway obstruction include significant intrathoracic pressure variations [3], increased transmural pressure load on the heart [4], and sharp blood pressure spikes induced by cortical arousal [5]. In addition, continuous tissue vibration can damage the upper airway and surrounding tissues, including the carotid artery [6].

Polysomnography (PSG) is the gold standard for diagnosing OSA, using multiple sensors to comprehensively record electroencephalogram (EEG), electrooculogram (EOG), electromyogram (EMG), electrocardiogram (ECG), oral-nasal airflow, blood oxygen saturation, respiratory effort and body position [7]. However, interpretation of PSG results requires experienced sleep specialists to manually score the Apnea-Hypopnea Index (AHI), a process that is both time consuming and costly [8]. Therefore, there is an urgent need for the development of convenient, affordable diagnostic devices for sleep disorders to facilitate personalised screening at home.

Existing interventions for snoring include the use of mandibular advancement devices (MADs) [9] and nasal devices [10]. Anti-snoring mouthpieces work by holding the lower jaw in a forward position during sleep, effectively preventing airway narrowing and eliminating vibrations in the throat. Nasal devices, on the other hand, physically dilate the nostrils, increasing airflow and reducing resistance in the nasal cavity. These devices have been shown to have a positive effect on relieving the symptoms of sleep apnoea. However, they often compromise wearer comfort and disrupt normal sleep patterns.

To overcome these limitations, this study introduces for the first time a snoring detection network based on a lightweight attention mechanism, LightSnore-Net, and on this basis designed and developed a smart pillow. Compared to existing anti-snoring devices, the intelligent pillow based on LightSnore-Net significantly improves in terms of intelligence and comfort without interfering with natural oral or nasal functions. This innovative solution offers new ways to screen and manage sleep disorders at home, potentially improving the quality of sleep and overall health of OSA patients.

2 Related Work

With the rapid development of wearable devices and Internet of Things (IoT) sensor technologies, deep learning-based snore detection models have been developed for long-term sleep monitoring in home environments. Hu et al. [12] developed a unified snore detector based on raw waveforms, using a multi-scale strategy that integrates different levels of feature maps and an adaptive receptive field (ARF) module to mimic the response of the human auditory system to sound frequencies. This method achieved a mean accuracy (mAP) of 85.19% in the SPC2018 snoring sound challenge. Song et al. [13] introduced a model combining a soft voting strategy of XGBoost, Mel-spectrogram and CNN, and Mel-spectrogram and Residual Networks (ResNet), which achieved an accuracy of 83.44%, demonstrating the effectiveness of multi-model fusion in snoring sound detection. Furthermore, Ye et al. [14] developed a snoring detection algorithm based on multi-channel spectrograms and CNN, which was tested on sleep recordings of 30 subjects. By extracting four different types of feature maps as model inputs, their accuracy reached 94.18%.

As attention mechanisms continue to evolve, snoring monitoring models are increasingly being adapted for embedded devices. Researchers are focusing on developing lightweight attention mechanisms to achieve efficient detection in resource-constrained environments. For example, Peng et al. [15] developed a dual-modal feature fusion CNN model for OSA detection by concatenating two types of extracted features. To improve the classification accuracy, they introduced a lightweight ECA channel attention mechanism in the dual-modal CNN module to dynamically weight the extracted feature maps, achieving an accuracy of 95.91% on relevant datasets. Similarly, Chen et al. [16] proposed a multiscale fusion network, SE-MSCNN, for SA detection using single-lead ECG signals collected from wearable devices. Experimental results on the PhysioNet Apnea-ECG dataset showed that SE-MSCNN achieved a segment-best accuracy of 90.64% and a record-best accuracy of 100%, with fast response times and lightweight parameters, making it suitable for embedding in wearable devices.

Recent research in snoring monitoring has increasingly focused on integrating deep learning networks with anti-snoring devices. This approach aims to enhance the accuracy of snoring sound detection and classification, as well as to provide physical interventions to mitigate snoring. However, many current solutions fail to prioritize user comfort and the preservation of natural sleep states. Although deep learning networks offer high accuracy, they often lack environmental adaptability, and the interventions from anti-snoring devices can negatively impact sleep quality. To address these issues, we present a novel deep neural network named LightSnore-Net. LightSnore-Net incorporates a lightweight attention mechanism, making it suitable for deployment on embedded devices. This network not only optimizes snore monitoring accuracy and usability but also ensures user comfort and the maintenance of natural sleep states. Our results demonstrate that LightSnore-Net effectively balances these critical factors, providing a more holistic solution to snoring monitoring and intervention.

3 Methods

To facilitate deployment on embedded devices, we are considering enhancements based on the MobileNetV3 framework. The project is divided into three parts: data processing, network training and hardware transplantation. First, the LightSnore-Net network is trained on a dataset, then the network is built on the STM32 by parameter optimisation, thus achieving hardware transplantation.

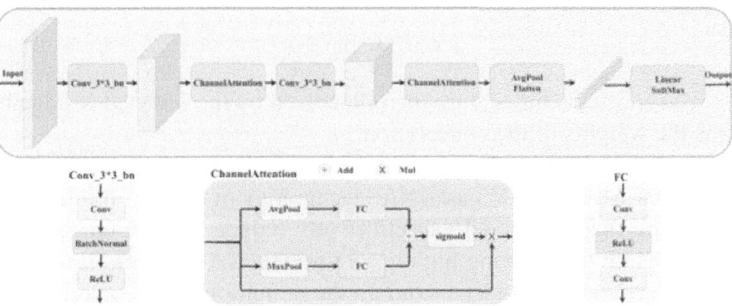

Fig. 1. LightSnore-Net Network Structure.

We present LightSnore-Net, a lightweight convolutional neural network architecture, as shown in Fig. 1. By eliminating the substantial transformation layers, specifically the self.conv3 and self.se3 convolution blocks, we reduce the depth and number of parameters of the model, thereby speeding up the inference process. To improve the feature extraction capabilities while minimising the computational cost, we adjusted the step in the conv_3x3_bn layer of the depth-separable convolution from 2 to 1, reducing the downsampling of the feature map to preserve more detail. Furthermore, using a lightweight attention mechanism, we modified the reduction ratio and replaced the fully connected layers with more efficient 1x1 convolution layers, further reducing the model parameters. At the same time, by integrating strategies of global average pooling and global max pooling, and by improving the channel recalibration process using ReLU and sigmoid functions, we have refined the architecture to efficiently capture channel-wise statistical information while ensuring model compactness and performance efficiency.

3.1 MobileNetV3

MobileNetV3 is a lightweight deep learning model optimised for mobile and embedded devices. Its design philosophy is to maintain high accuracy by reducing computational and model parameters [17].

Depthwise separable convolution is a process that decomposes standard convolution into depthwise convolution and pointwise convolution. The convolution is applied to each channel of the input independently. For each channel C_{in} of the input feature map F, depthwise convolution is performed with a kernel size of $D_K \times D_K$, where D_K is the kernel size. The computational cost of this step is $D_K \times D_K \times C_{in} \times H_{out} \times W_{out}$, where H_{out} and W_{out} are the height and width of the output feature map, respectively. The point convolution combines the channels of the output feature map from the depth convolution using 1×1 kernels. The computational load for this step is $C_{in} \times C_{out} \times H_{out} \times W_{out}$, where C_{out} is the number of channels in the output feature map.

The inverted residual structure, introduced in MobileNetV2 and continued in MobileNetV3, is a key structure that expands the channels of the feature map (using an expansion factor t), applies a depth-separable convolution, and then compresses it back to the original dimensions, forming a residual connection.

For an input feature map x, the output y of an inverted residual block can be expressed as

$$y = F(x, W) + x \tag{1}$$

where $F(x, W)$ represents the feature map processed by the inverted residue block and W represents the weights of the convolution.

NAS (Neural Architecture Search) is used to automatically search for optimal network architectures. MobileNetV3 uses NAS technology to further optimise the network structure, ensuring that the model can effectively reduce the amount of computation and parameters while maintaining high accuracy. The NAS process involves complex algorithms and formulas, typically including the definition of the search space, search strategy and performance evaluation criteria.

3.2 Squeeze-and-Excitation Block

The Squeeze-and-Excitation (SE) block is a mechanism that improves network performance by explicitly modelling the interdependencies between feature channels. For an input feature map U, the SE block first compresses each channel through global average pooling, then adjusts the weight of each channel through two fully connected layers (compression and activation layers), and finally outputs a weighted feature map by scaling the channels of the input feature map.

The channel c's feature compression z_c obtained by global average pooling is:

$$z_c = \frac{1}{H \times W} \sum_{i=1}^{H} \sum_{j=1}^{W} u_c(i,j) \tag{2}$$

With the output of the fully connected layers as s, the output $x_c\prime$ of the SE block is:

$$x_c\prime = u_c \cdot \sigma(s_c) \tag{3}$$

where u_c is channel c of the input feature map U, σ is the sigmoid activation function, and s_c is the processed z_c through two fully connected layers.

3.3 Hardware Design

PCB Design. Figure 2 illustrates the use of the STM32F103 chip as the main board's central processing unit. This chip is responsible for regulating the activity of six micro-solenoid valves within the pillow, which adjust the airbag inflation levels in a timely manner according to the user's sleeping posture. Connected to the Sound Collection Sensor Module via the UART interface, the system effectively monitors and analyses the intensity of snoring and ambient noise before feeding it into the LightSnore-Net for analysis and adjustment of the anti-snore function. To meet user needs, an external flash memory and a Bluetooth module interface are added to improve storage capacity and data transmission speed, enabling communication with external devices such as smartphone applications, supporting remote control and data exchange.

A power management module is introduced to ensure a stable power supply to the system, integrating overload protection and voltage stabilisation functions to ensure system stability and reliability. The inclusion of an ST-Link interface facilitates firmware upgrades and data transfer, simplifying maintenance processes and improving system serviceability and scalability. Circuit layout optimisation focuses on minimising the length of signal and power lines and adopting a strategy of separating ground lines from power lines to reduce interference. With regard to external interference, measures such as star grounding, power line filtering and differential mode filtering are implemented to improve the system's noise immunity.

Overall Design of Pillow. This study presents the design of an intelligent anti-snoring pillow (Fig. 3), which has a latex outer shell embedded with six inflatable airbags made of PTU composite fabric. The pillow uses a silent air pump and deceleration mechanism system for inflation and deflation, and aims to provide a comfortable, antibacterial, breathable and durable support experience without disturbing the user's sleep.

Fig. 2. Shows the design and physical comparison of the PCB, divided into two parts. The image on the left is the schematic of the PCB. The right image shows the physical PCB.

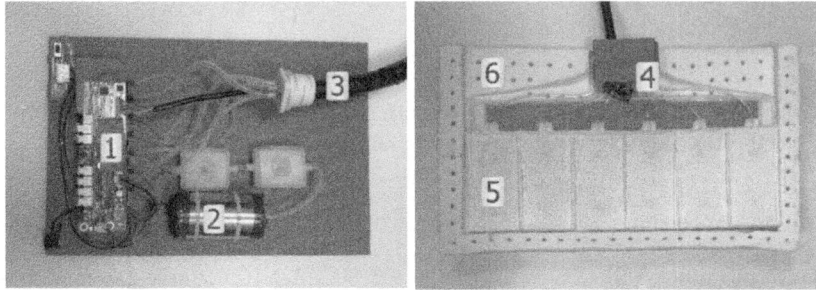

Fig. 3. The overall structural design of the pillow comprises the following six parts: 1) PCB board, serving as the electronic control center of the entire system, responsible for processing audio signals and issuing commands; 2) Silent air pump, providing air pressure to the inflatable airbags, adjusting the height of the pillow; 3) Twisted pair cables, used for the transmission of electronic signals, ensuring stable communication between parts of the system; 4) Sound collection sensor module, responsible for collecting external audio signals for subsequent snoring detection; 5) Inflatable airbags, adjusting the shape and firmness of the pillow according to commands; 6) Latex casing, offering a soft and comfortable appearance and touch to the pillow.

When users make snoring sounds, an acoustic signal collection sensor module on the pillow captures the snoring signals and transmits them, along with audio signals mixed with ambient noise, via a twisted pair cable to the Central Processing Unit (CPU). The CPU converts these signals into analogue signals. The LightSnore-Net model on the STM32F103 microcontroller unit (MCU) then performs a detailed analysis of the signals to identify background noise and quantify the intensity of the snoring. Based on the voltage values feedback from the model, the smart pillow implements a differentiated inflation adjustment strategy to reduce snoring.

The inflation adjustment strategy is divided into three levels based on the intensity of the snoring:

First level: For snoring below 30 dB, the pillow remains unchanged and no adjustments are made.

Second level: If the snoring intensity is between 30 dB and 40 dB, a primary inflation protocol is initiated, slightly raising the pillow to improve the user's airway.

Third stage: If the snoring intensity is between 40 dB and 60 dB, an intermediate inflation mode is activated, which further raises the pillow to facilitate smoother breathing.

Fourth stage: For snoring above 60 dB, an advanced inflation strategy is activated that significantly raises the pillow or adjusts its shape to guide the user's head to naturally turn to the side for maximum snoring reduction.

3.4 Software Design

At the same time, as shown in Fig. 4, we have developed an application called "sleep-monitoring", for which we have successfully obtained software copyright. The application is designed with several features reserved for functionality, including snoring monitoring, sleep scoring, and sleep duration tracking. Users can use this app in conjunction with the smart pillow, allowing them to access their sleep status and receive tailored sleep advice through the app. Future developments are planned to further enhance and expand the functionality of the app to provide an even more convenient experience for users.

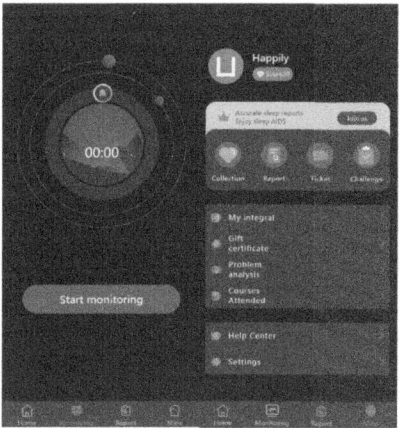

Fig. 4. Schematic Diagram of the "sleep-monitoring" App.

4 Experiments

4.1 Baseline

This section reviews related work in recent years to establish the baseline of this study and highlight its contributions. Li et al. [18] designed a non-contact device to capture sleep breathing sounds and developed a hybrid model combining one-dimensional and two-dimensional convolutional neural networks (1DCNN and 2DCNN) for automatic snore detection, achieving a classification accuracy of 89.3%. González-Martínez et al.

[19] discriminated snoring from non-snoring sounds by analysing the harmonic content of sounds and applying Harmonic/Percussive Sound Separation (HPSS) technology, achieving an accuracy of 88% with a CNN. Dogan et al. [20] proposed an automatic snoring classification method using Maximum A Posteriori Probability (MAP), non-linear patterns and Neighbourhood Component Analysis techniques and achieved an accuracy rate of 97.1% using a k-Nearest Neighbours (kNN) classifier and Leave-One-Out Cross-Validation (LOOCV). Tuncer [21] achieved an accuracy rate of 95.53% in the kNN classification stage with LOOCV using an SSC method based on LDOP. Khan [22] developed a CNN model for snore detection which, after training, was successfully applied to an embedded system "listener module" and achieved a recognition accuracy rate of 96%.

The significant contribution of this study lies not only in outperforming the afore-mentioned existing works but also in successfully implementing the model on an STM32 microcontroller. This implementation has enabled the development of an intelligent snore monitoring system that integrates deep learning-based hardware and software. Our system features low power consumption and real-time response, demonstrating the potential of deep learning technology in the field of medical health monitoring.

4.2 Datasets

This study uses a publicly available snoring dataset published by Khan (2021) on the Kaggle platform for experimentation. The dataset consists of 1,000 audio samples, evenly divided into snoring (label 1) and non-snoring (label 0) categories, each with 500 samples and a duration of 1 s per sample. The snoring samples include snoring sounds of children, adult males and adult females without background noise, while the non-snoring samples include various background noises. The dataset is divided into training, validation and test sets in a 7:2:1 ratio, as shown in Table 1, to ensure that the model is tested on unseen data.

Table 1. Snore dataset setup.

	Snoring	Non-snoring	Total
Training	350	350	700
Validation	100	100	200
Test	50	50	100

4.3 Data Preprocessing

In this study, a series of precise data pre-processing and optimisation techniques are used to preserve the original quality of the audio data and ensure that the deep learning model can accurately recognise snoring characteristics. First, the original audio data is loaded without changing the sampling rate to preserve the original quality of the sound.

Then, by computing Mel Frequency Cepstral Coefficients (MFCCs), we transform the audio signal into a set of coefficients that effectively represent the characteristics of the sound. During this process, 32 MFCCs are set with a hop length of 512 and a Fast Fourier Transform (FFT) window size of 512, ensuring adequate frequency resolution. In order to improve the generalisation ability of the model across different recording conditions, the generated MFCC images are subjected to min-max normalisation, scaling the coefficient values between 0 and 1. This normalisation process reduces the sensitivity of the model to recording volume differences, focusing more on the structural features of snoring rather than its volume.

Given the computational resources and memory limitations of embedded devices, this study uses TensorFlow Lite Converter to transform the trained TensorFlow model into TensorFlow Lite format. By implementing quantization techniques, we significantly reduce the size of the model and optimise its execution speed and efficiency on embedded devices. The quantization process converts model weights from floating-point to integer format, which not only reduces the model's memory requirements, but also increases inference speed, making the model more suitable for real-time applications such as snore detection in smart anti-snoring pillows.

4.4 Evaluation Indicators

To evaluate the overall performance of the LightSnore-Net model in detecting anomalies, we used accuracy, precision, recall and F1 score as evaluation metrics:

$$Accuracy = \frac{TP + TN}{TP + TN + FN + FP} \tag{4}$$

$$Precesion = \frac{TP}{TP + FP} \tag{5}$$

$$Recall = \frac{TP}{TP + FN} \tag{6}$$

$$F_1 = 2 \times \frac{Precision \times Recall}{Precision + Recall} \tag{7}$$

where TP is the number of anomalous samples correctly detected, FP is the number of samples falsely detected as anomalous but actually normal, FN is the number of anomalous samples not detected, and TN is the number of samples correctly detected as normal.

4.5 Experimental Results

In experimental comparisons with the commonly used microcontroller model STM32F407, our optimized system demonstrated a reduction in energy consumption by 13% to 20%. As shown in Fig. 5, a segment of random snoring audio from the internet was used to test the pillow's operational effectiveness in a mixed audio environment with sound levels between 40 and 50 decibels. The pillow inflated its airbags to the

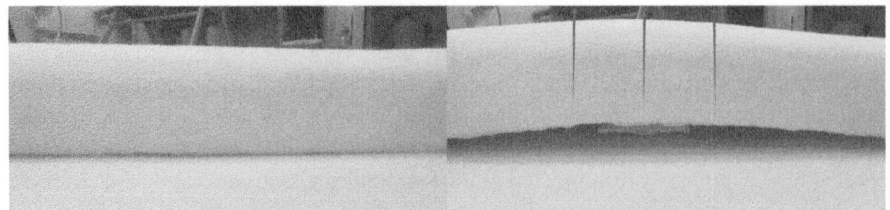

Fig. 5. Operational Demonstration of the Smart Anti-Snoring Pillow.

pre-set levels as expected, demonstrating the integrated coordination of the system and the effectiveness of the LightSnore-Net model.

This paper presents LightSnore-Net, a novel snore detection network that achieves 99% accuracy, outperforming both traditional and recent deep learning models, as detailed in Table 2. Using MFCC features and a lightweight attention mechanism, LightSnore-Net improves model performance in terms of accuracy, precision, recall and F1 score. The model's superiority is due to its efficient network structure and strategic placement of convolutional layers, which not only increases classification accuracy, but also maintains the model's compactness and speed.

Table 2. Comparison of the available studies with our work.

Authors	Method	Features	Snoring Detection Accuracy(%)
Li et al.	1D–2D CNN [18]	Raw audio+Mapped Images	89.30
González-Martínez et al.	CNN [19]	STFT+CQT+Mel-scaled	88.00
Dogan et al.	kNN+LOOCV [20]	SBox+NCAINCA	97.10
Tuncer et al.	DWT+LDOP [21]	LDOP+RFINCA	95.53
Khan	CNN [22]	MFCCs	96.00
Ours	Baseline	MFCCs	92.00
Ours	LightSnore-Net	MFCCs	**99.00**

Ablation studies on LightSnore-Net underlined the central role of its lightweight attention mechanism and convolutional layer configuration. The inclusion of the attention mechanism significantly improved model metrics, yielding a 7% increase in accuracy, 7.47% in precision, 5.75% in recall, and a 6.93% increase in F1 score, as detailed in Table 3. Examining the impact of the convolutional layer output channels revealed that halving them significantly reduced performance, while increasing them by 50% did not significantly improve performance, suggesting an optimal balance between efficiency and effectiveness in the current setup. The critical nature of specific convolutional layers was further demonstrated by performance drops following the removal of either conv1 or conv2, confirming the effectiveness of the lightweight attention mechanism and convolutional architecture in fine-tuning model performance.

Table 3. Ablation experiment

Method	Accuracy(%)	Precision(%)	Recall(%)	F1 Score(%)
Baseline	92.00	92.53	92.12	91.99
Decreased Channels (−50%)	93.00	93.00	93.16	92.99
Increased Channels (+50%)	98.00	97.83	98.21	97.98
Decreased conv1	90.00	91.25	89.66	89.85
Decreased conv2	89.00	88.78	91.13	88.81
LightSnore-Net	**99.00**	**100.00**	**97.87**	**98.92**

5   Conclusion

This work presents LightSnore-Net, a deep neural network enhanced with a lightweight attention mechanism, deployed on the STM32F103C8T6 microcontroller for snoring detection. By integrating smart hardware with deep learning, we propose a novel smart pillow for snoring monitoring, using Mel Frequency Cepstral Coefficients (MFCC) for sound analysis. Incorporating a lightweight attention mechanism significantly increases the network's ability to detect snoring sounds, achieving a remarkable classification accuracy of 99%, outperforming current methods.Ablation studies highlight the critical role of the attention mechanism and optimised convolutional layers in improving model efficiency while maintaining system compactness.

Future directions aim to extend monitoring capabilities to multimodal signals, including ECG, EEG, airflow and SpO2, to improve prediction accuracy and system robustness. In addition, advances in model compression and acceleration will be pursued to meet real-world application requirements and optimise detection efficiency.Ultimately, this research aims to provide an affordable, effective tool for the management of sleep-disordered breathing, improving sleep quality and health outcomes.

Acknowledgments. This work was supported by the National Natural Science Foundation of China (Grant No. 62377036), Tianjin Municipal Training Program of Innovation and Entrepreneurship for Undergraduates (202310057383).

References

1. Lechat, B., Naik, G., Appleton, S., et al.: Regular snoring is associated with uncontrolled hypertension. NPJ Digit. Med. **7**(1), 38 (2024)
2. Korkalainen, H., Töyräs, J., Nikkonen, S., et al.: Mortality-risk-based apnea–hypopnea index thresholds for diagnostics of obstructive sleep apnea. J. Sleep Res. **28**(6), e12855 (2019)

3. Pinsky, M.R.: Discovering the clinical relevance of heart-lung interactions. Anesthesiology **140**(2), 284–290 (2024)
4. Wan, Y.: Craniofacial morphology and dental characteristics in children and adolescents with sleep disorders: a systematic review. Boston University (2024)
5. Dai, Y., Vgontzas, A.N., Chen, L., et al.: A meta-analysis of the association between insomnia with objective short sleep duration and risk of hypertension. Sleep Med. Rev. **75**, 101914 (2024)
6. Govindharaj, K., Manoharan, M., Muthumalai, K., et al.: Interconnected SnO2 nanoflakes decorated WO3 composites as wearable and ultrafast sensors for real-time wireless sleep quality tracking and breath disorder detection. Chem. Eng. J. **482**, 148759 (2024)
7. Duarte, M., Pereira-Rodrigues, P., Ferreira-Santos, D.: The role of novel digital clinical tools in the screening or diagnosis of obstructive sleep apnea: systematic review. J. Med. Internet Res. **25**, e47735 (2023)
8. Yue, H., Chen, Z., Guo, W., et al.: Research and application of deep learning-based sleep staging: data, modeling, validation, and clinical practice. Sleep Med. Rev. **74**, 101897 (2024)
9. Parmenter, D., Millar, B.J.: How can general dental practitioners help in the management of sleep apnoea? Br. Dent. J. **234**(7), 505–509 (2023)
10. Alrejaye, N.S., Al-Jahdali, H.: Dentists' role in obstructive sleep apnea: a more comprehensive review. Sleep Epidemiol. **4**, 100073 (2024)
11. Cay, G., Ravichandran, V., Sadhu, S., et al.: Recent advancement in sleep technologies: a literature review on clinical standards, sensors, apps, and AI methods. IEEE Access **10**, 104737–104756 (2022)
12. Hu, X., Sun, J., Dong, J., et al.: Auditory receptive field net based automatic snore detection for wearable devices. IEEE J. Biomed. Health Inform. (2022)
13. Song, Y., Sun, X., Ding, L., et al.: AHI estimation of OSAHS patients based on snoring classification and fusion model. Am. J. Otolaryngol. **44**, 103964 (2023)
14. Ye, Z., Peng, J., Zhang, X., et al.: Snoring sound recognition using multi-channel spectrograms. Archives of Acoustics (2024)
15. Peng, D., Yue, H., Tan, W., et al.: A bimodal feature fusion convolutional neural network for detecting Obstructive Sleep Apnea/hypopnea from nasal airflow and oximetry signals. Artif. Intell. Med. **150**, 102808 (2024)
16. Chen, X., Chen, Y., Ma, W., et al.: SE-MSCNN: a lightweight multi-scaled fusion network for sleep apnea detection using single-lead ECG signals. In: 2021 IEEE International Conference on Bioinformatics and Biomedicine (BIBM), pp. 1276–1280. IEEE (2021)
17. Howard, A., Sandler, M., Chu, G., et al.: Searching for MobileNetV3. In: Proceedings of the IEEE/CVF International Conference on Computer Vision, pp. 1314–1324 (2019)
18. Li, R., Li, W., Yue, K., et al.: Automatic snoring detection using a hybrid 1D–2D convolutional neural network. Sci. Rep. **13**(1), 14009 (2023)
19. González-Martínez, F.D., Carabias-Orti, J.J., Cañadas-Quesada, F.J., et al.: Improving snore detection under limited dataset through harmonic/percussive source separation and convolutional neural networks. Appl. Acoust. **216**, 109811 (2024)
20. Dogan, S., Akbal, E., Tuncer, T., et al.: Application of substitution box of present cipher for automated detection of snoring sounds. Artif. Intell. Med. **117**, 102085 (2021)
21. Tuncer, T., Akbal, E., Dogan, S.: An automated snoring sound classification method based on local dual octal pattern and iterative hybrid feature selector. Biomed. Signal Process. Control **63**, 102173 (2021)
22. Khan, T.: A deep learning model for snoring detection and vibration notification using a smart wearable gadget. Electronics **8**(9), 987 (2019)

Malaria Cell Images Classification with Deep Ensemble Learning

Qi Ke[1]($\boxtimes$), Rong Gao[1], Wun She Yap[2], Yee Kai Tee[2], Yan Chai Hum[2], and YuJian Gan[3]

[1] School of Big Data and Artificial Intelligence, Guangxi University of Finance and Economics, Nanning 530003, China
keqikeruzhen@126.com
[2] Lee Kong Chian Faculty of Engineering and Science, Universiti Tunku Abdul Rahman, 43000 Kajang, Malaysia
teeyeekai@gmail.com
[3] School of Electronic Engineering and Computer Science, Queen Mary University of London, London 1 4NS, UK

Abstract. Malaria is a deadly infectious disease and a major threat to global health. Efficient and accurate malaria detection is crucial for timely identification of patients and subsequent treatment. Most traditional classification models use a single network to extract features, but each single model can only extract a limited number of image features for classification. To overcome this limitation and to improve the classification performance, this paper proposes a classification study of malaria cell pathology images based on a deep ensemble learning model. Transfer learning technique is employed to pretrain multiple networks, transferring the learned knowledge to the target dataset. The pretrained networks with optimal performance are selected as the basic classifiers for the ensemble model. A weighted voting strategy is used to integrate multiple pretrained networks for the final classification of malaria cell images. To validate the effectiveness of the proposed model, the classification performance is evaluated on a publicly available malaria cell images dataset. Experimental results demonstrate that the proposed deep ensemble model achieves excellent classification performance on the target dataset, with a classification accuracy of 98.49%, outperforming the classification performance of a single CNN model using transfer learning. The proposed deep ensemble learning technique proves to be feasible for classifying malaria cell pathology images.

Keywords: Deep Learning · Transfer Learning · Ensemble Learning · Malaria Cell Images · Image Classification

1 Introduction

Malaria is an insect-borne infectious disease caused by the infection of Plasmodium parasites transmitted through the bite of an infected female Anopheles mosquitos or the introduction of blood containing the parasites [1]. To date, malaria remains a serious global health concern, with approximately 40% of the world's population living in

D.-S. Huang et al. (Eds.): ICIC 2024, LNBI 14881, pp. 417–427, 2024.
https://doi.org/10.1007/978-981-97-5689-6_36

malaria-endemic areas [2]. The World Malaria Report 2023, published by the World Health Organization, indicates an estimated 249 million malaria cases globally in 2022, an increase of 2 million cases compared to 2021.

Screening and diagnosing malaria are not an easy task. Professional pathologists determine the type and severity of malaria infection by observing various features in blood samples under a microscope [3].With the increasing maturity of deep learning technology, convolutional neural networks (CNNs) have emerged, enabling the automatic extraction of features and highlighting the advantages of computer-assisted medical diagnosis [4]. Researchers have applied CNNs to the medical field, leading to the gradual development of deep learning algorithms for medical image classification and segmentation [5]. Developing deep learning-based techniques for malaria pathology detection can alleviate the burden on humans and assist in making accurate diagnoses.

While many studies have successfully applied deep learning technology to medical image processing, there are current limitations in obtaining medical image dataset due to difficulties in acquisition, small dataset sizes, expensive and time-consuming annotations, among other constraints [6]. These limitations result in insufficient training data for deep learning models, leading to issues such as overfitting and inadequate generalization.

Each network structure has differences, with its unique strengths and limitations. Different networks extract distinct important features and provide varying interpretations of images [7]. Although a single network can make correct predictions, there are limitations in feature extraction that may lead to insufficient extraction of crucial features, making it challenging to achieve high-accuracy predictions [8]. Therefore, the accuracy and precision of medical image analysis based on deep learning technology are still in its infancy.

To address the challenge of insufficient medical image samples and to overcome the issue of incomplete feature extraction by a single pretrained network, this study investigates the problem in malaria diagnosis via blood smear images by leveraging transfer and deep ensemble learning. The effectiveness and classification performance of the proposed method is evaluated on a publicly available dataset, Malaria Cell Images Dataset.

2   Related Works

Utilizing machine learning methods for quantitative analysis of hematological data or blood smear images can mimic the medical professionals' expertise in malaria diagnosis, mitigating the impact of subjective factors [9]. Lately, algorithms employing deep learning for pathological image classification and diagnosis have gained significant traction, primarily due to their capacity to automatically extract useful image features [10]. Mitrovic et al. [11] used convolutional neural networks to classify thin blood smear images of the potentially malaria-infected cells. The study managed to obtain 82.7% accuracy in malaria classification. Rahman et al. [12] proposed a fast convolutional neural network architecture for the classification of malaria cell images and achieved a classification accuracy of 96.7%.

Deep learning models usually require a large and diverse dataset to learn features efficiently to achieve high performance of the model. However, medical image data are scarce, expensive and time-consuming to annotate. In addition to employing data augmentation, transfer learning technique is also a common approach to address the need for huge amounts of data [13]. Transfer learning involves pre-training a network on natural image datasets or other image datasets, and then transferring the learned knowledge to the target task, thereby enhancing the performance of the network model [14].

Barath et al. [15] applied transfer learning for malaria cell classification. The transfer learning model was compared with AlexNet, ResNet, VGG-16 and DensNet, achieving a classification accuracy of 96.7%. Tajbakhsh et al. [16] systematically investigated the knowledge transfer learning between natural images and medical images. The performance of transfer learning was tested in four different medical imaging applications.

Most of the state-of-the-art deep learning models for malaria cell images classification employs a single CNN model. Using a single model as the pre-trained network for transfer learning may overlook certain layer features, leading to insufficient or incomplete extraction of crucial features [17]. Ensemble learning involves aggregating the output decisions of multiple base classifiers (pre-trained networks) through ensemble strategies to make the final classification predictions [18]. Ensemble models have a broader coverage in extracting features, contributing to improved classification performance [19, 20].

For instance, Li C et al. [21] proposed an ensemble learning framework for classifying cervical tissue pathology images into well-differentiated, moderately differentiated, and poorly differentiated categories. The model achieved the highest accuracy of 96.05%. D Xue et al. [22] proposed an ensemble learning technique to achieve a more accurate classification result for three cervical cancer differentiation stages. Four models (Inception-V3, Xception, VGG-16, and Resnet-50) were utilized for classification. They obtained the highest overall accuracy of 98.61% on poorly differentiated staining. Zhu et al. [23] proposed a ResNet-Based Output Ensemble Net for malaria parasite classification. Three randomized neural networks including Random Vector Function Link, Schmidt Neural Network and Extreme Learning Machine were selected as classifiers of ensemble model. The accuracy was as high as 95.73%.

One of the pressing issues in malaria diagnosis is the limited public malaria datasets, it lacks diversity in both parasitized and uninfected categories. The existing malaria cell image datasets are either relatively small or lack of detailed annotations, limiting the widespread use of deep learning approaches to automate or assist in malaria diagnosis. The ability of ensemble model to aggregate features or decisions from multiple pre-trained models can overcome these limitations. In this paper, we will propose a deep learning-based ensemble learning in malaria cell image classification to facilitate malaria diagnosis via blood smear images.

3   Model and Methodology

3.1   Pretrained Networks as Base Models

This paper used three CNN architectures and their variants as pretrained networks, including VGG16, VGG19, ResNet50, ResNet101, DenseNet121, and DenseNet201. In total, six CNNs were used in the experiments, a few best performed pre-trained networks will be selected as the base classifiers for the ensemble model.

3.2   Transfer Learning

The advantage of transfer learning lies in overcoming challenges such as limited dataset, scarcity of labeled samples, and avoiding issues like overfitting during pre-training [24]. Transfer learning is well-suited for application in pathology image datasets with limited data. In this study, we employed network models trained on the large-scale ImageNet dataset. Through the fine-tuning technique of transfer learning, the acquired knowledge was transferred on the malaria cell images dataset, achieving effective feature extraction.

3.3   Ensemble Learning

The concept of ensemble learning involves combining multiple base classifiers through ensemble strategy to collectively accomplish a classification or prediction task. Since a single network has certain limitations in feature extraction, if multiple different networks are trained, they can complement each other's prediction results, and then an excellent prediction result can be obtained through a suitable ensemble strategy. These base classifiers will be selected from the pretrained models of transfer learning with the best classification performance.

The ensemble strategies usually include voting, averaging and stacking. The voting strategy selects the result with the most votes as the final prediction. For weighted voting, different base models have different voting weights, the final prediction is based on the voting weights multiple by the model outputs. The voting is the most commonly used and effective strategy in classification problems. Considering the varying performance advantages and contributions of each base model, this study opted for a weighted voting strategy.

The weighted voting method: corresponding weights were assigned to each classifier based on the performance differences of each pretrained network. The weights of the base classifiers are used to calculate the voting scores, which are aggregated and the category with the largest weighted sum is selected as the final prediction. The better-performing base classifiers received higher weights, exerting a greater influence on the final classification results. The formular is given in Eq. (1):

$$En(x) = C_{arg_j max} \sum_{i=1}^{N} w_i b_i^j(x) \tag{1}$$

where ensemble model includes N base classifiers $\{b_1, b_2, b_3, ..., b_n\}$, En is the ensemble model of the final prediction, w_i is the weight of b_i, $w_i \geq 0$ and $\sum_{i=1}^{N} w_i = 1$.

3.4 Evaluation Metrics

This paper utilized accuracy, precision, recall, and F1-score as evaluation metrics for assessing the classification performance of the model. TP represents true positives (positive samples correctly predicted as positive), TN represents true negatives (negative samples correctly predicted as negative), FP represents false positives (negative samples incorrectly predicted as positive), and FN represents false negatives (positive samples incorrectly predicted as negative). The calculation equations are described.

$$Accuracy = \frac{TP + TN}{TP + TN + FP + FN} \tag{2}$$

$$Precision = \frac{TP}{TP + FP} \tag{3}$$

$$Recall = \frac{TP}{TP + FN} \tag{4}$$

$$F1 - score = 2 \times \frac{Precision \times Recall}{Precision + Recall} \tag{5}$$

The accuracy of image classification is the most commonly used metric. We aim to accurately differentiate between parasitized and uninfected malaria cells. In this paper, the accuracy serves as the criterion for judging the performance of pretrained models. The pretrained models with the highest accuracy are selected as the base classifiers for the ensemble model. In medical image diagnosis, the cost of false negatives is significant, so ensuring a high recall for identifying positive samples is also a crucial parameter. Therefore, among recall, precision, and F1 score, we choose the recall of the pretrained model as the weight for the model.

The whole workflow of the ensemble learning model of this study is depicted in Fig. 1. The basic framework comprises the following four steps:

1. Data augmentation. All malaria cell images are divided into training and test sets. Various data augmentation techniques are applied to augment the training set.
2. Transfer learning. Multiple convolutional neural networks are employed for transfer learning. Through fine-tuning and pre-training, the well-pretrained network models are obtained.
3. Ensemble learning. The top k best performance pretrained networks are chosen as the base classifiers for the ensemble model. Obtain the final prediction results through ensemble strategy.
4. Model evaluation. Accuracy, precision, recall, and F1 score are used to measure the classification performance of the ensemble model. Model parameters will be detailed in the next section.

4 Experiment and Evaluation

4.1 Dataset

In this paper, an open dataset of cell images was sourced from the National Institutes of Health [25]. The dataset comprises images related to malaria screening research activities collected and photographed at Chittagong Medical College Hospital, Bangladesh.

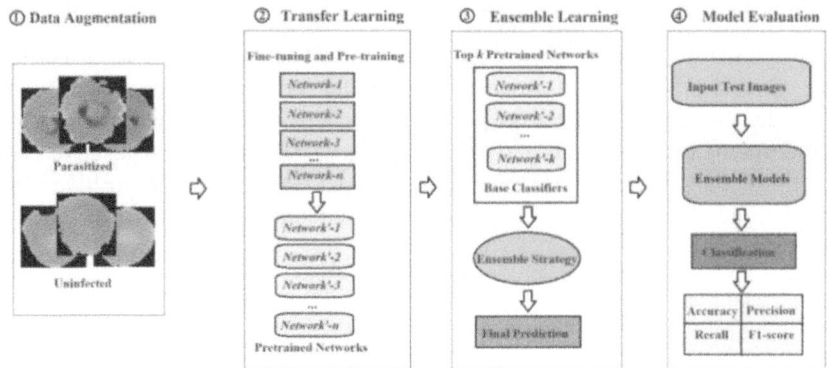

Fig. 1. Workflow of the proposed ensemble techniques for malaria diagnosis via blood smear images.

The images are Giemsa-stained thin blood smears of 150 patients infected with Plasmodium falciparum and 50 uninfected individuals. The dataset comprises a total of 27,558 cell images, divided into two balanced classes: parasitized cells and uninfected cells, each class comprises 13,779 images. The dataset was split into training and test sets at proportions of 80% and 20%, respectively. Examples of cell images from the dataset are shown in Fig. 2.

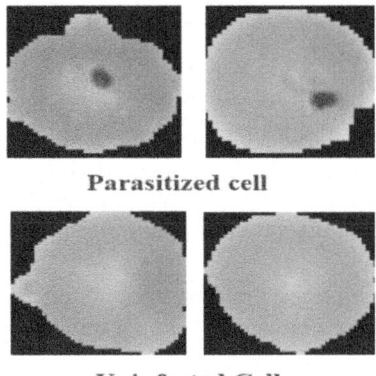

Fig. 2. Sample images of parasitized and uninfected cells.

4.2 Data Pre-processing

The purpose of data preprocessing is to generate a larger and more diversified image dataset to aid in model training. In this experiment, data augmentation techniques such as vertical flip, horizontal flip, rotation, and zoom were applied to augment the training samples of the dataset. The training set size after data augmentation is twice that of the original training set. Additionally, the images underwent Normalization to better respond to activation function, and improve the convergence speed of the model.

4.3 Experiment Setting

The experiments in this paper were conducted using a GeForce RTX 3090 GPU with 24GB of memory, and the deep learning framework employed was PyTorch. The initial learning rate was set to 0.0001, and the Adam was used as the optimizer. The batch size was 64, and the number of epochs was set to 30.

For the transfer learning part, 6 commonly used CNNs were selected for performance testing, including VGG16, VGG19, ResNet50, ResNet101, DenseNet121, and DenseNet201. The weights of each CNN in transfer learning were initialized with pretrained weights on the ImageNet dataset. The last layer of the CNNs was fine-tuned, and the remaining layers of the model were frozen. The last layer of each CNN was replaced with a new fully connected layer having 2 neurons and a SoftMax function to complete the classification performance test of the CNNs.

After pretraining the 6 CNNs through transfer learning, the top 3 best performing CNNs were selected as the basic classifiers for the ensemble model based on their accuracy on the test set. In this experiment, a weighted voting strategy was used as the ensemble strategy for the model to accomplish the final classification task of parasitized and uninfected cells in the dataset.

4.4 The Experimental Results and Analysis

Different CNNs exhibit varying classification performances when using transfer learning with fine-tuning technique on pathological image datasets. Table 1 presents the performance results of the 6 CNNs on the test set of the malaria cell image dataset, including accuracy, precision, recall, and F1 score values. In Table 1, among the classification performances of various CNNs, the top 3 models with the highest accuracy are: VGG16 with the best accuracy at 97.13%, DenseNet121 with an accuracy of 97.10%, and DenseNet201 with an accuracy of 97.08%.

The experimental results indicate that, with the same transfer learning technique and dataset, different CNNs achieve different classification accuracies. This demonstrates that different CNNs focus on distinct regions of interest in image features, leading to variations in the extracted image features and, consequently, different classification performances.

Table 2 displays the comparison of the accuracy of the ensemble learning model proposed in this paper with other advanced models from the literatures for classifying the same dataset. From Table 2, it is evident that the classification accuracy of the proposed ensemble learning model reaches 98.49%, achieving the best performance. Experimental verification indicates that the proposed ensemble learning technique exhibits superior classification performance.

Furthermore, as observed in Table 2, the classification accuracy of the malaria cell image dataset using the proposed ensemble model (98.49%) is higher than that achieved by transfer learning (97.13%) and other improved individual CNN models. The experimental results suggest that the classification performance of the proposed ensemble

Table 1. Classification performance of pre-trained networks with transfer learning on the malaria cell image dataset.

Models	Accuracy	Precision	Recall	F1-score
VGG16	97.13%	97.79%	96.44%	97.11%
VGG19	96.59%	98.38%	94.74%	96.52%
ResNet101	95.39%	97.03%	93.65%	95.31%
ResNet152	94.85%	97.65%	91.91%	94.69%
DenseNet121	97.10%	97.93%	96.23%	97.07%
DenseNet201	97.08%	98.00%	96.12%	97.05%

model surpasses that of transfer learning. When multiple basic classifiers are included in the ensemble model, the final classification decision depends on a broader and more comprehensive set of image features, resulting in improved classification performance. The feasibility and effectiveness of the proposed deep learning-based ensemble model for classifying malaria cell images are validated in this research.

Table 2. Comparison of classification accuracy between the proposed model and other existing models.

Models	Accuracy
InceptionV3 and SVM [26]	95.14%
Inceptionv3-GF [27]	97.83%
Improved ResNet50 [1]	95.47%
InceptionV3-RMSprop [28]	97%
ROENet [23]	95.73%
EL [29]	97.92%
Proposed transfer learning	97.13%
Proposed ensemble learning	98.49%

4.5 Extended Experiment

The proposed ensemble model achieves excellent performance in binary classification on the malaria cell dataset. To demonstrate the effectiveness and robustness of the proposed ensemble model and verify that the model is not restricted to dataset types, yielding good classification results in multi-class datasets, we conducted further experiment on different pathological image datasets. Therefore, in the extended experiment, we used the Cervical Cell Image dataset (Herlev-dataset). The Cervical Cell Image dataset is a multi-class pathological image dataset of cervical cells. This dataset comprises 917 single-cell images, with 242 belonging to the benign class and 675 to the malignant class. Each cell annotation is categorized into seven classes.

In the extended experiment, the experimental setup involves configuring the output SoftMax layer of the pretrained CNNs to have 7 nodes to accommodate the 7 classes in the Cervical Cell Image dataset. For the other experimental procedures, we utilized the same preprocessing steps and experimental settings as those used for classifying the malaria cell dataset.

Table 3 presents the ensemble learning model proposed in this paper achieves an excellent classification accuracy of 99.02% on Herlev-database dataset, which is the highest compared to the classification accuracies in the advanced models. The experimental results demonstrate that the proposed ensemble model exhibits good classification performance and generalization ability when dealing with multi-class classification tasks on pathological datasets from different organs.

Table 3. Comparison of classification accuracy between the proposed model and other existing models.

Models	Accuracy
TL–CNN [30]	92.60%
Deep Cervix [31]	98.32%
ETL-framework [22]	98.37%
Proposed ensemble learning	99.02%

5   Conclusion

This paper presents a deep ensemble learning model based on VGG, ResNet, and DenseNet architectures for the classification of malaria via blood smear cell images. High classification accuracy was achieved on the malaria cell pathology image dataset, reaching 98.49% with the proposed ensemble model. This can contribute towards early detection and diagnosis of malaria, providing effective information for early diagnostic intervention decisions and enhancing patient recovery and survival rates. Future research should consider the potential impact of feature types in the dataset on pretrained models to further enhance the performance of lesion detection in pathological images.

Acknowledgments. This study was funded by Guangxi first-class discipline statistics construction project fund, and Guangxi University Young and Middle-aged Teachers' Research Capacity Enhancement Project (2024KY0669).

References

1. Chima, J.S., Shah, A., Shah, K., Ramesh, R.: Malaria cell image classification using deep learning. Int. J. Recent Technol. Eng. (IJRTE) **8**, 5553–5559 (2020). https://doi.org/10.35940/ijrte.F9540.038620

2. Krishnadas, P., Chadaga, K., Sampathila, N., Rao, S., Swathi, K.S., Prabhu, S.: Classification of malaria using object detection models. Informatics. **9**(4), 76 (2022). https://doi.org/10.3390/informatics9040076

3. Vijayalakshmi, A.: Deep learning approach to detect malaria from microscopic images. Multimed Tools Appl. **79**, 15297–15317 (2020)

4. Ali, R., Hardie, R.C., Narayanan, B.N., Kebede, T.M.: IMNets: deep learning using an incremental modular network synthesis approach for medical imaging applications. Appl. Sci. (Switz.) **12**(11), 5500 (2022). https://doi.org/10.3390/app12115500

5. Li, X., et al.: A comprehensive review of computer-aided whole-slide image analysis: from datasets to feature extraction, segmentation, classification and detection approaches. Artif. Intell. Rev. **55**, 4809–4878 (2022). https://doi.org/10.1007/s10462-021-10121-0

6. Kloeckner, J., Sansonowicz, T.K., Rodrigues, Á.L., Nunes, T.W.N.: Multi-categorical classification using deep learning applied to the diagnosis of gastric cancer. J. Bras. Patol. Med. Lab. **56** (2020). https://doi.org/10.5935/1676-2444.20200013

7. Celik, Y., Talo, M., Yildirim, O., Karabatak, M., Acharya, U.R.: Automated invasive ductal carcinoma detection based using deep transfer learning with whole-slide images. Pattern Recognit Lett. **133**, 232–239 (2020). https://doi.org/10.1016/j.patrec.2020.03.011

8. Iizuka, O., Kanavati, F., Kato, K., Rambeau, M., Arihiro, K., Tsuneki, M.: Deep learning models for histopathological classification of gastric and colonic epithelial tumours. Sci. Rep. **10** (2020). https://doi.org/10.1038/s41598-020-58467-9

9. Hameed, Z., Zahia, S., Garcia-Zapirain, B., Javier Aguirre, J., María Vanegas, A.: Breast cancer histopathology image classification using an ensemble of deep learning models. Sensors. **20**, 4373 (2020). https://doi.org/10.3390/s20164373

10. Spanhol, F.A., Oliveira, L.S., Petitjean, C., Heutte, L.: Breast cancer histopathological image classification using convolutional neural networks. In: 2016 International Joint Conference on Neural Networks, pp. 2560–2567. IEEE (2016)

11. Mitrovic, K., Milosevic, D.: Classification of malaria-infected cells using convolutional neural networks. In: 2021 IEEE 15th International Symposium on Applied Computational Intelligence and Informatics (SACI), pp. 000323–000328 (2021). https://doi.org/10.1109/SACI51354.2021.9465636

12. Rahman, A., Zunair, H., Reme, T.R., Rahman, M.S., Mahdy, M.R.C.: A comparative analysis of deep learning architectures on high variation malaria parasite classification dataset. Tissue Cell **69**, 101473 (2021). https://doi.org/10.1016/j.tice.2020.101473

13. Alassaf, A., Sikkandar, M.Y.: Intelligent deep transfer learning based malaria parasite detection and classification model using biomedical image. Comput. Mater. Continua. **72**, 5273–5285 (2022). https://doi.org/10.32604/cmc.2022.025577

14. Sai Bharadwaj Reddy, A., Sujitha Juliet, D.: Transfer learning with RESNET-50 for malaria cell-image classification. In: Proceedings of the 2019 IEEE International Conference on Communication and Signal Processing, ICCSP 2019, pp. 945–949. Institute of Electrical and Electronics Engineers Inc (2019). https://doi.org/10.1109/ICCSP.2019.8697909

15. Narayanan, B.N., Ali, R.A., Hardie, R.C.: Performance analysis of machine learning and deep learning architectures for malaria detection on cell Images. In: Zelinski, M.E., Taha, T.M., Howe, J., Awwal, A.A., and Iftekharuddin, K.M. (eds.) Applications of Machine Learning, pp. 29. SPIE (2019). https://doi.org/10.1117/12.2524681

16. Tajbakhsh, N., Shin, J.Y., Gurudu, S.R., Hurst, R.T., Kendall, C.B., Gotway, M.B., Liang, J.: Convolutional neural networks for medical image analysis: full training or fine tuning? IEEE Trans. Med. Imaging. **35**, 1299–1312 (2016). https://doi.org/10.1109/TMI.2016.2535302

17. Liu, W., et al.: Preoperative prediction of Ki-67 status in breast cancer with multiparametric MRI using transfer learning. Acad. Radiol. **28**, e44–e53 (2021). https://doi.org/10.1016/j.acra.2020.02.006

18. Yong, M.P., et al.: Histopathological cancer detection using intra-domain transfer learning and ensemble learning. IEEE Access (2023)

19. Yong, M.P., et al.: Histopathological gastric cancer detection on GasHisSDB dataset using deep ensemble learning. Diagnostic. **13**, 1793 (2023)

20. Koo, J.C., et al.: Non-annotated renal histopathological image analysis with deep ensemble learning. Quant. Imageing Med. Surg. **13**, 5902 (2023)

21. Li, C., et al.: Cervical histopathology image classification using ensembled transfer learning. In: Advances in Intelligent Systems and Computing, pp. 26–37. Springer Verlag (2019). https://doi.org/10.1007/978-3-030-23762-2_3

22. Xue, D., et al.: An application of transfer learning and ensemble learning techniques for cervical histopathology image classification. IEEE Access. **8**, 104603–104618 (2020). https://doi.org/10.1109/ACCESS.2020.2999816

23. Zhu, Z., Wang, S., Zhang, Y.: ROENet: A ResNet-based Output Ensemble for Malaria Parasite Classification. Electron. (Switz.). **11**, 2040 (2022). https://doi.org/10.3390/electronics1113 2040

24. Pan, S.J., Yang, Q.: A Survey on transfer learning. IEEE Trans. Knowl. Data Eng. **22**, 1345–1359 (2010). https://doi.org/10.1109/TKDE.2009.191

25. Rajaraman, S., et al.: Pre-trained convolutional neural networks as feature extractors toward improved malaria parasite detection in thin blood smear images. PeerJ **6**, e4568 (2018). https://doi.org/10.7717/peerj.4568

26. Amrutha Reddy, M., Sai Siva Rama Krishna, G., Tanoj Kumar, T.: Malaria cell-image classification using inceptionV3 and SVM. Int. J. Eng. Res.Technol. **10** (2021)

27. Çinar, A., Yildirim, M.: Classification of malaria cell images with deep learning architectures. Ingenierie des Systemes d'Inf. **25**, 35–39 (2020). https://doi.org/10.18280/isi.250105

28. Minarno, A.E., Aripa, L., Azhar, Y., Munarko, Y.: Classification of malaria cell image using inception-V3 architecture. Int. J. Inf. Visual. **7**, 273–278 (2023)

29. Bhuiyan, M., Islam, M.S.: A new ensemble learning approach to detect malaria from microscopic red blood cell images. Sens. Int. **4**, 100209 (2023). https://doi.org/10.1016/j.sintl.2022.100209

30. Cho, B.J., et al.: Automated diagnosis of cervical intraepithelial neoplasia in histology images via deep learning. Diagnostics **12**, 548 (2022). https://doi.org/10.3390/diagnostics12020548

31. Rahaman, M.M., et al.: DeepCervix: a deep learning-based framework for the classification of cervical cells using hybrid deep feature fusion techniques. Comput. Biol. Med. **136**, 104649 (2021). https://doi.org/10.1016/j.compbiomed.2021.104649

Multi-source Unsupervised Domain Adaptation for Medical Image Recognition

Yujie Liu(✉) and Qicheng Zhang(✉)

Shanghai, People's Republic of China
{liuyujie,bababa_}@shu.edu.cn

Abstract. Medical image recognition is pivotal in intelligent healthcare, especially when addressing complex diseases and human anatomical structures. Intelligent models can be trained using a vast number of labeled medical images, which enables the automatic detection of lesions. However, existing models often overlook the domain gap between the training data and the actual clinical environment, resulting in poor performance in new clinical settings. In this paper, we propose an adaptive dynamic multi-stage pseudo-labeling mechanism based on multi-source domain samples for generating pseudo-labels of the unlabeled target domain images. Additionally, we introduce a multi-source domain adaptation (MSDA) framework for medical image recognition with limited labeled training samples for a specific dataset. Training with multi-source samples enhances the model's generalization and adaptability, in which the diverse samples enable the target model to learn more representative feature maps. Our method achieves high accuracy and robustness in medical image recognition, demonstrating strong adaptability and superior performance across various clinical scenarios.

Keywords: Deep learning · Domain adaptation · Pseudo-label · Medical image recognition

1 Introduction

Medical image recognition is essential for medical diagnostics [1]. Recent advances in imaging technology have provided clinicians with detailed insights into the structures and diseases of the human body. As shown in Fig. 1, there is variability in imaging data across medical institutions, devices, and patient demographics, and existing models struggle to handle this significant domain gap. This is because many models focus on improving feature extraction ability within a limited number of datasets to enhance test accuracy, neglecting the limited generalization ability across different domain. This results in a gap between the training environments and the actual application settings, leading to poor performance of these models in new clinical environments. Domain adaptation has great potential to address these issues because it can use knowledge from source domain to compensate for the lack of knowledge in the target domain. This method enables models to achieve great success in entirely new clinical environments.

Y. Liu and Q. Zhang—Independent Researcher.
Y. Liu and Q. Zhang—Contributing equally to this work.

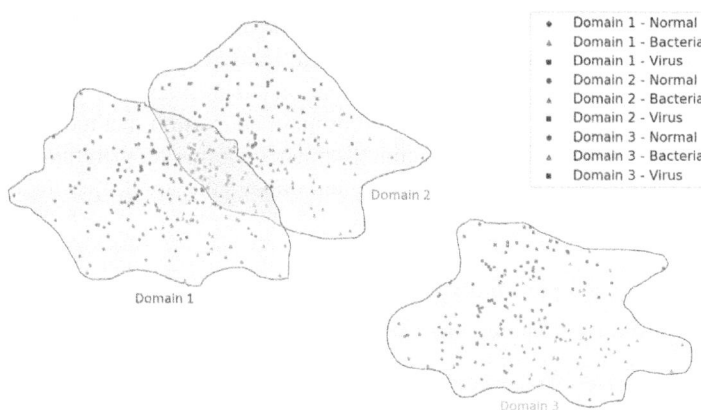

Fig. 1. This figure contains three domains, including two source domains and one target domain. The red and blue areas represent the source domains, while the green area represents the target domain. We can clearly observe that the datasets chosen for our study exhibit a marked domain shift. (Color figure online)

Domain adaptation is an important method that involves aligning the feature distributions of the source and target domains [3]. This alignment enables the model to use information from the source domain to enhance its performance in the target domain, particularly when the target domain lacks labeled data. A significant advantage of domain adaptation is its ability to mitigate the effects of domain gaps, which are the differences between data distributions in the training and deployment environments that impair the model's performance. Therefore, by aligning features across multiple domains, domain adaptation can facilitate the transfer of knowledge from one domain to another where such data is scarce, thus resolving issues caused by domain gaps that lead to decreased model performance.

In previous method that use domain adaptation to solve medical image recognition problems, the majority are supervised learning methods, using a certain amount of labeled target domain data to complete the adaptation process across multiple domains. However, in real-world medical scenarios, the target domain data often tends to be scarce [4]. Moreover, for simplicity, many previous studies focus on a single source domain, which restricts the applicability of the learned classification model to the target domain. Models trained on a single domain often face difficulties when adapting to the varied conditions present in real-world clinical environments [5].

In this paper, we propose an unsupervised multi-source domain adaptation method for medical image recognition. This method allows the model to capture a wider and more comprehensive array of medical images, thus providing a nuanced understanding of the diverse scenarios encountered in clinical settings. First, it generates pseudo-labels for target domain data based on a dynamic multi-stage pseudo-labeling strategy and trains based on these labels. By generating pseudo-labels corresponding to the features captured from the target domain data, it avoids dependence on labeled data from the target domain, thus addressing the problem of difficult-to-label datasets in real-world scenarios. Moreover, our method incorporates knowledge from multiple source domains to guide the training of the target domain, enabling the model to develop a more comprehensive understanding of the variations in medical images [4], significantly enhancing the model's adaptability as well as its generalization capabilities. The key aspect of our work is the development of an unsupervised domain adaptation method and the strategic integration of multiple source domains to bolster the model's ability, contributing significantly to the creation of more reliable and universally applicable medical image recognition models. The contributions of our work are summarized as follows:

– We propose a dynamic multi-stage pseudo-labeling strategy that dynamically generates the most suitable pseudo-labels based on the features of the target domain, avoiding the need for labeling target domain data.
– We design a multi-source unsupervised domain adaptation framework to solve the issue of poor model performance in real scenario with domain gap.
– Our method achieves outstanding performance with a correct identification probability of 98.96% compared to baseline methods.

2   Related Work

The Domain Gap Existing in Medical Image Recognition. In medical image recognition, a domain gap exists between datasets due to differences in medical institutions, devices, and patient demographics. One solution is to use transfer learning methods, starting with training on the ImageNet dataset and then adapting these models for medical imaging, achieving significant results. However, this method requires a large amount of labeled target domain data to achieve good results. Therefore, researchers have tried using domain adaptation methods, like dual-stream U-Net using Maximum Mean Discrepancy (MMD) [5] and Correlation Alignment, have been applied to blend source and target domain data, reducing reliance on labeled target data. Semi-supervised Generative Adversarial Networks (GANs) [6] and unsupervised methods like CycleGAN [7] for image style conversion have also been explored to address domain differences without needing many target domain labels. Despite these advances in reducing reliance on labeled target domain datasets, in real scenarios, target domain labeled data is often scarce or even nonexistent. In our research, we tackle this challenge by generating adaptively tailored pseudo-labels for unsupervised domain adaptation, thus ensuring elevated accuracy.

Multi-source Domain Adaptation. Multi-Source Domain Adaptation (MSDA) uses data from multiple sources to improve learning on a single target domain. Methods based on adversarial training play a significant role in multisource domain adaptation. Examples like Domain-Adversarial Neural Network (DANN) [8] and Adversarial Discriminative Domain Adaptation (ADDA) [9] minimize class label prediction errors while confusing domain classification to learn domain-invariant features. What's more, techniques such as Multi-Kernel Maximum Mean Discrepancy (MK-MMD) [10] and Joint Maximum Mean Discrepancy (JMMD) [11] align features and optimize discrepancies at both feature and classifier levels. However, in real medical scenarios, annotating the applied contexts is very challenging. Therefore, in this paper, we propose an unsupervised domain adaptation method that does not require target domain labels.

Generation of Pseudo-labels. In Unsupervised Domain Adaptation, generating pseudo-labels is pivotal for training models without direct supervision. A prevalent method is confidence thresholding, which selects pseudo-labels based on the highest predicted class probability. The others are cluster-based methods like CAN [12] and ClusterFit [13], which offer contrasting methodologies. However, these methodologies exhibit inherent limitations, often excelling in datasets with uniform distributions or simpler feature spaces, yet faltering otherwise. This paper introduces a novel multi-stage pseudo-label generation strategy. This strategy dynamically generates pseudo-labels from target domain image features and integrates results across training iterations, significantly enhancing the model's adaptability and understanding of target data.

3 Method

3.1 Framework

Our approach involves two source models and a target model. Each source model is trained on one of two labeled source datasets, while the target model is trained on the unlabeled target dataset. The source models generate pseudo-labels to guide the target model, leveraging a weight generation network to choose the most appropriate source model during target model training. We introduce a multi-stage pseudo-label generation strategy to enhance the quality of pseudo-labels for the unlabeled target samples. Figure 2 illustrates the setup: two source domain datasets ($\boldsymbol{D}_n^S = \{(\boldsymbol{x}_i^{Sn}, Y_i^{Sn})\}_{i=1}^{N_S}$, where $n = 1, 2$) and one target domain dataset $\boldsymbol{D}^T = \{\boldsymbol{x}_i^T\}_{i=1}^{N_T}$. Here, $\boldsymbol{x}_i^{Sn}$ represents an image from the n-th source domain, labeled as Y_i^{Sn}, while $\boldsymbol{x}_i^T$ is an image from the target domain.

The two-source and one-target model is trained using distinct samples. The weight generation network receives inputs from the feature extractor applied to the i-th target domain sample x_i^T from dataset D^T, producing a weight ω_s^n for each source domain sample θ_{sn}. For multi-stage pseudo-label generation, inputs include outputs from the feature extractor $f_\phi^{Sn}(x_i^T)$ and classifier $f_\theta^{Sn}(f_\phi^{Sn}(x_i^T))$ of the source domain model applied to the target domain dataset, where f_ϕ^{Sn} and f_θ^{Sn} represent the feature extractor and classifier, respectively. The output is the pseudo-label Y_T for the target samples.

Our approach harnesses knowledge from the source domain to produce pseudo-labels for the unlabeled target domain images using dual source models. These source models are trained on pairs (x_i^{Sn}, Y_i^{Sn}) with the training loss function $\mathcal{L}_{SD}$:

$$\mathcal{L}_{cross}^{SDk} = -\sum_{i=1}^{N_S} Y_i^k \log\left(f_\theta^{Sk}\left(f_\phi^{Sk}\left(\boldsymbol{x}_i^{Sk}\right)\right)\right), k = 1, 2. \tag{1}$$

After completing source domain pre-training, we apply the learned knowledge to the target domain, where the dataset $D^T = \{x_i^T\}_{i=1}^{N_T}$ is unlabeled. Differences between the target and source domains D_1^S and D_2^S necessitate feeding x_i^T into both source domain models, resulting in outputs $\left(f_\theta^{S1}\left(f_\phi^{S1}\left(x_i^T\right)\right)\right)$ and $\left(f_\theta^{S2}\left(f_\phi^{S2}(x_i^T)\right)\right)$. We design a variance loss to clearly delineate decision boundaries between the two source domain models, defined as:

$$\mathcal{L}_{variance} = \lambda\|\left(f_\theta^{S1}\left(f_\phi^{S1}\left(x_i^T\right)\right)\right) - \left(f_\theta^{S2}\left(f_\phi^{S2}\left(x_i^T\right)\right)\right)\|_1. \tag{2}$$

After completing the process of feeding x_i^T into the models of source domain 1 and source domain 2, we aim to use the knowledge from D_1^S and D_2^S to generate high-quality pseudo-labels for target domain training.

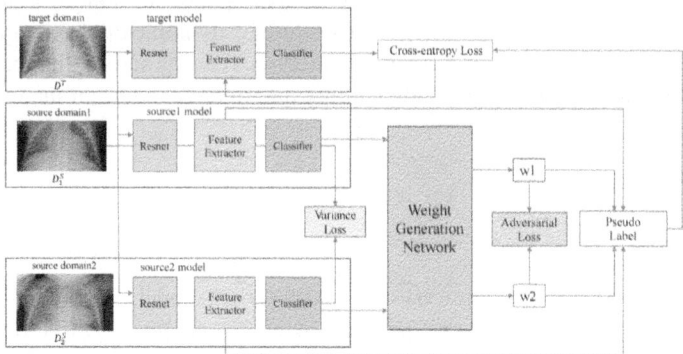

Fig. 2. The architecture of our model. Our model first combines two source domain datasets, D_1^S and D_2^S, with a target domain dataset D^T to train the corresponding source and target models. Then, through a weight generation network, the target model adaptively selects between the two source models. Pseudo-labels are generated based on the generated weights and the discrepancy loss between source domains to guide the supervised learning of the target model.

3.2   Weight Generation Network

Each source domain has distinct feature representations, so selecting the most suitable one enhances pseudo-label quality. The weight generation network accomplishes this by adaptively assigning weights to each source domain, guiding the creation of high-quality pseudo-labels. Its adaptive nature selects the optimal source domain based on feature differences, eliminating the need for manual annotation.

The Weight Generation Network outputs weights ω_s^1 and ω_s^2 to quantify the differences between two source domains. Representing the weight generation model as f_w and using ResNet for feature extraction from target domain data (x_i^T), the features $f_\phi^{Sn}(x_i^T)$ are processed through convolutions and pooling, and finally through a sigmoid-activated fully connected layer to produce weights ω_s^1 and ω_s^2 within the range [0, 1]. This yields the described formula.

$$\omega_s^1, \omega_s^2 = f_w\left(f_\phi^{Sn}\left(x_i^T\right)\right), \tag{3}$$

The weights, ranging from 0 to 1, are then utilized to highlight the distinctive characteristics of each source domain, facilitating targeted and effective pseudo-label generation. By weighting prediction probabilities $\{p_i^n\}$ and pseudo-labels Y_{Tn} from both domains, the network produces high-quality predictions $\{p_i'\}$ and pseudo-labels Y_{Ti}':

$$p_i' = \sum_{n=1}^{2} \omega_s^n \times p_i^n, \tag{4}$$

$$Y'_{Ti} = \sum_{n=1}^{2} \omega_s^n \times Y_{Ti}^n. \tag{5}$$

This approach improves prediction accuracy and enhances adaptability across different domains, leveraging their unique characteristics to produce better pseudo-labels.

3.3 Pseudo-label Generation

Given the differences between the two source domains, the above section allows us to learn domain-specific weights. This section focuses on designing robust pseudo-labels for model training. Our method improves pseudo-labeling with a multi-stage strategy that combines clustering with traditional techniques. Rather than solely relying on the highest-probability output, clustering identifies intrinsic features from model-generated vectors to establish class centers. By measuring the distances to these centroids, our approach generates accurate and nuanced pseudo-labels, mitigating the conventional limitation of focusing only on the most probable category.

Specifically, given the output of the source domain feature extractor $\phi_j^S(x)$ (j represents the category), and the corresponding source domain classification probability output $\theta_j^S(x)$, class centers μ_j for the source domain are computed through the feature map and classification probability output as follows:

$$\mu_j = \frac{\sum_{x_i^T \in D^T} \left(f_\phi^{Sn} (x_i^T) \right) \cdot \left(f_\theta^{Sn} \left(f_\phi^{Sn} (x_i^T) \right) \right)}{\sum_{x_i^T \in D^T} \left(f_\theta^{Sn} \left(f_\phi^{Sn} (x_i^T) \right) \right)}. \tag{6}$$

Single-stage pseudo-label generation methods often struggle with dataset complexity and heterogeneity, particularly when domain gaps are significant. This results in suboptimal pseudo-labels due to insufficient consideration of cross-domain feature disparities. To address this, we propose a multi-stage pseudo-label generation strategy, utilizing multiple training iterations to enhance the model's adaptability to the target domain. Each stage refines the model progressively, aligning it more closely with target domain characteristics. This iterative process improves pseudo-label quality and the model's ability to discern feature variations across domains, enhancing cross-domain generalization. The strategy uses N training stages, each generating a weighted class center $\mu_j^{(n)}$, where n is the stage number ($n = 1, 2, \ldots, N$).

$$Y_T = \arg\min_j \left(\sum_{n=1}^{N} e^{-\lambda(N-n)} \cdot |\left(f_\phi^{Sn} (x_i^T) \right) - \mu_j^{(n)}|^2 \right). \tag{7}$$

The decay function $e^{-\lambda(N-n)}$ prioritizes recent stages, incorporating historical data for a comprehensive pseudo-label generation. This approach enables finer adaptation to target domain characteristics, leading to improved pseudo-label quality.

4 Experiment

Dataset. To evaluate the performance of our algorithm, we conducted experiments on chest X-ray datasets from three distinct medical domains and the Office31 dataset.

- The Ped-pneumonia dataset from the Guangzhou Women and Children's Medical Center features frontal chest X-ray images of children aged 1 to 5 years. This dataset is pivotal for studying pediatric pneumonia, which differs significantly in its pulmonary structure and manifestation from adults, emphasizing Guangzhou's regional demographic and environmental influences [14].
- The Chest X-Ray Images (Pneumonia) dataset, adapted from data provided by the University of California San Diego, categorizes X-ray images as normal or pneumonia-afflicted, with pneumonia types (bacterial or viral) indicated in the filenames. This dataset illustrates how demographic and environmental factors affect pneumonia presentation in the U.S. [15].
- The Curated X-Ray Dataset COVID-19, developed collaboratively by several institutions in India and Concordia University, includes chest X-ray images pre and post COVID-19 infection. It emphasizes COVID-19's unique imaging characteristics among viral pneumonias, shaped by demographic and environmental contexts during the pandemic in India [16].
- The Office31 dataset, provided by Cornell University, consists of images of office items in 31 categories from three distinct office environments: Amazon, Webcam, and DSLR. It serves to explore the challenge of cross-domain adaptation, focusing on the transferability of model learning across different image-capture technologies [17].

To highlight the feature space distinctions across three datasets, we use a visualization technique shown in Fig. 1, we demonstrate feature space distinctions across three datasets, each forming a unique cluster due to differences in geography, population, and disease types. This highlights the importance of domain adaptation in pneumonia detection and classification to ensure model generalization across various conditions. In our experiments, we cycle through these datasets, assigning one as the target and the others as sources, without using source domain labels during training.

We introduce the **W**eighted **P**seudo-labels **N**etwork (WPN), a multi-source domain adaptation model specifically designed for medical image recognition tasks. It effectively utilizes data from various source domains through a weighted pseudo-label mechanism, enhancing the model's ability to learn and generalize to target domain data. In this section, we design experiments to validate the efficacy of the WPN model.

Baseline Method. We compare our method against a suite of domain adaptation baselines: DAN (ICML 2015) [18], DANN (ICML 2015) [8], ADDA (CVPR 2017) [9], MCD (CVPR 2018) [18], BSP (ICML 2019) [19], MCC (ECCV 2020) [20], SIG(NeurIPS 23) [21], MRDA(WWW 23) [22], DARE-GRAM(CVPR 23) [23], TranSVAE(NeurIPS 23) [24], PMTrans-ViT(CVPR 23) [25], MPA(NeurIPS 23) [26], HCLD(AAAI 24) [27] and CSR(AAAI 24) [28].

4.1 Comparative Experiments

Table 1 reports the accuracy of unsupervised domain adaptation on our custom dataset. We assessed the domain accuracy for all pairwise transfers among the three domains, as well as the overall average accuracy. Here, domain D1 corresponds to the Ped-pneumonia dataset, domain D2 to the Chest X-ray images dataset, and domain D3 to the Curated

Table 1. Results on the chest X-ray dataset. Our method outperforms all baseline models, achieving a performance improvement of 1.91% over the best-performing baseline.

Category	METHOD	D1 → D2	D2 → D3	D3 → D1	AVG
multi-source domain adaptation	SIG(NeurIPS'23) [21]	0.9668	0.9313	0.9761	0.9581
	MRDA(WWW'23) [22]	0.9239	0.9641	0.9527	0.9469
	CSR(AAAI'24) [28]	0.9751	0.9641	0.9672	0.9688
Unsupervised domain adaptation	DANN(ICML'15) [8]	0.5088	0.5894	0.5806	0.5596
	DAN(ICML'15) [18]	0.5336	0.5128	0.5350	0.5271
	ADDA(CVPR'17) [9]	0.6647	0.5889	0.6320	0.6285
	MCD(CVPR'18) [18]	0.7385	0.7381	0.7339	0.7368
	BSP(ICML'19) [19]	0.7905	0.7737	0.7253	0.7632
	DARE-GRAM(CVPR'23) [23]	0.9618	0.9031	0.9775	0.9475
	TranSVAE(NeurIPS'23) [24]	0.9453	0.9232	0.9482	0.9389
	PMTrans-ViT(CVPR'23) [25]	0.9512	0.9673	0.9662	0.9616
multi-source unsupervised domain adaptation	MCC(ECCV'20) [20]	0.7985	0.8079	0.7968	0.8011
	DSiT-B(ICCV'23) [29]	0.9490	0.9579	0.9642	0.9570
	MPA(NeurIPS'23) [26]	0.9626	0.9722	0.9547	0.9632
	HCLD(AAAI'24) [27]	0.9675	0.9688	0.9769	0.9711
	WPN(Ours)	**0.9889**	**0.9904**	**0.9895**	**0.9896**

X-ray dataset COVID-19. Our method is observed to outperform all current baseline methods.

We conducted tests on the public dataset Office-31 [30] to further test our method. The test results, as shown in Table 2, indicate that even outside of the medical domain, our model can achieve the best performance. To compare our method with others, we employed t-SNE for visualizing feature distribution across domains D1, D2, and D3. We analyzed: direct transfer using ResNet, DANN, MCC, and our method, depicted in Fig. 3. Results show that direct transfer (Fig. 3 (a)) leads to a disorganized distribution, DANN (Fig. 3 (b)) offers slight alignment improvement, MCC (Fig. 3 (c)) betters alignment yet remains flawed, and our method (Fig. 3 (d)) achieves superior alignment of domain D4 (weighted combination of D2 and D3) with D1, proving our method's efficacy in enhancing target domain adaptability through the weight generation network.

4.2 Ablation Experiments

Table 3 illustrates the impact of each component on adapting to the chest X-ray dataset. Here, no checkmarks indicates the method where parameters pre-trained on the source

Table 2. Results on the Office31 dataset. Our method outperforms all baseline models, achieving a performance improvement of 2.02% over the best-performing baseline.

Category	METHOD	→ A	→ D	→ W	AVG
multi-source domain adaptation	SIG(NeurIPS'23) [21]	0.7430	0.9445	0.9413	0.8763
	MRDA(WWW'23) [22]	0.7177	0.9713	0.9737	0.8876
	CSR(AAAI'24) [28]	0.7522	0.9723	0.9706	0.8984
unsupervised domain adaptation	DARE-GRAM(CVPR'23) [23]	0.7783	0.9710	0.9771	0.9088
	TranSVAE(NeurIPS'23) [24]	0.8073	0.9701	0.9734	0.9169
	PMTrans-ViT(CVPR'23) [25]	0.7891	0.9721	0.9792	0.9135
multi-source unsupervised domain adaptation	DSiT-B(ICCV'23) [29]	0.8280	0.9825	0.9820	0.9308
	MPA(NeurIPS'23) [26]	0.8122	**0.9847**	**0.9835**	0.9268
	HCLD(AAAI'24) [27]	0.8289	0.9574	0.9514	0.9126
	WPN(Ours)	**0.8388**	0.9837	0.9828	**0.9351**

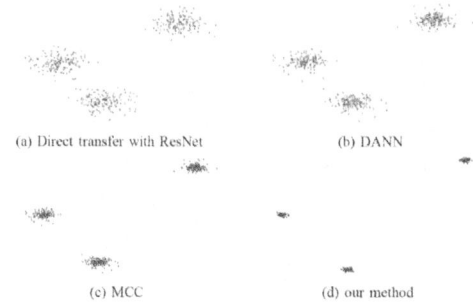

(a) Direct transfer with ResNet   (b) DANN

(c) MCC   (d) our method

Fig. 3. Feature Visualization using t-SNE. In Figures (a), (b), and (c), the red dots represent samples from domain D1, blue dots represent samples from domain D2, and green dots represent samples from domain D3. In Figure (d), red dots indicate samples from domain D1, while blue dots denote the weighted domain results of D2 and D3 post-application of the weight generation network. (Color figure online)

domain using ResNet are frozen and directly applied to the target domain without any transfer. Only 'Domain Adaptation' has checkmark means that the weights of two source domains are manually fixed at 0 and 1, respectively; while both 'Multi-Source Domain' and 'Domain Adaptation' checked indicates that the weights for both source domains are manually set to 0.5 each.

Contributions. Our framework introduces multi-source domain adaptation for medical image analysis, utilizing unsupervised learning and pseudo-label generation for the first time. We tackle dataset annotation challenges by creating a weight generation network that adaptively adjusts pseudo-label weights. We also propose a multi-stage method to enhance pseudo-label quality.

We tested transfer learning by applying ResNet pre-trained parameters directly to the target domain without adaptation, and compared single-source domain training (by manually setting source domain weights to 0 and 1) with conventional multi-source domain training (by setting weights to 0.5). To validate our weight generation network's effectiveness, we enabled it for adaptive weight and pseudo-label control. We further improved our method with a multi-stage weight generation method, demonstrating the strategy's validity.

Weight Generation Network. In this work, we use a Weight Generation Network to assign specific weights to each source domain, creating weighted pseudo-labels for detailed analysis. We fix feature extractors and focus on training the weight generation network, result (in Fig. 4.) shows the network's ability to determine the more relevant source domain for the target data by allocating higher weights, highlighting its effectiveness in enhancing model performance.

Multi-Stage Pseudo-Label Generation. In this paper, contrary to selecting pseudo- and historical data, generates more nuanced pseudo-labels. As shown in Fig. 3, our multi-stage pseudo-label generation method, which accounts for more complex features, results in a faster decline in training loss. The model achieves convergence more rapidly and consistently outperforms the traditional method in the long term (Fig. 5).

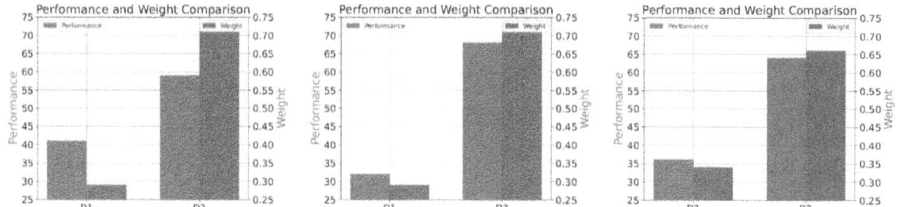

Fig. 4. Weight Distribution and Source Domain Accuracy Relationship Chart.

Table 3. Ablation study results on the adaptation of chest X-ray datasets.

Domain Adaptation	Multi-Source	Adaptive Pseudo-label	Multi-stage Generation	D1, D2 → D3	D1, D3 → D2	D2, D3 → D1	AVG
				D1 → D3 0.8697	D1 → D2 0.8739	D2 → D1 0.8815	0.8761
				D2 → D3 0.8850	D3 → D2 0.8694	D3 → D1 0.8772	
✓				D1 → D3 0.9462	D1 → D2 0.9484	D2 → D1 0.9481	0.9474
				D1 → D3 0.9462	D1 → D2 0.9484	D2 → D1 0.9481	
✓	✓			0.9514	0.9537	0.9589	0.9547
✓	✓	✓		0.9747	0.9751	0.9792	0.9763
✓	✓	✓	✓	**0.9889**	**0.9904**	**0.9895**	**0.9896**

5 Conclusions

We propose an unsupervised multi-source domain adaptation model, marking the first application of multi-source domain adaptation in the field of medical image processing. Our model adaptively selects the optimal source domain model, deriving target domain outcomes solely based on source domain information. We validated our model on the task of pneumonia detection, achieving remarkable results. However, our proposed multi-stage pseudo-label generation method, controlled by a decay function, lacks flexibility in generating pseudo-labels. In the future, we aim to develop more flexible and dynamic methods of pseudo-label generation in this field, to create more appropriate pseudo-labels.

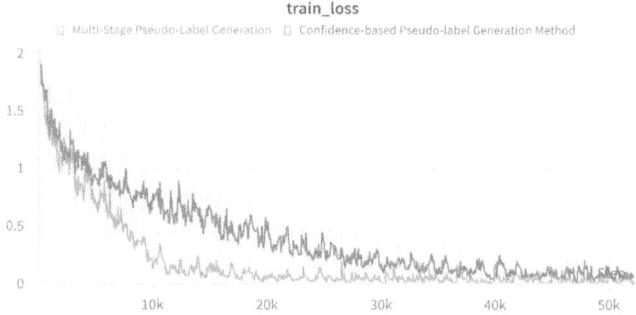

Fig. 5. Training Loss Comparison Chart. (Blue represents our pseudo-labels, while green indicates traditional pseudo-labels.) (Color figure online)

References

1. Long, M., Zhu, H., Wang, J., Jordan, M.I.: Deep transfer learning with joint adaptation networks. In: Proceedings of the 34th International Conference on Machine Learning, ICML 2017. PMLR (2017)
2. Sun, J., Wang, Z., Wang, W., Li, H., Sun, F.: Domain adaptation with geometrical preservation and distribution alignment. Neurocomputing **454**, 152–167 (2021)
3. Tajbakhsh, N., Jeyaseelan, L., Li, Q., Chiang, J.N., Wu, Z., Ding, X.: Embracing imperfect datasets: a review of deep learning solutions for medical image segmentation. Med. Image Anal. **63** (2020)
4. Li, H., et al.: Enhancing and adapting in the clinic: source-free unsupervised domain adaptation for medical image enhancement. IEEE Trans. Med. Imaging **43**(4), 1323–1336 (2024)
5. Dziugaite, G.K., Roy, D.M., Ghahramani, Z.: Training generative neural networks via maximum mean discrepancy optimization. In: Proceedings of the Thirty-First Conference on Uncertainty in Artificial Intelligence, UAI 2015. AUAI Press (2015)
6. Madani, A., Moradi, M., Karargyris, A., Syeda-Mahmood, T.F.: Semisupervised learning with generative adversarial networks for chest X-ray classification with ability of data domain adaptation. In: 15th IEEE International Symposium on Biomedical Imaging (2018)

7. Zhu, J., Park, T., Isola, P., Efros, A.A.: Unpaired image-to-image translation using cycle-consistent adversarial networks. In: IEEE International Conference on Computer Vision, ICCV 2017 (2017)

8. Ganin, Y., Lempitsky, V.S.: Unsupervised domain adaptation by backpropagation. In: Proceedings of the 32nd International Conference on Machine Learning, ICML 2015 (2015)

9. Tzeng, E., Hoffman, J., Saenko, K., Darrell, T.: Adversarial discriminative domain adaptation. In: 2017 IEEE Conference on Computer Vision and Pattern Recognition, CVPR (2017)

10. Xu, B., Wu, K., Wu, Y., He, J., Chen, C.: Dynamic adversarial domain adaptation based on multikernel maximum mean discrepancy for breast ultrasound image classification. Expert Syst. Appl. **207** (2022)

11. Zhang, W., Wu, D.: Discriminative joint probability maximum mean discrepancy (DJP-MMD) for domain adaptation. In: 2020 International Joint Conference on Neural Networks, IJCNN 2020. IEEE (2020)

12. Zhang, W., Ouyang, W., Li, W., Xu, D.: Collaborative and adversarial network for unsupervised domain adaptation. In: 2018 IEEE Conference on Computer Vision and Pattern Recognition, CVPR 2018. Computer Vision Foundation/IEEE Computer Society (2018)

13. Yan, X., Misra, I., Gupta, A., Ghadiyaram, D., Mahajan, D.: ClusterFit: improving generalization of visual representations. In: 2020 IEEE/CVF Conference on Computer Vision and Pattern Recognition, CVPR 2020. Computer Vision Foundation/IEEE (2020)

14. Kermany, D., Zhang, K., Goldbaum, M., et al.: Labeled optical coherence tomography (OCT) and chest X-ray images for classification (2018)

15. National Institutes of Health - Clinical Center. ChestXray-NIHCC. Accessed 10 Jan 2022

16. Sait, U., et al.: Curated dataset for COVID-19 posterior-anterior chest radiography images (X-Rays) (2020)

17. Saenko, K., Kulis, B., Fritz, M., Darrell, T.: Adapting visual category models to new domains. In: Daniilidis, K., Maragos, P., Paragios, N. (eds.) Computer Vision – ECCV 2010. Springer, Heidelberg (2010). https://doi.org/10.1007/978-3-642-15561-1_16

18. Saito, K., Watanabe, K., Ushiku, Y., Harada, T.: Maximum classifier discrepancy for unsupervised domain adaptation. In: 2018 IEEE Conference on Computer Vision and Pattern Recognition, CVPR (2018)

19. Chen, X., Wang, S., Long, M., Wang, J.: Transferability vs. discriminability: batch spectral penalization for adversarial domain adaptation. In: Proceedings of the 36th International Conference on Machine Learning, ICML 2019, vol. 97 (2019)

20. Jin, Y., Wang, X., Long, M., Wang, J.: Minimum class confusion for versatile domain adaptation. In: Vedaldi, A., Bischof, H., Brox, T., Frahm, J.M. (eds.) Computer Vision - ECCV 2020, vol. 12366, pp. 464–480. Springer, Cham (2020). https://doi.org/10.1007/978-3-030-58589-1_28

21. Li, Z., Cai, R., Chen, G., Sun, B., Hao, Z., Zhang, K.: Subspace identification for multi-source domain adaptation. In: Advances in Neural Information Processing Systems 36: Annual Conference on Neural Information Processing Systems 2023, NeurIPS 2023 (2023)

22. Zhou, J., Fu, C., Zhang, X.: Multi-source domain adaptation via latent domain reconstruction. In: Companion Proceedings of the ACM Web Conference 2023, WWW (2023)

23. Nejjar, I., Wang, Q., Fink, O.: DARE-GRAM : unsupervised domain adaptation regression by aligning inverse gram matrices. In: IEEE/CVF Conference on Computer Vision and Pattern Recognition, CVPR (2023)

24. Wei, P., et al.: Unsupervised video domain adaptation for action recognition: a disentanglement perspective. In: Advances in Neural Information Processing Systems 36: Annual Conference on Neural Information Processing Systems 2023, NeurIPS (2023)

25. Zhu, J., Bai, H., Wang, L.: Patch-mix transformer for unsupervised domain adaptation: a game perspective. In: IEEE/CVF Conference on Computer Vision and Pattern Recognition, CVPR 2023 (2023)

26. Chen, H., Han, X., Wu, Z., Jiang, Y.: Multi-prompt alignment for multi-source unsupervised domain adaptation. In: Advances in Neural Information Processing Systems 36: Annual Conference on Neural Information Processing Systems 2023, NeurIPS 2023 (2023)
27. Liu, X., et al.: UFDA: universal federated domain adaptation with practical assumptions. In: Thirty-Eighth AAAI Conference on Artificial Intelligence, AAAI 2024 (2024)
28. Zhou, C., Wang, Z., Du, B., Luo, Y.: Cycle self-refinement for multi-source domain adaptation. In: Thirty-Eighth AAAI Conference on Artificial Intelligence, AAAI 2024 (2024)
29. Sanyal, S., Asokan, A.R., Bhambri, S., Kulkarni, A.R., Kundu, J.N., Babu, R.V.: Domain-specificity inducing transformers for source-free domain adaptation. In: IEEE/CVF International Conference on Computer Vision, ICCV 2023 (2023)
30. Saenko, K., Kulis, B., Fritz, M., Darrell, T.: Adapting visual category models to new domains. In: Daniilidis, K., Maragos, P., Paragios, N. (eds.) Computer Vision - ECCV 2010. LNCS, vol. 6314, pp. 213–226. Springer, Heidelberg (2010). https://doi.org/10.1007/978-3-642-15561-1_16

PlantViQ: Disease Recognition Across Varied Environments with Vision Transformer and Quadrangle Attention

Shuting Li[1] , Baoyu Chen[2] , Feng Li[3] , Jingmei He[4] , Feiyong He[5,7] ,
Yingbiao Hu[5] , Jingjia Chen[6] , and Huinian Li[1(✉)]

[1] Macao University of Science and Technology, Macao 999078, China
3220001294@student.must.edu.mo
[2] Dongguan University of Technology, Dongguan 523808, China
[3] Guangdong Innovative Technical College, Dongguan 523960, China
[4] Guangdong University of Finance, Guangzhou 510521, China
[5] Macao Polytechnic University, Macao 999078, China
[6] Guangzhou University, Guangzhou 510006, China
[7] Guangdong Polytechnic of Science and Technology, Guangzhou 510640, China

Abstract. Accurate plant disease detection in real-world agricultural settings poses a significant challenge due to the complex and variable nature of field environments. To address this challenge, we introduce PlantViQ. This deep learning model leverages the strengths of both convolutional neural networks (CNNs) for local feature extraction and Transformers for capturing global context. This enables PlantViQ to overcome the limitations of traditional CNN-based approaches, particularly in scenarios with small-sample datasets and challenging environmental conditions. As demonstrated by its exceptional F1-score of 88.52% on cassava leaf disease and 97.21% on rice leaf disease, PlantViQ has the potential to revolutionize crop management practices by enabling early and accurate disease detection even in the most challenging real-world environments.

Keywords: Plant protection · QA · AAM · Precision agriculture · Deep learning

1 Introduction

Plant diseases threaten agricultural development, causing substantial economic losses if undetected and untreated [1]. Traditional manual detection methods could be more efficient and practical for large-scale monitoring [2], delaying disease identification and intervention and further exacerbating the losses. Therefore, implementing efficient and accurate disease detection methods is of paramount importance.

The advent of deep learning has revolutionized image recognition and classification tasks, offering powerful tools for addressing the challenge of plant disease detection. Convolutional neural networks (CNNs) have been particularly successful in this domain, achieving remarkable performance in various plant disease detection applications [3]. For instance, researchers have leveraged transfer learning with EfficientNet B4 and U-Net

© The Author(s), under exclusive license to Springer Nature Singapore Pte Ltd. 2024
D.-S. Huang et al. (Eds.): ICIC 2024, LNBI 14881, pp. 441–452, 2024.
https://doi.org/10.1007/978-981-97-5689-6_38

segmentation [4], integrated ResNet and Transformer architectures with FAMP-Softmax loss [5], and utilized CNNs for image-based cassava disease detection [6, 7].

Despite the success of CNNs, their restricted receptive field, constrained by the convolutional kernel size, limits their ability to capture global image information crucial for accurate disease detection. To address this limitation, transformers have emerged as a promising alternative, utilizing self-attention mechanisms to gather information from the entire image and effectively process irregular inputs [8]. However, the direct application of transformers in computer vision faces challenges, as image data is inherently more complex than text data, requiring transformers to be trained on extensive datasets to achieve optimal performance. Consequently, processing image information with transformers necessitates more complicated algorithms and robust hardware architectures.

To address these limitations, Zhang et al. [9] introduced QFormer, a deformable attention mechanism for visual transformers that includes a novel Quadratic Attention (QA) mechanism designed for image processing. QA enables the model to dynamically determine the position, size, orientation, and shape of different targets in the image, improving performance and flexibility. While our initial experiments with QFormer yielded better results than previous models, they still need to meet our expectations.

To leverage the strengths of QFormer while addressing its limitations, we propose a novel network architecture, PlantViQ, incorporating an Axial Attention Module (AAM) [10]. This design capitalizes on QFormer's efficiency and integrates AAM to enhance its capabilities. The aim is to effectively utilize Transformers while exploiting the observed benefits of combining CNNs with visual Transformers for image recognition and classification tasks. Additionally, considering the limited data availability in plant disease image datasets, we prioritize efficient Transformer utilization to maximize the utility of existing data. Furthermore, agricultural field images often contain noise that can impact model accuracy. Therefore, our proposed PlantViQ network prioritizes robustness to ensure reliable performance in real-world farm settings. The framework of the PlantViQ training method is illustrated in Fig. 1.

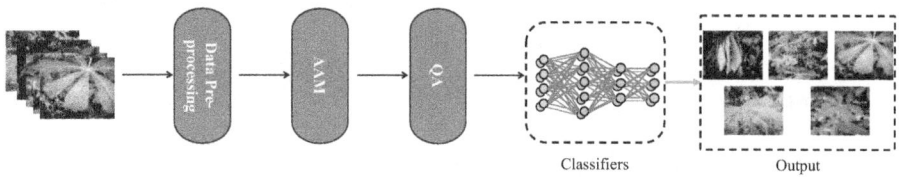

Fig. 1. Framework of PlantViQ Training Method

The main contributions of this study are summarized as follows:

(1) Deformable Attention Mechanism Integration: PlantViQ combines QFormer's deformable attention mechanism with the Axial Attention Module, enhancing model performance and flexibility by dynamically determining target positions, sizes, orientations, and shapes.

(2) Efficient Transformer Usage: PlantViQ efficiently utilizes the Transformer architecture to address limited sample challenges in plant disease image datasets, maximizing the impact of available data.

(3) Environmental Robustness: PlantViQ excels in diverse environments, maintaining reliable performance amidst real-world complexities and disturbances, particularly in agricultural fields.

2 Materials and Methods

2.1 Dataset Collection

PlantViQ was trained and evaluated on two public datasets: the Cassava Leaf Disease Dataset and the Rice Leaf Disease Dataset. These datasets were generated under different capture conditions to comprehensively assess the model's performance in real-world scenarios. Figure 2 shows some image samples of the two datasets.

(a)Cassava Leaf

(b)Rice Leaf

Fig. 2. Sample Images of Dataset

Cassava Leaf Disease Dataset - To establish a practical and widely applicable model for identifying cassava leaf diseases, we constructed a cassava leaf disease recognition dataset. This dataset consists of 21,367 original images and their corresponding disease category labels. We included four typical and common cassava leaf diseases and healthy cassava leaves. The cassava leaf disease dataset encompasses diverse cassava leaves captured under various lighting conditions and from different camera angles. Notably, the distribution of diseased leaves is uneven, with most images containing natural environmental backgrounds, including soil and ground. To ensure the robustness of the model, we deliberately included these diverse background conditions.

Rice Leaf Disease Dataset - This dataset integrates publicly available rice leaf disease image datasets from Kaggle with images independently captured by our team. It encompasses images representing four common disease categories and one healthy category. Employing data augmentation techniques to enrich the training set's scale and diversity, we compiled a final dataset comprising 8,080 images. The categorized diseases include bacterial leaf rot, brown spot, rice hispa, and leaf spot. All images are in JPG format with a resolution of 128x128 pixels, captured under controlled lighting conditions against a white background.

2.2 AAM

The Axial Attention module efficiently captures long-range dependencies within input features by leveraging their inherent spatial structure. The architecture of AAM is illustrated in Fig. 3. It achieves this by decomposing the input feature map into separate tensors representing the height, denoted as $\mathbf{X_h}$, and width, denoted as $\mathbf{X_w}$, of the feature map.

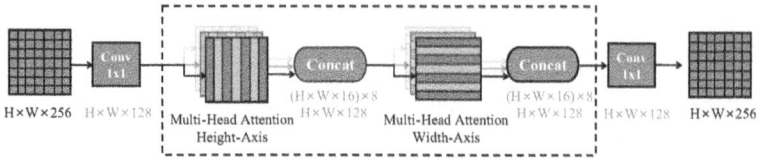

Fig. 3. Architecture of AAM

Next, row attention is applied along the height dimension. This process involves three linear transformations: one each for queries ($\mathbf{Q_h}$), keys ($\mathbf{K_h}$), and values $\mathbf{V_h}$ on the decomposed height feature map, $\mathbf{X_h}$. These transformations project the features into lower-dimensional spaces suitable for attention calculation. The dimensions of these transformed features are represented as (h, C_q), (h, C_k), and (h, C_v), respectively, where h represents the height and $C_{q'}$, $C_{k'}$, C_v represent the channel dimensions for each element. The calculation is as in Eq. (1).

$$\mathbf{Q_h} = \mathbf{W_q X_h}$$
$$\mathbf{K_h} = \mathbf{W_k X_h} \tag{1}$$
$$\mathbf{V_h} = \mathbf{W_v X_h}$$

where $\mathbf{W_{q'}}$, $\mathbf{W_{k'}}$, $\mathbf{W_v}$ represent the parameter matrices for the linear transformations. Like row attention, column attention is applied along the width dimension of the decomposed feature map, $\mathbf{X_w}$. The same steps, including linear transformations, attention score calculation, and weighted sum, are followed, with the output denoted as $\mathbf{Z_w}$.

After obtaining the outputs from both row and column attention, $\mathbf{Z_h}$ and $\mathbf{Z_w}$, they are concatenated along the channel dimension to form the concatenated tensor $\mathbf{Z}$. The calculation is as in Eq. (2).

$$\mathbf{Z} = [\mathbf{Z_h}; \mathbf{Z_w}] \tag{2}$$

The attention scores, denoted as $\mathbf{A_h}$, are calculated using the scaled dot-product operation between the projected queries $\mathbf{Q_h}$ and keys $\mathbf{K_h}$. The calculation is as in Eq. (3).

$$\mathbf{A_h} = \frac{\mathbf{Q_h K_h^T}}{\sqrt{d_k}} \tag{3}$$

where, $\mathbf{Q_h}$ represents the query tensor obtained by applying a linear transformation to the decomposed height feature map $\mathbf{X_h}$, $\mathbf{K_h}$ represents the critical tensor obtained through a similar linear transformation, and d_k represents the dimension of the essential vectors (i.e., C_k). The dot product is divided by the square root of d_k to prevent the attention scores from growing too large.

The attention scores $\mathbf{A_h}$ indicate the relevance of each element in $\mathbf{X_h}$ to the current element being processed. These scores are then normalized using the softmax function to obtain the normalized attention weights, $\mathbf{A'_h}$. The calculation is as in Eq. (4).

$$\mathbf{A'_h} = \text{softmax}(\mathbf{A_h}) \tag{4}$$

Finally, the row attention output, $\mathbf{Z_h}$, is obtained by performing a weighted sum of the value vectors, $\mathbf{V_h}$, using the normalized attention weights, $\mathbf{A'_h}$. The calculation is as in Eq. (5).

$$\mathbf{Z_h} = \mathbf{A'_h}\mathbf{V_h} \tag{5}$$

The concatenated tensor $\mathbf{Z}$ is then passed through a 1×1 convolutional layer to adjust its channel dimension to match the original input feature map $\mathbf{X}$. This produces the final attention-weighted feature map $\mathbf{X_{AA}}$. The calculation is as in Eq. (6).

$$\mathbf{X_{AA}} = \text{Conv}(\mathbf{Z}) \tag{6}$$

The convolutional operation ensures that the attention-weighted feature map $\mathbf{X_{AA}}$ has the same channel dimension as the input feature map $\mathbf{X}$, allowing for seamless integration with subsequent layers or operations in the neural network.

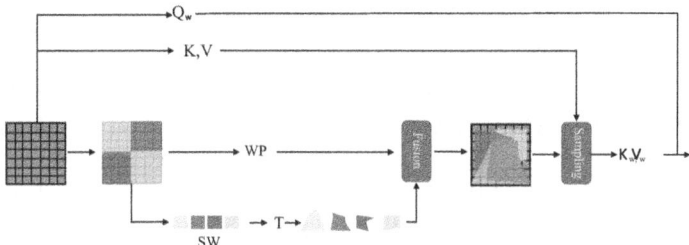

Fig. 4. Architecture of QA

2.3 QA

Quadrangle Attention (QA) [9] is a novel attention mechanism for image processing. The architecture of QA is illustrated in Fig. 4. In the QA architecture diagram, WP and SW are 'window positions' and 'Standard Windows. It empowers models to dynamically determine the location, size, orientation, and shape of different targets in an image, leading to enhanced performance and flexibility. At the core of QA lies the utilization of learnable projection transformations to convert image patches into target quadrilaterals, enabling precise recognition and localization of diverse targets. To achieve this, QA divides the image into multiple non-overlapping patches and applies the transformation above to each patch. This approach demonstrates remarkable adaptability to targets of varying scales and shapes, thus improving the model's generalization capability.

The process begins with patch generation. Given an input feature map $X \in \mathbb{R}^{H \times W \times C}$, QA divides it into $H \times W$ non-overlapping patches of size $w \times w \times C$. Next, a linear transformation is applied to each patch to extract queries Q, keys K, and values V:

$$Q, K, V = Linear(X) \tag{7}$$

Subsequently, each patch is converted into a target quadrilateral via a projection transformation. An eight-parameter transformation matrix represents this transformation:

$$T = \begin{pmatrix} a_1 & a_2 & b_1 \\ a_3 & a_4 & b_2 \\ c_1 & c_2 & 1 \end{pmatrix} \tag{8}$$

where $\alpha = [\alpha_1, \alpha_2; \alpha_3, \alpha_4]$ defines the transformation of scaling, rotation, and shearing. $b = [b1; b2]$ defines the translation. $c = [c1, c2]$ is a projection vector that describes how the perceived objects change when the observer's viewpoint changes in the depth dimension.

By predicting the parameters of each elementary transformation, we can accurately convert the patches into target quadrilaterals. Precisely, the parameters t of the elementary transformations are predicted first. Then, the transformation matrices for each elementary transformation are computed based on t and multiplied in sequence to obtain the final projection transformation matrix T. This completes the construction of a quadrilateral regression module that is **trainable end-to-end**. It converts the default patches into target quadrilaterals for label sampling and attention computation. Consequently, the model can model diverse targets with different shapes and orientations and capture rich contextual information, significantly enhancing the flexibility of Transformer models in adapting to objects of varying sizes, shapes, and orientations.

Combining Qformer and Axial Attention Modules yields a comprehensive attention mechanism that empowers the model with enhanced capabilities. Qformer captures long-range dependencies, facilitating the model's understanding of context and relationships between distant words or elements. Meanwhile, the Axial Attention Module focuses on local dependencies, enabling the model to capture fine-grained details and structural information within sentences.

3   Experiments and Results

3.1   Data Augmentation and Pre-processing

All images were uniformly resized to 224×224 pixels to meet the model's input requirements. Various image augmentation techniques were applied to enrich the datasets and prevent overfitting. These techniques included RandomRotation, GaussianBlur, ColorJitter, and Mixup.

The impact of Mixup [11] compared to other augmentation techniques is illustrated in Table 1 and Table 2 for the Cassava Leaf Disease Dataset and Rice Leaf Disease Dataset, respectively. As the tables show, Mixup achieved the highest F1-score (88.52% for cassava and 97.21% for rice) compared to other techniques. Figure 5 visually demonstrates the application of these augmentation techniques. The original images (left column) are transformed (middle column) to enhance diversity. Mixup amplifies this effect by creating novel samples with blended features (rightmost column). These augmentation strategies improve the model's ability to learn robust and generalizable features, leading to superior performance and resilience to real-world variations.

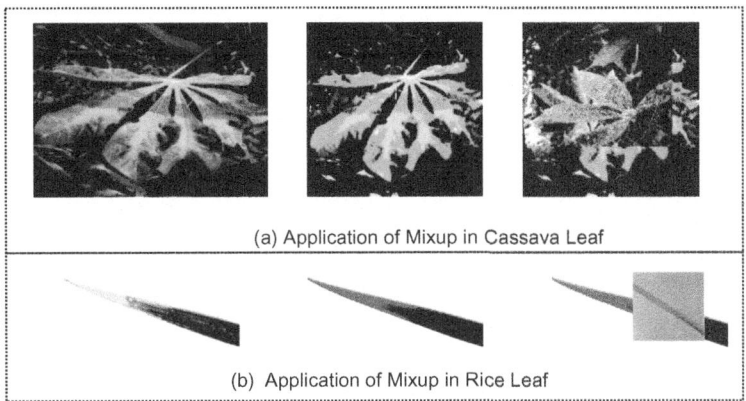

(a) Application of Mixup in Cassava Leaf

(b) Application of Mixup in Rice Leaf

Fig. 5. Image augmentation technology effects

Table 1. Mixup vs. Other Techniques: Cassava Leaf Disease Dataset

Technique	Precision (%)	Recall (%)	F1-Score (%)	Accuracy (%)
Cutout [12]	76.12	72.85	85.73	74.4
Random Erasing [13]	78.02	76.03	87.42	77.01
Flip [14]	79.22	77.67	88.31	78.44
Random Crop [15]	79.53	77.9	88.39	78.71
Mixup	80.00	78.00	88.52	79.00

By integrating a diverse range of images with effective data augmentation techniques, PlantViQ is well-equipped to handle the complexities of real-world agricultural settings and deliver accurate plant disease detection.

3.2   Experimental Results and Analysis

This section evaluates the performance of the PlantViQ model on both the Cassava Leaf Disease Dataset and the Rice Leaf Disease Dataset. The key findings are summarized below:

Table 2. Mixup vs. Other Techniques: Rice Leaf Disease Dataset

Technique	Precision (%)	Recall (%)	F1-Score (%)	Accuracy (%)
Cutout	79.23	76.19	93.24	76.49
RandomErasing	80.67	78.7	94.26	79.69
Flip	81.33	79.67	94.87	80.67
RandomCrop	82.26	80.13	95.36	81.00
Mixup	83.00	80.99	97.21	82.00

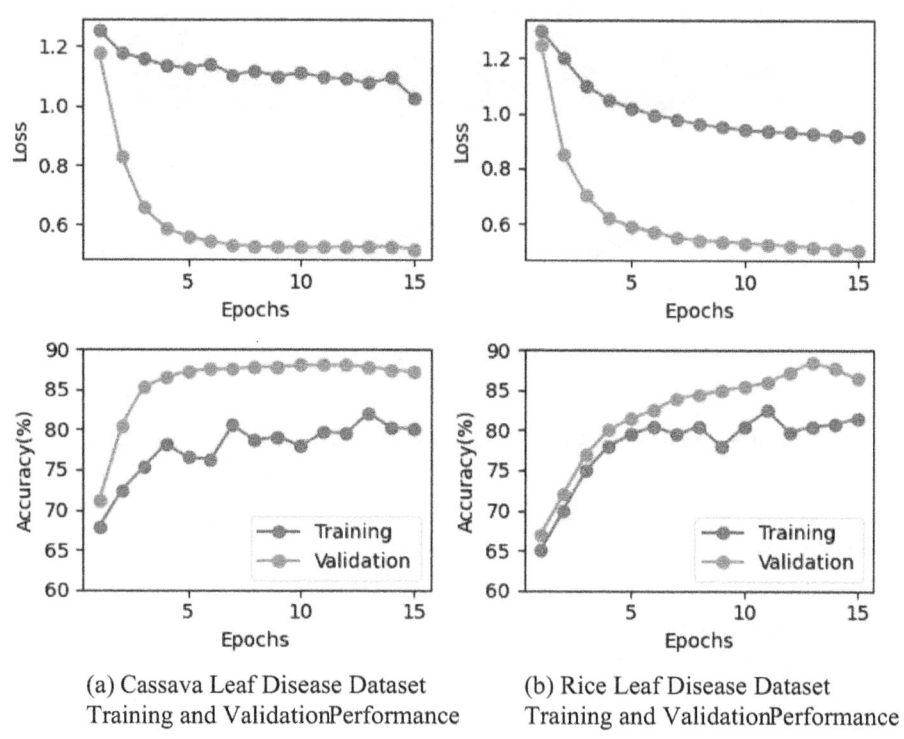

(a) Cassava Leaf Disease Dataset
Training and ValidationPerformance

(b) Rice Leaf Disease Dataset
Training and ValidationPerformance

Fig. 6. Loss and Accuracy for training and validation dataset

Figure 6 illustrates the training and validation loss and accuracy curves for the Cassava Leaf Disease Dataset. The model exhibited a consistent improvement in both training and validation accuracy throughout the training process. Training accuracy started at 67.9% and gradually increased to 82.0% by the 13th epoch. Validation accuracy followed a similar trend, starting at 71.1% and reaching a maximum of 88.1% at the 12th epoch. Following the peak, training and validation accuracy stabilized, indicating that the model had converged to an optimal performance level. Training loss steadily decreased from 1.26 to 1.03 during training, while validation loss dropped from 1.18 to 0.52. The continuous decline in training and validation loss demonstrates the model's practical learning and generalization capabilities on the Cassava Leaf Disease Dataset.

Figure 6 illustrates the training and validation loss and accuracy curves for the Rice Leaf Disease Dataset. The model's performance on this dataset exhibited a pattern similar to that observed on the Cassava Leaf Disease Dataset. The training accuracy surged from 70.0% to 99.0%, reaching a peak of 98.8% by the 14th epoch. Similarly, validation accuracy showed steady improvement, starting at 84.0% and reaching a maximum of 97.2% by the 11th epoch. The training loss declined from 1.30 to 0.915, while the validation loss decreased from 1.25 to 0.505. The convergence of training and validation losses suggests that the model effectively learned the discriminative features of rice leaf diseases, resulting in robust performance on the validation set.

3.3   Contrasting Various Models with the PlantViQ

To assess the performance of our proposed PlantViQ model thoroughly, we conducted experiments on two publicly available datasets: the Cassava Leaf Disease Dataset and the Rice Leaf Disease Dataset. The detailed results of these experiments are presented in Table 3 and Table 4, respectively.

Table 3. Results on Cassava Leaf Disease dataset

Author	Method	Accuracy (%)
Maurya et al. [16]	VIT	82.60
Surya et al. [17]	CNN	84.96
Zhong et al. [18]	ResNet	86.90
Singh et al. [19]	DenseNet169	87.86
Ours	PlantViQ	88.52

Table 4. Results on Rice Leaf Disease dataset

Author	Method	Accuracy (%)
Singh et al. [20]	CNN	93.33
Shanuja et al. [20]	XceptionNet	94.33
kukreja et al. [21]	CNN + LSTM	95.25
Saini et al. [22]	CNN	95.72
Ours	PlantViQ	97.21

Table 3 showcases the performance metrics on the Cassava Leaf Disease Dataset. Our PlantViQ model achieved the highest accuracy of 88.52% on this dataset, surpassing the previous state-of-the-art methods, including IVT, CNN, ResNet, and DenseNet169.

This outcome underscores the effectiveness of our model's architecture, which integrates the Qformer and Axial Attention Module, in accurately identifying cassava leaf diseases.

Table 4 provides a comprehensive overview of the results obtained from the experiments conducted on the Rice Leaf Disease Dataset. Notably, our PlantViQ model demonstrated an impressive accuracy of 97.21%, showcasing significant superiority over the previous state-of-the-art techniques, such as CNN, XceptionNet, CNN + LSTM, and CNN. This remarkable performance underscores the generalization capability of our model and its proficiency in capturing the distinctive features of rice leaf diseases.

4 Discussion

PlantViQ, a groundbreaking deep learning model, has demonstrated exceptional capabilities in accurately identifying plant diseases across two widely used datasets: the Cassava Leaf Disease Dataset and the Rice Leaf Disease Dataset. Its remarkable performance, surpassing previous state-of-the-art methods, stems from its innovative architectural design, which effectively integrates the Qformer architecture and the Axial Attention Module. This combination enables PlantViQ to capture both local and global features within plant leaf images, leading to more precise and nuanced disease diagnoses.

The model's effectiveness is further solidified by its consistent improvement in training and validation accuracy, coupled with a steady decline in training and validation loss. These results underscore PlantViQ's robust learning and generalization capabilities. On the Cassava Leaf Disease Dataset, PlantViQ achieved an impressive accuracy of 88.52%, outperforming established techniques such as IVT, CNN, ResNet, and DenseNet169. Similarly, on the Rice Leaf Disease Dataset, PlantViQ attained an exceptional accuracy of 97.21%, surpassing methods like CNN, XceptionNet, CNN + LSTM, and CNN.

Despite its overall excellence, PlantViQ's accuracy for certain specific leaf disease categories was relatively lower. This limitation is primarily attributed to these diseases' subtle or complex visual characteristics, making their symptoms more challenging for the model to extract and recognize accurately.

To address this limitation and further enhance PlantViQ's robustness, future work will optimize the attention mechanism design to improve its ability to capture key disease characteristics. Additionally, exploring the fusion of multi-source information, such as leaf images and leaf spectral data, will be investigated to compensate for the limitations of single-modal visual details and improve the recognition accuracy for complex diseases. These improvements aim to enhance PlantViQ's applicability in complex real-world scenarios, providing a more comprehensive and reliable solution for agricultural production.

5 Conclusion

In conclusion, the results of this study demonstrate the exceptional performance of the PlantViQ model in accurately identifying plant diseases across diverse datasets. The model's innovative architectural design, which combines the Qformer and Axial Attention Module, enables it to outperform previous state-of-the-art approaches, showcasing its potential to revolutionize agricultural practices through early and reliable disease detection.

The model's robust and generalizable performance highlights its suitability for practical deployment in real-world settings, paving the way for improved crop management and increased agricultural productivity. By providing an accurate and automated tool for plant disease identification, the PlantViQ model has the potential to significantly enhance the sustainability and efficiency of the agricultural sector, ultimately leading to improved food security and economic outcomes for farmers and communities worldwide.

References

1. Faithpraise, F., Birch, P., Young, R., Obu, J., Faithpraise, B., Chatwin, C.: Automatic plant pest detection and recognition using k-means clustering algorithm and correspondence filters. Int. J. Adv. Biotechnol. Res. **4**, 189–199 (2013)
2. Mohanty, S.P., Hughes, D.P., Salathé, M.: Using deep learning for image-based plant disease detection. Front. Plant Sci. **7**, 215232 (2016)
3. Arnal Barbedo, J.G.: Digital image processing techniques for detecting, quantifying and classifying plant diseases. Springerplus **2**, 660 (2013)
4. Maryum, A., Akram, M.U., Salam, A.A.: Cassava leaf disease classification using deep neural networks. In: 2021 IEEE 18th International Conference on Smart Communities: Improving Quality of Life Using ICT, IoT and AI (HONET), pp. 32–37 (2021)
5. Sambasivam, G., Opiyo, G.D.: A predictive machine learning application in agriculture: cassava disease detection and classification with imbalanced dataset using convolutional neural networks. Egypt. Inf. J. **22**, 27–34 (2021)
6. Paiva-Peredo, E.: Deep learning for the classification of cassava leaf diseases in unbalanced field data set. In: Woungang, I., Dhurandher, S.K., Pattanaik, K.K., Verma, A., Verma, P. (eds.) ANTIC 2022. CCIS, vol. 1798, pp. 101–114. Springer, Cham (2022). https://doi.org/10.1007/978-3-031-28183-9_8
7. Singh, R., Sharma, A., Sharma, N., Sharma, K., Gupta, R.: A deep learning-based InceptionResNet V2 model for cassava leaf disease detection. In: Rathore, V.S., Piuri, V., Babo, R., Ferreira, M.C. (eds.) ICETEAS 2023. LNNS, vol. 682, pp. 423–432. Springer, Singapore (2023). https://doi.org/10.1007/978-981-99-1946-8_38
8. Thai, H.-T., Tran-Van, N.-Y., Le, K.-H.: Artificial cognition for early leaf disease detection using vision transformers. In: 2021 International Conference on Advanced Technologies for Communications (ATC), pp. 33–38 (2021)
9. Zhang, Q., Zhang, J., Xu, Y., Tao, D.: Vision transformer with quadrangle attention. IEEE Trans. Pattern Anal. Mach. Intell. (2024)
10. Ho, J., Kalchbrenner, N., Weissenborn, D., Salimans, T.: Axial attention in multidimensional transformers. arXiv preprint arXiv:1912.12180 (2019)
11. Zhang, H., Cisse, M., Dauphin, Y.N., Lopez-Paz, D.: mixup: beyond empirical risk minimization. arXiv preprint arXiv:1710.09412 (2017)
12. DeVries, T., Taylor, G.W.: Improved regularization of convolutional neural networks with cutout. arXiv preprint arXiv:1708.04552 (2017)
13. Zhong, Z., Zheng, L., Kang, G., Li, S., Yang, Y.: Random erasing data augmentation. Proc. AAAI Conf. Artif. Intell. **34**, 13001–13008 (2020)
14. Flipped Learning Network: The four pillars of FLIP (2014)
15. Wang, M., Luo, C., Hong, R., Tang, J., Feng, J.: Beyond object proposals: random crop pooling for multi-label image recognition. IEEE Trans. Image Process. **25**, 5678–5688 (2016)
16. Maurya, R., Pandey, N.N., Singh, V.P., Gopalakrishnan, T.: Plant disease classification using interpretable vision transformer network. In: 2023 International Conference on Recent Advances in Electrical, Electronics & Digital Healthcare Technologies (REEDCON), pp. 688–692 (2023)

17. Surya, R., Gautama, E.: Cassava leaf disease detection using convolutional neural networks. In: 2020 6th International Conference on Science in Information Technology (ICSITech), pp. 97–102 (2020)
18. Zhong, Y., Huang, B., Tang, C.: Classification of cassava leaf disease based on a non-balanced dataset using transformer-embedded ResNet. Agriculture **12**, 1360 (2022)
19. Singh, R., Sharma, A., Sharma, N., Gupta, R.: Automatic detection of cassava leaf disease using transfer learning model. In: 2022 6th International Conference on Electronics, Communication and Aerospace Technology, pp. 1135–1142 (2022)
20. Singh, G., Guleria, K., Sharma, S.: Fine-tuned convolutional neural network model for rice leaf disease prediction. In: 2023 2nd International Conference on Futuristic Technologies (INCOFT), pp. 1–6 (2023)
21. Kukreja, V., Sharma, R., Vats, S.: Revolutionizing rice farming: automated identification and classification of rice leaf blight disease using deep learning. In: 2023 Third International Conference on Secure Cyber Computing and Communication (ICSCCC), pp. 586–591 (2023)
22. Saini, A., Guleria, K., Sharma, S.: Multiclass classification of rice leaf disease using deep learning based model. In: 2023 3rd Asian Conference on Innovation in Technology (ASIANCON), pp. 1–6 (2023)

Unsupervised Anomaly Detection in Tongue Diagnosis with Semantic Guided Denoising Diffusion Models

Hongbo Huang[✉], Xiaoxu Yan, Longfei Xu, Yaolin Zheng, and Linkai Huang

Beijing Information Science and Technology University, Beijing, China
hhb@bistu.edu.cn

Abstract. Tongue diagnosis is one of the core diagnostic methods in Traditional Chinese Medicine (TCM), primarily involving the visual inspection of tongue images to assess a patient's health status. However, the subjectivity and environmental differences in tongue diagnosis may lead to potential errors and limitations. In this paper, we introduce an unsupervised tongue coating anomaly detection model based on diffusion models, aiming to address the limitations of traditional supervised learning and existing anomaly detection models. Our approach combines the semantic classification ability of the cross-attention module within the diffusion model with score-based conditional guidance to achieve high-quality image reconstruction and precise identification of discriminative regions. Experimental results have demonstrated that our anomaly detection model exhibits state-of-the-art performance, surpassing the accuracy of existing models.

Keywords: Diffusion models · Anomaly detection · Tongue diagnosis

1 Introduction

In modern medicine, tongue diagnosis is regarded as a non-invasive, simple, and cost-effective diagnostic method, capable of rapidly assessing and diagnosing a patient's health condition preliminarily. Since the World Health Organization (WHO) included traditional Chinese medicine (TCM) in its latest global medical guidelines, TCM has increasingly gained the attention and recognition of medical experts. However, a major drawback of TCM's tongue diagnosis is the variability in diagnostic outcomes among different physicians, leading to higher subjectivity and uncertainty in diagnoses, which also poses certain challenges for clinical treatment. TCM doctors rely on their professional knowledge and past experience to identify tongue characteristics, which are easily influenced by environmental factors, resulting in unstable and inaccurate diagnoses.

In recent years, as deep learning has gradually matured, many scholars and researchers have begun to apply deep learning techniques to the field of traditional TCM tongue diagnosis. The goal is to leverage the efficiency and accuracy of deep learning to enhance the precision and standardization of TCM tongue diagnostics.

H. Huang, X. Yan and L. Xu—Contributing equally to this work.

© The Author(s), under exclusive license to Springer Nature Singapore Pte Ltd. 2024
D.-S. Huang et al. (Eds.): ICIC 2024, LNBI 14881, pp. 453–465, 2024.
https://doi.org/10.1007/978-981-97-5689-6_39

In the aspect of image recognition, Xue et al. [1] conduct various optimization methods on the AlexNet to enhance the accuracy of tongue image classification. They train the AlexNet using fissured and non-fissured tongue data, successfully extracting high-level features specific to the fissured region. Chang et al. [2] focused on the study of fissured tongues. They utilize the ResNet-50 network coupled with the Grad-CAM [3] technique, which enables the localization and diagnosis of fissured tongues. Tang et al. [4] employee the ResNet network to extract tongue coating features and utilize the multi-instance support vector machine [5] for classification, addressing the issue of inconsistent model performance caused by variations in the size or position of the tongue coating regions. In the aspect of image detection, Hu et al. [6] design a neural network framework for prescription construction, utilizing convolutional channels and fully connected layers to encode tongue features and generate herbal prescriptions. They also introduce an auxiliary treatment loss mechanism to mitigate the interference of sparse output labels on the diversity of results. Li et al. [7] employee a multi-label classification method to identify the complex composition of tongue images and compared the feature extraction capabilities of different networks. Zhuang et al. [8] leverage the ResNet34 network to automatically extract tongue image features and accomplish automatic tongue image classification tasks.

Deep learning applications in traditional TCM tongue diagnosis have made some progress, but still face several challenges. TCM tongue diagnosis data is relatively small, and there are discrepancies in diagnosis results among different doctors. The annotation process is time-consuming and labor-intensive, with high costs and potential subjectivity and inconsistencies in the annotations. To overcome these challenges, we propose an innovative architecture (see Fig. 1) that leverages the remarkable semantic recognition capability provided by the cross-attention modules and introduces score-based conditional guidance to generate high-quality normal tongue coating images.

In the experimental phase, our model utilize the collected normal tongue coating image data for unsupervised training, eliminating the need for manual feature descriptions and expert comments throughout the training process. By using the trained model for inference, the tongue coating images to be diagnosed are compared with generated normal tongue coating images to identify the affected areas. Additionally, we conduct performance tests on our proposed model, utilizing a dataset of tongue coating cases from 100 patients undergoing tongue diagnosis and consensus labels provided by clinical professionals to validate the performance of our model. The experimental results have showcased the preeminence of our model when compared to traditional rule-based approaches and other cutting-edge anomaly detection methods.

In summary, we have made the following key contributions:

- We propose an unsupervised tongue coating anomaly detection framework based on the diffusion model.
- By utilizing the semantic masks generated by cross-attention modules and score-based conditional guidance, our proposed framework is able to effectively reconstruct high-quality images and accurately identify pathological areas, thereby achieving automatic grading of pathological tongue coating.

- The experimental results show that our model outperforms existing models in terms of accuracy, making it a promising tool for improving the practice of TCM tongue diagnosis.

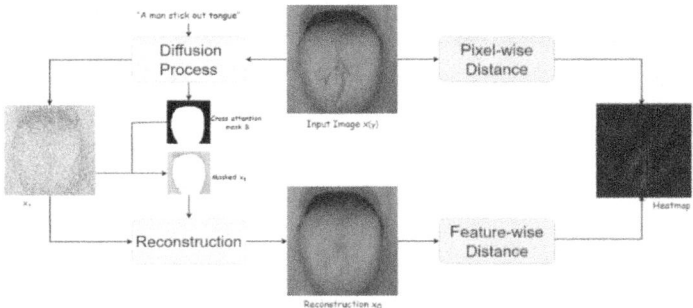

Fig. 1. Framework of our proposed method. During sampling process, semantic masks are applied to the input image to conditionally denoise it, achieving reconstruction. Subsequently, accurate anomaly localization is generated by comparing the reconstructed image with the input image through pixel matching and feature matching.

2 Method

2.1 Attention Mask Generation

Text-guided Diffusion generative models merge visual and textual embedding via spatial cross-attention, employing a text prompt P to direct the content-related image I creation from a random gaussian image noise z. A text encoder, a variational autoencoder (VAE), and a U-shaped network comprise Stable Diffusion [9]. The U-Net architecture integrates text and vision through the use of cross-attention mechanisms, producing spatial attention maps for each textual token by synthesizing visual and textual embeddings at every time step. In a formal sense, at each step t, the visual features of the noisy image $\varphi(z_t) \in \mathbb{R}^{H \times W \times C}$ are flattened and linearly transformed into a query vector $Q = \mathcal{L}_Q(\varphi(z_t))$. The text prompt P is then encoded into a textual embedding by the text encoder $\tau_\theta(P) \in \mathbb{R}^{N \times d}$ (where d denotes the dimension of the latent space and N represents the length of the text sequence). Subsequently, this embedding is transformed into a Value matrix $V = \mathcal{L}_V(\tau_\theta(P))$ and a Key matrix $K = \mathcal{L}_K(\tau_\theta(P))$ via learned projections $\mathcal{L}_Q, \mathcal{L}_K$, and $\mathcal{L}_V$. One method for calculating the cross-attention mappings involves the following steps:

$$\mathcal{A} = \text{Softmax}\left(\frac{QK^T}{\sqrt{d}}\right) \tag{1}$$

where $\mathcal{A} \in \mathbb{R}^{H \times W \times N}$. The relevant weight $\mathcal{A}_j \in \mathbb{R}^{H \times W}$ for the j-th text character may be found on the visual map $\varphi(z_t)$. Lastly, $\hat{\varphi}(z_t) = \mathcal{A}V$ may be utilized to obtain the cross-attention output, which is subsequently utilized to refine the spatial features $\varphi(z_t)$.

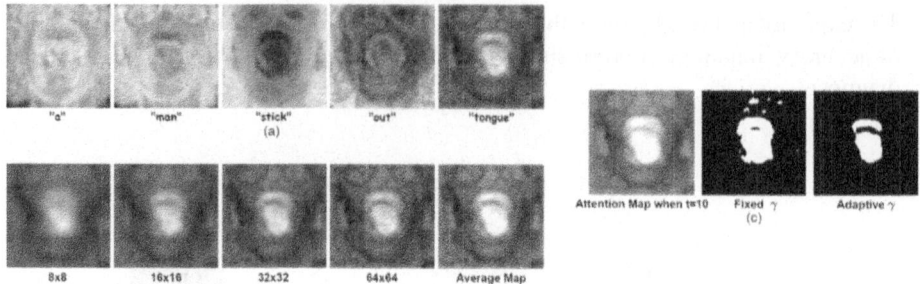

Fig. 2. (a) Cross attention maps of different text tokens. (b) Cross attention maps of different resolutions. (c) Binarization Mask with different γ.

Equation 1 allows us to derive the matching cross-attention map $\mathcal{A}_j^{s,t}$. Whereas s stands for the attention map from U-Net's s-th layer and corresponds to four distinct resolutions: 8×8, 16×16, 32×32, and 64×64, t stands for t-th time step. In practical terms, greater resolution maps (64×64) offer a more fine-grained grouping of components in larger items and may be superior for detecting small things. Lower resolution maps (8×8) provide a better grouping of large objects as a whole. In this paper, we design to calculate the average cross-attention map by consolidating the attention maps from multiple layers (see Fig. 2(b)), as illustrated below:

$$\hat{\mathcal{A}}_j = \sum_{s \in S} \frac{\mathcal{A}_j^s}{max\left(\mathcal{A}_j^s\right)} \tag{2}$$

where S denotes the number of layers(four in the case of U-Net). Normalization is required as the attention map's value in the Softmax output isn't a probability between 0 and 1. Subsequently, it is crucial to transform an average attention map (a probability map) $M \in \mathbb{R}^{H \times W}$ for the j-th text word generated by the cross attention into a binary map, where pixels with 1 represent the foreground region (such as "tongue"). Usually, a fixed threshold value γ is used for binarization procedure and refine using DenseCRF [10] in the manner shown below:

$$B = \text{DenseCRF}\left(\left[\gamma; \hat{\mathcal{A}}_j\right]_{argmax}\right) \tag{3}$$

However, as can be observed from Fig. 2(c), when the time step t is relatively small, the regional division within the attention map is not yet distinct. Applying a fixed threshold γ in this case would result in an inaccurate binary mask B (see Fig. 2(c)). Accordingly, in this paper, we propose an adaptive threshold γ setting based on the time step t.

$$\gamma = \gamma_T \left(1 + \frac{t}{T}\right) \tag{4}$$

wherein T is number of total time step. And γ_T is the optimal parameter value for γ when T time step. We replace γ in Eq. 3 with the expression of γ from Eq. 4 to obtain B. After then, this mask B was used in the sampling process.

2.2 High-Quality Reconstruction

In this paper, we train the diffusion model according to the method proposed in the original DDPM work. That is, by employing the latent variable model $p_\theta(x_0) := \int p_\theta(x_{0:T}) dx_{1:T}$, we add noise $\epsilon \sim \mathcal{N}(0, I)$ at each time step t to transform the latent variable x_0 into x_T, which has the same dimension as x_0, and ultimately x_T follows a Gaussian distribution $q(x_T) \sim \mathcal{N}(0, I)$. Forward diffusion persists for T time steps, with Gaussian noise added at each step according to a predetermined variance table $\beta_1, ..., \beta_T$. This process can be represented by a fixed Markov chain:

$$q(x_{1:T}|x_0) = \prod q(x_t|x_{t-1}) \tag{5}$$

$$q(x_t|x_{t-1}) = \mathcal{N}\left(x_t; \sqrt{1 - \beta_t}x_{t-1}, \beta_t I\right) \tag{6}$$

During the forward diffusion process, it is permissible to sample x_t in a closed-form sampling at any specified time step t. By setting $\alpha_t := 1 - \beta_t$ and $\overline{\alpha}_t := \prod_{s=1}^{t} \alpha_s$, the formula can be expressed as:

$$q(x_t|x_0) = \mathcal{N}\left(x_t; \sqrt{\overline{\alpha}_t}x_0, (1 - \overline{\alpha}_t)I\right) \tag{7}$$

On the contrary, the reverse process aims to restore the original data distribution, progressively eliminating noise. It can be represented as: $p_\theta(x_{t-1}|x_t) = \mathcal{N}(x_{t-1}; \mu_\theta(x_t, t), \beta_t I)$, within this equation, the mean is computed from a learnable function $\epsilon_\theta^{(t)}(x_t)$, where the primary learning goal is to minimize the value of $\left\| \epsilon_\theta^{(t)}(x_t) - \epsilon \right\|^2$. In DDIM, a deterministic non-Markovian sampling method is utilized, which significantly expedites the sampling process without causing significant degradation in image quality. More precisely, DDIM introduces a novel variance schedule and incorporates a forward process $q_\delta(x_{t-1}|x_t, x_0)$. The variable x_t can be expressed as a linear combination of x_0 and a noise variable, represented as $x_t = \sqrt{\alpha_t}x_0 + \sqrt{1 - \alpha_t}\,\epsilon$, which is a unique characteristic of the forward diffusion process. This facilitates the straightforward extraction of the denoised observation from x_t:

$$f_\theta^{(t)} := \left(x_t - \sqrt{1 - \alpha_t}\,\epsilon_\theta^{(t)}(x_t)\right) / \sqrt{\alpha_t} \tag{8}$$

Next, we can delineate a generative process incorporating a predetermined prior $p_\theta(x_T) = \mathcal{N}(0, I)$ and $p_\theta^{(t)}(x_{t-1}|x_t) = q_\sigma(x_{t-1}|x_t, f_\theta^{(t)}(x_t))$. Consequently, x_{t-1} can be written as:

$$x_{t-1} = \sqrt{\alpha_{t-1}}f_\theta^{(t)}(x_t) + \sqrt{1 - \alpha_{t-1} - \sigma_t^2}\,\epsilon_\theta^{(t)}(x_t) + \sigma_t\,\epsilon_t \tag{9}$$

In this case, σ_t determines how stochastic the sampling process is. First developed by Song et al., who introduced a score-based function to predict the needed deviation at each time step for creating a less noisy image, Song and Ermon established the relationship between diffusion models and score matching. This may be said in the following way:

$$\nabla_{x_t} \log p_\theta(x_t) = -\frac{1}{\sqrt{1 - \alpha}}\,\epsilon_\theta^{(t)}(x_t) \tag{10}$$

Dhariwal and Nichol [11] subsequently used this strategy to develop conditional sampling for DDIM.

Algorithm 1: Proposed Sampling Process

Input: Disease tongue coating image x (y), Prompt P: 'a man stick out tongue', Encoder τ_θ, Adaptive threshold γ.

$x_T \leftarrow \sqrt{\alpha_T} x + \sqrt{1 - \alpha_T} \epsilon_t$;

for t $\leftarrow$ T **do**:

    //Obtain semantic mask B

    $K \leftarrow \mathscr{L}_K(\tau_\theta(P)), Q \leftarrow \mathscr{L}_Q(\tau_\theta(x_T))$;

    $\mathcal{A}_4 \leftarrow \text{Softmax}(\frac{QK_4^T}{\sqrt{d}})$;

    $\hat{\mathcal{A}}_4 \leftarrow \sum_{s \in S} \frac{\mathcal{A}_4^s}{max(\mathcal{A}_4^s)}$;

    $B \leftarrow \text{DenseCRF}([\gamma; \widehat{\mathcal{A}}_4]_{argmax})$;

    //Conditional-guided reconstruction

    $y_t \leftarrow \sqrt{\alpha_t} y + \sqrt{1 - \alpha_t} \epsilon_\theta^{(t)}(x_t)$;

    $\hat{\epsilon} \leftarrow \epsilon_\theta^{(t)}(x_t) - \omega \sqrt{1 - \alpha_t}(y_t[1 - B] - x_t[1 - B])$;

    $\hat{f}_\theta^{(t)}(x_t) \leftarrow (x_t - \sqrt{1 - \alpha_t} \hat{\epsilon})/\sqrt{\alpha_t}$;

    $x_{t-1} \leftarrow \sqrt{\alpha_{t-1}} f_\theta(x_t) + \sqrt{1 - \alpha_{t-1} - \sigma_t^2} \epsilon^2 + \sigma_t \epsilon_t$;

End for

Output: Reconstructed tongue coating image x_0

With the target image y and the noisy image x_t, our goal is to progressively denoise x_t to obtain an image that closely matches the regions beyond the target area (e.g., tongue) of y, and ensures that the target area is normally generated. To accomplish this, we propose to construct a posterior score function $\nabla_{x_t} \log p_\theta(x_t[1 - B]|y[1 - B])$ (B is the binary mask obtained from the previous section.) by conditioning the score function on the target image. However, because x_t and y do not have the same signal-to-noise ratio, it is difficult to calculate this posterior score function directly. To address this challenge, we consider that the goal is to make the reconstructed image x_0 similar to y. Therefore, we attempt to add noise to y that is the same as the noise composed of x_t, obtaining y_t. At each denoising step, we guide $x_t[1 - B]$ to approach $y_t[1 - B]$, which indirectly ensures that the final obtained $x_0[1 - B]$ is similar to $y[1 - B]$. We add the predicted $\epsilon_\theta^{(t)}(x_t)$ from the trained diffusion model to y to calculate y_t. Subsequently, we use $y_t[1 - B]$ as a guiding condition to obtain $\nabla_{x_t} \log p_\theta(x_t[1 - B]|y_t[1 - B])$ to direct the denoising process. According to the Bayes' rule, it can be decomposed into:

$$\nabla_{x_t} \log p_\theta(x_t[1 - B]|y_t[1 - B]) = \nabla_{x_t} \log p_\theta(x_t[1 - B]) \\ + \nabla_{x_t} \log p_\theta(y_t[1 - B]|x_t[1 - B]) \tag{11}$$

It makes sense to think of the likelihood $\nabla_{x_t} \log p_\theta(y_t[1 - B]|x_t[1 - B])$ as a correction score for each denoising step's deviation in x_t from y_t. Given that the noise in x_t and y_t is identical, this deviation can only exist at the image (signal) level. As a result, $y_t[1 - B] - x_t[1 - B]$ can be used to calculate the divergence, and the adjusted noise term $\hat{\epsilon}$ is updated in this way:

$$\hat{\epsilon} = \epsilon_\theta^{(t)}(x_t) - \omega\sqrt{1 - \alpha_t}(y_t[1 - B] - x_t[1 - B]) \quad (12)$$

whereby ω regulates the conditioning's power. Ultimately, the denoising procedure yields the following calculation for the less-noisy image x_{t-1}:

$$x_{t-1} = \sqrt{\alpha_{t-1}}f_\theta(x_t) + \sqrt{1 - \alpha_{t-1} - \sigma_t^2}\,\epsilon^2 + \sigma_t\,\epsilon_t \quad (13)$$

Algorithm 1 displays a summary of our proposed sampling process.

2.3 Anomaly Scoring

In its most straightforward application, we can identify and pinpoint anomalies by conducting a pixel-by-pixel comparison between the original input and its reconstructed version. Nevertheless, relying solely on the pixel distances between two images might fail to detect certain anomalies, in cases of fissured tongue symptoms, protrusions or indentations may occur, particularly when there are no accompanying visible changes in color. Therefore, we calculate the distance between image features extracted by the Speeded Up Robust Features (SURF) algorithm extra, which has faster computational speed and good robustness to capture perceptual similarity.

With the target image i and a reconstructed image j, to create the anomalous heatmap, we establish two distance functions: a feature-wise distance function D_f and a pixel-wise distance function D_p. Based on the L1 norm in pixel space, D_p is computed. The calculation of feature point differences adopts cosine similarity, for each pair of matched feature points (p_i, p_j), we calculate the cosine similarity between their descriptors:

$$cos\_sim(d_i, d_j) = \frac{d_i \cdot d_j}{|d_i| \cdot |d_j|} \quad (14)$$

$$D_f = 1 - \frac{\sum_{i=1}^{N} cos\_sim(d_i, d_j)}{N} \quad (15)$$

wherein, d_i and d_j respectively represent the descriptors of feature points p_i and p_j, $\cdot$ denotes the dot product of vectors, $\|\cdot\|$ denotes the magnitude (or length) of a vector. N represents the number of matched feature point pairs. As a result, the pixel and the feature distance combine to form the final anomaly score function:

$$D_{ano} = D_f + \left(v\frac{\text{Mean}(D_f)}{\text{Mean}(D_p)}\right)D_p \quad (16)$$

where the pixel-wise distance's significance is controlled by v.

3 Experiment

3.1 Implementation Details

Every experiment is run on a computer that has two GeForce RTX Titan graphics cards installed. PyTorch 1.12.1 is utilized for model training and sampling. As input for our model, every image is painstakingly resized to a resolution of 256×256 and then normalized. 20,000 steps are covered in the training phase, using the Adam optimizer with a learning rate of 0.0001 and a batch size of 8. The γ_T ($T = 999$) for DenseCRF is 0.3. v is set to 0.6. We calculate anomaly scores depending on the reconstruction distance. We assess the model's efficacy utilizing the Area Under the Curve (AUC) metric.

3.2 Dataset

In our experiment, we gather patient data from hospitals, focusing on those who have undergone tongue diagnosis in recent years. Specifically, we utilize 437 unlabeled image data from 200 patients with normal tongue fur and various diseased tongue images, such as fissured tongue, tongue with black coating, thick and greasy tongue etc. (see Table 1, hereafter in the text, we refer to symptom 1, 2, 3 for simplicity). We employ the images of normal tongue fur to train a diffusion model for abnormal scoring. To evaluate the performance of the method we proposed, we use the diseased images as the evaluation dataset, in which the photos are annotated by professional doctors to mark the defective areas.

Table 1. Details of datasets.

Dataset	Unlabled	Labled			Total
		Symptom 1	Symptom 2	Symptom 3	
Train	437	-	-	-	437
Test	-	50	50	50	150

3.3 Anomaly Scores for Tongue Coating Symptom Groups

Figure 3 presents the anomaly scores for three types of tongue coating symptoms as determined by our proposed model. The distribution of anomaly scores between different groups is not uniform. The median scores of these three symptoms are as follows: median 5.075 (range from 1.321–8.139) for the fissured tongue group, median 14.152 (range from 10.419–18.778) for the tongue with black coating group, median 12.024 (range from 8.312–16.231) for the thick and greasy tongue group, It is noteworthy that there are significant differences in scores between these groups. We believe that the increased anomaly scores are due to the larger lesion areas in the other two symptoms compared to tongue fissures.

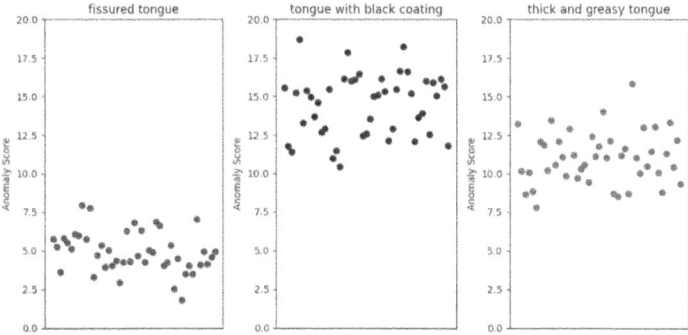

Fig. 3. Comparison of anomaly scores between three types of tongue coating symptoms.

3.4 Model Performance Evaluation

We next proceeded to assess the efficacy of our proposed model in the context of anomaly detection. To thoroughly evaluate its efficacy, we conducted a comparative analysis of our model against the state-of-the-art (SOTA) feature-based and reconstruction-based methods employed in anomaly detection. Table 2 summarizes the performance metrics of diverse anomaly detection models, inclusive of the methodology recently introduced by us. Conspicuously, the reconstruction-based anomaly detection frameworks yielded more favorable outcomes than their feature-based counterparts. It is noting that our proposed model significantly outperformed all other competitors, a superiority that is statistically robust. Figure 3 illustrates the effectiveness of our model.

Table 2. Performance comparison with anomaly detection models.

Methods	Symptom 1	Symptom 2	Symptom 3	Average
Feature-based				
PaDiM [13]	0.528	0.566	0.557	0.550
FastFlow [14]	0.556	0.588	0.579	0.574
CFA [15]	0.483	0.526	0.512	0.507
PatchCore [16]	0.682	0.722	0.706	0.703
Reconstruction-based				
Ganomaly [17]	0.705	0.756	0.736	0.732
DRAEM [18]	0.647	0.676	0.664	0.663
Ours	**0.846**	**0.881**	**0.868**	**0.865**

Table 3. Ablation study of condition guidance.

Methods	Symptom 1	Symptom 2	Symptom 3	Average
No condition guidance	0.816	0.839	0.827	0.827
With condition guidance	**0.846**	**0.881**	**0.868**	**0.865**

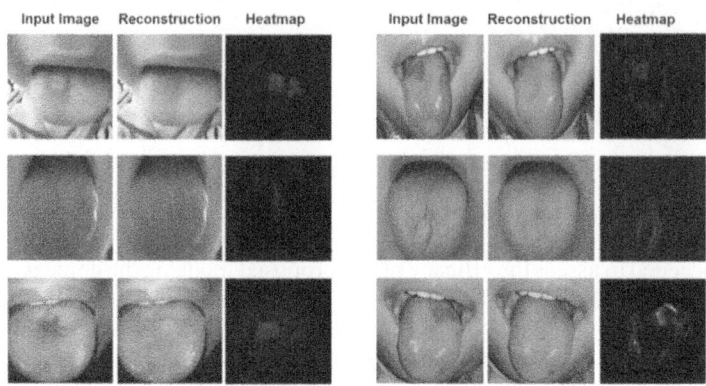

Fig. 4. Our method produces defect-free reconstruction of anomaly-free input images. A precise heatmap for anomaly detection is generated.

3.5 Ablation Study

We investigate the performance improvements conferred by our proposed framework. In the absence of our framework, that is, with conditional guidance excluded, the sampling in the reverse process is carried out in full accordance with the DDIM approach. As detailed in Table 3 and Fig. 4, without conditional guidance, the images generated by DDIM exhibit a higher number of discrepancies when compared to the original images, and the detection performance experiences a notable decline. Figure 5 also qualitatively analyzes the ablation comparison of using pixel distance features and combining feature distances when calculating the distance between the reconstructed image and the original image. It can be observed that the effect of combining feature distances is better in cases where there are prominent or indented fissured tongue symptoms.

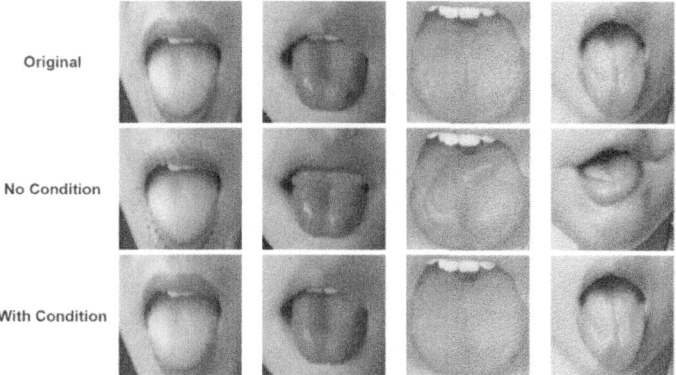

Fig. 5. Compared to the original image, the reconstructed image without conditional guidance lacks precision in many detail aspects, which will affect the calculation of subsequent anomaly scores.

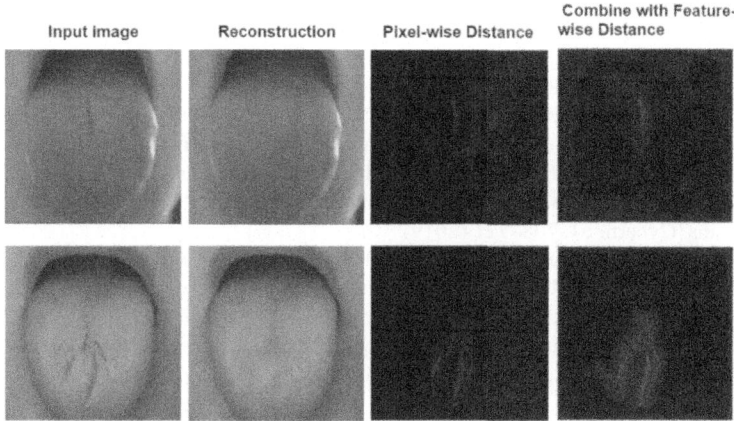

Fig. 6. When calculating the anomaly score, in the case of fissured tongue symptoms, the effect of combining feature-wise distance is better.

4   Conclusion

This study demonstrates the application of diffusion models in the core diagnostic method of Traditional TCM - tongue coating diagnosis. We aim to address the subjectivity in tongue coating diagnosis and eliminate the interference of environmental factors by proposing an unsupervised tongue coating anomaly detection framework based on the diffusion model. Our framework first utilizes the semantic classification capability of cross-attention module and conditional guidance based on scoring function, effectively reconstructing high-quality images and accurately identifying pathological regions, enabling the automatic scoring of pathological tongue coatings. Compared to traditional rule-based methods, our fully automated approach eliminates the need for manual annotation and provides an objective assessment. Moreover, experimental results indicate that our model outperforms existing models in terms of accuracy.

Acknowledgments. This work was supported by National Natural Science Foundation of China (62376286).

References

1. Xue, Y., Li, X., Cui, Q., Wang, L., Wu, P.: Cracked tongue recognition based on deep features and multiple-instance SVM. In: Hong, R., Cheng, W.-H., Yamasaki, T., Wang, M., Ngo, C.-W. (eds.) Advances in Multimedia Information Processing – PCM 2018. LNCS, vol. 11165, pp. 642–652. Springer, Cham (2018). https://doi.org/10.1007/978-3-030-00767-6_59
2. Chang, W.-H., Chu, H.-T., Chang, H.-H.: Tongue fissure visualization with deep learning. In: 2018 Conference on Technologies and Applications of Artificial Intelligence (TAAI), pp. 14–17. IEEE (2018)
3. Selvaraju, R.R., Cogswell, M., Das, A., Vedantam, R., Parikh, D., Batra, D.: Grad-cam: visual explanations from deep networks via gradient-based localization. In: Proceedings of the IEEE International Conference on Computer Vision, pp. 618–626 (2017)
4. Tang, Y., Sun, Y., Chiang, J.Y., Li, X.: Research on multiple-instance learning for tongue coating classification. IEEE Access **9**, 66361–66370 (2021)
5. Andrews, S., Tsochantaridis, I., Hofmann, T.: Support vector machines for multiple-instance learning. In: Advances in Neural Information Processing Systems, vol. 15 (2002)
6. Hu, Y., Wen, G., Liao, H., Wang, C., Dai, D., Yu, Z.: Automatic construction of Chinese herbal prescriptions from tongue images using CNNs and auxiliary latent therapy topics. IEEE Trans. Cybern. **51**, 708–721 (2019)
7. Li, T., Wu, C., Ma, Y.: Multi-label constitution identification based on tongue image in traditional Chinese medicine. In: 2021 China Automation Congress (CAC), pp. 1617–1622. IEEE (2021)
8. Zhuang, Q., Gan, S., Zhang, L.: Human-computer interaction based health diagnostics using ResNet34 for tongue image classification. In: Computer Methods and Programs in Biomedicine, vol. 226, p. 107096 (2022)
9. Rombach, R., Blattmann, A., Lorenz, D., Esser, P., Ommer, B.: High-resolution image synthesis with latent diffusion models. In: Proceedings of the IEEE/CVF Conference on Computer Vision and Pattern Recognition, pp. 10684–10695 (2022)
10. Krähenbühl, P., Koltun, V.: Efficient inference in fully connected CRFs with Gaussian edge potentials. In: Advances in Neural Information Processing Systems, vol. 24 (2011)
11. Song, J., Meng, C., Ermon, S.: Denoising diffusion implicit models. arXiv preprint arXiv: 2010.02502 (2020)
12. Defard, T., Setkov, A., Loesch, A., Audigier, R.: Padim: a patch distribution modeling framework for anomaly detection and localization. In: Del Bimbo, A., et al. (eds.) Pattern Recognition. ICPR International Workshops and Challenges. LNCS, vol. 12664, pp. 475–489. Springer, Cham (2021). https://doi.org/10.1007/978-3-030-68799-1_35
13. Yu, J., et al.: FastFlow: unsupervised anomaly detection and localization via 2D normalizing flows. arXiv preprint arXiv:2111.07677 (2021)
14. Lee, S., Lee, S., Song, B.C.: CFA: coupled-hypersphere-based feature adaptation for target-oriented anomaly localization. IEEE Access **10**, 78446–78454 (2022)
15. Roth, K., Pemula, L., Zepeda, J., Schölkopf, B., Brox, T., Gehler, P.: Towards total recall in industrial anomaly detection. In: Proceedings of the IEEE/CVF Conference on Computer Vision and Pattern Recognition, pp. 14318–14328 (2022)

16. Akcay, S., Atapour-Abarghouei, A., Breckon, T.P.: GANomaly: semi-supervised anomaly detection via adversarial training. In: Jawahar, C.V., Li, H., Mori, G., Schindler, K. (eds.) Computer Vision – ACCV 2018. LNCS, vol. 11363, pp. 622–637. Springer, Cham (2019). https://doi.org/10.1007/978-3-030-20893-6_39
17. Zavrtanik, V., Kristan, M., Skočaj, D.: DRAEM-a discriminatively trained reconstruction embedding for surface anomaly detection. In: Proceedings of the IEEE/CVF International Conference on Computer Vision, pp. 8330–8339 (2021)
18. Ho, J., Jain, A., Abbeel, P.: Denoising diffusion probabilistic models. In: Advances in Neural Information Processing Systems, vol. 33, pp. 6840–6851 (2020)

Author Index

© The Editor(s) (if applicable) and The Author(s), under exclusive license
to Springer Nature Singapore Pte Ltd. 2024
D.-S. Huang et al. (Eds.): ICIC 2024, LNBI 14881, pp. 467–470, 2024.
https://doi.org/10.1007/978-981-97-5689-6

GPSR Compliance

The European Union's (EU) General Product Safety Regulation (GPSR) is a set of rules that requires consumer products to be safe and our obligations to ensure this.

If you have any concerns about our products, you can contact us on ProductSafety@springernature.com

In case Publisher is established outside the EU, the EU authorized representative is:

Springer Nature Customer Service Center GmbH
Europaplatz 3
69115 Heidelberg, Germany

The manufacturer's authorised representative in the EU is Springer
Nature Customer Service Centre GmbH, Europaplatz 3, 69115 Heidelberg,
Germany. If you have any concerns regarding our products, please
contact ProductSafety@springernature.com

Printed and bound by CPI Group (UK) Ltd, Croydon, CR0 4YY
05/05/2026
02102981-0010